T0302092

Bird Migration Across the Himalayas

Birds migrating across the Himalayan region fly over the highest heights in the world, facing immense physiological and climatic challenges. The authors show the different strategies diverse species use to cope. Many wetland avian species are seen in the high-altitude lakes of the Himalayas and the adjoining Tibetan Plateau, including Bar-headed Geese, one of the highest-flying species known.

Ringing programmes have generated information about origins and destinations, but this book is the first to present information on the birds' exact migratory paths. Capitalizing on knowledge generated through satellite telemetry, the authors describe the migratory routes of a multitude of birds flying over or skirting the Himalayas.

The myriad of threats to migratory birds and the wetland system in the Central Asian Flyway are discussed, with ways to mitigate them. This is a volume to inform and persuade policy-makers and conservation practitioners to take appropriate measures for the long-term survival of this unique migration system.

Herbert H.T. Prins is Professor in Resource Ecology at Wageningen University. He is known for savanna ecology and has investigated wild goose ecology in Europe, on Spitsbergen and in Siberia. For his conservation efforts, he received the Aldo Leopold Award and was appointed Officer in the Order of Oranje Nassau and Officer in the Order of the Golden Ark.

Tsewang Namgail heads the Snow Leopard Conservancy India Trust. After completing his higher education in Europe, he moved to the United States and worked on migratory birds. He has done pioneering ecological work on mammals in the Himalayas, and serves on the editorial boards of the *Ecological Research* and *Pastoralism: Research, Policy and Practice* journals.

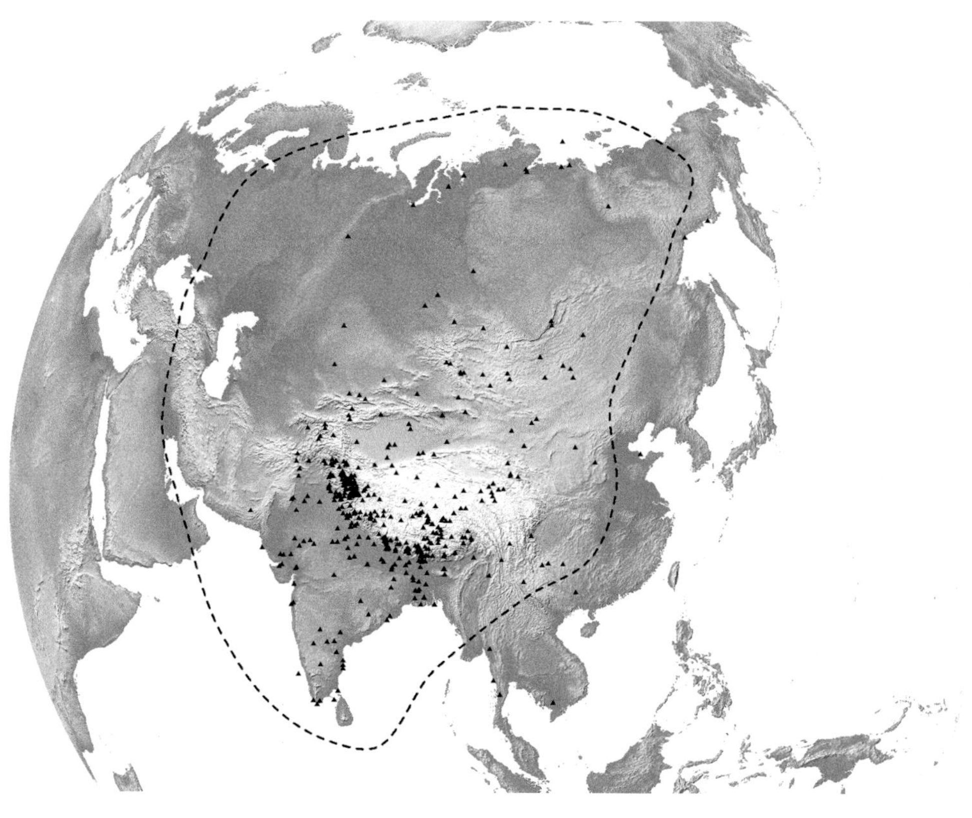

Map of the Central Asian Flyway (outer limits indicated by dashed line). Triangles refer to all localities mentioned in this book as presented in the Gazetteer at the end of the book. The map was produced by Yorick Liefting in ArcGIS 10.2.2 on the 'Gray Earth' base-map from www.naturalearthdata.com

Bird Migration Across the Himalayas

Wetland Functioning Amidst Mountains and Glaciers

EDITED BY

HERBERT H.T. PRINS

Wageningen University, The Netherlands

TSEWANG NAMGAIL

Snow Leopard Conservancy India Trust, India

CAMBRIDGE
UNIVERSITY PRESS

University Printing House, Cambridge CB2 8BS, United Kingdom

One Liberty Plaza, 20th Floor, New York, NY 10006, USA

477 Williamstown Road, Port Melbourne, VIC 3207, Australia

314-321, 3rd Floor, Plot 3, Splendor Forum, Jasola District Centre, New Delhi - 110025, India

79 Anson Road, #06-04/06, Singapore 079906

Cambridge University Press is part of the University of Cambridge.

It furthers the University's mission by disseminating knowledge in the pursuit of education, learning and research at the highest international levels of excellence.

www.cambridge.org
Information on this title: www.cambridge.org/9781107114715

First published 2017

A catalogue record for this publication is available from the British Library

Library of Congress Cataloging in Publication data
Prins, H. H. T. (Herbert H. T.), editor. | Namgail, Tsewang, 1973– editor.
Bird migration across the Himalayas : wetland functioning amidst mountains and glaciers / edited by Herbert Prins, Affiliation Wageningen University, The Netherlands, and Tsewang Namgail, Snow Leopard Conservancy India Trust, India.
Cambridge : Cambridge University Press, 2016. | Includes bibliographical references and index. LCCN 2016047007 | ISBN 9781107114715 (alk. paper)
LCSH: Waterfowl – Migration – Himalaya Mountains Region. | Birds – Migration – Himalaya Mountains Region. | Migratory birds – Himalaya Mountains Region. | Flyways – Himalaya Mountains Region. | Himalaya Mountains Region.
LCC QL698.9 .B56 2016 | DDC 598.4/1568095496–dc23
LC record available at https://lccn.loc.gov/2016047007

ISBN 978-1-107-11471-5 Hardback
ISBN 978-1-107-53525-1 Paperback

Contents

Colour plates are to be found between pp. 174 and 175

Contributors

Monisha Ahmed
Ladakh Arts and Media Organisation, India

Sumanta Bagchi
Centre for Ecological Sciences, Indian Institute of Science, India

Sivananinthaperumal Balachandran
Bombay Natural History Society, India

Nyambayar Batbayar
Wildlife Science and Conservation Center of Mongolia, Mongolia

Keith L. Bildstein
Acopian Center for Conservation Learning, USA

Charles M. Bishop
School of Biological Sciences, University of Bangor, UK

Bodo Bookhagen
Institute of Earth and Environmental Science, University of Potsdam, Germany

Laurianne Bruneau
Centre de recherché sur les civilisations de l'Asie orientale – CRCAO Collège de France, France

Patrick J. Butler
School of Biosciences, University of Birmingham, UK

Simon Delany
Delany Environmental, The Netherlands

Andrew Dixon
International Wildlife Consultants Ltd, UK and Environment Agency-Abu Dhabi, UAE

Peter B. Frappell
Office of the Dean of Graduate Research, University of Tasmania, Australia

David Garbutt
Garbutt Consult, Switzerland

Sunetro Ghosal
Stawa, Chamshenpa, India

Thomas A. Groen
Department of Natural Sciences, University of Twente, The Netherlands

Ekta Gupta
Centre for Ecological Sciences, Indian Institute of Science, India

Lucy A. Hawkes
Centre for Ecology and Conservation, University of Exeter, UK

René Heise
OSTIV Mountain Wave Project, Germany

Hiroyoshi Higuchi
Graduate School of Media and Governance, Keio University, Japan

Blaise Humbert-Droz
Independent Researcher, Wildlife and Environment, Bangalore, India

Rob J. Jansen
Resource Ecology Group, Wageningen University, The Netherlands

Matías A. Juhant
Acopian Center for Conservation Learning, Hawk Mountain Sanctuary, Orwigsburg, PA, USA

Hansoo Lee
Korea Institute of Environmental Ecology, Republic of Korea

Ze Luo
Computer Network Information Center (CNIC), Chinese Academy of Sciences, China

Tracy McCracken
Emergency Prevention System (EMPRES) for Transboundary Animal and Plant Pests and Diseases, Wildlife Health and Ecology Unit, Italy

Jessica U. Meir
Department of Anesthesia, Massachusetts General Hospital, Harvard Medical School, Boston, MA, USA

William K. Milsom
Department of Zoology, University of British Columbia, Vancouver, Canada

Jason Minton
Wild Bird Society of Japan, CA, USA

Charudutt Mishra
Nature Conservation Foundation, India. Snow Leopard Trust, Seattle, WA, USA

Taej Mundkur
Wetlands International, The Netherlands

Karthik Murthy
Centre for Ecological Sciences, Indian Institute of Science, India

Tsewang Namgail
Snow Leopard Conservancy India Trust, India. US Geological Survey, NV, USA

Tseveenmyadag Natsagdorj
Wildlife Science and Conservation Center of Mongolia, Mongolia

Scott H. Newman
Emergency Centre for Transboundary Animal Diseases, Vietnam

John Norton
John Norton Ecology, UK

Klaus Ohlmann
Serres Airfield, La Bâtie-Montsaléon, France

Lewis A. Owen
Department of Geology, University of Cincinnati, Cincinnati, OH, USA

Eric C. Palm
US Geological Survey, Beltsville, MD, USA

Herbert H.T. Prins
Resource Ecology Group, Wageningen University,
The Netherlands

Diann J. Prosser
US Geological Survey, Beltsville, MD, USA

Lutfor Rahman
International Wildlife Consultants Ltd, UK. Environment Agency-Abu Dhabi, UAE

Gopal S. Rawat
Wildlife Institute of India, India

Ponnusamy Sathiyaselvam
Bombay Natural History Society, India

Graham S. Scott
Department of Biology, McMaster University, Hamilton, ON, Canada

Michael Searle
Department of Earth Sciences, University of Oxford, UK

T.R. Shankar Raman
Nature Conservation Foundation, India

Navinder J. Singh
Department of Wildlife, Fish and Environmental Studies, Swedish University of Agricultural Sciences, Sweden

Aleksandr Sokolov
Ecological Research Station of Institute of Plant and Animal Ecology, Russian Academy of Sciences, Russia

Vasiliy Sokolov
Ecological Research Station of Institute of Plant and Animal Ecology, Russian Academy of Sciences, Russia

Clare Sulston
Earth Trust, Oxfordshire, UK

Kulbhushansingh R. Suryawanshi
Nature Conservation Foundation, India. Snow Leopard Trust, Seattle, WA, USA

John Y. Takekawa
US Geological Survey, CA, USA. National Audubon Society, Science Division, San Francisco, CA, USA

Víctor Martín Vélez
Resource Ecology Group, Wageningen University, The Netherlands

Martin Vernier
Rue du Village 23, Switzerland

Sipke E. van Wieren
Resource Ecology Group, Wageningen University, The Netherlands

Charles Williams
Natural England, UK

Xiangming Xiao
Department of Botany and Microbiology, University of Oklahoma, OK, USA. Institute of Biodiversity Science, Fudan University, China.

Ron C. Ydenberg
Centre for Wildlife Ecology, Simon Fraser University, Canada. Resource Ecology Group, Wageningen University, The Netherlands

THE DALAI LAMA

FOREWORD

Bird migration is one of the natural wonders of the world. As they traverse the earth, with no regard for national borders, birds exercise a freedom to which could be the envy of many people. Every spring and autumn, the lakes and rivers, plains and forests of Tibet teem with migratory birds. As a child, they fascinated me, whether they were red-billed choughs in the crevices of the Potala, the elegant black-necked cranes landing and dancing on the marshes around the Norbulingka, or the majestic vultures soaring in the skies above Lhasa. These birds enchanted me.

When I was in Tibet, bird life across the Tibetan Plateau was rich. They brought life and beauty to the stark Tibetan landscape. Not only were there laws to protect nature and the environment, but also the Tibetan Government assigned guards to protect birds and their eggs at nesting time. In the years following my arrival in India, Tibetans from Tibet and non-Tibetans who have traveled there have told me about the steep decline in all kinds of wildlife, including birds. They say the habitats are being destroyed by reckless activities, including mining, leading to depletion of forest cover and pollution, etc.

As much as we human beings have right to the natural habitat of our mother earth, other inhabitants on this earth also have the same right to thrive peacefully. Although we need development and modernization, their purpose is to create joy and alleviate suffering. In doing so, if we forget to consider the wellbeing of other living creatures, how could we justify our human intelligence? Birds and other living creatures hardly endanger the lives of humans; conversely, our actions have detrimental consequences on their lives. Therefore, I hope that this book on 'Bird Migration Across the Himalayas' would help the readers to understand the lives of tens of thousands of birds across the Himalayas, and inspire them to extend their compassion towards other living species as well.

February 2, 2017

Preface

This book grew out of our deep fascination for the Himalayas and their wildlife. The Himalayas (meaning 'the abode of snow' in Sanskrit) include the highest mountains on our planet, and the region is sometimes considered 'the third pole' because of its massive ice and snow deposits. The Himalayas and the adjoining Tibetan Plateau also have innumerable high-altitude lakes. Visiting some of these wetlands, teeming with birds, and trekking to reach them was exciting and adventurous. We crossed high passes, traversed vast expanses of dry plateaus, waded through strong Himalayan torrents, walked across the steepest slopes and ploughed through deep snow. We saw Lammergeyers flying high, Robin Accentors flitting from boulder to boulder, Horned Larks feeding their downy young on tawny slopes and flocks of Yellow-billed Choughs indulging in high-altitude acrobatics. We also heard passerines such as Rosefinches singing the most melodious songs.

While trekking through the high Himalayas, we always envied the birds cruising overhead, leaving us behind, slogging across steep slopes. We often looked up at them wistfully and wondered: Where did they come from and where are they headed? Sitting on a high pass, thinking about the next trough and the crest to be scaled (people who have flown from New Delhi to Leh in Ladakh on a clear morning would recognize these landscape features), we heard a flock of Bar-headed Geese calling on their way north. After hours of plodding across a desolate plateau, we reached a high-altitude lake, where we observed some geese, touted as the highest-flying bird in the world, foraging on the first grass blades of the season.

The migratory birds visit these high-altitude lakes every year, come what may, in both autumn and spring. Needless to say, they face a lot of hardships on their way to these wetlands. The most prominent of these is the effort required to cross some of the highest mountains and plateaus in the world, but others include pesticide pollution and loss of habitat in breeding, staging and wintering areas. While crossing the high mountains and plateaus, the birds face snowstorms, rains, raptors and terrestrial predators preying on them at the staging sites. Some birds ultimately succumb to these threats. Indeed, we have seen carcasses of Golden Oriole and Common Kestrel in areas higher than 4500 m above sea level. Some of these migratory birds arrive at high-altitude wetlands in spring, feed and raise their chicks during the short summer and then return to wintering areas in autumn. Others cross the mountains from their breeding grounds far to the north, in the Arctic or in the taiga of Siberia, on their way to spend the winter in the Indian subcontinent or even Africa. During our treks, we have seen Horned Larks and

Redshanks feeding on narrow strips of sedge meadows, sometimes no bigger than 20 m by 5 m. Some species also stop at small, ephemeral pools of water no bigger than 10 m^2, formed as a result of short bursts of rain, to feed on aquatic invertebrates.

Given their often wide ranges of distribution, the future of these birds is uncertain because their survival depends not only on the proper functioning of wetland ecosystems in one country, but also on the functioning of ecosystems in many countries spread over more than one biogeographic realm. Thus, the long-distance migratory birds in the Central Asian Flyway will survive and continue to amaze with their flights only if the small, dispersed wetlands along the route are protected alongside wetlands in the subarctic and the Indian subcontinent, and if other protection measures within this flyway are taken soon. For this to be accomplished, government agencies in Central Asia and the Indian subcontinent need to come forward and collaborate on local and international conservation efforts. Such efforts, also transcending national boundaries, are essential to ensure the survival of the spectacular migrations across the highest mountain range in the world.

After spells of trekking in the mountains, we got back to our offices and tried to learn more about the migratory birds we had observed, but, to our disappointment, there was very little to be found, especially when it came to understanding the migratory routes these birds take. Although ringing programmes had generated some information about the origins and destinations of migratory birds, there was only very limited data on their exact migratory paths.

We even tried to tap the deep knowledge of the holy men who live in the great Buddhist monasteries of the Himalayas. In the Ki Monastery (founded in about 1040 CE) there was a monk walled up in his room. It had only a small window from which he could observe the outside world, and through which he got his food. He had profound knowledge of the Blue Sheep (*Bharal*) and their comings and goings on the slope he could monitor. He spent all his waking hours observing the small fraction of the sky that he could see to study the movements of Red-billed Choughs, Ravens and Golden Eagles, but he could not give us information about the great bird migrations across the Himalayas. When the abbot of the monastery, His Eminence Lochen Rinpoche, asked us whether we were in search of enlightenment, we denied it from a spiritual point, but asked him whether he could help us understand the past. His Eminence is believed to be one of the most frequently reincarnated people on earth, but he explained to us that it was the subtle consciousness (roughly translated as 'soul'), not the mind, that was involved. In other words, his memory only stretched as far as the youth of his present body and he could not tell us whether nature and its climate were different now as compared to centuries ago. He advised us to delve deeply into science to better understand wildlife and the migration of birds across the Himalayas.

The idea of this book was conceived during a visit to the Grand Canyon, Arizona, on a foggy day in 2013. The Canyon was filled with mist from rim to rim, and although our view of the canyon was obscured, we did begin to visualize a book on bird migration across the Himalayas. We discussed possibilities, opportunities and challenges at length. This book is aimed at stimulating further research into and conservation of migratory birds in Asia. We learned much from the different authors, and we are certain that much more can be learned. The Central Asian Flyway encompasses some of the most exciting places on earth, and the vast stretches of land between the Arctic Ocean and the Indian Ocean are awe-inspiring. The Central Asian Flyway is one of the most vulnerable flyways in the world, and it needs urgent protection measures. This book, we hope, will inform and persuade policy-makers and conservation practitioners to take appropriate measures to safeguard the bird migration systems in the Himalayas and beyond. Finally, we hope that the Himalayas remain well known not only for their geological and geographical wonders, but also for their avian populations and migrations.

Introduction

Herbert H.T. Prins and Tsewang Namgail

> *We behold the face of nature bright with gladness, we often see superabundance of food: we do not see, or we forget, that the birds which are idly singing round us mostly live on insects or seeds, and are thus constantly destroying life; or we forget how largely these songsters, or their eggs, or their nestlings, are destroyed by birds and beasts of prey; we do not always bear in mind, that food may be now superabundant, is not so at all seasons of each recurring year. I should premise that I use the term Struggle for Existence in a large and metaphorical sense, including dependence of one being on another.*
>
> Charles Darwin (1859, p. 116)[1]

Perhaps the seasonal migrations of birds across the highest mountains on Earth, the Himalayas, represent the ultimate 'struggle for existence' that Charles Darwin referred to in 1859 when *The Origin of Species* saw the light. Imagine yourself being born somewhere in the northern Siberian taiga. Autumn approaches and soon the area freezes to a mind-numbing −50°C. Gales howl from the Arctic at wind speeds of more than 100 km.h^{-1}. The ground freezes and food is difficult to find, but somewhere in the Deep South along the coasts from Myanmar (Burma) to Iran, or even across the Indian Ocean along the East African coasts, there is much better weather and much more food. Would you not rather migrate too if you had wings to take you there? However, between the Deep South and the Frozen North lies Central Asia, a region replete with high mountains and deserts. Central Asia is also bordered to the south by the Himalayas, with the Hindu Kush to the west and the Hengduan Shan to the east. The highest peaks in this mountain range frequently reach 8000 m and above (more than 100 peaks are higher than 7300 m); more important, though, the few lowest valleys that present a path through this barrier are still higher than 5000 m, with more in the east than in the west. (Figure 0.1). In the field of ornithology, this connection between North Asia and South Asia and beyond is named the 'Central Asian Flyway' (Figure 0.2).

The Central Asian Flyway covers a vast area, encompassing about 15 countries between the Arctic Ocean and the Indian Ocean. After leaving Siberia, the migratory birds in this flyway first cross the Altai Mountains, then the Gobi and Taklamakan Deserts and then the Tian Shan (Mountains). If migrating birds were to decide to circumvent these mountains that rise up to 6000 m, then they could not benefit from the lush meadows and deciduous forests before they have to cross the Taklamakan Desert. After flying over this desert, they have to cross the Tibetan Plateau, after which loom the high Himalayas, straddling almost

[1] Darwin, C. (1859). *The Origin of Species by Means of Natural Selection or the Preservation of Favoured Races in the Struggle for Life*. Reprinted by Penguin Books Ltd, London (1968).

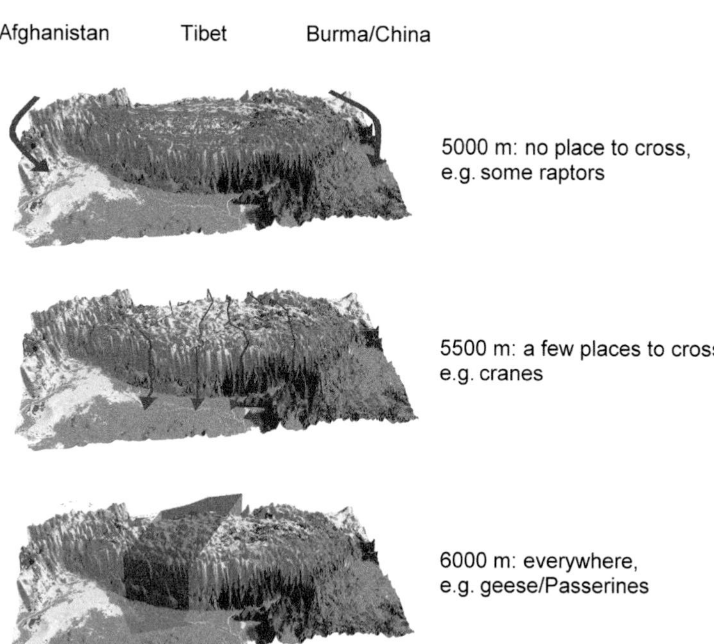

Figure 0.1. A digital elevation model of the area between the Pamirs in the west and the Hengduan Shan mountains of Sichuan in the east (vertical scale exaggerated 1000x) shows that there are no routes through this vast barrier of the Himalayas and the Tibetan Plateau for birds that cannot fly higher than 5000 m above sea level (or about 540 millibar). If the maximum ceiling is 5500 m (500 mb), then there are only a few routes open for passage, with more in the east than in the west, but if their ceiling is 6000 m (475 mb), nearly the whole area can be crossed even though peaks have to be circumvented. (A black-and-white version of this figure will appear in some formats. For the colour version, please refer to the plate section.)

across the breadth of the flyway. The Himalayas stretch more than 2400 km between Namche Barwa in the east and Nanga Parbat in the west. The width of the Himalayas ranges from about 400 km in the west to about 150 km in the east. This young mountain range is bounded to the north by the vast Tibetan Plateau and to the south by the plains of the rivers Indus, Ganges, Brahmaputra and Irrawaddy, but between these southern plains and the Greater Himalaya and the Lesser Himalaya there is still another mountain range to be passed by birds migrating south, namely the Siwalik Range, which runs for the entire length of the western and central Himalayas. Further to the west, the Himalayas merge into the Zanskar Range, the Ladakh Range, the Pir Panjal, the Karakoram, the Hindu Kush and the Pamirs. These far western mountain ranges include many peaks that are higher than 7000 m and even 8000 m.

Birds go to incredible altitudes: we have seen Horned Larks foraging in snow-clad scree in mid-summer at 6200 m above sea level (a.s.l.); we found nests of Black Redstarts at 5200 m a.s.l. and of Common Redshanks at 4800 m a.s.l. We have seen Golden Eagles being attacked by Bar-headed Geese, and vice versa, hundreds of metres above our heads while we ourselves were at some 4800 m. Resting at a pass (5670 m a.s.l.), we have

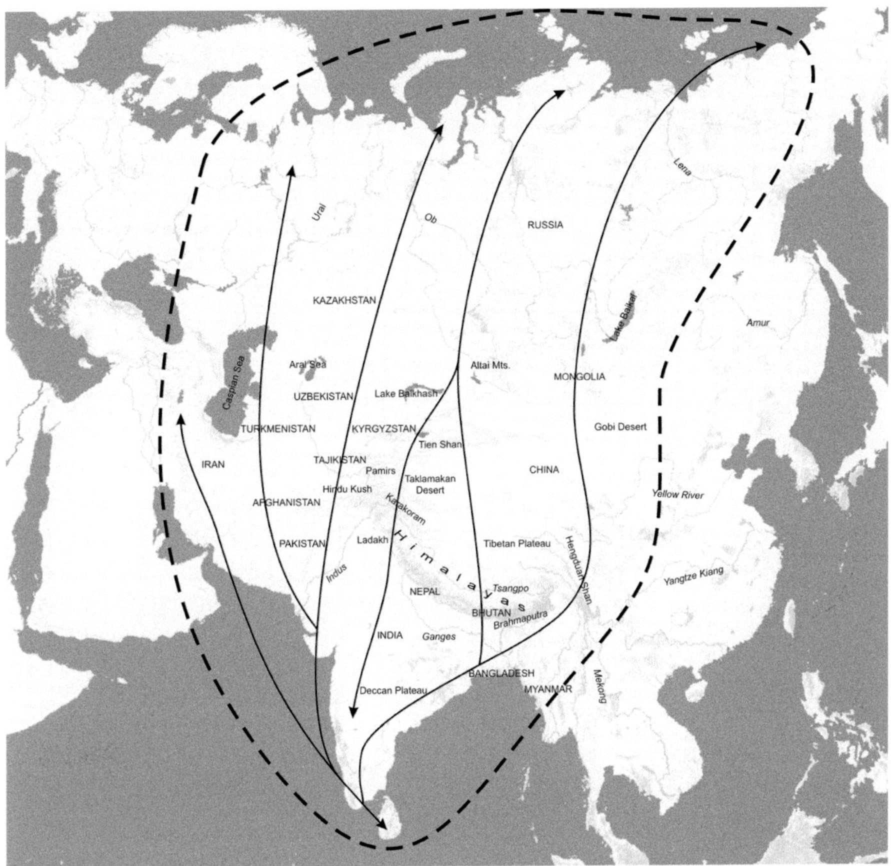

Figure 0.2. The Central Asian Flyway covers a vast area, encompassing about 15 countries between the Arctic Ocean and the Indian Ocean. Birds migrating in this flyway have to cross many obstacles (from north to south): the Altai Mountains, the Gobi and Taklamakan Deserts, the Tian Shan Mountains, the Tibetan Plateau, the Karakoram, the Pamirs, the Zanskar Range, the Ladakh Range, the Great Himalaya, the Lesser Himalaya, the Pir Panjal Range and finally the Siwalik Range. Many of the mountain peaks are higher than 7000 m and even 8000 m, and mountain passes are typically between 4000 m and 6000 m.

witnessed flocks of Ruddy Shelducks flying across to the myriad wetlands of eastern Ladakh. We have seen Tibetan Snowfinches feeding at 6000 m. Telemetry studies have also shown that Bar-headed Geese and Steppe Eagles fly higher than 7000 m in the Himalayas. Yes, indeed, birds can go to incredible altitudes – so high that earlier observers put forward the notion that they went so high that they could not fly any more: when in 1827 the American Quaker Joziah Harlan led an Afghan army across the Hindu Kush into what is now Turkmenistan, he observed 'storks walking over the passes' because putatively these passes were too high for the storks to cross on the wing, and he derived the ancient Greek name *petiamplus* for what we now call the Hindu Kush ('slayer of Indians') from the Persian *petipluampus*, meaning (according to Harlan) 'peaks over which eagles cannot soar'. Even though we are not aware of any recent observation of pedestrian bird crossings of these highest mountain passes on Earth, the point is that people have been standing in awe of these mountains and have queried the possibility that birds could fly over them.

How high is 'high' actually? This question is not easy to answer, but to give the reader a feeling for these incredible altitudes: one of the highest permanently occupied villages is Kibber, in the district of Lahul-Spiti, at 4200 m a.s.l. Possibly the highest permanent camp by pastoralists is close to Kibber at 5200 m (the head of the family had eight living children – of these one was living in Germany, one in Japan and one in New Delhi when we interviewed him). At this altitude, the partial oxygen pressure is only about 55 per cent of that at sea level. More telling, perhaps, is the fact that even though arterial human blood still contains oxygen, at this altitude the oxygen saturation of the venal blood is reaching only 30 per cent. In the perpetual state of semi-war between Pakistan and India, troops on both sides of the frontier try to occupy ever-higher permanent posts to give their howitzers the advantage, but soldiers have discovered at a high cost that neither Indian nor Pakistani can live longer than about six weeks in camps higher than about 5200 m a.s.l. But the barrier to bird migration is higher and goes up into the eternal ice, which starts here at about 6000 m. The higher one gets into these mountains, the shorter the window of opportunity to find food; at the lowest places the growing season for plants is still some five months at 3000 m a.s.l., but at 5500 m this is reduced to just a month (Figure 0.3). This book focuses

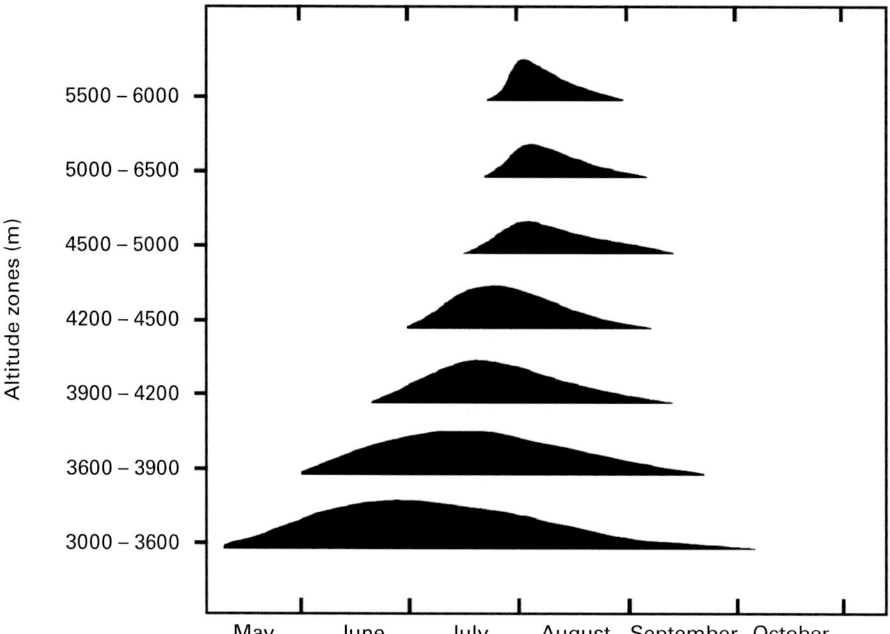

Figure 0.3. The growing season decreases with altitude in the Himalayas, and the onset of spring advances, but the end of the growing season does not get much delayed. This could imply that there is much less suitable habitat available in spring for birds migrating north, then in autumn for birds migrating south (after Figure 12 in M.S. Mani, 1978. *Ecology and Phytogeography of High Altitude Plants of the Northwest Himalaya*. London: Chapman & Hall).

on how birds are able to pass this nearly unsurmountable barrier in the Central Asian Flyway.

We, the editors, have been struck with awe every time we were observing these birds doing the nearly impossible, and out of that awe arose our wish to better describe what is perhaps much more of an accomplishment than the well-featured annual migration of the zebra and wildebeest over the Serengeti or the sardine run along the Wild Coast of South Africa. The annual struggle of the ducks, geese, cranes, raptors, waders and passerines over the Himalayas is a triumph of evolution. We thus invited many specialists to write chapters for this book, because without their deep knowledge of the Himalayas in the wider sense of the word, we could not construct the portrayal of this triumph. Together as a team of authors, we represent approximately 2000 man-years of observation and thought, give or take 500 years. That is minute as compared to the accumulated knowledge with regard to the European or American Flyways, but is not minute in comparison to the study of the Serengeti migrations referred to earlier. To the best of our knowledge, we are the first to bring together this information concerning the bird migration across the Himalayas, not under the assumption that this will be the ultimate volume describing this seasonal trek, but in the hope that it will focus and stimulate research and the description of the natural history of one of the most challenging places for natural selection to act.

In this book, we brought on board not only ornithologists and ecologists, but also geologists, climatologists, glaciologists, sociologists, archaeologists and aviators. Before the Himalayas came into existence, the ancestors of the present-day birds may have flown across the Tethys Sea to reach wintering areas on the Indian Island, which was then drifting northward. The migration must have become easier as the island came closer to the Eurasian land mass. Once the Indian Plate collided with the Eurasian Plate some 45 million years ago, the Tibetan Plateau and the Himalayas rose up, creating the most massive physical barrier in the world for present-day migratory birds. To come to grips with this, we included chapters on understanding the historical processes that created this joint barrier of the Tibetan Plateau and Himalayas. Birds have been carrying out this epic journey probably for tens of millions of years, as alluded to earlier, and apparently adapted as the mountains grew. When the barrier arose, the climate changed. In this day and age of climate change, of course we must pay attention to the changes that took place in the Himalayas. The formation of the Himalayas, and especially the rise of the Tibetan Plateau in the recent 20 million years, altered the monsoon regimes that dominate the current climates of Asia. The rise of the Himalayas and the associated mountain formation made the climate of the land beyond progressively drier, forming desert habitats in Central Asia. But the Himalayas also protected the Indian subcontinent from subarctic winds. Despite the differences in opinion over the exact pattern, consensus exists that the uplift of Central Asia influenced the current monsoon regimes in Asia during the Miocene. The rise of the Tibetan Plateau may even have been one of the causes of the ice ages. To grasp what drives the weather patterns (needed to appreciate air movements that birds use to help them across these mountains) we included chapters that help the reader to understand weather and climate. We even included work by glider pilots to help explain how air movements may help (or indeed, hinder) birds in their efforts to cross the massive mountains.

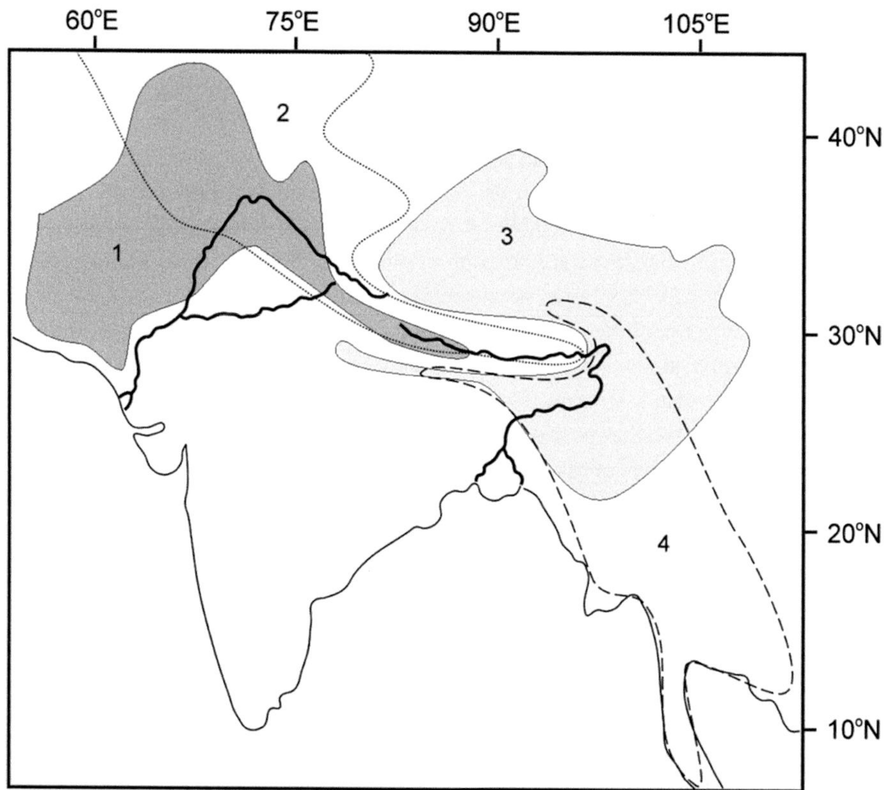

Figure 0.4. The major phytogeographical areas for the high-altitude flora of the 'Roof of the World': 1 – Mediterranean affinity, 2 – Central Asian affinity, 3 – Indo-Chinese affinity and 4 – Malayan affinity (after Figure 47 in M.S. Mani, 1978. *Ecology and Phytogeography of High Altitude Plants of the Northwest Himalaya*. London: Chapman & Hall). The high mountains truly represent a crossroads between East and West.

Since the end of the Pleistocene, these climate changes are quite well known, although there is still controversy over the extent of ice cover on the Tibetan Plateau and the Himalayas. Some contend that the Tibetan Plateau was covered with an ice sheet, while others argue that ice sheets were restricted only to the mountain ranges. In any case, it is thought that the extent of ice in the Himalayan region was relatively restricted, generally extending less than 10 km from contemporary ice margins. It is apparent that glaciers in the Himalayan region reached their maximum extent early in the last glacial maximum (about 20,000 years ago). These glacial advances significantly influenced the hydrology and the wetlands of the Himalayas. The Himalayas hold about 15,000 glaciers, which store about 12,000 km^3 of fresh water. They cover an area of about 1.1 million km^2, and include the territory of eight countries. They are the source of 10 major rivers that together provide irrigation and drinking water to one-third of the world's population. The vast icecaps of the Himalayas are perhaps the world's largest freshwater reserve after the polar icecaps.

Before the advent of humans in the Himalayas, the availability of wetlands for migratory birds must have been much higher than today. Much of the water melted from the glaciers then fed the wetlands in the Himalayan valleys. The rapidly growing human populations and their ceaseless activity claimed innumerable wetlands and converted them for human use. All the valley bottom wetlands at altitudes lower than ~3500 m a.s.l. have been converted to agricultural land. But the region still has a myriad of wetlands, including glacial lakes, marshes, wet grasslands, rivers, streams, ponds and artificial water reservoirs, some of which are relatively undisturbed, at higher altitudes. And still the high lands of the Himalayas, Pamirs and other mountain ranges and the Tibetan Plateau (Figure 0.2) form a botanical Valhalla, because here four major phytogeographical regions intermingle (Figure 0.4), providing a high diversity of seed sources for migratory birds. Lakes in the Himalayas play a crucial role in driving the hydrological regime of mighty rivers such as the Ganges, Brahmaputra and Indus, and act as buffers between glacial meltwaters and outflows to smaller rivers and streams. Apart from supporting countless migratory birds, these wetlands also support unique ecosystems and services that sustain the livelihoods of almost a quarter of the world's human population. They are thus crucial in sustaining biological and cultural diversity. Many of these wetlands also play important roles in the religious and social fabric of life in the Himalayan region.

The geographical features and climatic vagaries have limited the production systems to agriculture and livestock rearing in the Himalayan-Tibetan region. Agriculture in the region is believed to have started in the second millennium BCE. Retrieval of carbonized seeds from Neolithic sites, and the seeds of wild wheat and barley from the aceramic and ceramic levels, provides evidence that the Neolithic settlers in the Himalayas cultivated both wild and domestic plants. Recent excavations of the remains of a horse from a megalithic site in the Garhwal Himalayas indicate that pastoral communities had arrived at the beginning of the second millennium BCE. The transhumant livestock grazing in the Himalayas was, however started by the Aryans in 1500 BCE. Much knowledge concerning the advent of man in these mountains can also be learned from petroglyphs that can be found at many places in these mountain ranges.

This was only the beginning of land use changes driven by man. Wetlands continue to disappear as the needs of humans increase and their aspirations change. The most prominent factors that are threatening the wetlands are land reclamation for agriculture, and extraction of fuel wood, fodder and timber resources from the wetlands. Since the disappearance of wetlands is threatening to disrupt bird migration systems in the Himalayan region, it is crucial to understand the dynamics of the extent of wetlands in relation to other land use in the region. The focus of agriculture in the Himalayan region is slowly shifting from traditional cereal crops to more lucrative cash crops such as fruits and vegetables. This transformation from subsistence to commercial agriculture poses new threats to migratory birds. Since the demand for cash crops is increasing, to cater to the needs of a burgeoning human population, wetlands are being drained and reclaimed for agriculture, which reduces the extent of foraging areas available for migratory birds. Modern agriculture also makes indiscriminate use of government-sponsored, subsidized, artificial fertilizers, insecticides and herbicides, which pollutes the wetlands and reduces the abundance of insects and aquatic invertebrates that the migratory birds rely on. Likewise, governments build roads in the Himalayan-Tibetan region, also for geopolitical reasons, thus making agriculture economically viable at the household level. Therefore, modern agriculture, characterized by monoculture and extensive use of chemicals and pesticides, threatens the bird migration system in the Himalayan-Tibetan region. However, it is not only arable agriculture that is changing in these vast lands straddling the Central Asian Flyway of migratory birds.

Modern livestock production systems are also threatening migratory birds and their wetland habitats. The wetlands on the Tibetan Plateau and surrounding regions are scattered among the grazing lands of pastoral nomads. Government agencies provide subsidized hay in winter, thus boosting domestic livestock numbers at ever higher altitudes. The livestock then encroach on the grasslands and wetlands in summer. The meadows around lakes, rivers and streams are therefore intensively grazed by domestic livestock. Since the needs of the pastoral communities on these rangelands are increasing and their aspirations are changing, the livestock population is increasing steadily, which is putting immense pressure on the wetland resources. For example, the population of Pashmina goats is increasing rapidly. Within India, there is high pressure from the Kashmiri government to increase Pashmina production to absorb about 50,000 families, who lost their livelihoods due to the ban on *Shahtoosh* ('king of wool' in Persian – made from the now Globally Threatened Tibetan antelopes). In addition, with globalization, some nomads look down on their pastoral lifestyle as primitive. Thus, they tend to settle down, especially near water bodies such as lakes and streams, thereby increasing the pressure on the wetland resources.

The Himalayas form a unique and popular tourist destination. Tens of thousands of people from all over the world visit the natural and cultural sites in the Himalayan region. An increasing number of tourists also visit various lakes for observing migratory birds, and insensitive tourists often disturb and scare these birds away. Furthermore, indiscriminate camping at the edge of high-altitude lakes pollutes the wetlands, thereby affecting the abundance of aquatic insects. Military activities close to the wetlands also cause damage and disturb the birds. Since the Himalayan range straddles the boundaries of many countries with strained relationships, these issues are becoming more serious. The military also feeds dogs around the military camps. Some of these feral dogs wander off into the mountains, and predate the eggs and chicks of migratory birds. As if these issues are not problematic enough for bird migration, global climate change is increasingly seen as a major threat to migratory birds in their breeding and/or wintering areas. Increased glacial melting is resulting in the retreat of nearly two-thirds of the 15,000 glaciers in the Himalayas. With more than 2 billion people dependent on this glacial melt, the wise use of wetlands is critical for maintaining steady water flows and reducing the risks of floods. Apart from being water banks, the wetlands in the Himalayan region play a crucial role in supporting millions of migratory and resident birds that forage and nest around these lakes and in these high-altitude wetlands.

So, there may be problems enough to threaten the long-term viability of the Central Asian Flyway. Are there solutions to protect this migratory thoroughfare? There is a glimmer of hope due to the socio-religious protection of birds by some communities in the Himalayas. Bottom-up bird conservation is not uncommon in the Himalayan region. For example, the Chinese government has initiated a major bottom-up community-based conservation programme to train local people to monitor the health of the wildlife habitat. In India, people have managed wetlands in the Himalayan mountains for centuries, for subsistence and livelihood. Several communities across the Indian Himalayas are actively engaged in managing lakes and rivers. For instance, local people in the Ladakh Trans-Himalaya discourage tourists from camping along the shores of lakes and encourage them to camp only at the designated sites. Communities have also freed areas from grazing so that wild animals can thrive. But is there a role for the state governments too? States, not local communities, are signatories to the Ramsar Convention. We hope this book will serve as a clarion call for the governments of Afghanistan, Bangladesh, Bhutan, Burma (Myanmar), China, India, Kazakhstan, Kyrgyzstan, Mongolia, Nepal, Pakistan, the Russian Federation, Tajikistan and Uzbekistan to take a hard look at the effectiveness of their conservation efforts.

Aldo Leopold already taught us that conservation is tightly linked to ethics. Although, perhaps too often, we may think that conservation boils down to 'it is the economy, stupid', there is still the lingering thought that ethics may supplant – at least partially – money when it comes to sustaining conservation. It is in the Himalayas that Buddhism took strong roots. Compassion towards sentient beings is at the heart of Buddhism. Furthermore, birds have a special place in Buddhism. Garuda is a mythological bird of Buddhists and Hindus, the main inhabitants of the Himalayas. Hindus even worship this mythical bird. The Black-necked Crane is another bird that features in folklore, as well as myths, legends and songs in the Himalayan region. Furthermore, people regard most of the lakes in the Himalayas as sacred. The most sacred of all is Mansarovar Lake in Tibet, revered by both Buddhists and Hindus. Of course, we do not claim that Buddhism and Hinduism are the only ethical systems that could uphold conservation in this part of the world: we are only too aware that in this part of the world ethical systems such as atheism, Christianity, Islam, Marxism and Sikhism all compete to offer guidance to people living with the great migration across the Himalayas in stepping up their conservation efforts. Indeed, time is running out.

Wetlands form the backbone for this great migration in the dry grasslands and scree plains that cover vast tracts of the high-altitude lands, but high-altitude wetlands are increasingly being degraded due to intense grazing by livestock and resource extraction by farmers, whose needs are increasing apace. In other words, the 'fuelling stations' for the migratory birds are disappearing, making migration to and from the south increasingly difficult, which may ultimately lead to the extinction of these migratory birds. Increasing human populations and associated developmental initiatives are diverting water away from these wetlands. Furthermore, many of the wetlands associated with the Himalayan river systems are small and widely scattered, but they offer suitable habitat for the migratory birds. These small wetlands are fed by glaciers, but because of their small size and dispersed nature, they are generally ignored by conservation planners. These wetlands do come under the purview of the Ramsar Convention, but are insufficiently distinct and individually important for bureaucrats to act on; however, these ephemeral or dispersed wetlands are crucial for the success of avian migration across the Himalayas and need to be delineated and conserved.

Isolated studies have been carried out on different aspects of the ecology of migratory birds and wetlands in the Himalayan region, but there has not yet been an apparent effort to synthesize the existing knowledge. This book is an attempt to extract relevant information from the existing literature, invite current researchers to share their new findings and generate new knowledge that is relevant to the conservation of migratory birds and wetlands in the Himalayan region.

Many chapters in this book focus on Ladakh in the western part of the Himalayas and in the Trans-Himalaya of India. Anyone knowledgeable about the vast Himalayan range from the West to the East knows that Ladakh does not represent that whole range at all. However, in the past 20 to 40 years, access to most of this vast barrier to study avian migration was difficult: in the West, the wars in Afghanistan and Tajikistan prevented most if not all ornithologists from going to the mountains, and civil unrest in northern Pakistan led to low numbers of birders in the region. Similarly, in the Central Himalayas, in Nepal and in the Far East, the lack of permits given by the Indian authorities and the then pariah state of Myanmar also led to a dearth of observation opportunities for serious birdwatchers. Only in Ladakh was there a free window for study and research. Hence the predominance of studies on Ladakh in the current book. We do not apologize for that: on the contrary, we thank the Ladakhi people and authorities for being so open-minded in facilitating overseas scientists but also local ones to study their mountainous paradise.

A Note on the Names of Places and Geographical Features in this Book

Many geographical features have been mentioned in this book. That is, of course, a consequence of the subject of migration ('The periodic passage of groups of animals from one region to another for feeding or breeding' as defined by www.biology-online.org/dictionary/ accessed 31 May 2016). The editorial team has taken great pains to find and verify the names and coordinates of these locations. All these names and coordinates can be found in the gazetteer at the end of this book. We had some difficulty in finding standardized names, but took as a starting point *The Times Comprehensive Atlas of the World*, eleventh edition. If we could not find a name there, we chose to follow Wikipedia. If we could not find it there, we chose to follow names as found on the Internet through the Google search engine (last accessed 1 June 2016). The Appendix includes more than 400 locations within the Central Asian Flyway, and gives names in Tibetan too.

Acknowledgements

A book such as this could not of course have been made by us alone. First we want to acknowledge our indebtedness to Dominic Lewis who, as Editor Life Sciences of the Press was willing to propose our book outline to the Syndicate. Equally important was Patricia Meijer of Wageningen University, who assisted us with all administrative details, but who also vigorously kept track of the progress of all authors, checked references and chased them if necessary. Herman van Oeveren spent much time making sure that figures and graphics were up to standard, and re-drew numerous figures to ensure a uniform style. Simon Delany spent countless hours on editing tasks, checking information and honing the language of texts, making the chapters as clear as possible without compromising individual style. TN wishes to acknowledge the Graduate School Production Ecology & Resource Conservation (PE&RC) for supporting his stay at Wageningen University during the early phase of the writing. He is deeply indebted to Herbert and Sip van Wieren for the opportunity to join them on many hikes in the mountains and to study the fascinating assemblage of birds in Ladakh. He is also grateful to John Takekawa for giving the opportunity to work on migratory ducks in India. He is thankful to Tsewang Dolma and Odkar Namgail for their love and support. HP is grateful to Wageningen University for allowing him the time to work on this book and so often to go to Central Asia and elsewhere on this fascinating Earth. He acknowledges his deep indebtedness to Charu Mishra for first taking him to the high mountains, to Sip van Wieren for so often being his companion in these mountains, and of course to Tsewang, who taught him so much about his homeland. He also acknowledges the Fulbright Centre for granting him a Fulbright Scholarship to start working on this book, and the Department of Ecology and Evolutionary Biology of Princeton University for hosting him as Visiting Professor. He is equally grateful to the Members of the Welgevonden Game Reserve (South Africa), and in particular its chairman, Mr. François Spruyt, for welcoming him in their reserve and hosting him for writing. François: thank you so much for introducing me to your family doctor, Theo Louridos, and, Theo, thank you so much for ensuring a very good gastroenterologist to operate on me when I was in the African bush. Dr Ian Berkowitz and your colleagues and nurses – perhaps you did not save my life, but at least you brought me back to be a working ecologist again! We both thank all the authors for forming such a wonderful team.

Part I

Migratory Routes and Movement Ecology

1 Goose Migration across the Himalayas: Migratory Routes and Movement Patterns of Bar-headed Geese

John Y. Takekawa, Eric C. Palm, Diann J. Prosser, Lucy A. Hawkes, Nyambayar Batbayar, Sivananinthaperumal Balachandran, Ze Luo, Xiangming Xiao and Scott H. Newman

True Geese in Asia

True geese (subfamily Anserinae) are large-bodied wildfowl with long necks and legs adapted for grazing, and they are represented by more than 30 species or subspecies worldwide. Most species form strong and persistent breeding pair bonds (Owen, 1980; del Hoyo *et al.*, 2001) and migrate long distances between southern wintering areas and northern breeding grounds in temperate or Arctic regions of the Northern Hemisphere (no true geese occur in the Southern Hemisphere with the exception of one introduced population). True geese are much more abundant in North America (16 million individuals) and Europe (4 million individuals) than in Asia, where fewer than 1 million individuals winter, and numbers continue to decline (Kear, 2005; Fox *et al.*, 2013). Asian goose populations include: Dark-bellied Brent *Branta bernicla bernicla*; Aleutian Cackling Goose *B. hutchinsinii asiatica*, Red-breasted Goose *B. ruficollis*, Bar-headed Goose *Anser indicus*, Taiga Bean Goose *A. fabalis*, Tundra Bean Goose *A. serrirostris*, Emperor Goose *A. canagicus*, Greater White-fronted Goose *A. albifrons*, Greylag Goose *A. anser rubrirostris*, Lesser Snow Goose *A. caerulescens*, Lesser White-fronted Goose *A. erythropus* and Swan Goose *A. cygnoides*.

Three flyways encompass the migration of geese across Asia: the West Asian–Eastern African Flyway, the Central Asian Flyway and the East Asian–Australasian Flyway (Boere & Stroud, 2006). The Central Asian Flyway (Figure 1.1 inset) is uniquely defined by the 200–400-km-wide barrier of the Himalayan mountain range (hereafter Himalayas) that extends 2300 km from Nanga Parbat in Pakistan in the west to Namche Barwa in China in the east, rising to an average elevation of 6100 m above sea level (a.s.l.). The Himalayas present an obstacle to migration of waterbirds wintering on the Indian subcontinent from northern breeding grounds (Figure 1.1). Only two populations of true geese numbering in the thousands, the Bar-headed Goose and the eastern population of the Greylag Goose, regularly winter on the Indian subcontinent.

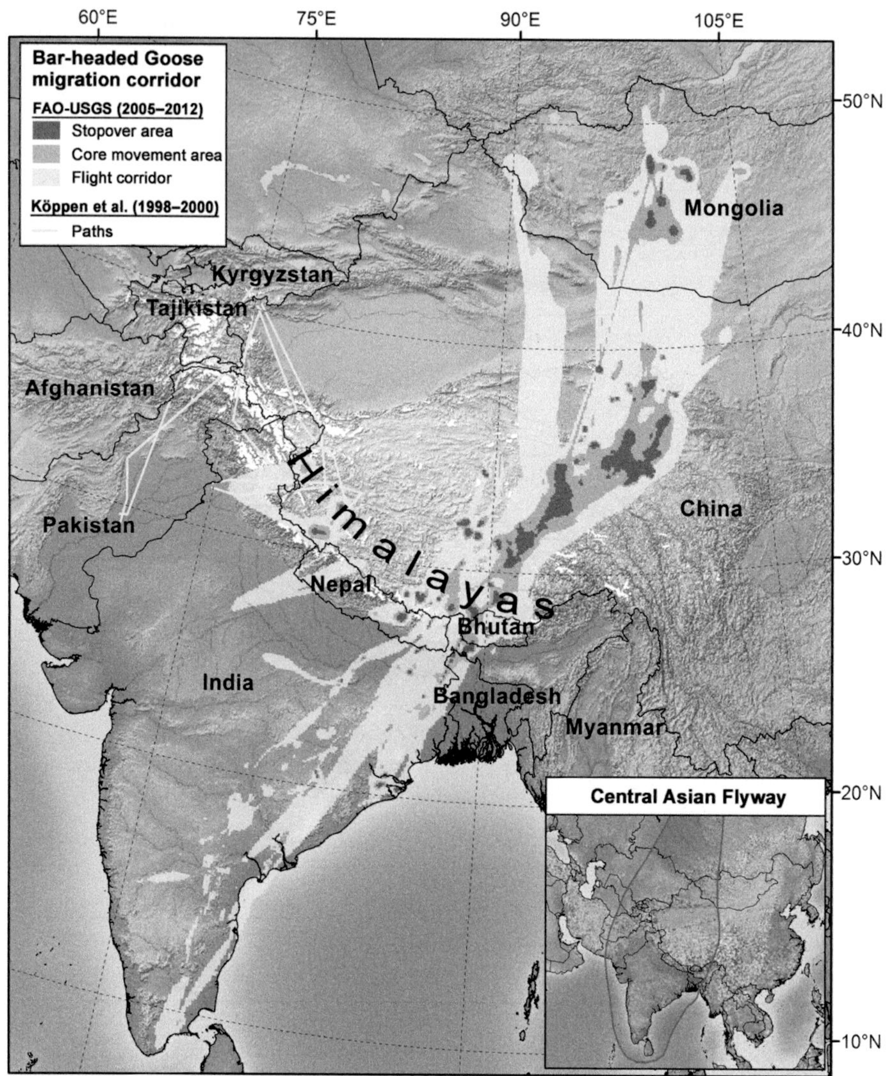

Figure 1.1. Extent of the Central Asian Flyway (inset) and migration routes and relative use of stopover areas for 44 satellite-marked Bar-headed Geese (BHGO). Colours, from darkest to lightest, represent 50%, 75% and 99% cumulative probability contours from a dynamic Brownian Bridge movement model. Adapted from Palm *et al.* (2015) and Köppen (2010). (A black-and-white version of this figure will appear in some formats. For the colour version, please refer to the plate section.)

Although the western subspecies of Greylag Goose found in Europe is abundant, widespread and well-studied, the eastern subspecies are little studied and represent a small population distributed in few areas. Recent estimates from India suggest 6000–8000 geese with the largest congregations of 3500 at Pong Dam, Himachal Pradesh (S. Balachandran, pers. obs.). Until recently, movements of the Bar-headed Goose were also poorly known (see Javed *et al.*, 2000; Takekawa *et al.*,

2009; Köppen *et al.*, 2010), but in 2005, an outbreak of highly pathogenic avian influenza (HPAI), H5N1, killed more than 3000 individuals at Qinghai Lake in west-central China (Chen *et al.*, 2005). After that outbreak, several studies (see Prosser *et al.*, 2009; Prosser *et al.*, 2011; Newman *et al.*, 2012) were undertaken to examine the movement ecology of the species and to determine its potential role in disease transmission as well as its physiological adaptations for crossing the Himalayas (Hawkes *et al.*, 2011; Hawkes *et al.*, 2013; Bishop *et al.*, 2015; see also Hawkes *et al.*, Chapter 16). Thus, in this chapter, we focus on the migration of the Bar-headed Goose as the most studied goose species that crosses the Himalayas.

Bar-headed Geese are medium-sized geese with males (2370 g, 95% CI = 68 g, N = 16) generally larger than females (2100 g, 95% CI = 93 g, N = 17; (J.Y. Takekawa, unpubl. data).

Population Size

Global population estimates vary widely, but the Bar-headed Goose is not considered vulnerable to extinction due to its extensive range (BirdLife International, 2015). Miyabayashi and Mundkur (1999) had estimated a total world population of 32,300–35,400 individuals, with two distinct subpopulations based on wintering areas comprising 16,800–18,900 geese in South Asia to Myanmar and 15,500–17,500 birds in China. Wetlands International (2015) reported a worldwide population of 52,000–60,000; however, recent estimates of large populations (> 30,000 individuals) wintering or staging in the Lhasa River Valley of the southern Tibetan Autonomous Region (China) and 60,000–70,000 in India, with 30,000–40,000 at Pong Dam alone (S. Balachandran, unpubl. data) suggest that this total may be an underestimate.

Breeding Ecology

The Bar-headed Goose breeds at high altitudes (3000–5000 m a.s.l.) in central Asia. Many geese visit breeding lakes prior to the spring thaw in April and then retreat before returning to breed in May. They nest primarily on islands in high-altitude lakes, 0.5 to 2.0 m above the water level, in dense colonies of up to 530 nests per ha (Owen, 1980). In northwest China, colonial island nests may be located 0.4 to 1.5 m apart with 0.6 nests per m^2, which is among the highest reported nesting densities of any true goose species (Ma & Cai, 1997).

Breeding birds remain at the nest site and nearby feeding areas during the nesting period. Females incubate the eggs while males remain vigilant nearby, but pairs may share nesting islands with gulls and cormorants (Batbayar *et al.*, 2014). Clutch size averages 3–4 eggs (range: 2–10 eggs; Owen, 1980; Kear, 2005; Batbayar *et al.*, 2014), and incubation lasts 27–30 days (Owen, 1980; del Hoyo *et al.*, 2001; Batbayar *et al.*, 2014). Smaller numbers of geese nest on river islands, in old raptor or raven nests in trees, or on cliffs where nest survival may be higher. They use a variety of nesting substrates, including pondweed, sand and soil (Owen, 1980; Batbayar *et al.*, 2014).

Bar-headed Geese forage on natural wetlands, including freshwater and brackish lakes and riverine habitats (del Hoyo *et al.*, 2001; Kear, 2005; Prosser *et al.*, 2011). Parents and goslings eat sedges, grasses and aquatic plants, including pondweed, but goslings also appear to forage on insect larvae. Goslings feed on lake edges with their parents and typically reach 400–500 g body mass within one month, fledging in 60 days (Owen, 1980). Families are thought to remain together throughout the autumn migration and the winter, and birds reach sexual maturity at two to three years old (del Hoyo *et al.*, 2001).

Wintering Numbers and Ecology

Bar-headed Geese spend most of the winter on grasslands, pastures or agricultural areas, but they also use habitats in nearby natural wetlands, rivers and lakes (Prosser *et al.*, 2011). Recent counts in Tibet and China suggest an increase in the population size of Bar-headed Geese, but it is unclear if these estimates reflect a true increase in population size, or shifts from India to southern Tibet to winter, or changes in survey coverage to include previously undocumented wintering areas. The largest wintering population in China appears to be in southern Tibet near Lhasa, where numbers have increased markedly in the past two decades from 13,000 to 14,500 in 1990–1991 and 1995–1996 (Bishop *et al.*, 1997) to an estimated 32,000 by 2006 (Bishop & Tsamchu, 2007). Zhang *et al.* (2011) estimated 50,000 birds in 2009 in the Yarlung River Basin, including the Lhasa watershed and much of south-central Tibet. Rivers in Shanxi, Sichuan and Yunnan support several thousand wintering birds, and populations on the Yungui Plateau in northern Yunnan increased from 1800 in 2002 to 5300 in 2004 (Liu *et al.*, 2014).

South of the Himalayas on the Indian subcontinent, the estimated wintering population included 5000 birds in Myanmar (Van der Ven & Gole, 2010) and 15,000–17,000 in India during the 1990s (Miyabayashi & Mundkur, 1999), but numbers of Bar-headed Geese wintering in India appear to have increased (S. Balachandran, pers. comm.). For example, between 19,000 and 40,000 birds were counted at Pong Dam (Rahmani & Zafar-ul Islam, 2009; S. Balachandran, unpubl. data). Wintering birds forage and roost in flocks of hundreds to a few thousand, and feed on river flood plains at night, but roost on mid-stream sand-banks during the day. Winter diets consist of grasses and sedges, but they may feed on seaweed, crustaceans and invertebrates on coasts (Owen, 1980). In recent decades, they have been recorded feeding on wheat, barley and rice stubble in agricultural areas (Bishop *et al.*, 1997). In Keoladeo National Park, wheat *Triticum aestivum* L. and native knotgrass *Paspalum distichum* L. are important food sources in mid-winter, while peas *Pisum sativum* L. and chickpeas *Cicer arietinum* L. constitute a large portion of diets in March and April (Middleton & Van der Valk, 1987).

Table 1.1. Season, marking sites (origin), major stopover sites, destination, departure and arrival dates, duration of migration and number of Bar-headed Geese marked with satellite transmitters in the Central Asian Flyway

Season	Origin	Major stopover site(s)	Destination	Median departure date	Median arrival date	Duration (days)	n
Spring	Koonthankulam, India	Torsa River, India	Zhaling-Eling Lakes, China	26 Mar	25 May	60	1
	Koonthankulam, India	Dochen Lake, Tibet	W Mongolia: Darkhad Valley, Tuul River, Terhiyn Tsagaan Nuur, Gegeen Lake	17 Mar	7 May	51	4
	Chilika Lake, India	Dochen Lake and Yamdrok Lake, Tibet	Qinghai Lake, China	6 Mar	25 Apr	50	2
	Chilika Lake, India	Dochen Lake, Tibet	Sagsai River, Mongolia	14 Mar	24 May	71	1
	Chilika Lake, India	Ganges River, India	Yamdrok Lake, Tibet	26 Feb	16 Mar	18	1
	Chilika Lake, India	Jaldhaka River, India	Elephant Lake, India	21 Mar	10 May	50	1
	NE Bihar, India	Yeru River, Tibet	Zhaling-Eling Lakes, China	20 Apr	27 May	37	1
	Lhasa and Yarlung Rivers, Tibet	Zhaling-Eling Lakes, China	Qinghai Lake and Qilian Mountains, China	5 Apr	27 Apr	22	7
	Lhasa River, Tibet	Wetlands SW of Nagqu, Tibet	Sanjiangyuan National Nature Reserve, China	5 Apr	22 May	47	1
	Indus River, Pakistan	-	Son Lake, Kyrgyzstan	1 Apr	2 Apr	1	1
	Pong Dam, India	River valleys near Rutog and Rabang, Tibet	Chatyr Lake, Kyrgyzstan	13 Apr	7 May	24	1
	Pong Dam, India	-	Tibe: Tsokar Lake, Bangong Lake, Indus River Headwaters	13 Apr	9 May	26	3
	Nepal: Chitwan National Park India: Keoladeo National Park	-	S Tibet: Lake Ngangla Rinco, Dangquezangbu River, Argog Lake, Yoqag Lake, Urru Lake, Ngoin Lake	28 Mar	9 Apr	12	7
Post-breeding (moulting)	Qinghai Lake and Qilian Mountains, China	-	Qinghai-Tibetan Plateau: Zhaling-Eling Lakes, Huangheyuan Lake, Donggeicuona Lake, Alake Lake, Cuolongka Lake, Kuhai Lake	20 Jun	27 Jun	7	10

Table 1.1. (cont.)

Season	Origin	Major stopover site(s)	Destination	Median departure date	Median arrival date	Duration (days)	n
Autumn	W Mongolia: Terkhiin Tsagaan Lake, Gegeen Lake	China: Zhaling-Eling Lakes, Alake Lake, Sanjiangyuan National Nature Reserve; NW Qilian Mountains; Tibet: Damxung	E India: Tilaiya Dam Reservoir, Chilika Lake, Ganges River, Tel River	3 Sep	2 Dec	90	5
	W Mongolia: Darkhad Valley, Terkhiin Tsagaan Lake, Sagsai River	China: Chuma'er River, Zhaling-Eling Lakes; Tibet: Namtso Lake	S India: Koonthankulam, Tungabhadra Valley, Almatti Reservoir	25 Aug	3 Dec	100	3
	Qinghai Lake, China	Sanjiangyuan National Nature Reserve, China; Yamdrok Lake, Tibet	E India: Chilika Lake, Mahananda River	4 Sep	8 Dec	95	2
	Qinghai Lake, China	Zhaling-Eling Lakes, China Zhamucuo Wetland, Tibet	Tibet: Lhasa, Yarlung, Nyang Chu Rivers	9 Sep	10 Nov	62	8
	Qinghai-Tibetan Plateau: Alake Lake, Bayan Har Mts., Upper Yellow River Basin, Dou Lake, Donggeicuona Lake	China: Zhaling-Eling Lakes, Bayan Har Mts.; Zhamucuo Wetland; Tibet: Selincuo Reserve	Tibet: Lhasa, Yarlung Rivers	5 Oct	7 Nov	33	6
	Yonghong Lake, China	Selincuo Reserve, Tibet	Chilika Lake, India	14 Sep	22 Nov	69	1
	Zhaling-Eling Lakes, China	-	Nyang Chu River, Tibet	1 Oct	10 Nov	40	1
	Yamdrok Lake	-	Chilika Lake, India	5 Nov	21 Nov	16	1
	Elephant Lake, India	-	Tungabhadra Valley, India	20 Nov	23 Nov	3	1
	Yoqag Lake, Tibet	East Tingri, Tibet	Dhasan River, India	3 Nov	6 Dec	33	1
	Chatyr Lake, Kyrgyzstan	Mindam Lake, river valleys near Rutog and Rabang, Tibet	Pong Dam, India	28 Sep	15 Nov	48	1
	Son Lake, Kyrgyzstan	Oksu and Pyandsh Rivers, Tajikistan-Afghanistan-China border area	Indus River, Pakistan	3 Oct	14 Nov	42	2

Migration Routes and Stopover Areas

Several migration routes have been described for individual Bar-headed Geese based on satellite telemetry studies from 1999 to 2012 (Table 1.1; Figure 1.2), and a broad-front migration strategy has been identified (Javed *et al.*, 2000; Takekawa *et al.*, 2009). To the west, geese marked in Kyrgyzstan followed three routes: (1) south over the Pamir Range to the Tajikistan-Afghanistan-China border and wintering areas in Punjab, Pakistan; (2) southwest across the Tien Shan Range and Turkmenistan to the Vakhsh River Valley in Uzbekistan; and (3) southeast to western Tibet crossing the Himalayas to wintering areas in northwestern India, including the Ravi, Degh and Chenab River Valleys and the Beas Reservoir (Köppen *et al.*, 2010). Geese marked at Pong Dam in northwest India migrated short distances northeast across to Tsokar and TsoMoriri in Ladakh, where some geese breed (Prins & Van Wieren, 2004; Namgail *et al.*, 2009), while one goose continued towards high-altitude lakes in southwestern Tibet. Marked geese wintering in northwest India and Nepal (n = 11) flew short distances north to breeding areas in south-central Tibet (Takekawa *et al.*, 2009, Kalra *et al.*, 2011), including saline lakes on the Qinghai-Tibetan Plateau such as Taro Tso northwest of Lhasa (Javed *et al.*, 2000).

Not all populations crossed the Himalayas, however, and many marked birds wintered in southern Tibet. In fact, there may be a trade-off for geese wintering north and south of the Himalayas; geese wintering in Tibet must withstand harsher winters but offset this cost with a shorter migration distance and earlier arrival at the breeding grounds (Takekawa *et al.*, 2009). Geese wintering in the milder climates of central and southern India migrate longer distances and cross the highest mountain barrier on the planet to breed in northern China and Mongolia. Breeding colonies at Qinghai Lake were strongly associated with wintering grounds in the Lhasa River Valley in southern Tibet (Prosser *et al.*, 2011), as nearly all marked geese from the Qinghai Lake breeding population (94%, n = 16) wintered there. The remaining geese crossed the Himalayas to winter in Chilika Lake, Odisha, India (Takekawa *et al.*, 2009). Overall, geese from Qinghai Lake flew twice as far (800 km) as geese wintering in northwestern India and Nepal, but geese from the latter group crossed the Himalayas.

Bar-headed Geese marked in southern and eastern coastal India followed a 3000–5000 km leapfrog migration (Figure 1.3) past Qinghai Lake, crossing the Himalayas, the Qinghai-Tibetan Plateau and the Gobi Desert to breeding areas in Mongolia (Takekawa *et al.*, 2009, Hawkes *et al.*, 2013). For these geese, Dochen Lake and wetlands southeast of Lake Como Chamling were the first spring stopovers north of the Himalayas (J.Y. Takekawa, unpubl. data).

Altogether, 65% (11 of 17) of geese wintering in south and east India spent the breeding season at sites across western Mongolia, including Terkhiin Tsagaan Lake, the Sagsai River Valley, the southern Darkhad Valley and the Tuul River. The remaining marked birds flew to Qinghai Lake (n = 2), the eastern Qinghai-Tibetan Plateau (n = 2) or near Lhasa (n = 2). Most geese marked in Mongolia crossed to northeast India and

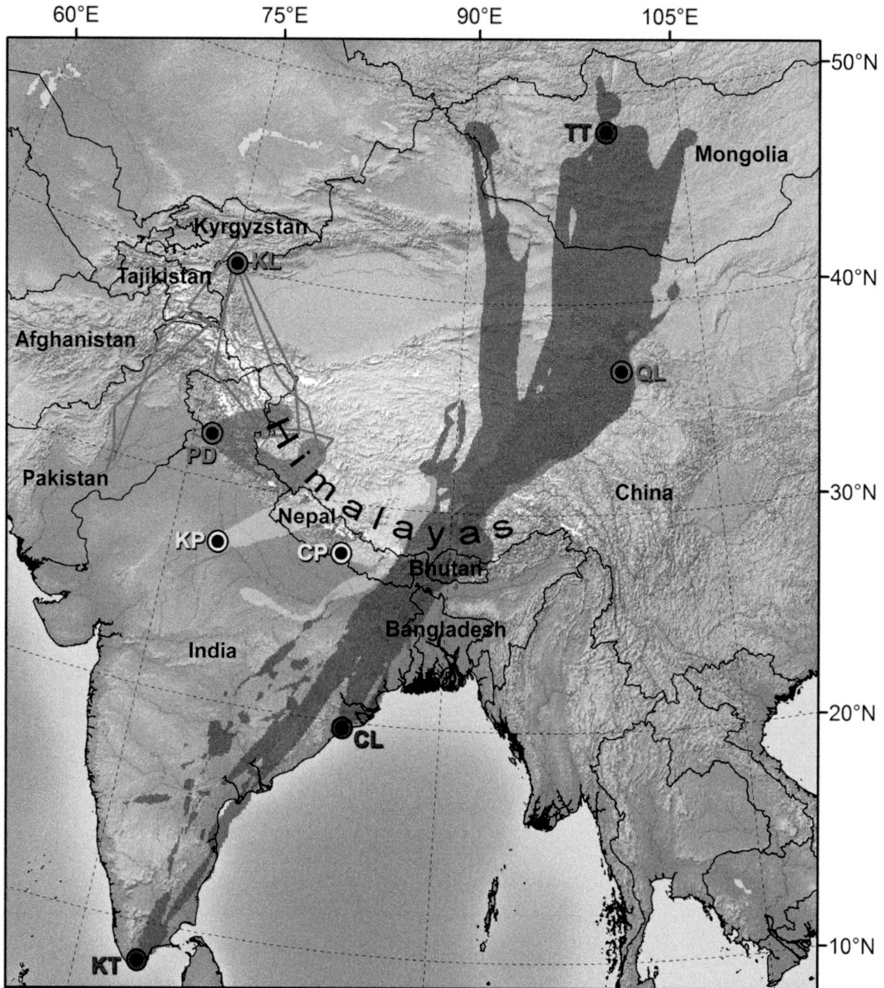

Figure 1.2. Distinct migration patterns of Bar-headed Geese from different marking sites of Terkhiin Tsagaan Lake, Mongolia (TT), Kyrgyzstan Lakes, Kyrgyzstan (KL), Qinghai Lake, China (QL), Chitwan National Park, Nepal (CP), Pong Dam, India (PD), Keoladeo National Park, India (KP), Chilika Lake, India (CL) and Koonthankulam, India (KT). Colours of marking site labels correspond to shaded areas highlighting separate migration routes. Shaded areas represent 99% cumulative probability contours of a dynamic Brownian bridge movement model. (A black-and-white version of this figure will appear in some formats. For the colour version, please refer to the plate section.)

Bhutan near the Yarlung-Brahmaputra River, although some (31%, n = 16) wintered near Lhasa. Many geese (70%, n = 20) crossed the Himalayas through Nathu Pass; 4,300 m) on the border between Sikkim, India and Tibet, China, an area south of Dochen Lake was often used as a stopover site. Zhamu Co Wetland and Sanjiangyuan Nature Reserve were major autumn stopover sites. The Nam Co region 115 km northwest of Lhasa also was an important stopover site in the late autumn (Prosser *et al.*, 2011). The

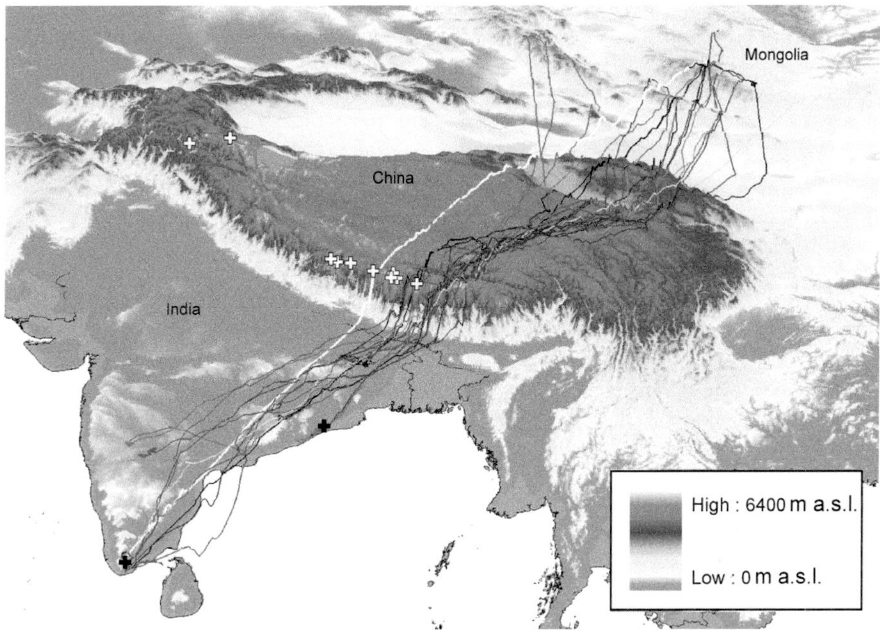

Figure 1.3. Three-dimensional tracks of 16 Bar-headed Geese from CL, KT and TT crossing the Himalayas. Solid white line shows the great circle route, and white plus signs indicate locations of peaks > 8000 m in elevation. Adapted from Hawkes *et al.* (2011). (A black-and-white version of this figure will appear in some formats. For the colour version, please refer to the plate section.)

Lhasa River Valley, Yamdrok Lake and Dochen Lake were all important autumn stopovers for birds migrating from Mongolia to India.

Timing of Migration

Bar-headed Geese migrating over the Himalayas typically climbed 4000–6000 m in 7–8 h (Figure 1.3), but did not rely on tailwinds to complete the crossing (Hawkes *et al.*, 2013). They usually varied their elevation by following the terrain (Bishop *et al.*, 2015) and had an average maximum flight elevation of 5600 m a.s.l. However, the duration and speed of migration and the stopovers used varied by population segment (Figure 1.2). During the spring and autumn migration, geese flew 1280–1550 km from Kyrgyzstan breeding areas to a major staging area over 14–49 days. Average stopover was twice as long during autumn (32–46 days) versus spring migration (16–23 days), and maximum distance covered in a day was 680 km (Köppen *et al.*, 2010). Average migration between Qinghai Lake and southern Tibet was 51 days (1–4 stops) in the autumn and 25 days (0–2 stopovers) in the spring. Geese that moulted southwest of Qinghai Lake averaged 40 days (1–2 stopovers) en route to southern Tibet (~850 km) during autumn migration (Prosser *et al.*, 2011). Migration from western Mongolia to India (3400–4850 km) averaged 62 days (2–5 stopovers) in the spring and 80 days (2–7 stopovers) in the autumn.

Wintering – Most geese arrived on wintering grounds from mid-October to early December (Köppen *et al.*, 2010; Prosser *et al.*, 2011). The median arrival date was 8 November in southern Tibet (n = 20) and 28 November (n = 14) in India (J.Y. Takekawa, unpubl. data), including wetlands and rivers in southern Bihar and northern Jharkhand.

Spring migration – Median departures from wintering areas were 14 March (n = 11) from Chilika Lake, Odisha in east India, 17 March (n = 8) from Koonthankulam, Tamil Nadu in south India, and 5 April (n = 9) from the Lhasa River Valley (Prosser *et al.*, 2011).

Spring stopovers – Major stopovers (Figure 1.1) included the Zhaling-Eling Lakes area at the source of the Yellow River, and Siling Co Lake, the largest lake in Tibet and the second largest saline lake in China after Qinghai Lake.

Breeding arrival – Most geese arrive at their breeding grounds in late April to early May (Kear, 2005). The median arrival date was 4 May (n = 11) at Qinghai Lake and 18 May (n = 5) in Mongolia (J.Y. Takekawa, unpubl. data). Egg-laying began from 3–20 May but started in mid-June at higher altitudes or in years when snow cover extended late into the spring (Owen, 1980; Prins & Van Wieren, 2004).

Moulting – Non-breeding geese begin moulting earlier than breeders (Kear, 2005). Adults began wing moult 24–28 days after the young hatched and lost all flight feathers within 24 h. Geese regained their ability to fly 32–35 days after the start of wing moult and moult was fully completed after 40–45 days. Body moult starts 21–24 days after wing moult and lasts 10 weeks (Kear, 2005). Geese that bred at Qinghai Lake remained there during the moult. Unsuccessful breeders (about 50% of marked geese) moulted in several locations southwest of Qinghai Lake. The median departure date to moulting areas for non-breeders was 20 June, and the median arrival date at Qinghai-Tibetan Plateau moulting areas was 27 June (n = 10). Zhaling-Eling Lakes and Huangheyuan wetlands received the most use, but other sites included Donggeicuona and Alake Lake, Cuolongka Lake and Kuhai Lake (Cui *et al.*, 2011). Geese remained at moulting sites for 28–42 days and sometimes flew short distances southward before autumn migration (Cui *et al.*, 2011; Prosser *et al.*, 2011).

Autumn migration – Most geese began autumn migration between late August and early October. Geese moulting in western Mongolia departed on 5 September (n = 33; J.Y. Takekawa, unpubl. data) and geese from Qinghai Lake on 10 September (n = 20; Prosser *et al.*, 2011; Zhang *et al.*, 2011); the departure date was 5 October for geese moulting southwest of Qinghai Lake, including the Zhaling-Eling Lakes, Galalacuo and Donggeicuona Lakes and the Huangheyuan wetland (n = 10).

Conservation Threats and Priorities

Outside of a few notable exceptions, most goose populations in North America and Western Europe have been increasing, while many East Asian populations have decreased. Geese face numerous threats in East Asia, but the primary concerns are changing land use patterns, including damming of rivers for electricity generation, illegal hunting and climate change (Kear, 2005; De Boer *et al.*, 2011). Satellite telemetry has been instrumental in identifying important habitats for conservation. Of potential conservation targets, high-elevation lakes and wetlands in the Qinghai-Tibetan Plateau are critical areas for Bar-headed Geese. These lakes provide wintering, moulting, staging and breeding habitat for several populations (Takekawa *et al.*, 2009; Prosser *et al.*, 2011; Palm *et al.*, 2015; *cf.* Groen & Prins, Chapter 17). They provide autumn and spring staging habitats for geese breeding in western Mongolia, northern China and Qinghai Lake. High-elevation lakes and wetlands in the Qinghai-Tibetan Plateau also serve as breeding habitat for geese wintering in northwest India and Nepal and provide moulting habitat for Qinghai Lake breeders.

Land use change – In many areas across the range of Bar-headed Geese, riverine habitat is being lost to channelization. In Tibet, hydroelectric development threatens to disrupt river flows and alter important roosting and feeding areas. Power lines from these projects are already a major source of injury and mortality to geese in some areas, including along the Yarlung River in Tibet, which is part of a large wintering area in the flyway (Li *et al.*, 2011). Elsewhere, power lines near roosting sites and migratory corridors are a significant threat. Around Lhasa, valuable wetlands and agricultural habitats are being converted to greenhouses and other developments (Lang *et al.*, 2007). Availability of agriculture may have resulted in geese short-stopping in this region rather than following traditional migration routes crossing the Himalayas (Elmberg *et al.*, 2014). Wetland reclamation, urbanization and water pollution have led to population declines at wintering areas in Guizhou Province. Illegal gold mining, overgrazing, poaching and other anthropogenic disturbances have also degraded goose habitats (Wu *et al.*, 2007) and may have decreased breeding success in the Altun Mountain Nature Reserve (Zhang *et al.*, 2012).

Climate change– Changing climatic conditions are evident in the Qinghai-Tibetan Plateau, where temperature increases have been greater than Northern Hemisphere and global averages (Liu & Chen, 2000) and are more pronounced during the winter and at higher elevations. In the northern Qinghai-Tibetan Plateau and Mongolia, higher temperatures and decreased precipitation coupled with higher water demand by local communities have altered semiarid habitats in recent decades (but see Bagchi *et al.*, Chapter 13). Shrinking lakes and decreasing wetlands have reduced suitable nesting areas and increased the risk of nest trampling on lake islands (Batbayar *et al.*, 2014). In the southern Qinghai-Tibetan Plateau, warming temperatures and higher humidity have led to an increase in vegetative cover (Xu *et al.*, 2008). Breeding phenology of Arctic nesting geese has been sensitive to climatic variation during the spring and the summer (e.g. Dickey *et al.*, 2008). In single-brooded species such as the Bar-headed Goose, a temporal mismatch between the brood rearing period and peak food availability could negatively affect reproductive success (Jiguet *et al.*, 2007).

Disease threats– Bar-headed Geese are susceptible to outbreaks of highly pathogenic avian influenza H5N1 virus. The largest recorded wild bird outbreak at Qinghai Lake killed 3282 Bar-headed Geese in May 2005. It coincided with the arrival of thousands of northward-migrating geese and with onset of breeding (Takekawa *et al.*, 2009). Brown *et al.* (2008) found that geese experimentally infected with H5N1 virus survived up to eight days. Thus, during the asymptomatic period, infected geese could fly hundreds of kilometres and disseminate the virus. The wintering population in southern Tibet overlaps with high-density poultry production, including domesticated Bar-headed Geese (Prosser *et al.*, 2011). Numerous highly pathogenic avian influenza H5N1 outbreaks have occurred in this area (Prosser *et al.*, 2011) and have the potential to spread the disease (*cf.* Si *et al.*, 2009).

Predation and harvest – Island nests in central Mongolia are at risk of nest predation by Mongolian Gulls *Larus vegae mongolicus*, Common Ravens (*Corvus corax*) and domestic dogs as well as being susceptible to trampling (Batbayar *et al.*, 2014). In India, China and Russia, Black Kites *Milvus migrans* and Common Ravens also predate nests (Gole, 1982; Baranov, 1991; Ma & Cai, 1997). Over-hunting and egg collecting also have contributed to population declines across much of their range (del Hoyo *et al.* 2001). The Bar-headed Goose is officially protected in Tibet, but hunting still occurs there, and it is not protected in the rest of China (Bishop *et al.*, 1997).

Summary of Bar-headed Goose Migration

Only two species of geese number in the thousands in the Indian subcontinent (Bar-headed Goose and Greylag Goose), and until recently, not much was known about the migration of either species. However, recent satellite telemetry studies provide a better picture of the movement ecology of Bar-headed Geese wintering from southern China to the southern tip of India with a population that seems to be increasing. In contrast, movement ecology of the Greylag Goose is still little known, but it is often found in relatively small flocks of tens to low thousands, often in the same wintering areas as the Bar-headed Goose.

Different populations of Bar-headed Geese may employ different strategies to balance the costs and benefits associated with variation in migration distance, arrival date at breeding areas and weather conditions at wintering grounds. In general, geese wintering in harsh conditions north of the Himalayas migrate shorter distances and arrive on their breeding grounds earlier than the geese that winter in the warm climate of central and southern India and breed farther north in Mongolia (Takekawa *et al.*, 2009). Populations that winter immediately south of the Himalayas in Nepal and far northern India have the shortest migrations (usually lasting less than one week), but must still cross the mountains twice a year.

The most heavily used migration corridor for Bar-headed Geese breeding in Qinghai Lake and Mongolia extends southwest from the Qinghai Lake region towards Lhasa, Tibet, and southward to the northernmost extent of Bangladesh. Geese breeding south of this core area stop more frequently and for longer durations than when migrating through the margins of the range (Palm *et al.*, 2015). Furthermore, geese migrated across a broad front but travelled within a relatively narrow corridor while migrating through the eastern Qinghai-Tibetan Plateau. Satellite marking studies are helping to provide more details about the connectivity of their wintering, breeding and migration areas in the eastern (Myanmar; Yunnan Province, China) and western (Pakistan, northwest India, Tajikistan, Kyrgyzstan) areas of their range. In addition, studies on the eastern population of Greylag Goose may provide insights into the migration strategies of this population, and whether it also has adapted to crossing the Himalayas.

Acknowledgments

We thank the many colleagues who cooperated in the collection of the field data on the geese of the Central Asian Flyway. We appreciate the assistance of T. Natsagdorj, Baoping Yan, Zhi Xing, Yongsheng Hou, David Douglas and Bill Perry. The use of trade, product or firm names in this publication is for descriptive purposes only and does not imply endorsement by the US government.

References

Baranov, L.A. (1991). *Rare and little-studied birds of Tuva*. M.S. Thesis. Krasnoyarsk, Russia: Krasnoyarsk State University.

Batbayar, N., Takekawa, J.Y., Tseveenmyadag, N., Spragens, K.A. & Xiao, X. (2014). Site selection and nest survival of the Bar-headed Goose (Anser indicus) on the Mongolian Plateau. *Waterbirds*, **37**, 381–393.

BirdLife International (2015). Species factsheet: Anser indicus. Retrieved from www.birdlife.org on 12/02/2015.

Bishop, C.M., Spivey, R.J., Hawkes, L.A., *et al.* (2015). The roller coaster flight strategy of Bar-headed Geese conserves energy during Himalayan migrations. *Science*, **347**, 250–254.

Bishop, M.A. & Tsamchu, D. (2007). Tibet Autonomous Region January 2007 survey for Black-necked Crane, Common Crane, and Bar-headed Goose. *China Crane News*, **11**, 24–26.

Bishop, M.A., Yanling, S., Zhouma, C. & Binyuan, G. (1997). Bar-headed Geese Anser indicus wintering in south-central Tibet. *Wildfowl*, **48**, 118–126.

Boere, G.C. & Stroud, D.A. (2006). The flyway concept: what it is and what it isn't. In G.C. Boere, C.A. Galbraith & D.A. Stroud., eds., *Waterbirds around the world*. Edinburgh, UK: The Stationery Office, pp. 40–47.

Brown, J.D., Stallknecht, D.E. & Swayne, D.E. (2008). Experimental infection of swans and geese with highly pathogenic avian influenza virus (H5N1) of Asian lineage. *Emerging Infectious Diseases*, **14**, 136–142.

Chen, H., Smith, G.J., Zhang, S.Y., *et al.* (2005). Avian flu: H5N1 virus outbreak in migratory waterfowl. *Nature*, **436**, 191–192.

Cui, P., Hou, Y., Tang, M., *et al.* (2011). Movement patterns of Bar-headed Geese Anser indicus during breeding and post-breeding periods at Qinghai Lake, China. *Journal of Ornithology*, **152**, 83–92.

De Boer, W.F., Cao, L. Barter, M., *et al.* (2011). Comparing the community composition of European and eastern Chinese waterbirds and the influence of human factors on the China waterbird community. *Ambio*, **40**, 68–77.

del Hoyo, J., Elliott, A. & Christie, D. (2001). *Handbook of the birds of the world, vol. 1.* Ostrich to ducks, 2nd ed. Barcelona, Spain: Lynx Editions.

Dickey, M.H., Gauthier, G. & Cadieux, M.C. (2008). Climatic effects on the breeding phenology and reproductive success of an arctic-nesting goose species. *Global Change Biology*, **14**, 1973–1985.

Elmberg, J., Hessel, R., Fox, A.D. & Dalby, L. (2014). Interpreting seasonal range shifts in migratory birds: a critical assessment of 'short-stopping' and a suggested terminology. *Journal of Ornithology*, **155**, 571–579.

Fox, A.D., Lei, C., Barter, M., *et al.* (2013). The functional use of East Dongting Lake, China, by wintering geese. *Wildfowl*, **58**, 3–19.

Gole, P. (1982). Status of Anser indicus in Asia with special reference to India. *Aquila*, **89**, 141–149.

Hawkes, L.A., Balachandran, S., Batbayar, N., *et al.* (2011). The trans-Himalayan flights of Bar-headed Geese (Anser indicus). *Proceedings of the National Academy of Sciences of the United States of America*, **108**, 9516–9519.

Hawkes, L.A., Balachandran, S., Batbayar, N., *et al.* (2013). The paradox of extreme high-altitude migration in Bar-headed Geese (Anser indicus). *Proceedings of the Royal Society: Biological Sciences*, **280**, 1–8.

Javed, S., Takekawa, J.Y., Douglas, D.C., *et al.* (2000). Tracking the spring migration of a Bar-headed Goose (Anser indicus) across the Himalaya with satellite telemetry. *Global Environmental Research*, **4**, 195–205.

Jiguet, F., Gadot, A.-S., Julliard, R., Newson, S.E. & Couvet, D. (2007). Climate envelope, life history traits and resilience of birds facing global change. *Global Change Biology*, **13**, 1672–1684.

Kalra, M., Kumar, S., Rahmani, A.R., *et al.* (2011). Satellite tracking of Bar-headed Geese Anser indicus wintering in Uttar Pradesh, India. *Journal of the Bombay Natural History Society*, **108**, 79.

Kear, J. (Ed.). (2005). *Ducks, geese and swans volume 1: general chapters; species accounts (Anhima to Salvadorina)*. Oxford: Oxford University Press.

Köppen, U., Yakovlev, A.P., Barth, R., Kaatz, M. & Berthold, P. (2010). Seasonal migrations of four individual Bar-headed Geese Anser indicus from Kyrgyzstan followed by satellite telemetry. *Journal of Ornithology*, **151**, 703–712.

Lang, A., Bishop, M.A. & Le Sueur, A. (2007). An annotated list of birds wintering in the Lhasa River watershed and Yamzho Yumco, Tibet Autonomous Region, China. *Forktail*, **23**, 1–11.

Li, F., Bishop, M.A. & T. Drolma. (2011). Power line strikes by Black-necked Cranes and Bar-headed Geese in Tibet Autonomous Region. *Chinese Birds*, **2**, 167–173.

Liu, Q., Li, F. & Yang, F. (2014). Winter distribution and abundance of Bar-headed Geese Anser indicus on the Yun-Gui Plateau, China. *Goose Bulletin*, **18**, 3–6.

Liu, X.D. & Chen, B.D. (2000). Climatic warming in the Tibetan Plateau during recent decades. *International Journal of Climatology*, **20**, 1729–1742.

Ma, M. & Cai, D. (1997). Aggregated distribution of Anser indicus nest in Bayinbuluke of Tianshan Mountains and its breeding ecology. *Chinese Journal of Applied Ecology*, **8**, 287–290.

Middleton, B. & van der Valk, A.G. (1987). The food habits of Greylag and Bar-headed Geese in the Keoladeo National Park, India. *Wildfowl*, **38**, 93–102.

Miyabayashi, Y. & Mundkur, T. (1999). *Atlas of Key Sites for Anatidae in the East Asian Flyway*. Kuala Lumpur: Wetlands International. www.jawgp.org/anet/aaa1999/aaae ndx.html.

Namgail, T., Mudappa, D. & Shankar Raman, T.R. (2009). Waterbird numbers at high altitude lakes in eastern Ladakh, India. *Wildfowl*, **59**, 135–142.

Newman, S.H., Hill, N.J., Spragens, K.A., *et al.* (2012). Eco-virological approach for assessing the role of wild birds in the spread of avian influenza H5N1 along the Central Asian Flyway. *PloS One*, **7**(2), e30636.

Owen, M. (1980). *Wild geese of the world: their life history and ecology*. London, UK: Batsford.

Palm, E.C., Newman, S.H., Prosser, D.J., *et al.* (2015). Mapping migratory flyways in Asia using dynamic Brownian bridge movement models. *Movement Ecology*, **3**, 3.

Prins, H.H.T. & van Wieren, S.E. (2004). Number, population structure and habitat use of Bar-headed Geese Anser indicus in Ladakh (India) during the brood-rearing period. *Acta Zoologica Sinica*, **50**, 738–744.

Prosser, D.J., Cui, P., Takekawa, J.Y., *et al.* (2011). Wild bird migration across the Qinghai-Tibetan plateau: a transmission route for highly pathogenic H5N1. *PloS One*, **6**, e17622.

Prosser, D.J., Takekawa, J.Y., Newman, S.H., *et al.* (2009). Satellite-marked waterfowl reveal migratory connection between H5N1 outbreak areas in China and Mongolia. *Ibis*, **151**, 568–576.

Rahmani, A.R.& Zafar-ul Islam, M. (2009). *Ducks, geese, and swans of India: their status and distribution*. Mumbai: Bombay Natural History Society.

Si, Y., Skidmore, A.K., Wang, T., *et al.* (2009). Spatio-temporal dynamics of global H5N1 outbreaks match bird migration patterns. *Geospatial Health*, **4**, 65–78.

Takekawa, J.Y., Heath, S.R., Douglas, D.C., *et al.* (2009). Geographic variation in Bar-headed Geese Anser indicus: connectivity of wintering areas and breeding grounds across a broad front. *Wildfowl*, **59**, 100–123.

Van der Ven, J. & Gole, P. (2010). Bar-headed Geese Anser indicus: notes from breeding and wintering areas. *Goose Bulletin*, **10**, 7–17.

Wetlands International (2015). 'Waterbird Population Estimates'. Retrieved from wpe.wetlands.org.

Wu, G., Leeuw, J. de., Skidmore, A.K., Prins, H.H.T. & Liu, Y. (2007). Concurrent monitoring of vessels and water turbidity enhances the strength of evidence in remotely sensed dredging impact assessment. *Water Research*, **41**, 3271–3280.

Xu, X.K., Chen, H. & Levy, J.K. (2008). Spatiotemporal vegetation cover variations in the Qinghai-Tibet Plateau under global climate change. *Chinese Science Bulletin*, **53**, 915–922.

Zhang, G., Liu, D., Hou, Y., *et al.* (2011). Migration routes and stop-over sites determined with satellite tracking of Bar-headed Geese Anser indicus breeding at Qinghai Lake, China. *Waterbirds*, **34**, 112–116.

Zhang, T., Ma, M., Ding, P., *et al.* (2012). Population ecology and current status of Bar-headed Goose (Anser indicus) in autumn at the Altun Mountain Natural Reserve, Xinjiang, China. *Goose Bulletin* **14**, 27–34.

2 Himalayan Thoroughfare: Migratory Routes of Ducks over the Rooftop of the World

Tsewang Namgail, John Y. Takekawa, Sivananinthaperumal Balachandran, Eric C. Palm, Taej Mundkur, Víctor Martín Vélez, Diann J. Prosser and Scott H. Newman

Migratory Ducks in the Himalayas

Every year, millions of ducks migrate to their breeding grounds in Central Asia and Siberia in the spring and back to their wintering areas on the Indian subcontinent in the autumn (Rahmani & Islam, 2008). The use of productive wetlands in the north, however, comes at a great cost to these ducks: they must cross the Himalayas, the most formidable mountain barrier in the world (Groen & Prins, Chapter 17). Migratory ducks cross this massive barrier twice a year. How they accomplish this epic journey and which routes they take has intrigued scientists for a long time. Owing to their size and public appeal, the geese and swans and their migratory systems are relatively better studied than ducks in the Central Asian Flyway (Rahmani & Islam, 2008).

Much of our knowledge of the migratory habits of ducks in the Central Asian Flyway has come from information collected over the 1850s–1980s. During the 1850s–1940s, British army and civil service personnel were stationed in the hilly regions of Afghanistan, Pakistan and India. Many of these personnel were avid hunters of ducks and other game birds, and they documented the early evidence of the migratory movements of wildfowl across mountain passes of the Trans-Himalayan region (see Ali & Ripley, 1983). The Bombay Natural History Society (BNHS) under the Migratory Animal Pathology Survey also undertook extensive ringing of ducks at selected sites in India, notably Bharatpur (Keoladeo National Park) (BNHS 1998; Rahmani & Islam, 2008), which provided information on the breeding ranges and migratory routes of some of the long-distance migrants in the Central Asian Flyway. There is, however, a gap in our knowledge of migration routes of ducks from India to China and Mongolia, because of a dearth of ringing programmes in Nepal, Bhutan, Bangladesh and Sri Lanka.

Although ringing projects (McLure, 1974; Rogacheva, 1992; BNHS, 1998) and observations by naturalists during their expeditions to the high mountains (e.g. Swan, 1970) revealed that these ducks migrate over some of the highest peaks in the Himalayas, very little was known about their exact migratory routes until recently (Takekawa *et al.*, 2009; Takekawa *et al.*, 2010; Hawkes *et al.*, 2011). Satellite telemetry and remote sensing helped in determining the migratory routes of these ducks, when researchers fitted more than 50 ducks in India with satellite transmitters (Takekawa *et al.*, 2010; Palm *et al.*, 2015).

In this chapter, we describe the current state of knowledge of the migratory routes of ducks across the Himalayas during their annual migrations. Although we fitted satellite transmitters to more than 50 individual ducks belonging to seven species, only 15 individuals of six species crossed the Himalayas before they stopped transmitting, and their routes are described here in detail.

Wetland Configuration and Ease of Migration

Crossing a barrier as massive as the Himalayas is energetically expensive (Groen & Prins, Chapter 17). It may, however, become possible largely because of the presence of numerous lakes and rivers on either side of this mountain range. The Tibetan Plateau to the north has numerous lakes of all sizes and shapes, most of which have lush meadows along their shores. Hundreds of thousands of migratory ducks breed at these wetlands or use them to replenish their fat reserves during their northward and southward migrations (Prins *et al.*, Chapter 18). Furthermore, the Indus flowing west and the Tsangpo River flowing east from the Mount Kailash area form a linear wetland continuum just north of the Himalayas. These rivers have innumerable mudflats and meadows along their banks, which serve as important staging sites for migratory ducks just before they ascend the Himalayan range during their southward migration (Jun *et al.* 2004). On the southern side, the Ganga River flowing east and the Brahmaputra flowing west also form a linear set of floodplains and associated lakes and marshes with various mudflats and meadows, vital stopover sites for migratory ducks crossing the Himalayas during their northward migration (Dutta & Konwar, 2013). Moreover, the southern slopes of the Himalayas receive extremely high precipitation and have various seasonal wetlands (Burbank *et al.*, 2012).

The routes migratory ducks take through the Himalayas to reach these wetlands on either side were not known until recently. Referring to ducks, Lawrence wrote: 'These birds undoubtedly migrate across the Himalaya, but it is not known whether they fly over the peaks, keep to the passes or follow the gorges' (Lawrence, 2000). Answers to this question have begun to emerge since more than 100 migratory ducks and geese were tagged with satellite transmitters in the Central Asian Flyway (Takekawa *et al.*, 2010; Prosser *et al.*, 2011; Newman *et al.*, 2012) in a quest to understand the role of migratory birds in the spread of the highly pathogenic avian influenza (HPAI) virus (H5N1). These studies took place after the deaths of more than 1000 Bar-headed Geese *Anser indicus* and gulls at Qinghai Lake in western China (Liu *et al.*, 2005). The wildfowl marked in these studies indicated that migrant ducks probably adopt different migration strategies to cope with high energy demands while flying across the Himalayas (Hawkes *et al.*, 2011). Some of these strategies include clear trade-offs between distance and altitude (Groen & Prins, Chapter 17), while others include morphological, behavioural and physiological adaptations (Hawkes *et al.*, Chapter 16).

Satellite Telemetry and Migratory Birds

In the twentieth century, ringing was a widely used method to identify the destinations and longevity of migratory birds, but this technique could not adequately determine the migratory paths (Balachandran, 2012), despite extensive ringing efforts (e.g. Wernham *et al.*, 2002). Radar has also been used to track migratory birds (Bruderer, 1997a, 1997b; Sjöberg & Nilsson, 2015), but one of the major drawbacks of this technology during its early years was that it could not identify species. Conventional radio-telemetry was used in the 1990s only for larger birds with limited movements. Today, satellite telemetry is the most efficient, although relatively expensive technology to study the migratory routes of birds.

The first satellite telemetry study to reveal the migratory routes of birds in the Himalayan region started in the 1990s, when Higuchi *et al.* (1994) fitted a Common Crane *Grus grus* in India with a Platform Terminal Transmitter (PTT) and tracked it to Siberia. Subsequently, two Bar-headed Geese were fitted with transmitters in Keoladeo National Park in India (Javed *et al.*, 2000). Out of these, one stopped transmitting before the onset of the spring migration, while the other bird departed on migration on 24 March. It flew across the Himalayas to Taro Tso in Tibet, and remained there through the breeding season. Following this, a number of studies fitted transmitters to Bar-headed Geese, ducks, gulls and Great Cormorants *Phalacrocorax carbo*, although largely to study the role of migratory birds in spreading H5N1 (e.g. Muzaffar *et al.*, 2008; Zhang *et al.*, 2008; Prosser *et al.*, 2009; Gaidet *et al.*, 2010; Takekawa *et al.*, 2010; Prosser *et al.*, 2011; Newman *et al.*, 2012; Takekawa *et al.*, 2013; Zhang *et al.*, 2014). Information derived from these efforts was also used to study the habitat use and home ranges of seven species of ducks (Namgail *et al.*, 2011; Namgail *et al.*, 2014).

Migratory Routes and Stopover Sites in the Himalayas

We tracked migratory ducks that had been trapped in their wintering areas in lowland India. The ducks were fitted with Argos and GPS solar-powered Platform Terminal Transmitters (PTTs; Microwave Telemetry Inc., Columbia, MD, USA) between 2007 and 2009 in the states of Tamil Nadu, Odisha and West Bengal. In Tamil Nadu, ducks were captured and fitted with transmitters at Puthalam and Koonthankulum. In Odisha, ducks were marked at Chilika Lake, which is the second largest coastal lagoon in the world, and supports nearly 1 million ducks and other waterbirds each year (Islam & Rahmani, 2008). In West Bengal, ducks were marked at Purbasthali and Ahiron. Capture, handling and marking procedures were approved by the US Geological Survey Patuxent Wildlife Research Center Animal Care and Use Committee (2007–2001).

We used the Brownian bridge movement model (BBMM) to estimate the migratory routes of ducks crossing the Himalayas (Figure 2.1). BBMM is a continuous-time stochastic model of movement in which the probability of an individual duck being in an area is conditioned on starting and ending locations (Horne *et al.*, 2007). We also

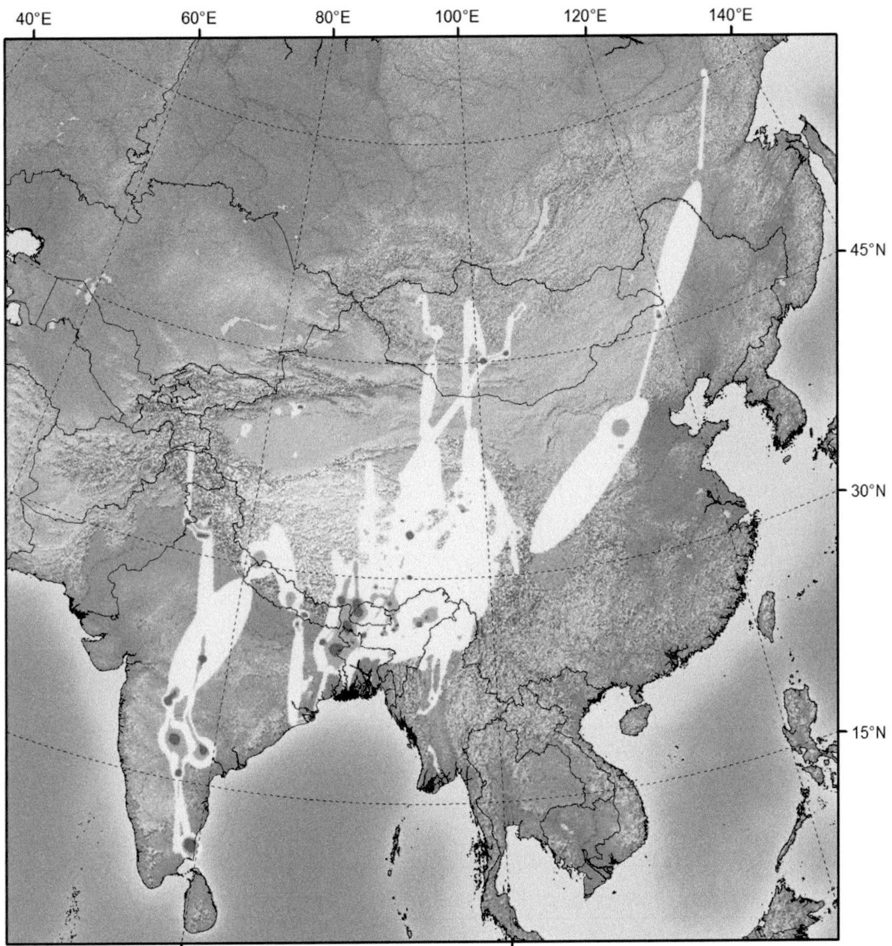

Figure 2.1. Estimated migration routes and utilization distributions of ducks in the Central Asian Flyway. From darkest to lightest, colours represent 50%, 75% and 99% cumulative probability contours (adapted from Palm *et al.*, 2015). Note that ringing locations in eastern and central India introduce bias into the mapped distributions, which do not include migration routes from western lowland India. Map data: Google earth. (A black-and-white version of this figure will appear in some formats. For the colour version, please refer to the plate section.)

showed the approximate migratory paths of individual ducks (Figures 2.2 to 2.7). Most of the ducks wintering in India pass over, and not around the Himalayas (Figures 2.1 to 2.7), which is in line with Groen and Prins (Chapter 17).

Eurasian Wigeon

The Eurasian Wigeon (EW) *Anas Penelope* is a relatively large duck (600–1000 g) with an extensive range, breeding in Northern Europe and Asia, and spending the non-breeding period in West and South Asia, Southeast and East Asia, Africa and Europe (Carboneras

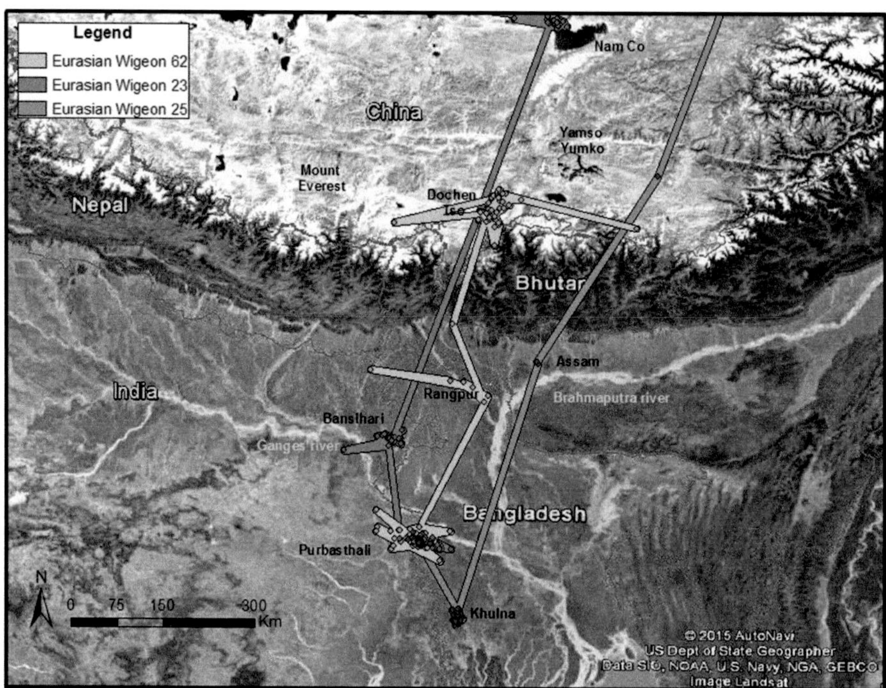

Figure 2.2. Movement paths of Eurasian Wigeon migrating between Central Asia and the Indian Peninsula, crossing the Himalayas. Map data: Google earth. (A black-and-white version of this figure will appear in some formats. For the colour version, please refer to the plate section.)

et al., 2016) with an estimated population of about 250,000 in South Asia (Wetlands International, 2012). A total of 110 rings was recovered during the Bombay Natural History Society (BNHS) ringing programme between 1980 and 1992. Of these, 100 were recovered from Russia, Kazakhstan, Kyrgyzstan, Turkmenistan and Afghanistan, nine from Pakistan and one from India (BNHS, 1998). The recoveries from their known breeding grounds in Russia indicate that the South Asian non-breeding populations originate from a large part of temperate Russia, from the Ural Mountains across to the Far East (Rahmani & Islam, 2008).

During the present study, seven individuals were tracked, of which three crossed the Himalayas before they stopped transmitting, and their routes are described here. We captured Eurasian Wigeon number 62 (EW #62) and fitted it with a transmitter at Purbasthali in West Bengal (Figure 2.2). It left Purbasthali on 13 April, flew northwards and stopped at the banks of the Teesta River in Bangladesh for a day, and then flew to Dochen Lake in south China just north of Bhutan, where it spent the summer. EW #23 was also tagged at Purbasthali, and it departed on 2 April. It flew north to Bhalgapur near Farakka Barrage on the River Ganga near the Indo–Bangladesh border, and spent 39 days there. It then flew to Nam Co, north of Lhasa, flying over the Nathu Pass (4310 m a.s.l.) in northern Sikkim. EW #25 was tagged at Ahiron in West Bengal, and it departed on 21 April. It flew north and spent three days at the banks of the Brahmaputra in Assam. It then continued northward, and stopped for two days at Dreku Lake and then flew north to Mongolia (Figure 2.2).

Gadwall

Gadwall (GD) *Anas Strepera* is a relatively small duck (550–1000 g) with a wide global distribution. It breeds in Northern Europe, Asia and North America (Carboneras *et al.*, 2016), with an estimated population of 300,000 in South Asia (Wetlands International, 2012). It is widely distributed across India during the non-breeding period (Rahmani & Islam, 2008). During the Bird Ringing Project from 1980 to 1992, the BNHS and the US Fish and Wildlife Service ringed more than 1500 Gadwalls at different wetlands in India, and 45 ducks were recovered, mostly from the western part of the Asian breeding range in Kazakhstan and central Russia (BNHS, 1998; Rahmani & Islam, 2008), although the species breeds across temperate Asia.

During the present study, nine individuals were fitted with transmitters and tracked, but only two crossed the Himalayas before they stopped sending signals. GD #35 was marked at Purbasthali in West Bengal. It flew north and spent 11 days on the Ganga River near Farakka Barrage, then flew north to the northwestern tip of Torsa Nature Reserve in Bhutan, and then returned to the Teesta River near the border between India and Bangladesh (Figure 2.3). GD #61 was tagged at Ahiron, West Bengal. From there, it flew eastward and spent three days near Guwahati on the banks of the

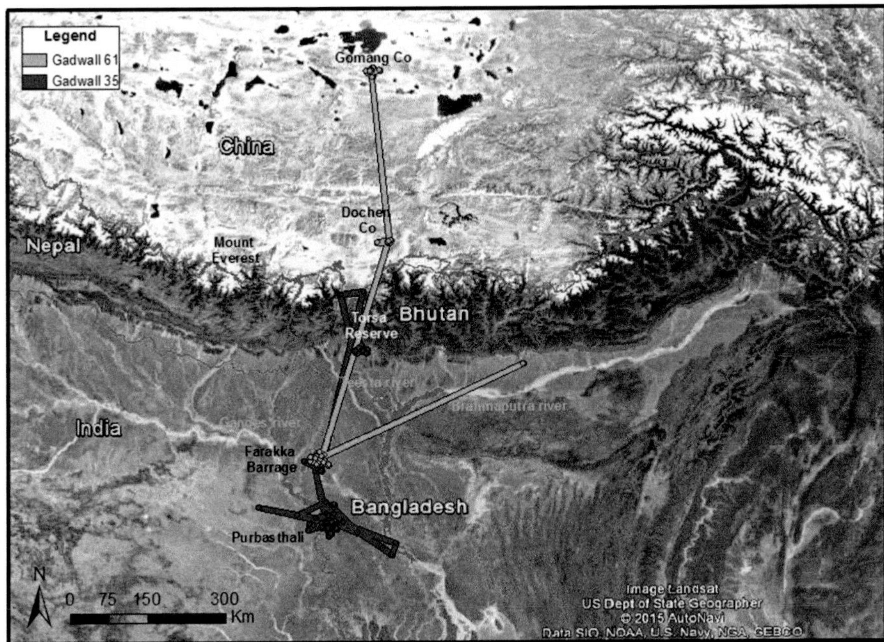

Figure 2.3. Movement paths of Gadwall migrating between Central Asia and the Indian Peninsula, crossing the Himalayas. Map data: Google earth. (A black-and-white version of this figure will appear in some formats. For the colour version, please refer to the plate section.)

Brahmaputra and returned to Ahiron. It then flew northwards to Dochen Lake via Nathu Pass (4310 m a.s.l.). It spent 28 days there, then flew north to Gomang Co in Tibet (Figure 2.3).

Northern Pintail

Northern Pintail (NP) *Anas acuta* is a relatively large duck (730–1030 g), and it breeds in northern parts of Europe, Asia and North America. The northern populations are migratory, but there are small sedentary populations on southern Indian Ocean islands (Carboneras & Kirwan, 2016a). The species is widespread in Asia, with an estimated population of 1 million in South Asia (Wetlands International, 2012). During the Bird Ringing Project (BNHS, 1998), 4050 Northern Pintails were ringed in India, of which 157 were recovered, mostly from Russia, Kazakhstan, Kyrgyzstan, Uzbekistan and western China (Rahmani & Islam, 2008).

During the present study, 10 individuals were tracked, but only two crossed the Himalayas before they stopped transmitting. NP #37 was tagged at Purbasthali. It left on its spring migration on 8 March, and flew northwards to the Yumthang Valley in northern Sikkim; it spent two days there and then flew to Dochen Co in southern Tibet (Figure 2.4). NP #75 was tagged at Purbasthali. It left on 27 February and flew northwards to Como Chamling in Tibet. This duck remained

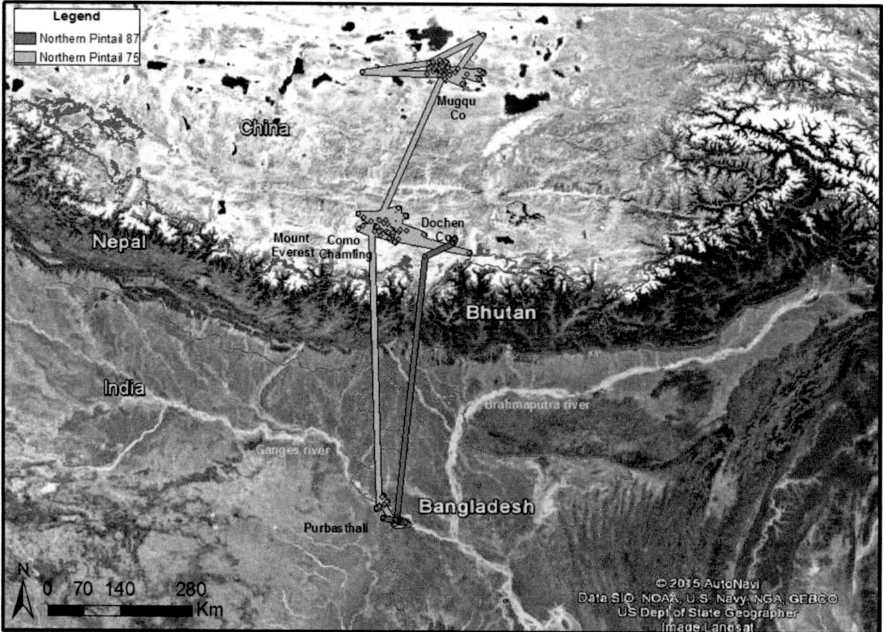

Figure 2.4. Movement paths of Northern Pintail migrating between Central Asia and the Indian Peninsula, crossing the Himalayas. Map data: Google earth. (A black-and-white version of this figure will appear in some formats. For the colour version, please refer to the plate section.)

around this lake for 52 days, flying between various wetlands, including Qiangzuo Co. It then flew to a place near Dochen Co, where it stayed for a few hours and then flew back. Finally, it flew north to the wetland complex of Mugqu Co, Ringco Gongma and Ringco Yokma, and spent the breeding season there.

Northern Shoveler

The Northern Shoveler (NS) *Anas clypeata* is a relatively large duck (410–1100 g). It is a widespread Holarctic species, and is highly migratory, although there are some non-migratory populations in parts of Europe (Carboneras & Kirwan, 2016b). Within South Asia, it has an estimated population of 500,000 (Wetlands International, 2012). During the Bird Migration Project (BNHS, 1998), a total of 69 individuals was recovered. Of these, 53 were recovered from Russia, Kazakhstan, Kyrgyzstan, Uzbekistan and Tajikistan, 10 from Pakistan, two from China and four from India. The ring recoveries revealed that the population wintering in India comes from an extensive breeding area in the north (38° to 63° N and 66° to 133° E; Rahmani & Islam, 2008). The recoveries also showed that ducks often move together in flocks between their breeding and wintering areas.

During the present study, six individuals were tracked, but only two crossed the Himalayas before they stopped transmitting. NS #38 was tagged at Chilika Lake. From there, it appeared to fly straight to an area south of Como Chamling in southern Tibet, crossing the Himalayan range near Chomo Yummo (6830 m) on the border between Sikkim and Tibet (Figure 2.5). NS #67 flew to an area near Mohanpur in Rajshahi District in Bangladesh, where it remained for four days. It then flew south to the bank of the Rupsa River near Khulna in Bangladesh. From there it flew north to the Mathabhanga River near Chuadanga in Bangladesh, where it remained for six days. It then went to the Teesta Barrage at Gajoldoba in northeastern West Bengal for 26 days before flying to the confluence of the Teesta and Brahmaputra Rivers, where it stopped transmitting for unknown reasons.

Ruddy Shelduck

Ruddy Shelduck (RS) *Tadorna ferruginea* is a large duck (1200–1640 g), which breeds in Central Asia, and scarcely in southeastern Europe and North Africa (Carboneras & Kirwan, 2016c). It breeds at high-altitude wetlands in the mountains of Central Asia in Mongolia and China as far south as the Himalaya in Ladakh (Rahmani & Islam, 2008), Sikkim, western Arunachal Pradesh (Mazumdar *et al.*, 2011), Nepal and Bhutan (BirdLife International, 2016). The Asian population is largely migratory, and moves to southern latitudes in the winter, with an estimated population of 50,000 in South and Southeast Asia (Wetlands International, 2012). During the Bird Migration Project between 1980 and 1992 (BNHS, 1998), 51 Ruddy Shelducks were ringed, of which three were recovered: one from Bangladesh and two from India (Rahmani & Islam, 2008).

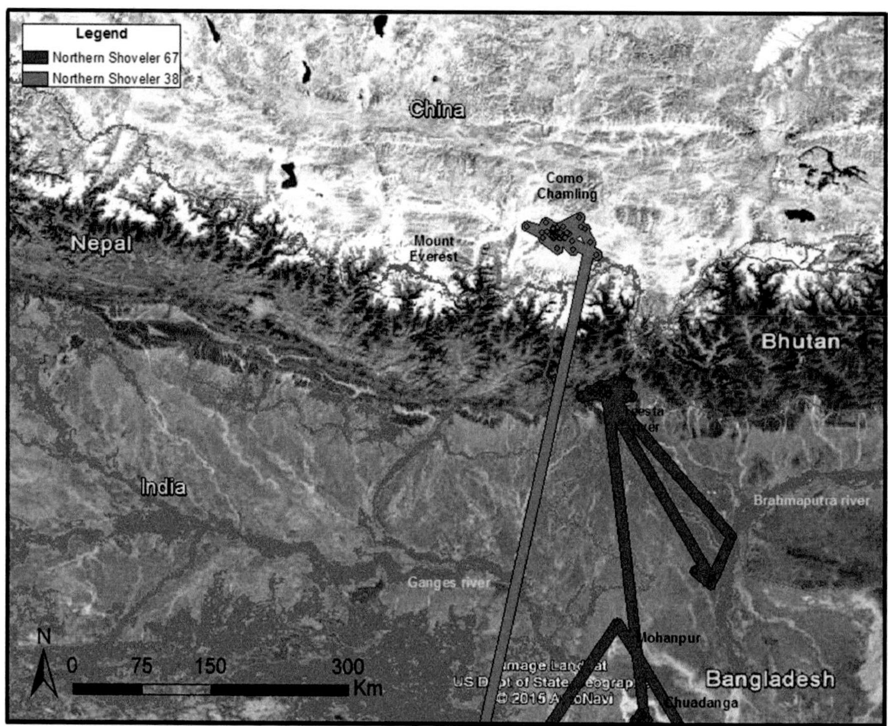

Figure 2.5. Movement paths of Northern Shoveler migrating between Central Asia and the Indian Peninsula, crossing the Himalayas. Map data: Google earth. (A black-and-white version of this figure will appear in some formats. For the colour version, please refer to the plate section.)

During the present study, six individuals were tracked, and only three crossed the Himalayas before they stopped signalling. RS #81 was tagged at Deepor Beel Wildlife Sanctuary in Assam, and it left on 27 April. It flew north and crossed into Tibet through the Zemithang Valley of western Arunachal Pradesh. It stopped at Shijianlame Co for 18 days and then flew north before stopping at a small lake near Quo Yangqu for three days, and then continued to a meadow near Zhamu Co, where it presumably fed for two days before flying north to the surroundings of Zhaling Lake. RS #83 was also tagged at Deepor Beel, which it left on 26 February. It flew north along the Kulong Chhu Valley in the Bumdeling Wildlife Sanctuary in eastern Bhutan (Figure 2.6). It stopped briefly at the northern tip of the sanctuary before flying to the Tsangpo River just south of Lhasa for the summer. RS #21 was tagged at Chilika Lake from where it left on 6 May. It flew northwards and stopped for a day at the confluence of the Murti, Sipsu and Jaldhaka Rivers in northeastern West Bengal. It then flew north and stopped very briefly at Chumba Yumco. From there it went to a small wetland north of Chomo Yummo peak in northern Sikkim north of the Kanchenjunga National Park, halted there for a day and then flew to Dadu Co just northwest of the Dochen Co, and stayed there for 32 days. It again flew north to Simi-la Lake, where it stopped for three days. It then

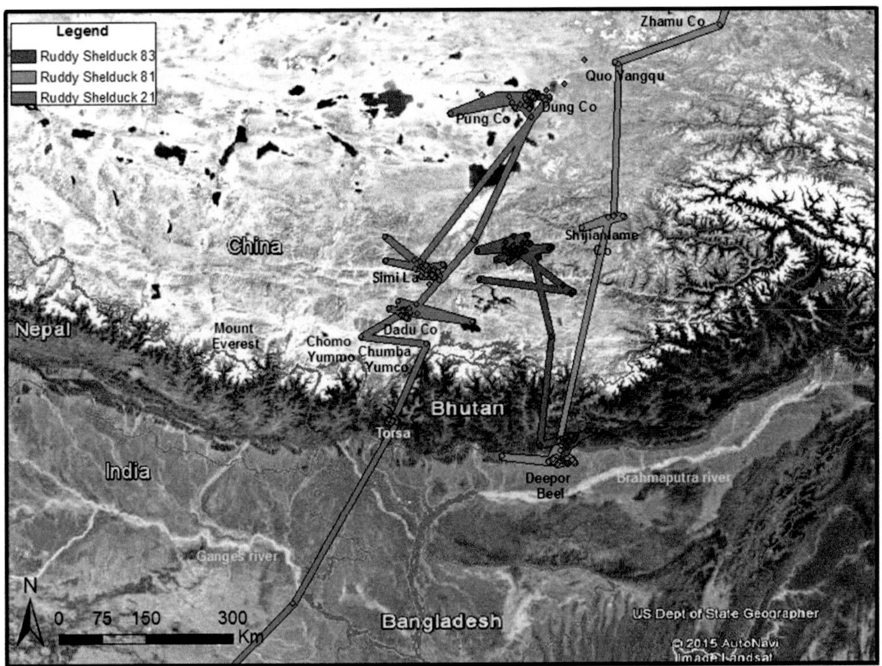

Figure 2.6. Movement paths of Ruddy Shelduck migrating between Central Asia and the Indian Peninsula, crossing the Himalayas. Map data: Google earth. (A black-and-white version of this figure will appear in some formats. For the colour version, please refer to the plate section.)

continued to a wetland complex comprising Pung and Dung Co, and spent the summer on these lakes and the Simi-la Lake.

Garganey

Garganey (GG) *Anas querquedula* is the smallest duck studied (240–585 g), and it has a wide distribution. It breeds across Northern Europe and Asia, and spends the non-breeding period in Africa and in South and Southeast Asia (Carboneras & Kirwan, 2016d). It has an estimated population of 350,000 in South Asia (Wetlands International, 2012). During the ringing project between 1980 and 1992 (BNHS, 1998), more than 700 Garganeys were ringed, and 27 individuals were recovered: 20 from the former USSR, five from India and two from Pakistan. The recoveries from their known breeding grounds in Russia indicate that the South Asian non-breeding populations originate from a large part of temperate Russia from the Ural Mountains in the west to Lake Baikal in the far east (Rahmani & Islam, 2008).

A total of 13 individuals was tracked during this study, and three crossed the Himalayas, while two stopped transmitting after reaching the Himalayan foothills. GG #12 was marked at Purbasthali in West Bengal. It left this wetland on 14 May for the Farakka Barrage on the River Ganga near the border between India and Bangladesh. From there it flew to Karatoya River near Debiganj in Bangladesh.

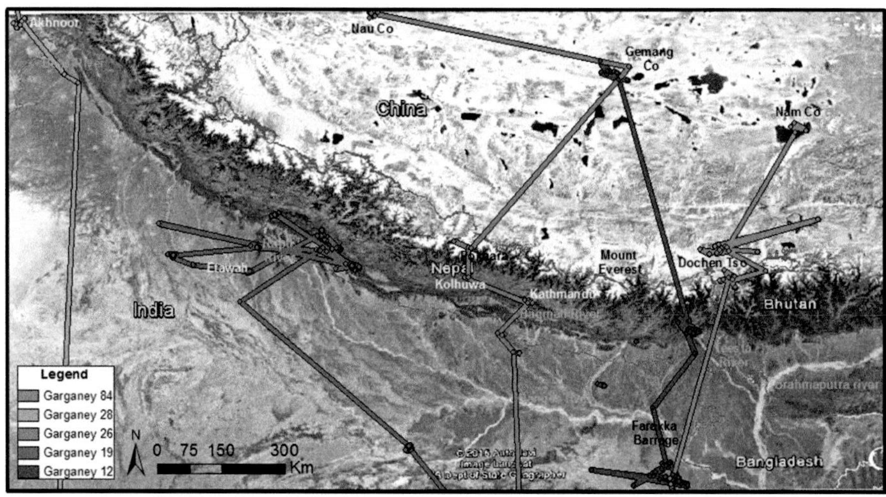

Figure 2.7. Movement paths of Ruddy Garganey migrating between Central Asia and the Indian Peninsula, crossing the Himalayas. Map data: Google earth. (A black-and-white version of this figure will appear in some formats. For the colour version, please refer to the plate section.)

After a three-day halt there, it flew further north to the bank of the Teesta River near the Indo–Bangladesh border for 29 days (Figure 2.7). From there it continued north along the Zemu Valley, and crossed the Himalaya near the Chorten Nyima Ri (6930 m) in northern Sikkim. It spent the breeding season around the Gemang Co. GG #26 was marked at Chilika Lake, from where it left on 26 April. It flew north before resting on a meadow along the Gandak River west of Muzzafarpur in Bihar for six days. It then continued north to the bank of the Bagmati River south of Kathmandu, Nepal, where it spent 10 days. From there the duck flew to the banks of Narayani River near Kolhuwa, Nepal, where it stayed for two days, and then flew to Phewa Lake near Pokhara, Nepal and stayed there for four days. Thereafter, it flew to Dagze Co in Tibet, where it stopped for two days before flying to Nau Co, where it spent the summer. GG #84 was tagged at Purbasthali, which it left on 5 March. It flew north to the Paro Valley in Bhutan, and then to the Phobjika Valley, where it halted for 45 days before flying to Dochen Co and then Nam Co in Tibet. GG #19 was fitted with a transmitter at Chilika Lake. It first flew to a wetland near Etawah in Uttar Pradesh, from where it proceeded to the banks of the Rapti River at the border between Nepal and India. GG #28, marked at Koonthankulum, left this wetland on 25 April. It flew north to Chohal Dam in Punjab, where it stopped for 48 days. It then flew northwest to the Beas River, stayed there for two days and then flew to the Chenab River near Akhnoor in Jammu and Kashmir before it stopped transmitting.

Over the Mountains or through the Valleys?

The question of whether migratory ducks fly over high mountains, follow gorges and valleys or take a detour to circumvent the Himalayas challenged ornithologists and bird conservationists for a long time (Rahmani & Islam, 2008). Although several naturalists and members of expeditions to the Himalayas observed migratory ducks flying over high passes, no major effort was made to understand their migratory routes in the Himalayas. Conventional methods of ringing and recovering rings by hunters and bird watchers did not shed much light on this crucial aspect of their migration in the Himalayan region (Javed *et al.*, 2000). Radar technology was also not available to study the ducks in the Himalayas. The present study using satellite telemetry work in the Central Asian Flyway accurately revealed the migratory routes of six duck species across the Himalayas for the first time.

Given the low oxygen levels at high altitude, one would expect the migratory ducks to cross the Himalayas by flying through gorges that traverse the Himalayan range (Groen & Prins, Chapter 17). But none of the ducks tracked in this study flew through valleys like the Kali Gandaki or the Sutlej River gorges to get through the Himalayas. Most of the ducks studied flew over relatively low passes such as Niti Pass (5070 m) in Uttarakhand and Nathu Pass (4310 m) in northern Sikkim to cross the Himalayas into China. Before flying over these passes, the ducks stopped at wetlands on the Indo-Gangetic plain and at floodplains along the Brahmaputra River and the foothills of the Himalayas to replenish their fat reserves (Choudhury & Gupta, 2015; Jha & McKinley, 2015). Thus, this study has shown that although migratory ducks cross the Himalayas, they do not fly over the highest peaks, but rather make use of the 'low' passes to get past this mountain barrier.

Acknowledgements

We thank William Perry, Ben Gustafson, David Douglas, Shane Heath and Bridget Collins at the US Geological Survey for their help and support. We also thank our partners and government officials in India (Yogendrapal Singh, R. Sundararaju, P. Gangaimaran, Dr Anmol Kumar, Dr Lal Krishna, Devojit Das, A.K. Pattnaik, Dr S. Nagarajan, Dr G. Chetri, D.M. Singh, Mr. Mukherji and V. Thirunavukarasu). The use of trade, product or firm names in this publication is for descriptive purposes only and does not imply endorsement by the US government.

References

Ali, S. & Ripley, S.D. (1983). *Compact Handbook of the Birds of India and Pakistan.* Bombay: Oxford University Press.

Balachandran, S. (2012). Avian diversity in coastal wetlands of India and their conservation needs. Proceedings of International Day for Biological Diversity, 22 May. http://www.upsbdb.org/pdf/Souvenir2012/ch-19.pdf

BirdLife International (2016). Species factsheet: Tadorna ferrugenia. Downloaded from www.birdlife.org on 2 March 2016.

BNHS (1998). Studies on the movement and population structure of Indian avifauna. Phase I: 1980–1986. P. 177 and Phase II:1987–1992. P. 142. Final Reports. Mumbai: Bombay Natural History Society

Bruderer, B. (1997a). The study of bird migration by radar part 1: the technical basis. *Naturwissenschaften*, **84**, 1–8.

Bruderer, B. (1997b). The study of bird migration by radar part 2: major achievements. *Naturwissenschaften*, **84**, 45–54.

Burbank, D.W., Bookhagen, B., Gabet, E.J. & Putkonen, J. (2012). Modern climate and erosion in the Himalaya. *Comptes Rendus Geoscience*, **344**, 610–626.

Carboneras, C., Christie, D.A. & Kirwan, G.M. (2016). Eurasian Wigeon (Mareca penelope). In del Hoyo, J., Elliott, A., Sargatal, J., Christie, D.A. & de Juana, E., eds., *Handbook of the Birds of the World Alive*. Barcelona: Lynx Edicions (retrieved from www.hbw.com/node/52862 on 2 March 2016).

Carboneras, C. & Kirwan, G.M. (2016a). Northern Pintail (Anas acuta). In del Hoyo, J., Elliott, A., Sargatal, J., Christie, D.A. & de Juana, E., eds., *Handbook of the Birds of the World Alive*. Barcelona: Lynx Edicions, (retrieved from www.hbw.com/node/52 884 on 11 April 2016).

Carboneras, C. & Kirwan, G.M. (2016b). Northern Shoveler (Spatula clypeata). In del Hoyo, J., Elliott, A., Sargatal, J., Christie, D.A. & de Juana, E., eds., *Handbook of the Birds of the World Alive*. Barcelona: Lynx Edicions (retrieved from www.hbw.com /node/52896 on 2 March 2016).

Carboneras, C. & Kirwan, G.M. (2016c). Ruddy Shelduck (Tadorna ferruginea). In del Hoyo, J., Elliott, A., Sargatal, J., Christie, D.A. & de Juana, E., eds., *Handbook of the Birds of the World Alive*. Barcelona: Lynx Edicions, (retrieved from www.hbw.com /node/52835 on 2 March 2016).

Carboneras, C. & Kirwan, G.M. (2016d). Garganey (Spatula querquedula). In del Hoyo, J., Elliott, A., Sargatal, J., Christie, D.A. & de Juana, E., eds., *Handbook of the Birds of the World Alive*. Barcelona: Lynx Edicions, (retrieved from http://www .hbw.com/node/52890 on 2 March 2016).

Choudhury, D. & Gupta S. (2015). Aquatic insect community of Deepor beel (Ramsar site), Assam, India. *Journal of Entomology and Zoology Studies*, **3**, 182–192.

Dutta, P. & Konwar, M. (2013). Morphological aspects of floodplain wetlands with reference to the upper Brahmaputra River Valley. *International Journal of Scientific and Research Publications*, **3**, 1–6.

Gaidet, N., Cappelle, J., Takekawa, J.Y., *et al.* (2010). Potential spread of highly pathogenic avian influenza H5N1 by wildfowl: dispersal ranges and rates determined from large-scale telemetry. *Journal of Applied Ecology*, **47**, 1147–1157.

Hawkes, L.A., Balachandran, S., Batbayar, N., *et al.* (2011). The Trans-Himalayan flights of Bar-headed Geese (Anser indicus). *Proceedings of the National Academy of Sciences*, **108**, 9516–9519.

Higuchi, H., Nagendran, M., Sorokin, A.G. & Ueta, M. (1994). Satellite tracking Common Cranes *Grus grus* migrating north from Keoladeo National Park, India. In H. Higuchi & J. Minton, eds., *The Future of Cranes and Wetlands*. Tokyo: Wild Bird Society of Japan, pp 26–31.

Horne, J.S., Garton, E.O., Krone, S.M. & Lewis, J.S. (2007). Analyzing animal movements using Brownian bridges. *Ecology*, **88**, 2354–2363.

Islam, M.Z. & Rahmani, A.R. (2008). *Existing and Potential Ramsar Sites of India.* Mumbai, India: Oxford University Press.

Javed, S., Takekawa, J.Y., Douglas, D.C., *et al.* (2000). Tracking the spring migration of a Bar-headed Goose (Anser indicus) across the Himalaya with satellite. *Global Environmental Research*, **4**, 195–205.

Jha, K.K. & McKinley, C.R. (2015). Composition and dynamics of migratory and resident avian population in wintering wetlands from northern India. *Notulae Scientia Biologicae*, **7**, 1–15.

Jun, W., Zhongbo, S. & Yaoming, M. (2004). Reconstruction of a cloud-free vegetation index time series for the Tibetan Plateau. *Mountain Research and Development*, **24**, 348–353.

Lawrence, S.W. (2000). *Tales of the Himalayas: Adventures of a Naturalist.* La Crescenta, California: Mountain N' Air Books.

Liu, J., Xiao, H., Lei, F., *et al.* (2005). Highly pathogenic H5N1 influenza virus infection in migratory birds. *Science*, **309**, 1206.

Mazumdar, K., Maheswari, A., Dutta, P.K., Borah, P.J. & Wange, P. (2011). High altitude wetlands of western Arunachal Pradesh: new breeding ground for Ruddy Shelduck (Tadorna ferruginea). *Zoo's Print*, **26**, 9–10.

McLure, H.E. (1974). Migration and Survival of the Birds of Asia. U.S. Army Medical Component, South East Asia Treaty Organisation (SEATO) Medical Project, Bangkok.

Muzaffar, S.B., Takekawa, J.Y., Prosser, D.J., *et al.* (2008). Seasonal movements and migration of Pallas's Gulls Larus ichthyaetus from Qinghai Lake, China. *Forktail*, **24**, 100–107.

Namgail, T., Takekawa, J.Y., Balachandran, S.B., Sathiyaselvam, P., Mundkur, T. & Newman, S. (2014). Space use of wintering waterbirds in India: influence of trophic ecology on home-range size. *Current Zoology*, **60**, 616–621.

Namgail, T., Takekawa, J.Y., Sivananinthaperumal, B., *et al.* (2011). Ruddy Shelduck Tadorna ferruginea home range and habitat use during the non-breeding season in Assam, India. *Wildfowl*, **61**, 182–193.

Newman, S.H., Hill, N.J., Spragens, K.A., *et al.* (2012). Eco-virological approach for assessing the role of wild birds in the spread of avian influenza H5N1 along the Central Asian Flyway. *PloS One*, **7**(2), e30636.

Palm, E.C., Newman, S.H., Prosser, D.J., *et al.* (2015). Mapping migratory flyways in Asia using dynamic Brownian bridge movement models. *Movement Ecology*, **3**, 1–10.

Prosser, D.J., Cui, P., Takekawa, J.Y., *et al.* (2011). Wild bird migration across the Qinghai-Tibetan plateau: a transmission route for highly pathogenic H5N1. *PloS One*, **6**, e17622.

Prosser, D.J., Takekawa, J.Y., Newman, S.H. *et al.* (2009). Satellite-marked waterfowl reveal migratory connection between H5N1 outbreak areas in China and Mongolia. *Ibis*, **151**, 568–576.

Rahmani, A.R. & Islam, M.Z. (2008). *Ducks, Geese and Swans of India: Their Status and Distribution*. Mumbai, India: Oxford University Press.

Rogacheva, H. (1992). *The Birds of Central Siberia.* Husum: Husum Druck- und Verlaggesellschaft.

Sjöberg, S. & Nilsson, C. (2015). Nocturnal migratory songbirds adjust their travelling direction aloft: evidence from a radiotelemetry and radar study. *Biology Letters*, **11**(6), 20150337.

Swan, L.W. (1970). Goose of the Himalayas. *Natural History*, **79**, 68–75.

Takekawa, J.Y., Heath, S.R., Douglas, D.C., *et al.* (2009). Geographic variation in Bar-headed Geese Anser indicus: connectivity of wintering areas and breeding grounds across a broad front. *Wildfowl*, **59**, 100–123.

Takekawa, J.Y., Newman, S.H., Xiao, X., *et al.* (2010) Migration of waterfowl in the East Asian Flyway and spatial relationship to HPAI H5N1 outbreaks. *Avian Diseases*, **54**, 466–476.

Takekawa, J.Y., Prosser, D.J., Douglas, D.C., *et al.* (2013). Movements and space use of Ruddy Shelducks in the Central Asian Flyway and HPAI H5N1 risk. *Viruses*, **5**, 2129–2152.

Wernham, C.V., Toms, M.P., Marchant, J.H., Clark, J.A., Siriwardena, G.M. & Baillie, S.R., eds. (2002). *The Migration Atlas: Movements of the Birds of Britain and Ireland.* London: T. & A. D. Poyser.

Wetlands International (2012). 'Waterbird Population Estimates'. Retrieved from wpe .wetlands.org on 20 Jan 2016.

Zhang, G.G., Liu, D.P., Hou, Y. Q., *et al.* (2014). Migration routes and stopover sites of Pallas's Gull Larus ichthyaetus breeding at Qinghai Lake, China determined by satellite tracking. *Forktail*, **30**, 104–108.

Zhang G.G., Liu D.P., Jiang, H.X., Hou, Y.Q., Dai, M., Qian, F.W. & Chu, G.Z. (2008). Movement of four breeding waterbirds at Qinghai Lake, China. *Biodiversity Science*, **16**, 279–287. (In Chinese with English summary.)

3 Migratory Routes across the Himalayas Used by Demoiselle Cranes

Hiroyoshi Higuchi and Jason Minton

Long-Distance Migration of Demoiselle Cranes

Demoiselle Cranes *Anthropoides virgo* breed on wetlands across Eurasia, from Eastern Europe to northeastern China, and they winter on the Indian subcontinent (Ali, 1941). Therefore, the birds which breed in the central and eastern portion of that range have to traverse the Himalayan Mountains and intervening deserts twice per year as they travel between their winter and summer ranges. The migration of these cranes over such an epic route has fascinated ornithologists for many decades. In the early 1970s, members of the Alpine Club in Japan presented ornithologists with photos, taken in the autumn of 1969, of cranes in flight over the Himalayas. Although these photos, and others taken during the 1970s, were at the time considered Siberian Cranes *Leucogeranus leucogeranus*, they were eventually proven to be Demoiselle Cranes (Matsuda, 1999). Research on bird migration, including cranes, over the Himalayas has included numerous reports (e.g. Donald, 1953; Bijlsma, 1991; Inskipp & Inskipp, 1991) recorded during a variety of expeditions, but systematic migration counts were becoming more regular in the 1970s (Martens, 1971; Beaman, 1973; Thiollay, 1979). With the advent of relatively small satellite telemetry units in the 1980s, the phenomenon of bird migration could finally be researched in detail, and a study of Demoiselle Cranes and their Himalayan migration was soon proposed. Migration is itself a challenge for ornithologists to study because of the long distances covered and high altitudes reached by the migrants, and migration across the Himalayas obviously provides an even greater challenge due to the remote locations and extreme terrain. Study of Demoiselle Crane migration by satellite tracking was expected to confirm that the birds do pass over the Himalayan Mountains, providing confirmation of that physiological feat. In addition, detailed migration route data would reveal the crucial stopover sites along the route, which are expected to be important links in the chain of migration, made all the more important because of the challenging elevations of the route. Successful satellite tracking in 1995 demonstrated the migratory routes of three cranes flying across the Himalayas to reach India from their breeding areas in Mongolia, and one crane illustrated an alternate route by avoiding a direct crossing of the Himalayas during its migration between Kazakhstan and India.

During the migration between breeding and wintering areas, the Demoiselle Cranes must navigate some of the harshest physical barriers, such as the Taklamakan and Gobi Deserts, and the mountains of the Hindu Kush, Pamirs, Great Himalaya, Tien Shan and

Figure 3.1. A flock of Demoiselle Cranes in flight around Mount Nilgiri (western Himalayas) on 7 October 2013. Photo by Rajendra Suwal. (A black-and-white version of this figure will appear in some formats. For the colour version, please refer to the plate section.)

Karakoram, and the Tibetan Plateau. Despite these apparent barriers to their migration, several thousand Demoiselle Cranes fly over the Himalayas every year (Figure 3.1) after the monsoon season ends in late September or early October (Ali, 1941; Bijlsma, 1991; Inskipp & Inskipp, 1991). Since the Demoiselle Cranes fly long distances (up to 4000 km), crossing some of the most inaccessible regions on earth, satellite tracking was ideal for gathering data to identify their migratory routes and timings, adding detailed information to the various reports of observations and numbers that had been recorded over many years.

Satellite tracking is a useful method for gathering ornithological and ecological information for long-distance migrants with a body size large enough to carry a transmitter, and it was successfully used to delineate migratory routes of several species of cranes prior to its deployment on Demoiselle Cranes crossing the Himalayas (Higuchi *et al.*, 1991; 1992; 1994a; 1994b; 1996; 1998; Kanai *et al.*, 1997; Ueta *et al.*, 1998; Higuchi, 2012). At the beginning of the research project, it was anticipated that satellite tracking would not only show the migratory routes in the Himalayas, but also identify important stopover sites along those routes.

It was predicted that cranes could have multiple routes between their northern breeding areas and wintering sites in India. Since Demoiselle Cranes have a wide longitudinal (east-west) distribution across Eurasia, their routes to the wintering sites in India may vary depending on the longitude of the breeding area. Cranes are large wetland obligate birds, requiring large areas for breeding, stopover and wintering (Johnsgard, 1983), so migration routes should ultimately be determined by the distribution of relatively large wetlands in the arid regions that the cranes cross. Therefore, identifying and conserving wetland habitats used by cranes could also conserve habitats for numerous other wildlife species dependent on wetland resources. Cranes are very recognizable and beautiful species that many cultures and individuals hold in high esteem, and ornithologists and conservationists proposed to use cranes as flagship species to both identify specific wetlands for conservation and to serve as an educational tool that could introduce conservation issues to the public. Since cranes are relatively large species, they are also capable of carrying satellite-telemetry equipment, and thus Demoiselle Cranes provided an opportunity to examine the distribution of wetlands over a very large area of Eurasia, and to serve as a flagship species that conservationists could reference when implementing educational programmes in the region.

Capture of the Cranes and PTT Deployment

Satellite transmitters (Platform Transmitter Terminals or PTTs) were deployed on a total of 21 Demoiselle Cranes within the breeding range in Kazakhstan, Mongolia and Russia (Kanai *et al.*, 2000). Following deployment of the transmitters, the entire autumn migration was successfully recorded and the migratory routes were mapped for three cranes from Mongolia, and for one crane from Kazakhstan, thus illustrating important routes across the Himalayas and the Hindu Kush, the world's highest mountains (Figure 3.2). The remaining birds stopped transmitting signals before reaching the Himalayas, and the relatively low success rate was attributed to an unexpectedly high rate of battery discharge. The high temperatures potentially encountered at several stopover sites in Central Asia may have caused batteries to fail at this unexpectedly high rate.

In 1995, Demoiselle Cranes were captured near the city of Khovd, western Mongolia, which is located near Har Us Lake, an important habitat for waterbirds in this arid steppe region. A second capture site was located further west, near the town of Kopa, Kazakhstan. The primary method of capture was by ingestion of alpha-chloralose, an oral tranquilizer, applied to a grain bait (whole corn or wheat; Nesbitt, 1976, 1984).

Cranes at the Mongolian site were captured on 1 July 1995, and they were kept in captivity until their release between 27 and 30 August, when they were fitted with satellite transmitters (IDs 25319, 25322 and 25324) and released along the shore of Har Us Lake. All birds were captured as adults from a non-breeding flock located on a flat plain to the northwest of the lake, where the Buyant River and its delta form many wet meadows surrounded by the desert. The vegetation was characterized by short grass with bushes scattered across the landscape. Some of the meadow areas had been irrigated to support small-scale agriculture and grazing, and it was in one of these areas that cranes were captured using baits of wheat coated with alpha-chloralose. At that time of the year, the cranes readily consumed the grain as if it was remnants from a harvest. Later in the

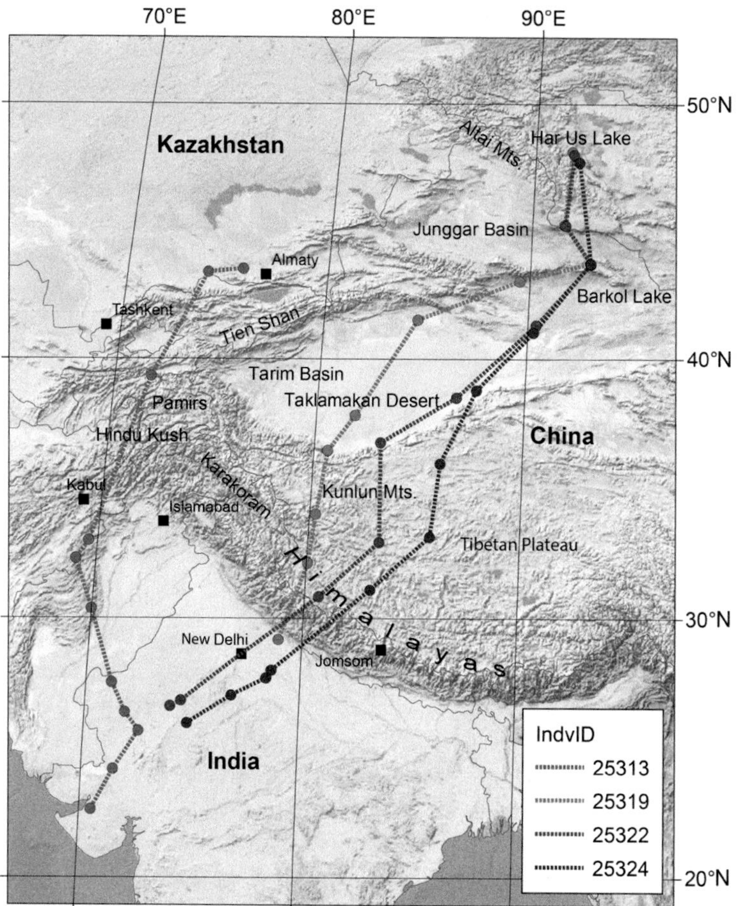

Figure 3.2. Map showing the satellite-tracked migration routes of Demoiselle Cranes from India onto the western Tibetan Plateau. Based on Kanai *et al.*, 2000. (A black-and-white version of this figure will appear in some formats. For the colour version, please refer to the plate section.)

summer, the capture technique was repeated unsuccessfully. Cranes ignored the proffered grain, and foraged on abundant grasshoppers that had emerged. The cranes captured in July were kept in temporary captivity near Khovd and cared for by students of biology at the National University of Mongolia in Khovd. They were fed twice daily on wheat and wild-caught insects, and a temporary water diversion from a natural stream was directed through the enclosure to provide water.

In Kazakhstan, a crane (ID 25313) was captured on 5 July 1995 near the small town of Kopa, approximately 100 km west of Almaty, the capital city of Kazakhstan. The breeding territories of cranes in that area were extremely arid with steppe-type vegetation dominated by *Artemisia* spp., which were grazed heavily by sheep and cattle. Temperatures in June–July were as high as 45–50°C. Cranes in the area formed a flock of non-breeding birds that were attracted to a small artesian seep that had been developed for livestock.

Satellite transmitters seemed to be an ideal method to research the migration of Demoiselle Cranes because the supposed routes were highly inaccessible to traditional radio telemetry methods. Despite being one of the smallest species of cranes, they are large enough to carry a satellite transmitter. The satellite transmitters used were Platform Transmitter Terminals (PTT; model T-2050 II) developed by the Nippon Telegraph and Telephone Corporation (NTT) and Toyo Communication Equipment Co., Ltd. (TOYOCOM). The size of the transmitter-case was 55 x 34 x 25 mm, with an antenna of 18 cm, weighing 43 g. The transmitter represented approximately 2% of the 1985 g body weight reported for adult female Demoiselle Cranes (Dementiev & Gladkov, 1968, as cited in Johnsgard, 1983). The transmitter was attached dorsally with Teflon-treated ribbon, which was used because the low friction of the Teflon coating was minimally abrasive to the skin and feathers. The ribbon was put through holes in flanges on the transmitter case, and crossed at the crane's breast. The ends of ribbon were sewn together with absorbable surgical suture (Nagendran *et al.*, 1994) so that the transmitter-harness would eventually fail due to degradation of the absorbable sutures, and release the crane from the expired transmitter.

The deployment dates were selected carefully, keeping in view the duty cycle and life expectancy of the transmitters, and the accessibility of potential capture sites. The limited information about the movement of the cranes throughout the season on the summering range made it advisable to capture birds as early as possible in the season to ensure that individuals were available to carry transmitters. However, the expected battery life averaged four to five months, so it was preferable to deploy transmitters immediately prior to the migration period. September deployment was considered optimum to ensure sufficient battery life during the migration period, but earlier dates were necessary to ensure that cranes could be located for capture.

Satellite Tracking of Routes over and around the Himalayan Mountains

Four individual Demoiselle Cranes were tracked from the summering range to their wintering grounds (Figure 3.2, Table 3.1). The crane from Kopa, Kazakhstan was tracked to western India via a route that avoided the Himalayan Mountains; the bird first flew to the east of the Hindu Kush Mountains and then to west of the Pamir Mountains. Three cranes tracked from Mongolia, however, passed over the Himalayas after migrating from Har Us Lake to western India via the Tarim Basin, the Tibetan Plateau and the Himalaya Mountains.

Cranes began migration from Kazakhstan in mid-August. The Demoiselle Crane fitted with a satellite transmitter in Kazakhstan was tracked from 7 July to 17 November, transmitting 111 locations. The crane appeared to depart from the general area of capture and release on 25 August. At the same time, visual observations indicated that hundreds of cranes had already begun their autumn migration in the area. The tracked crane from Kazakhstan departed Kopa and travelled 2000 km to India in seven days. The typical stopover duration was only one or two days at any one site, and the crane arrived in India on 31 August. The daily average distance travelled was 290 km.

The location data for the crane's route were mapped to reveal the topography of the route, and its changes in elevation as it migrated towards India. From Kopa, the tracked crane immediately entered the most mountainous section of its migration route, and crossed its maximum elevation (4700 m a.s.l.) at 460 km from the capture site, near Mount Kommunizma. At a distance of 1200 km to 1600 km from Kopa, the crane veered west through Afghanistan to avoid the northern end of the Pamir Highland and the Hindu Kush Mountains. The diversion along that leg of the migration took the crane over elevations of 1000 m to 2000 m, and it then descended to the lowlands of Pakistan, with altitudes of less than 200 m.

Most of the migration routes had little vegetation, as indicated by a vegetation index calculated from a five-minute grid-cell around each location of the marked cranes. Near Kopa and Mount Kommunizma the global vegetation index was 80 to 90 (out of 255), which is nearly barren. The vegetation index along the latter half of the route, 1000 km to 2000 km from Kopa, were even lower, ranging from 40 to 60, as the bird crossed stretches of desert. The crane did traverse the Indus in Pakistan at the south terminus of the Sulaiman Mountains, but it rested little along the floodplains. The crane continued south through the Thar Desert and arrived in a rest area south of Jodhpur in Rajasthan, India, at the end of August. It remained in Rajasthan until late September, when it moved to Gujarat, near the Rann (Gulf) of Kutch, a region with thousands of wintering migratory birds, including thousands of Demoiselle and Eurasian Cranes (*Grus grus*).

The cranes tracked from Mongolia began migration between 14 and 22 September, which is about one month later than the crane from Kazakhstan. Regional variation in climatic patterns may have affected this difference in the start date of migration. The Himalayas experience a strong influence from the monsoon of India, unlike the western Pamirs and the Hindu Kush. The lack of monsoon influence in Kazakhstan may make it easier for cranes to migrate in late summer to early autumn, while cranes in Mongolia must delay their departure until after the peak of the monsoon air flow.

After starting the migration by departing Har Us Lake, each of the three Demoiselle Cranes flew directly south, covering 500 km in one to two days, and rested at Barkol Lake for the following one to two weeks. After leaving Barkol Lake between 29 September and 2 October, the cranes crossed the Himalayas without substantial rest and reached India in seven days. The distances between Barkol Lake and India along the routes followed by the cranes were 2200 km to 2800 km, which the cranes covered in daily movements of 300 km to 400 km.

Demoiselle Cranes from Mongolia demonstrated that a primary section of the migration route led south from Har Us Lake. All three cranes moved from Har Us Lake, at an elevation of approximately 1000 m a.s.l., south to Barkol Lake at an elevation of approximately 1500 m. From Barkol Lake, two cranes (IDs 25322 and 25324) flew south towards Lop Nur, a remnant saline lake in Xinjiang Province, China. This route took them east of the Taklamakan Desert, and to the northern slopes of the Kunlun

Mountains, after travelling approximately 1500 km from the start of migration at Har Us Lake.

The third marked crane (ID 25319) flew over the Taklamakan desert for almost 500 km. It did so by departing Barkol Lake and descending southwest through the Turpan Basin to an elevation of approximately 150 m, and then climbing along the southern edge of the Tien Shan Mountains, increasing in elevation from 2000 m to 3000 m before descending to the Tarim River basin at 1200 m to 380 m. The crane then appears to have flown into the desert, and to have joined the Keriya River and followed it south and out of the desert to the foot of the Kunlun Mountains, where it rested for a single day at a small lake on the southern edge of the desert at almost 2000 m elevation.

From the northern limits of the Kunlun Mountains, each of the three marked cranes took different routes across the Tibetan Plateau. At that point, they were approximately 2000 km to 3000 km from the starting points of their migrations. Elevations increased to more than 5000 m as they migrated south across the Tibetan Plateau towards the Himalayas. Each of the three birds crossed the Tibetan Plateau within three days.

The marked cranes took three discrete routes across the Tibetan Plateau. These routes were about 600 km apart from one another as they climbed onto the Tibetan Plateau. As they continued south towards the Himalayan Mountains, the cranes' routes converged towards the border of the Plateau with Uttarakhand, India, and with Nepal. River gorges reach high into the Himalayas in that region, including the Baspa Valley in India, and the Kali Gandaki Gorge in Nepal.

After crossing the Himalayas, two cranes arrived at the wintering area, near Ajmer in India on 11 October (ID 25322) and 13 October (ID 25324), where they stayed until the transmitters' batteries expired. The third crane (ID 25319) ceased transmitting shortly after arriving in northern India (at 29°.48'N 76°.54'E), after flying over the Himalayas on 9 October. The three cranes that left Barkol Lake travelled 1800 km without substantial rest to the wintering grounds in India.

A survey of the cranes' destination sites in India was carried out in late October, 1995, in an attempt to locate the marked individuals. A Demoiselle Crane that was tracked from Mongolia (ID 25322) was observed in Rajasthan, India. In late November 1995, more than 17,000 Demoiselle Cranes were counted between Jodia and Ahmedabad in Gujarat, India (Nagendran, pers. obs.).

Ajmer is typified by dry steppe, most of which is cultivated wheat and peanut fields. During a field reconnaissance on 22 October, a marked crane (ID 25322) was observed in a flock of 500 Demoiselle Cranes at an irrigation pond. They fed in agricultural fields during the day, and loafed and roosted at the pond. The birds arrived in India just after the summer monsoon season. In October, western India has just finished harvesting its agricultural fields, so bare ground was widespread and the vegetation index was low. Harvested fields provide foraging sites for cranes because fallen grain can be gleaned from the soil.

Declining Field Observations of Cranes Crossing the Himalayas

The Kali Gandaki River flows through the Himalayas from north to south, and cranes have often been observed flying along this route between late September and mid-October after the monsoon season (Bijlsma, 1991; Inskipp & Inskipp, 1991). Since the 1970s, ornithologists have conducted intermittent counts of Demoiselle Cranes on autumn migration over Jomsom, by the Kali Gandaki River. Over the decades, migrant crane numbers have been counted during the same general two to three-week period in late September and early October during different years. Although the peak of that migration is dependent on the passing of the monsoon season, and therefore some counts vary widely, the results have generally trended downward. In 1969 and 1978, counts tallied more than 31,000 and 63,000 cranes, respectively (Martens, 1971; Thiollay, 1979). The potential for variability in counts, probably due to shifts in timing of the peak migration, is shown by a similar observation of only 2200 cranes in 1973 (Beaman, 1973). Those high numbers have not been repeated in recent counts. Surveys in 1995, 2012 and 2013 have resulted in only 20,870 cranes, 805 cranes and 5500 cranes, respectively (Higuchi & Kan, pers. obs.).

Any particular day of observation can result in widely varying numbers. For example, during the 13 days of field observations for the count in 1995, a total of 20,870 cranes was counted migrating past the study site in Jomsom. The peak of the count was between 6 and 10 October (Figure 3.3), when 77% of the cranes were observed during the study period, a total of 16,000 cranes, with daily totals ranging from 3300 to 6000 cranes. The satellite-tracked individuals crossed the Himalayas during this period, although they were not specifically identified during the field observations in Jomsom, which is further east than the crossings by the marked cranes. Strong southerly winds blew every afternoon (Ohlmann, Chapter 14). Cranes flew from north to south against this wind, and some flocks were blown back and stayed one night on the riverside. The following day, the stranded flocks regained altitude on updraughts along the steep gorge, and proceeded towards India.

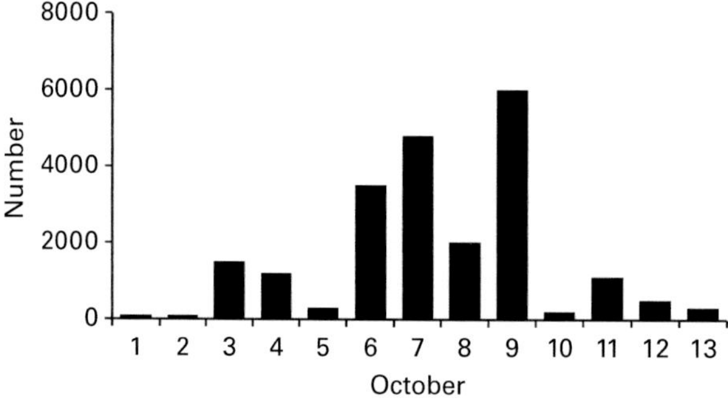

Figure 3.3. The number of Demoiselle Cranes observed to fly over the Himalayas at Jomsom, Nepal between 6 and 10 October 1995. Based on Kanai *et al.*, 2000.

The Kali Gandaki has also been proposed as a potentially important route for raptors, which appear to use this natural corridor during soaring flights over the Himalayan Mountains from the Tibetan Plateau (Bijlsma, 1991). Although it may not be the species primary route across the Himalayan Mountains, Bar-headed Geese (*Anser indicus*) also use the Kali Gandaki route on spring migration towards the Tibetan Plateau, and satellite-tracked geese rested in similar riverine areas to those used by marked Demoiselle Cranes (Javed *et al.*, 2000).

In 2012, the environment along the Kali Gandaki River was found to have been developed on a large scale, with housing, buildings and a large, modernized airport. In 1995, the cranes landed along riverbeds when they had difficulty in flying over the high mountains, but only a few cranes were observed to do so in 2012 and 2013. Many cranes now seem to avoid this area, and it may be a result of the construction impacts on this crucial link in the migration route.

Important Migration Routes and Staging Sites

There appear to be two migration routes for the Demoiselle Cranes in the Central Asian Flyway; one passes through the Himalayas, while the other skirts the highest ranges to the west (Figure 3.2, Table 3.1). Both routes included transitional areas between basins and surrounding highlands, which may benefit migrating cranes because they can use the updraughts along the mountain slopes, and can rest and feed at oases along the foot of the mountains.

A characteristic feature of Demoiselle Crane migration is that they cover long distances within short periods of time. Since these cranes migrate over mountains and deserts, they probably do not have many suitable stopover sites. Those that are used provide crucial rest and nourishment before scaling the rugged Himalayas. We believe that wetland stopover sites on the margins of deserts and high-altitude ascents are very important for Demoiselle Cranes and other wetland birds that migrate long distances through rugged terrain.

For cranes migrating from Mongolia, Har Us Lake itself is an important stopover site for migrating cranes. Although the marked cranes originated at this point, field observations and general knowledge indicated that it is a major staging area for Demoiselle Cranes migrating to and from areas further north. Several cranes also rested for up to five days in other areas of Khovd Province, but their first important stopover site after leaving Har Us Lake was Barkol Lake. Barkol Lake in China was used the longest of all the stopover sites, followed by Lop Nur.

All other stopover sites on the route to the east of the Taklamakan Desert were used for only one or two days, but some were used by several cranes. For the crane which was tracked through the Taklamakan, one stopover site stands out for its apparently crucial location, the Keriya River in the centre of the Desert, where the crane halted for only one day, but it is difficult to imagine any alternative rest area other than the river, as it descends into the desert basin. The crane apparently crossed the centre of the Taklamakan, joined the Keriya River and followed it out of the desert.

Table 3.1. Details of migration and stopover sites of tracked cranes.

ID	Latitude	Longitude	Stopover sites	Country	Stopover period
25319	45.232 N	91.577 E	E Baytag Bogd Uul	Mongolia	23 Sep
	43.721–43.771 N	92.951–93.085 E	Barkol Lake	China	24 Sep–2 Oct
	43.030 N	89.511 E	E Turpan	″	3 Oct
	41.522 N	84.613 E	N Ts'ao Lake	″	4 Oct
	37.814 N	81.883 E	Keriya River	″	5 Oct
	36.451 N	80.670 E	Hanko-la	″	6 Oct
	34.009–34.019 N	80.285–80.249 E	Doman River	″	7 Oct
	32.124–32.145 N	80.061–80.249 E	Gar Zangbo River	″	8 Oct
	29.180–29.821 N	78.888–78.920 E	Kaphrauri	India	9 Oct–9 Dec
25322	47.996 N	91.944 E	W Har Us Lake	Mongolia	22 Sep
	45.214 N	91.633 E	E Baytag Bogd	″	23 Sep
	43.703–43.762	92.987–93.181	Barkol Lake	China	24 Sep–3 Oct
	41.315 N	90.392 E	Lop Nur	″	4 Oct
	38.495 N	86.723 E	Yakatograk	″	5 Oct
	36.763 N	83.184 E	Yeyik	″	6 Oct
	32.926 N	83.380 E	N Oma	″	7 Oct
	30.837 N	80.673 E	Mandi	″	8 Oct
	26.828 N	74.480 E	N Ajmer	India	10 Oct
	26.602–26.694 N	74.002–74.303	W Ajmer	″	11–25 Oct
25324	47.704 N	92.219 E	S Har Us Lake	Mongolia	14 Sep
	43.729–43.765	92.964–93.106	Barkol Lake	China	15 Sep–1 Oct
	41.041 N	90.296 E	Lop Nur	″	2 Oct
	38.771 N	87.663 E	Waxxari	″	3 Oct
	35.949 N	86.095 E	W Chainjoin Lake	″	4 Oct
	33.132 N	85.776 E	SE Ch'a-k'o-ch'a-mu Lake	″	5 Oct
	31.092 N	83.063 E	S Jeu-ch'ing Hsiu-pu Ts'o Lake	″	6 Oct
	27.986–27.990	78.640–78.650	N Kasuganji	India	7–10 Oct
	27.687 N	78.435 E	Sikandra Rao	″	11 Oct
	27.027–27.031	76.843–76.850	W Bhasawar	″	12–13 Oct
	25.953–26.293	74.809–75.006	SE Ajmer	″	14 Oct–24 Nov
25313	43.374 N	74.021 E	Kokkaymar	Kazakhstan	10 Aug or 20–25 Aug
	39.334 N	71.792 E	E Karashura	″	26 Aug
	32.990 N	69.583 E	W Miram Shah	Pakistan	28 Aug
	32.292 N	69.040 E	Ruins	Afghanistan	29 Aug
	30.366 N	69.990 E	Khart	″	30 Aug
	27.520 N	71.184 E	Reservoirs	India	31 Aug
	26.371 N	71.886 E	Rateu	″	1 Sep
	25.660–25.844 N	72.580–72.757 E	E Mokalsar	″	2–21 or 26 Sep
	24.177 N	71.469 E	Tharad	″	26 Sep
	22.626–22.953 N	70.551–70.743 E	Dahinsara	″	28 Sep–17 Nov

The stopover sites on the Tibetan Plateau, and at high elevations within the Himalayas, were used for short durations of one to two days each, which is likely to indicate that they were used solely as resting locations and did not provide significant foraging opportunities. The marked cranes approached the Plateau along a broad front, approximately 500 km across. During the first day of ascent to the Plateau, the cranes travelled between 200 and 250 km and stopped over near aquatic features, including Chainjoin Lake, Yeyik and the Doman River. In general, each crane appeared to seek out the shores of lakes and rivers as they moved south across the Plateau. The detailed ecological conditions at these sites are unknown. Over the course of two to three days, each crane stopped over at several sites on the Plateau, including the Gar Tsangpo River, the north side of Oma Lake and a chain of lakes, including Zhari Namco. The marked cranes used numerous brief stopover sites, and generally converged toward the border of Tibet with India at Uttarakhand, from where they quickly descended to the south.

Conservation Prospects

The availability of stopover sites may partially define migration routes for Demoiselle Cranes migrating over or around the Himalayan Mountains. Conservation of these vulnerable wetland sites, which are few and far between, is essential to ensure that the migration routes remain open to the cranes. As has been seen with the satellite tracking results from Asia, collaborative projects are needed to identify more important areas across many geopolitical boundaries, and will help in efforts to protect rapidly declining species and their habitats.

Acknowledgements

This project was organized with Yomiuri Newspaper Co. Ltd., in collaboration with NTT Wireless Systems Laboratories, Applied Biology Co. Ltd., International Crane Foundation, Mongolian National University, Saitama Children's Zoo, Tama Zoological Park Tokyo, Toyo Communication Equipment Co., Ltd., and Yamashina Institute for Ornithology, and sponsored by NEC Corporation. We would like to thank our Japanese, Mongolian, Kazakh, Russian and American colleagues who collaborated on tracking Demoiselle Cranes: Yutaka Kanai, Mini Nagendran, Mutsuyuki Ueta, Bold Auyrsana, Oleg Goroshko, Anatoliy F. Kovhsar, Nagahisa Mita, Rajendra N. Suwal, Kunio Uzawa and Vladimir Krever. We would also like to thank Y. Ono and S. Kawashita of NHK, Tsuneo Hayashida, Dr Munkhtogtokh, J. Dagbadorj and T. Buyant of the National University of Mongolia in Khovd; M. Shigehara and N. Sukhbalsan of Mongolian Television; B. Tsendsuren, B. Hulan and Gamba of Meridian, Co., Ltd.; Y. Kakimoto of Yamaha Motor Co., Ltd.; A. Gistov, N. Berezovikov, E. Lapschin and L. Prunchenko of Institute of Zoology NAS; T. Sakamoto, K. Terauchi, T. Tada Masumitsu and N. Shigeta of Yomiuri Newspaper for fieldwork; and Dr T. Mizutani, Dr A. Kondho of Chiba University and Dr N. Mita of Kumamoto for advice on this study.

References

Ali, S. (1941). *The Book of Indian Birds*. Mumbai: The Bombay Natural History Society.

Beaman, M.A.S. (1973). Report of the ornithological Cambridge expedition to the Himalayas, 1973. Unpublished. As cited in Matsuda, Y. (1999). Cranes that cross the Himalaya. *The Himalayan Journal* 55.

Bijlsma, R.G. (1991). Migration of raptors and demoiselle cranes over Central Nepal. *Birds of Prey Bulletin*, **4**, 73–80.

Donald, C.H. (1953). Bird migration across the Himalayas. *The Journal of the Bombay Natural History Society*, **51**, 269–271.

Higuchi, H. (2012). Bird migration and the conservation of the global environment. *Journal of Ornithology*, **153**(Suppl. 1), S3–S14.

Higuchi, H., Nagendran, M., Sorokin, A.G. & Ueta, M. (1994b). Satellite tracking common cranes *Grus grus* migrating north from Keoladeo National Park, India. In H. Higuchi & J. Minton, eds., *The Future of Cranes and Wetlands*. Tokyo: Wild Bird Society of Japan, pp. 26–31.

Higuchi, H., Ozaki, K., Fujita, G., Soma, M., Kanmuri, N. & Ueta, M. (1992). Satellite tracking of the migration routes of cranes from southern Japan. *Strix*, **11**, 1–20.

Higuchi, H., Ozaki, K., Fujita, G., *et al.* (1996). Satellite-tracking white-naped crane *Grus vipio* migration, and the importance of the Korean DMZ. *Conservation Biology*, **10**, 806–812.

Higuchi, H., Ozaki, K., Golovuskin, K., *et al.* (1994a). The migration routes and important rest-sites of cranes satellite tracked from south-central Russia. In H. Higuchi & J. Minton, eds., *The Future of Cranes and Wetlands*. Tokyo: Wild Bird Society of Japan, pp. 15–25.

Higuchi, H., Sato, F., Matsui, M., Soma, M. & Kanmuri, N. (1991). Satellite tracking of the migration routes of whistling swans Cygnus columbianus. *Journal of the Yamashina Institute for Ornithology*, **23**, 6–12.

Higuchi, H., Shibaev, Y., Minton, J., *et al.* (1998). Satellite tracking the migration of red-crowned cranes Grus japonensis. *Ecological Research*, **13**, 273–282.

Inskipp, C. & Inskipp, T. (1991). *A Guide to the Birds of Nepal* (2nd edition). London: Christopher Helm.

Javed, S., Takekawa, J., Douglas, D.C., *et al.* (2000). *Documenting Trans-Himalayan Migration through Satellite Telemetry: A Report on Capture, Deployment and Tracking of Bar-headed Goose (Anser indicus) from India*. Dehradun: Department of Wildlife Sciences, AMU Aligarh and Wildlife Institute of India.

Johnsgard, P.A. (1983). *Cranes of the World*. Bloomington: Indiana University Press.

Kanai, Y., Minton, J., Nagendran, M., *et al.* (2000). Migration of demoiselle cranes in Asia based on satellite tracking and fieldwork. *Global Environmental Research*, **4**, 143–153.

Kanai, Y., Sato, F., Ueta, M., *et al.* (1997). The migration routes and important rest sites of whooper swans satellite tracked from northern Japan. *Strix*, **15**, 1–13.

Martens, J. (1971). Zur Kenntnis des Vogelzuges im nepalischen Himalaya. *Vogelwarte*, **26**, 113–128.

Matsuda, Y. (1999). Cranes that cross the Himalaya. *The Himalayan Journal*, 55.

Nagendran, M., Higuchi, H. & Sorokin, A. (1994). A harnessing technique to deploy transmitters on cranes. In H. Higuchi & J. Minton, eds., *The Future of Cranes and Wetlands*. Tokyo: Wild Bird Society of Japan, pp. 57–60.

Nesbitt, S. (1976). Capturing sandhill cranes with oral tranquilizer. In J.C. Lewis, (ed.). *Proceedings of the 1975 International Crane Workshop*. Stillwater: Oklahoma State University, pp. 296–298.

Nesbitt, S. (1984). Effects of an oral tranquilizer on survival of sandhill cranes. *Wildlife Society Bulletin*, **12**, 387–388.

Thiollay, J. M. (1979). La migration des grues â travers l'Himalaya et la prédation par les aigles royaux. *Alauda*, **47**, 83–92.

Ueta, M., Sato, F., Lobkov, E.G. & Mita, N. (1998). Migration route of white-tailed sea eagle Haliaeetus albicilla in northeastern Asia. *Ibis*, **140**, 684–686.

4 Passerine Migration across the Himalayas

Simon Delany, Charles Williams, Clare Sulston, John Norton and David Garbutt

History of Research

The avifauna of the Tibetan-Himalayan region has been well studied, but mostly in the summer months, outside the main migration seasons. Many papers have been published since the 1830s (Hodgson, 1833), and Ladakh has been a popular focus of interest, with a detailed avifauna published (Pfister, 2004). Current information about the migration of passerines in the Himalayan region is scarce, and is dispersed widely in the literature. Useful summary information is provided in regional avifaunas such as Vaurie (1972), Ripley (1982), Inskipp and Inskipp (1985), Cheng (1987), Ali and Ripley (1987–1999), MacKinnon and Phillips (2000), Grimmett *et al.* (2008) and Rasmussen and Anderton (2012).

Graduate students from Southampton University conducted a detailed, systematic study of bird migration in the Himalayas, involving comprehensive fieldwork in Ladakh through four autumn migration seasons, one winter and one spring, between 1976 and 1981–1982 (Denby & Phillips, 1977; Delany *et al.*, 1982; Williams & Delany, 1979, 1983, 1985, 1986). In this chapter we draw strongly on the results of this work, which are only now being published in widely accessible form.

Moreau (1972) observed that a majority of migrant passerine species breeding in the mid-Palearctic region follow an indirect, westerly migration route through Central and Southwest Asia to winter in Africa, in preference to the shorter, but more arduous route to South Asia. Dolnik (1990) described the vast scale of this migration between the Palearctic and Africa through Central Asia, to the north and west of the Tibetan-Himalayan region. Moreau concluded that while 74 Palearctic passerine species wintered regularly in Africa, only 27 species had at that time been recorded tackling the shorter route to South Asia, and concluded that there was 'virtually no evidence' of migration by passerines across the Tibetan Plateau. He observed that most of these 27 species filtered through the Pamir and Hindu Kush ranges to the west of the Tibetan Plateau, or flew around the eastern edge of the Tibetan Plateau. A year after Moreau's statement, Beaman (1973) provided some of the hitherto missing evidence, and recorded 19 passerine species on autumn migration through the Great Himalaya Range via the Kali Gandaki Valley, observations that strongly suggested small-scale passerine migration across the Tibetan Plateau.

A majority of migratory passerine species breeding in Europe migrate to Africa for the winter, but Moreau (1972) listed 10 that migrate in a southeasterly direction to South Asia. The most straightforward route for these birds migrating from the western Palearctic to South Asia would appear to be over the Hindu Kush range in Afghanistan, where the Kurram Valley and the Salang Pass (3880 m a.s.l.) are major migration routes (Akhtar, 1955). Because of war and political instability, there have still been no systematic studies to describe the nature and extent of migration along this potentially extremely important route. The recent tracking of five Red-spotted Bluethroats *Luscinia svecica* from two European populations to winter quarters in India (Lislevand *et al.*, 2015) suggests that much remains to be learned about what they term the 'Indo-European Flyway'.

Waterbirds and raptors dominate accounts of trans-Himalayan migration (e.g. Chapters 1–3 and 5–8), being more conspicuous than passerines, and perhaps presenting easier targets for collectors with guns, who provided a majority of observations before the use of optical equipment and mist nets came to predominate in the field in the 1960s. Because of their small body size, passerines have relatively low ring recovery rates, and cannot be fitted with satellite transmitters. Geolocators have become available in the twenty-first century which are small enough to deploy on passerines (Stutchbury *et al.*, 2009), but such studies have so far mainly been undertaken in Europe and North America. A factor that introduces bias into the recording of passerines on migration is the tendency of some groups, particularly flycatchers, chats, thrushes and warblers, to migrate at night, when they can only be observed by moon-watching or radar (e.g. Dokter *et al.*, 2011). On the other hand, larks, accentors, swallows, pipits, wagtails, finches, buntings, sparrows and corvids tend to migrate by day and so are more frequently recorded (Alerstam, 2009).

The Indomalayan and Palearctic Zoogeographic Regions

The crest of the Great Himalaya Range forms the southern boundary of the Palearctic zoogeographic region, and the northern limit of the Indomalayan (or Oriental) zoogeographic region. About 220 passerine species breed on the southern slopes of this range, mostly in subtropical forests. These include more than 60 species of babbler and more than 20 species each of old-world warbler and flycatcher, as well as Himalayan specialists such as nine rosefinch and five forktail species (del Hoyo *et al.*, volumes 9–16, 2004–2011). Most of these Himalayan southern slope species are non-migratory, and for very many, the Himalayan southern slopes form the northern and western extremities of their ranges of distribution. In this chapter we only consider passerine species which breed in the Palearctic zoogeographic region to the north of the Great Himalayan crest, and migrate over the Himalayas to the Indomalayan and Afrotropic zoogeographic regions, or to the southern parts of the Palearctic zoogeographic region, from breeding areas further to the north.

Passerines Recorded as Residents and Summer Visitors in the Tibetan-Himalayan Region

A total of about 45 passerine species visit the Palearctic part of the Tibetan-Himalayan region to breed in the summer, migrating south to the Indian lowlands for the winter (see Table 4.1, where scientific names are provided). Two species of summer visitor, Pied Wheatear and Common Rock Thrush, migrate to Africa for the winter. Table 4.1 also lists 79 resident passerine species which remain in the Tibetan-Himalayan region throughout the year, many of which are altitudinal migrants, moving relatively short distances to lower altitudes when winter arrives.

Passerines Recorded as Passage Migrants in the Tibetan-Himalayan Region

Long-distance passerine migrants were already recorded in the Himalayan region in the nineteenth century. A Dusky Warbler was collected in the Nubra Valley, Ladakh in October 1873 (Sharpe, 1891), and a Barred Warbler in Gilgit in September 1879 (Kinnear, 1931). Migrant passerines have been observed in the Brahmaputra (Tsangpo) Valley in the eastern Himalayas, and Siberian Rubythroat, Red-breasted Flycatcher, Brown Shrike, Black-faced and Yellow-breasted Buntings are also thought to follow this route through the Himalayas (Inskipp & Inskipp, 1985; Spierenburg, 2006). Further west, in Nepal, Inskipp (1985) noted that 19 passerine species had been recorded on autumn migration through the Great Himalaya Range via the Kali Gandaki Valley by Beaman (1973). The only numerous species were Greater and Hume's Short-toed Larks, White Wagtails, Black Redstarts and Tickell's Leaf Warblers. Bad weather did not increase the number of passerine migrants, suggesting that few used this route. Small numbers of passerines have been recorded crossing the Tibetan Plateau, including White Wagtail, Booted Warbler, Lesser Whitethroat and Common Chiffchaff (Inskipp, 1985). In the Kathmandu Valley, Proud (1949, 1952, 1955) noted 'thousands' of Common Stonechats on spring and autumn migration, and observed that the following species were also numerous on migration: Barn Swallow, White Wagtail, Citrine Wagtail, Yellow Wagtail, Bluethroat, Blyth's Reed Warbler, Greenish Warbler, Common Rosefinch and Yellow-breasted Bunting. Less common spring migrants were Brown Shrike and Paddyfield Warbler. Kinnear (1922) in the Mount Everest region reported Common Stonechats as well as House Martins and several pipits at 5200 m a.s.l., and even higher, up to 6100 m a.s.l., reported Richard's Pipit, Black Redstart, Short-toed Lark and Godlewski's Bunting (see also Prins *et al.*, Chapter 20). In Bhutan, Spierenburg (2006) also noted many high altitude records of Common Stonechat, and suggested that this was probably indicative of direct Himalayan passage.

A series of observations in the Gyangtse region of Tibet, south of Lhasa and adjacent to the border with Bhutan and Sikkim, at altitudes up to 4500 m a.s.l., was published by

Table 4.1. Passerine species resident in the Palearctic Tibetan-Himalayan region, and summer visitors which breed in the Tibetan-Himalayan region and migrate to the South Asian lowlands or to Africa for the winter. The species sequence in this and all tables follows del Hoyo *et al.* 2004–2011.

Residents in Tibetan-Himalayan Region (includes altitudinal migrants)		Summer visitors to Tibetan-Himalayan region
Tibetan Lark *Melanocorypha maxima*	White-winged Snowfinch *Montifringilla nivalis*	Hume's Short-toed Lark *Calandrella acutirostris*
Horned Lark *Eremophila alpestris*	Black-winged Snowfinch *Montifringilla adamsi*	Oriental Skylark[3] *Alauda gulgula*
Nepal House Martin *Delichon nipalense*	White-rumped Snowfinch *Montifringilla taczanovskii*	Plain Sand Martin *Riparia paludicola*
Upland Pipit *Anthus sylvanus*	Rufous-necked Snowfinch *Montifringilla ruficollis*	Pale Sand martin[1] *Riparia diluta*
White-throated Dipper *Cinclus cinclus*	Red-fronted Serin *Serinus pusillus*	Barn Swallow[1] *Hirundo rustica*
Brown Dipper *Cinclus pallasii*	Tibetan Siskin *Carduelis thibetana*	Wire-tailed Swallow[2] *Hirundo smithii*
Winter Wren *Troglodytes troglodytes*	Eurasian Goldfinch *Carduelis carduelis*	European Crag Martin *Hirundo rupestris*
Alpine Accentor *Prunella collaris*	Twite *Carduelis flavirostris*	Northern House Martin[1] *Delichon urbicum*
Altai Accentor *Prunella himalayana*	Eurasiann Linnet[2] *Carduelis cannabina*	Asian House Martin *Delichon dasypus*
Robin Accentor *Prunella rubeculoides*	Plain Mountain Finch *Leucosticte nemoricola*	Rosy Pipit[3] *Anthus roseatus*
Rufous-breasted Accentor[3] *Prunella strophiata*	Brandt's Mountain Finch *Leucosticte brandti*	White Wagtail[1] *Motacilla alba*
Brown Accentor *Prunella fulva*	Sillem's Mountain Finch[2] *Leucosticte sillemi*	Citrine Wagtail[1] *Motacilla citreola*
Black-throated Accentor *Prunella atrogularis*	Trumpeter Finch[2] *Bucanetes githagineus*	Grey Wagtail[1] *Motacilla cinerea*
Maroon-backed Accentor *Prunella immaculata*	Mongolian Finch *Eremopsaltria mongolicus*	Tickell's Thrush *Turdus unicolor*
Mistle Thrush[3] *Turdus viscivorus*	Blanford's Rosefinch *Carpodacus rubescens*	Common Rock Thrush[4] *Monticcola saxatilis*
Common Blackbird[2] *Turdus merula*	Dark-breasted Rosefinch *Carpodacus nipalensis*	Bluethroat[1] *Luscinia svecica*
Tibetan Blackbird[2] *Turdus maximus*	Beautiful Rosefinch *Carpodacus pulcherrimus*	White-tailed Rubythroat *Luscinia pectoralis*
Grey-winged Blackbird[2] *Turdus boulboul*	Dark-rumped Rosefinch *Carpodacus edwardsii*	Orange-flanked Bush Robin[1] *Tarsiger cyanurus*
White-backed Thrush *Turdus kessleri*	Pale Rosefinch[2] *Carpodacus synoicus*	Black Redstart *Phoenicurus ochruros*
Chestnut Thrush *Turdus rubrocanus*	White-browed Rosefinch *Carpodacus thura*	Hodgson's Redstart *Phoenicurus hodgsoni*
Blue Whistling Thrush *Myophonus caeruleus*	Red-mantled Rosefinch *Carpodacus rhodoclamys*	Common Stonechat[1] *Saxicola rubetra*
Grandala *Grandala coelicolor*	Streaked Rosefinch *Carpodacus rubicilloides*	Variable Wheatear[2] *Oenanthe picata*

Table 4.1. (cont.)

Residents in Tibetan-Himalayan Region (includes altitudinal migrants)		Summer visitors to Tibetan-Himalayan region
Blue Rock Thrush *Monticola solitarius*	Great Rosefinch *Carpodacus rubicilla*	Pied Wheatear[4] *Oenanthe pleschanka*
White-capped Water Redstart *Chaimarrornis leucocephalus*	Red-fronted Rosefinch *Carpodacus puniceus*	Desert Wheatear *Oenanthe deserti*
Blue-capped Redstart[2] *Phoenicurus caeruleocephala*	Crimson-browed Finch *Pinicola subhimachala*	Isabelline Wheatear *Oenanthe isabellina*
Guldenstadt's Redstart *Phoenicurus erythrogaster*	Scarlet Finch *Haematospiza sipahi*	Spotted Flycatcher[2] *Muscicapa striata*
Goldcrest *Regulus regulus*	Brown Bullfinch *Pyrrhula nipalensis*	Dark-sided Flycatcher *Muscicapa sibirica*
Great Tit *Parus major*	Red-headed Bullfinch *Pyrrhula erythrocephala*	Asian Paradise Flycatcher[2] *Terpsiphone paradisis*
Hume's Ground Jay *Pseudopodoces humilis*	White-winged Grosbeak *Mycerobas carnipes*	Blunt-winged Warbler *Acrocephalus concinens*
Long-billed Bushwarbler *Bradypterus major*	Rock Bunting[2] *Emberiza cia*	Clamorous Reed Warbler[3] *Acrocephalus stentoreus*
Smoky Warbler *Phylloscopus fuliginventer*	Godlewski's Bunting[2] *Emberiza godlewskii*	Mountain Chiffchaff *Phllosciopus sindianus*
White-browed Tit Warbler *Leptopoecile sophiae*	Tibetan Bunting *Emberiza koslowi*	Tickell's Leaf Warbler *Phylloscopus affinis*
Wallcreeper *Tichodroma muraria*	Reed Bunting[3] *Emberiza schoeniclus*	Sulphur-bellied Warbler *Phylloscopus griseola*
Long-tailed Shrike[2] *Lanius schach*		Brooks's Leaf Warbler[2] *Phylloscopus subviridis*
Black Drongo[2] *Dicrurus macrocercus*		Greenish Warbler[1] *Phylloscopus viridanus*
Black-billed Magpie *Pica pica*		Hume's Leaf Warbler *Phylloscopus humei*
Yellow-billed Chough *Pyrrhocorax graculus*		Tytler's Leaf Warbler[2] *Phylloscopus tytleri*
Red-billed Chough *Pyrrhocorax pyrrhocorax*		Western Crowned Leaf Warbler[2] *Phylloscopus occipitalis*
Eurasian Jackdaw[2] *Corvus monedula*		Hume's Whitethroat *Sylvia althaea*
Carrion Crow *Corvus corone*		Fire-capped Tit *Cephalopyrus flammiceps*
Large-billed Crow *Corvus macrohynchos*		Indian Golden Oriole *Oriolus kundoo*
Common Raven *Corvus corax*		Isabelline Shrike *Lanius isabellinus*
Common Myna *Acridotheres tristis*		Grey-backed Shrike[2] *Lanius schch*
House Sparrow[3] *Passer domesticus*		Common Rosefinch[1] *Carpodacus erythrynus*

Table 4.1. (cont.)

Residents in Tibetan-Himalayan Region (includes altitudinal migrants)	Summer visitors to Tibetan-Himalayan region
Spanish Sparrow[2, 3] *Passer hispaniolensis*	White-capped Bunting *Emberiza stewarti*
Eurasian Tree Sparrow *Passer montanus*	

[1] These species also have populations which breed further north and migrate to South Asia around or through the Tibetan-Himalayan region.
[2] These species have a peripheral distribution in the Tibetan-Himalayan region.
[3] These species are partial migrants. Some individuals remain in the Tibetan-Himalayan region in winter; some migrate south.
[4] These species winter in Africa.

Ludlow (1927, 1928, 1944, 1950). Common Stonechat was again identified as a numerous passage migrant in spring and autumn, and he also noted good numbers of White and Citrine Wagtail, Olive-backed and Blyth's Pipit, Bluethroat, Dusky Warbler and 'enormous flocks' of Short-toed Lark in September and October. One of the Olive-backed Pipits shot by Ludlow had heavy deposits of subcutaneous fat, indicating that it was at an early stage of a long migratory flight. Scarcer migrant passerines Ludlow recorded in Gyangtse included Greenish Warbler and Common Starling. Ludlow (1928) was one of the first to remark on the scale of altitudinal migration in the Himalayan region, and observed Guldenstadt's Redstarts 'swarming in every *Hippophaë* thicket' from the end of September onward, and also noted Brown and Robin Accentor, Winter Wren and Horned Lark.

In the western Himalayan region, Sillem (1934) recorded White, Citrine and Grey Wagtails, Tree Pipit, Common Rock Thrush and Common Rosefinch in the region of the Kara-tagh Pass at 5340 m a.s.l. in the eastern Karakoram in mid-August. Also in the Karakoram, Vaurie (1972) reported that the 1937 Shaksgam expedition had found the corpse of a Common Starling in snow at an altitude of 5490 m a.s.l. The most frequently observed passerines migrating through Pakistan are Red-breasted Flycatcher, Rosy Starling, Red-headed and Black-headed Buntings (Grimmett *et al.*, 2008). Several African-wintering migrants have also been recorded, including Common Whitethroat, Common Rock Thrush, Red-backed Shrike (Grimmett *et al.*, 2008) and Common Redstart (Christison, 1939). Alexander (1974) described how, in late May and early June, he often observed swallows, martins, swifts, warblers and Golden Oriole in the western Himalayan foothills above Dalhousie in Himachal Pradesh, at 2450 m a.s.l., flying directly towards the high ranges. He remarked on the readiness of some migratory birds, even small warblers, to follow a direct route to their destination, even over high obstacles.

Akhtar (1955) reported the considerable scale of passerine migration in the Hindu Kush of Afghanistan, especially in spring, with basketfuls of songbirds being sold in the bazaars of Kabul, many having been killed with catapults and blowpipes. Rosy Starlings, wagtails, Short-toed Larks and sparrows were particularly numerous.

Near-Passerines

A number of species closely allied to the passerines, notably Common Swift (*Apus apus*), Common Cuckoo (*Cuculus canorus*), Common Hoopoe (*Upupa epops*) and Eurasian Wryneck (*Jynx torquilla*), have been recorded as summer visitors and passage migrants in many parts of the Himalayas. One, or possibly two Hoopoes were recorded by Kinnear (1922) at 6400 m a.s.l. in the Mount Everest region, and Ludlow (1927) found them on migration in Gyangtse. Common Swift, Hoopoe and Wryneck are observed regularly on migration in Nepal (Proud, 1952; Inskipp & Inskipp, 1985), and Wryneck was mentioned as a spring migrant in the Hindu Kush (Akhtar, 1955). The Southampton University expeditions also recorded all these species in Ladakh.

Methods Used and Species Recorded by the Southampton University Ladakh Expeditions

On three expeditions between 1977 and 1981–1982, we established daily constant effort mist-netting at the Forestry Department Plantation at Thikse, a village on the River Indus at an altitude of 3300 m a.s.l., 18 km southeast of Leh, the capital of Ladakh. The plantation covers an area of about one square kilometre, and vegetation is dominated by young willow *Salix* and poplar *Populus* trees with extensive thickets of sea buckthorn *Hippophae rhamnoides* and tamarisk *Myricaria*. The plantation at Thikse is one of the most extensive areas of tree and scrub habitat in the Upper Indus Valley and its oasis-like position in a cultivated valley bottom, amid thousands of square kilometres of arid mountains, explains its attraction to passing migrants in need of rest, food and shelter. We set about 100 m of full-height mist nets daily and checked them every half hour from dawn to midday and from 16.00 to dusk. Bad weather occasionally curtailed these sessions. Additional observation routines covered a 13 km stretch of the valley bottom between Thikse and Choglamsar, on foot and bicycle, two to three times per week. We recorded 51 of the 79 resident and altitudinal migrant passerine species listed in Table 4.1 (65%) and 26 of the 45 breeding summer visitors (58%). Passerine species recorded in Ladakh which breed north of the Tibetan-Himalayan region included three species of numerous winter visitors, and no fewer than 45 species on spring and autumn passage (Table 4.2). We also recorded 90 species of non-passerines on migration, making 134 long-distance migrant species, more than half of which (71 species) had not previously been recorded in Ladakh, four of which were previously unrecorded in the Indian subcontinent and nine of which were new for India (Delany *et al.*, 2014). These long-distance trans-Himalayan migrants included Palearctic migrants en route to South Asia, to Southeast Asia and to Africa.

Table 4.2. Passerine species recorded in Ladakh with breeding ranges to the north of the Tibetan-Himalayan region, and wintering ranges in South Asia and/or Africa. For numbers mist-netted at Thikse, see Table 4.3. For scientific names, see previous table; previously unmentioned scientific names are given.

Trans-Tibetan-Himalayan passage migrants	Breeding range	Wintering range	Notes
Bimaculated Lark	Central & SW Asia	NW South Asia	One record, winter 1981–1982
Greater Short-toed Lark	Europe-Central Asia – Mongolia	Northern South Asia	Occasional on autumn passage
Sand Martin	Lowland Palearctic	Africa & SE Asia	Regular, spring and autumn
Barn Swallow[1]	Palearctic & Tibet-Himalaya		Numerous on spring and autumn passage
Red-rumped Swallow	S Europe, Southern Asia, Africa	Africa, S & SE Asia	Occasional on spring passage
Tree Pipit[1]	Palearctic & Tibet-Himalaya	S Asia & Africa	Regular on spring and autumn passage
Red-throated Pipit	Central & E Siberia	S & SE Asia	Many on spring passage
Water Pipit	N of Tibet-Himalaya	SW & SE Asia	Many on spring passage
Yellow Wagtail	Lowland Palearctic	SW, S & SE Asia & Africa	Two subspecies on spring passage
Citrine Wagtail[1]	Palearctic & Tibet-Himalaya	SW, S & SE Asia	Two subspecies on spring passage
Grey Wagtail[1]	Palearctic & Tibet-Himalaya	SW, S & SE Asia, Africa	Occasional migrants
White Wagtail[1]	Palearctic & Tibet-Himalaya	S & SE Asia & Africa	Three subspecies on spring passage
Black-throated Accentor	Ural & Altai Mountains	NW Indian subcontinent, SW Asia	Occasional migrants
Black-throated Thrush	Central Siberia	S W & S Asia	Winter visitor to Ladakh
Red-throated Thrush	Central Siberia	NE Indian subcontinent	Scarce winter visitor to Ladakh
Naumanns Thrush *Turdus naumanni*	Central & E Siberia	East Asia	Vagrant?
Dusky Thrush	Central & E Siberia	East Asia to Japan	Vagrant?
Song Thrush *Turdus philomelos*	W Palearctic to C Siberia	Europe, N Africa, SW Asia	One record, Nov–Dec 1981. Vagrant
Bluethroat[1]	Palearctic & Tibet-Himalaya	S& SE Asia, Africa	Many local breeders
Common Redstart	Europe to C Siberia	Africa	Three records, May 1982
Common Stonechat	Palearctic & Tibet-Himalaya	S & SE Asia, Africa	Northern forms observed regularly
Red-breasted Flycatcher	E Europe-E Siberia	S & SE Asia	One record, Dec 1981
Sedge Warbler *Acrocephalus schoenobanus*	Europe to Central Siberia	Africa	Two records, autumn 1980 & 1981 No other records in South Asia

Table 4.2. (cont.)

Trans-Tibetan-Himalayan passage migrants	Breeding range	Wintering range	Notes
Black-browed Reed Warbler *Acrocephalus bistrigiceps*	Far east Asia	SE Asia to Assam	One record, autumn 1980. Vagrant from East Asia
Great Reed Warbler *Acrocephalus arundinaceus*	Europe to Central Siberia	Africa	One record, August 1977
Paddyfield Warbler	SE Europe to Central Siberia	S Asia	Five records
Blyth's Reed Warbler	NE Europe to Central Siberia	S Asia & SE Asia	22 records
Common Chiffchaff	Europe to E Siberia	S Asia & Africa	20 records
Dusky Warbler	Central & E Siberia	SE & S Asia	One record, autumn 1980
Greenish Warbler[1]	Palearctic & Tibet-Himalaya	S Asia	17 records
Yellow-browed Warbler *Phylloscopus inornatus*	Central & E Siberia	SE & S Asia	5 records, autumn 1981;
Garden Warbler *Sylvia borin*	Europe to Central Siberia	Africa	Three records, autumn 1980 & 1981. No other records in South Asia
Barred Warbler	Europe to Xinjiang	East Africa	Two records, autumn 1980 & 1981.
Lesser Whitethroat	Europe to Central Asia	S & SW Asia	28 records of subspecies *blythi*
Desert Lesser Whitethroat	Central Asia to Xinjiang	S & SW Asia	Two records
Common Whitethroat *Sylvia communis*	Europe to Central Siberia	Africa	One record, June 1982
Isabelline Shrike *Lanius isabellinus*	Iran to Mongolia	Arabia, Africa, S Asia	Spring 1982
Red-backed Shrike *Lanius collurio*	Europe to Central Asia	Africa	Spring 1982
Lesser Grey Shrike *Lanius minor*	SE Europe-W Siberia	Southern Africa	Spring 1982. Two records
Rose-coloured Starling	SE Europe to Xinjiang	South Asia	Occasional migrant
Common Rosefinch[1]	N Palearctic & Tibet-Himalaya	South & SE Asia Asia	Many local breeders
Yellowhammer *Emberiza citrinella*	Europe to central Siberia	Europe to S central Siberia, central Asia	One record, Dec 1981
Pine Bunting	Siberia	Central Asia to Himalayas	Winter visitor to Ladakh
Little Bunting	N Palearctic	SE Asia to Himalayas	Two records, Oct 1980
Common Reed Bunting	Palearctic	S Palearctic	Scarce winter visitor to Ladakh

[1] These species have also established separate breeding populations within the Tibetan-Himalayan region, which migrate into South Asia for the winter. Both Siberian and Himalayan forms of White and Citrine Wagtails, Greenish Warbler and Common Stonechat were recorded.

Possible Routes Taken by Passerine Migrants in Ladakh

There are no obvious north-to-south migration routes through Ladakh. The Indus, and its northern tributary, the Shayok, flow in a southeast–northwest direction. The parallel massifs of the Karakoram, Ladakh, Zanskar and Great Himalaya ranges run parallel to these rivers, forming formidable southeast–northwest barriers to bird migration with only high-altitude passes. The Suru River, a southern tributary of the Indus, flows for 60 km in a northerly direction to join the Indus near Kharmang, and may offer a suitable route for migration (Holmes, 1986), but the river does not cross the Great Himalaya Range, whose northern wall still presents a barrier nearly 6000 m high towards the head of the valley. The very high and rugged Muztagh Pass (5380 m a.s.l.) through the Karakoram range lines up with the Suru Valley to provide a continuing potential route to the north, and Holmes (1986), based on numbers of birds mist-netted in the Suru Valley in 1978 and 1983, considered it a possibility that Yellow-browed Warblers and Common Rosefinches use this route in some numbers. Since the taxonomic split of

Figure 4.1. The recovery of a ringed Hume's Lesser Whitethroat suggests a direct migratory route across the Zanskar and Great Himalaya ranges from the Indus Valley. (A black-and-white version of this figure will appear in some formats. For the colour version, please refer to the plate section.)

Hume's Leaf Warbler *Phylloscopus humei* from Yellow-browed Warbler (Svensson, 1987), both species may have been involved in these observations as at Thikse.

Migrant passerines probably fly directly over the Zanskar and Great Himalaya ranges from breeding and stopover sites along the Indus such as the Thikse Forestry Plantation. The recovery of a ringed Hume's Lesser Whitethroat directly southwest of Thikse in Jammu and Kashmir State supports this view (Figure 4.1). This bird was ringed as a juvenile on 22 August 1980 and recovered 226 days later, 320 km away at Raipur, near Jammu, on 15 May 1981, shortly before (presumably) returning to Ladakh to breed. The recovery suggests a direct flight over ranges with a lowest transit point of about 5200 m a.s.l. DeLuca *et al.* (2015) used geolocators to track Blackpoll Warblers migrating from the northeastern United States to northern South America, and revealed non-stop flights averaging 2540 km, and lasting an average of 62 hours, an average speed of 41 $km.h^{-1}$. Assuming a similar flight speed, the Hume's Lesser Whitethroat would have covered the distance from Raipur to Thikse in less than eight hours.

Numbers of Migrants at Thikse

Our constant effort mist-netting activities at Thikse provided a sampling method for recording relative numbers of different species. The samples were, however, strongly biased in favour of species occurring in bushes and low trees, such as warblers, thrushes and chats (Figure 4.2, Figure 4.4), and against those that prefer open country, such as larks, swallows, pipits and wagtails (Figure 4.3). We redressed this bias by conducting almost continuous observations at Thikse of species that were seen but not trapped, and made regular counts of birds migrating through more open habitats downstream.

Table 4.3 shows that altitudinal migrant passerines were the most numerous type of migrants trapped at Thikse, making up 52% of the catch, with one species, Guldenstadt's Redstart, alone comprising 37% of birds caught. Brown Accentor and Eastern Great Rosefinch were the other numerous winter visitors at Thikse (Table 4.3). The sheer numbers of Guldenstadt's Redstarts and Brown Accentors that use the *Hippophae*-dominated scrub habitat of the Upper Indus Valley in winter suggest that birds come from a wide catchment, and that altitudinal migration might also involve considerable longitudinal and latitudinal components.

Summer visitors made up 45% of the birds trapped at Thikse, the bulk of the catch consisting of Mountain Chiffchaffs, House Sparrows, Common Rosefinches, Bluethroats, Hume's Lesser Whitethroats and Citrine Wagtails, all of which bred in and around the plantation (Table 4.3). These birds traverse the Great Himalaya range twice per year, and while their migrations into lowland South Asia can cover relatively short horizontal distances (e.g. Figure 4.1), these species are all widely distributed in South Asia, and some in Southwest and Southeast Asia in winter (Ali & Ripley, 1987–1999).

A few species that breed in Siberia winter in the Himalayan region and were recorded around Thikse. Groups of up to 30 or more Black-throated Thrushes and 20 Pine Buntings used mature trees adjacent to the plantation as a night-time roosting site, along with large numbers of resident or altitudinal migrant Black-billed Magpies and Eastern Great

Figure 4.2. Guldenstadt's Redstart, Thikse, Ladakh, October 1980. A sparsely distributed, high-altitude breeder throughout the Himalayas and ranges to the north, large numbers migrate altitudinally to lower-lying river valleys for the winter, where they are associated with thickets of Sea Buckthorn, *Hippophaë rhamnoides*. Photo, by John Norton. (A black-and-white version of this figure will appear in some formats. For the colour version, please refer to the plate section.)

Figure 4.3. Citrine Wagtail, Thikse, Ladakh autumn 1981. A common summer visitor to the Tibetan-Himalayan region, wintering widely in southern Asia. Photo, by Clare Sulston. (A black-and-white version of this figure will appear in some formats. For the colour version, please refer to the plate section.)

Figure 4.4. Sedge Warbler, Thikse, Ladakh, 11 October 1981. One of only two recorded in the Himalayan region, both by Southampton University expeditions. This was one of five warbler species discovered migrating between Asia and Africa via the northwest Himalayas. Photo, by Clare Sulston. (A black-and-white version of this figure will appear in some formats. For the colour version, please refer to the plate section.)

Rosefinches, and joined on occasion by scarcer thrushes and buntings. Small numbers of these birds were trapped, mostly during the winter of 1981–1982 (Table 4.3).

The long-distance passage migrant passerines made up a numerically small proportion of the catch (2.1%), but included the widest species diversity of any category of migrant, comprising 39% of the species trapped (Table 4.3). A considerable majority (19 out of 23 species, 83%) were warblers, thrushes and shrikes, all of which prefer to migrate at night, and all of which were stopping over in the northwest Himalayan region to rest on prodigious journeys.

Six (26%) of the long-distance migrant species trapped at Thikse breed in Palearctic Asia north of the Tibetan-Himalayan region, and winter in Africa. Of these, the two Sedge Warblers and three Garden Warblers trapped at Thikse remain the only records of these species in the Indian subcontinent. Barred Warbler has been recorded on at least two other occasions migrating through the northwest Himalayas, and Holmes (1986) suggested that it may be a regular migrant through these ranges. The other African wintering species comprised one Great Reed Warbler and one Common Whitethroat ringed, respectively in August 1977 and May 1982, and two Common Redstarts trapped (and a third observed) on subsequent days in May 1982.

The most numerous long-distance passage migrant species recorded at Thikse were Tree Pipit (35 ringed) Lesser Whitethroat (28), Blyth's Reed Warbler (22), Common Chiffchaff (20) and Greenish Warbler (17). All five species were trapped in every

Table 4.3. Numbers of altitudinal migrant, winter visitor, summer visitor and long-distance Palearctic migrant passerines trapped and ringed at Thikse, Ladakh, 1977–1982.

	Total ringed 1977–1982		Total ringed 1977–1982
Residents and altitudinal migrants		*Long-distance migrant winter visitors*	
Horned Lark	1	Black-throated Thrush	44
Brown Accentor	882	Red-throated Thrush	2
Robin Accentor	12	Song Thrush	1
Guldenstadt's Redstart	2817	Pine Bunting	9
Winter Wren	1	Reed Bunting	2
Goldcrest	1	**Total winter visitors**	**58**
White-browed Tit-warbler	7		
Great Tit	68	*Long-distance passage migrants*	
Black-billed Magpie	9	Red-rumped Swallow	1
Mongolian Finch	1	Tree Pipit	35
Red-fronted Serin	2	Black-throated Accentor	8
Eastern Great Rosefinch	154	Sedge Warbler[1]	2
Great Rosefinch	2	Black-browed Reed Warbler	1
Total residents and altitudinal migrants	**3957**	Great Reed Warbler[1]	1
		Paddyfield Warbler	5
		Yellow-browed Warbler	5
Summer visitors		Blyth's Reed Warbler	22
Hume's Short-toed Lark	20	Common Chiffchaff	20
Oriental Skylark	1	Dusky Warbler	1
European Crag Martin	1	Greenish Warbler	17
Barn Swallow	2	Garden Warbler[1]	3
Rose-breasted pipit	1	Barred Warbler[1]	2
White Wagtail	23	Lesser Whitethroat	28
Citrine Wagtail	98	Desert Lesser Whitethroat	2
Hume's Leaf Warbler	2	Common Whitethroat[1]	1
Mountain Chiffchaff	1199	Common Redstart[1]	2
Hume's Lesser Whitethroat	330	Common Stonechat	3
Bluethroat	342	Tickell's Thrush	1
Black Redstart	197	Dusky Thrush	1
Indian Golden Oriole	1	Isabelline Shrike	1
Long-tailed Shrike	25	Little Bunting	2
Common Rosefinch	559	**Total long-distance passage migrants**	**164**
House Sparrow	649		
Brahminy Starling	1		
Total Summer visitors	**3451**	**Overall total passerines ringed**	**7630**

[1] These species winter only in Africa.

migration season that was covered, in numbers that suggest annual migration through and over these ranges. All five of these species winter in the lowland Indian subcontinent, and Tree Pipit, Lesser Whitethroat and Blyth's Reed Warbler breed as close to the Himalayas as Central Asia. Common Chiffchaffs breed further north in Siberia, and Greenish Warblers of the form *viridanus* trapped at Thikse mainly breed from Eastern

Europe to West Siberia, being replaced in the northwest Himalayas by the form *ludlowi*. All of these species except Greenish Warbler also breed abundantly in Central Siberia (Rogacheva, 1992).

Four of the species trapped at Thikse, namely Black-Browed Reed Warbler, Dusky Warbler, Dusky Thrush and Little Bunting, winter mainly in Southeast Asia. There were two records of the latter species, and only single records of the others. Given the location of Thikse in the northwestern part of the Himalayan ranges, it was perhaps surprising to record any of these four species.

Visible Diurnal Migration in the Upper Indus Valley

We recorded detailed, systematic bird observations on each day of the expeditions to Ladakh, and these data complemented those collected by the ringing programme. We made visits every two to three days to more open grassland and marshland habitats downstream of Thikse at Shey and Choglamsar in 1980 and 1981–1982. Observations at these additional sites were less intense in 1977 because there were only two observers, and this reduced level of coverage affected observations of species that used these habitats in 1977. Numbers of visible diurnal migrants were often higher than numbers of mostly nocturnal migrants trapped at Thikse, and numbers of bird-days in the Upper Indus Valley of the conspicuous species are shown in Figures 4.5, 4.6 and 4.7.

Black Redstart

The Black Redstart is one of the most characteristic passerine summer visitors throughout the Tibetan-Himalayan region, occurring in most habitats from valley-bottom villages to the highest remote and desolate mountain passes. Our observations in the Upper Indus Valley showed a very regular autumn migration that was remarkably similar in 1980 and 1981, peaking in the week of 18–24 September, and an intense, rather short migration in the spring of 1982, peaking in the week of 14–20 May (Figure 4.5).

Pipits and Wagtails

Autumn migration of pipits and wagtails involved large movements of Tree Pipits, Citrine and White Wagtails (Figures 4.6 and 4.7). The wagtails may have been mostly Ladakhi breeders, but also included long-distance migrants from further north. The wagtails were in breeding plumage in spring and could usually be identified to subspecies level. Spring passage migrants included three subspecies of White and two of Citrine Wagtails, as well as two subspecies of Yellow Wagtail in lower numbers. Spring passage of pipits was more diverse than in autumn, with good numbers of Rosy Pipits (which occur throughout the year in Ladakh) and Red-throated Pipits, and fewer Water Pipits.

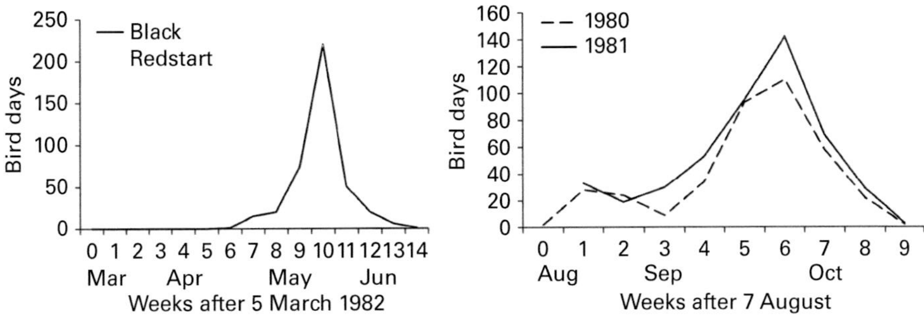

Figure 4.5. Migration pattern of Black Redstart in the Upper Indus Valley, spring 1982 and autumn 1980 and 1981. Weekly bird-day totals by seven-day periods

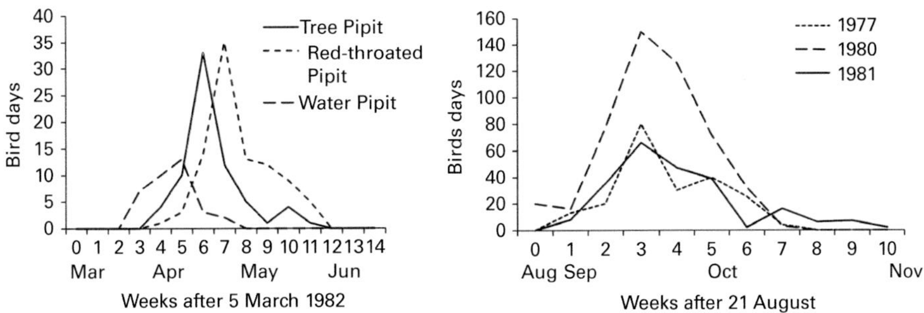

Figure 4.6. Migration patterns of pipits in the Upper Indus Valley. Weekly bird-day totals by seven-day periods. Left: Tree Pipit, Water Pipit, and Red-throated Pipit, spring 1982. Right: Tree Pipit, autumn 1977, 1980, 1981

Our observations of Tree Pipits in the Upper Indus Valley showed a very regular autumn migration that was similar in 1977, 1980 and 1981, peaking each year in the week of 11–17 September, and an intense migration involving fewer birds in the spring of 1982, peaking in the week of 16–22 April (Figure 4.6). The timing of the autumn migration peak is one week later than the timing of departure of birds from Central Siberia, where the autumn migration begins in the last third of August and ends in the first third of September (Rogacheva, 1992). The spring migration patterns of Water Pipit and Red-throated Pipit (species that did not occur on autumn migration) are also shown in Figure 4.6.

White and Citrine Wagtails were more numerous than any of the pipits, and White Wagtail numbers peaked later in the autumn and earlier in the spring than Citrine Wagtails (Figure 4.7). In the autumn of 1980, we recorded a total of 1644 bird-days for White Wagtail and 1411 for Citrine Wagtail, with 1981 totals being a little lower. The totals in 1977 were noticeably lower, but this was an effect of lower levels of coverage. White Wagtails were the first migrants to arrive on spring migration and peaked in number in the last week of March, after which numbers decreased steadily throughout

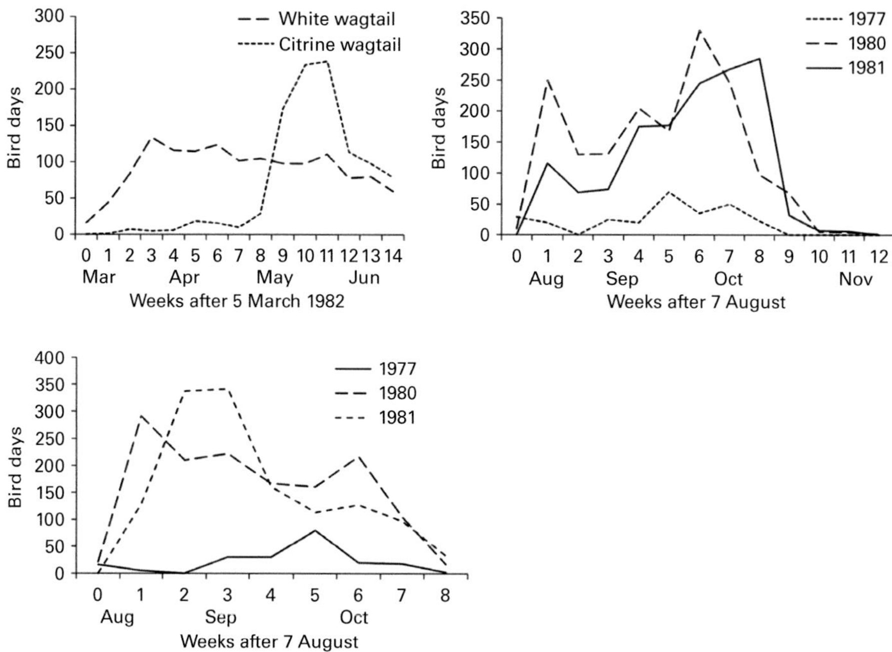

Figure 4.7. Migration patterns of wagtails in the Upper Indus Valley. Weekly bird-day totals by seven-day periods. Top left: White Wagtail and Citrine Wagtail, spring 1982. Right: White Wagtail, autumn 1977, 1980, 1981. Bottom left: Citrine Wagtail, autumn 1977, 1980, 1981.

the spring. Citrine Wagtail numbers remained low until the end of April, then built rapidly to a peak at the end of May before decreasing again. A total of 45 bird-days for Yellow Wagtail were also recorded in the spring of 1982, with passage peaking in the second half of May.

Swallows, Martins and Larks

Swallows and martins passed through in their thousands (but in smaller numbers than swifts, which are not passerines). While Barn Swallow and Northern House Martin breed in Ladakh, the number of birds observed suggests a more distant (and possibly much more distant) provenance, at least for some of the birds. Most recorded flocks were small, but in the last week of August 1980, the summed weekly bird-day counts attained a total of 1193 individuals. Red-rumped Swallows, Sand Martins and Plain Sand Martins were regularly seen in the flocks, as well as a single Wire-tailed Swallow.

Short-toed and Hume's Short-toed Larks were also regular autumn migrants over Thikse, often in flocks flying over at considerable altitude. Like the swallows and martins, these and other lark species appeared from time to time, usually during cloudy weather, with no clear pattern of occurrence.

Other Passerine Species

Other observed long-distance migrant and vagrant passerine species that hardly featured in the ringing programme included regular Black-throated Accentors and Common Starlings, occasional Brahminy and Rose-Coloured Starlings and single records of Bimaculated Lark, Forest Wagtail, Red-breasted Flycatcher, Common Jackdaw and House Crow. Shrikes were also regular on spring migration with Isabelline, Red-backed and Lesser Grey Shrikes all recorded in small numbers. The latter two species only winter in Africa.

Near-Passerines

Frequently recorded migrants in the Upper Indus Valley that are allied to passerines included Common Swift, Common Cuckoo, Common Hoopoe, Common Kingfisher *Alcedo atthis*, and Eurasian Wryneck. The first three are summer visiting breeders in Ladakh, and kingfishers probably also breed (Pfister, 2004). The numbers of Common Swifts observed, especially on days of low cloud, suggests that birds from a wide catchment were moving through. Small numbers of Pacific Swifts *Apus pacificus* were also seen in these flocks. Common Swift and Common Cuckoo winter in Africa, while the kingfisher, hoopoe and wryneck migrate the shorter distance to lowland India. European and Blue-cheeked Bee-Eaters (*Merops apiaster* and *Merops persicus*), two other African wintering species, were scarce migrants in the spring of 1982.

Weather Effects

Conspicuous visible migration almost invariably occurred during periods of low cloud. Migrants brought down by low cloud or rain often stayed in the valley bottom for a few days, even after the weather cleared. For example, cloudy and squally weather on 19 and 20 September 1981 resulted in the trapping and ringing at Thikse of a Sedge Warbler, five *tristis* Common Chiffchaffs, one Greenish Warbler, one Hume's Leaf Warbler (and three additional birds observed), one *maura* Common Stonechat (and at least two additional birds observed) and observations of one Isabelline Wheatear and one Desert Wheatear. More regular long-distance migrants recorded in this period included six Tree Pipits, 16 Citrine Wagtails and 40 White Wagtails. By 23 September the weather had cleared, but a Barred Warbler was trapped and several of the other Siberian species were still in the area. A total of 10 non-passerine long-distance migrant species was also observed in the same period. There was a correlation between cloudy weather and the variety of migrants trapped by the ringing programme, and it seems likely that in the clear weather conditions which are frequent in Ladakh during the spring and autumn, a high proportion of migrants pass over the Upper Indus Valley without stopping. This possibility is supported by the fact that several of the long-distance migrants trapped at Thikse were carrying substantial deposits of subcutaneous fat (visible as yellowish deposits in the abdominal, axillary and thoracic cavities, and registering as increases in weight). In particular a Barred Warbler trapped on 18 September 1980, and a Sedge Warbler trapped on 11 October 1981, both en route for Africa, weighed 27.3 g and 12.9 g, and were about 30% and 15% heavier, respectively, than the mean fat-free weights of these species (Cramp *et al.*, 1992). This suggests that these birds were interrupted by poor weather during a long flight, or, less likely, that they were engaging in pre-migratory fattening at Thikse or somewhere nearby.

More than half (26 out of 45 species) of the mostly Palearctic-breeding long-distance migrants listed in Table 4.2, and 83% (19 out of 23) of such species listed in Table 4.3 are warblers, thrushes, chats and shrikes. These species are nocturnal, high-altitude migrants whose juveniles migrate separately from adults, generally following migratory trajectories inherited from their parents (Berthold & Querner, 1981). These characteristics suggest the possibility that these birds pass over Ladakh as broad-front migrants, following long-haul, high-altitude routes paying little attention to the terrain below. Such a strategy risks catastrophe in the event of unfavourable weather, which occurs regularly in the Himalayan region in spring and autumn (Ohlmann, Chapter 14; Heise, Chapter 15). The larks, swallows, pipits, wagtails, starlings, finches and buntings in Table 4.2 mostly migrate by day, and they seem more likely to use migratory strategies involving the following of river valleys and mountain passes, although there are few such possibilities with appropriate orientation in Ladakh.

Insights from Studies in the European Alps and Israel

Bruderer (1996) summarized the results of 15 years of research on nocturnal migration in and around the European Alps using radar and moon-watching. These studies showed similarities with the Tibetan-Himalayan region in that the main migration stream passes to the north of the mountain range in a predominantly southwesterly direction. The influence of topography increased with increasing cloud cover and headwinds, which reduced flight altitudes, and high-flying birds flew on more direct, southerly trajectories. Birds above the Alps flew faster than those above the lowlands at the same latitude, and Alerstam (1990) showed that average flight speed increases by 5% per 1000 m increase in elevation (but energy requirements also increase) (Groen & Prins, Chapter 17). The advantages of shortened flight time and (possibly) access to stronger tailwinds (Heise, Chapter 15) seem sometimes to outweigh the disadvantages of greater energy requirements and higher risk of encountering bad weather.

Liechti and Schaller (1999) in a radar study over Israel, found that nocturnal migrants on spring migration flew at altitudes between 5000 m a.s.l. and nearly 9000 m a.s.l. to take advantage of optimal wind conditions. Advantages of migrating at night also include taking advantage of cooler, denser air with reduced turbulence, especially over mountains (Ohlmann, Chapter 14), and reduced risks of predation (Ydenburg, Chapter 19). Evidence that a high proportion of migrants pass around the Tibetan-

Himalayan region to the east and west suggests, however, that the advantages of direct, high-altitude migratory flights across the mountains and deserts are generally outweighed by the disadvantages.

Evolutionary Considerations

Irwin and Irwin (2005) suggested that the Tibetan-Himalayan region presents such a barrier that migration divides have evolved in several passerine species breeding in Siberia, with birds in the western part of the breeding ranges migrating to the west, and those in the eastern part of the breeding ranges migrating to the east. For one species, Greenish Warbler, this migration pattern has resulted in the evolution of two subspecies, and the existence of separate breeding populations in the eastern and western Palearctic and in the Tibetan-Himalayan region (Irwin *et al.*, 2001). Two other species, Common Rosefinch and Common Stonechat, have evolved a similar distribution consisting of a ring of populations encircling the Tibetan Plateau with a migratory divide between distinct subspecies in Siberia (Irwin & Irwin, 2005).

The juveniles of many nocturnal migrant passerines migrate separately from adults, and have been shown to use an inherited impulse to migrate in a particular direction (Berthold & Querner, 1981; Berthold, 2001). Furthermore, hybrids from populations which migrate in different directions have been shown to inherit an intermediate impulse (Helbig, 1991). This experimental work with captive birds has been confirmed in the wild by Delmore and Irwin (2014), who used geolocator tracking to demonstrate that hybrid populations of Swainson's Thrushes in a migratory divide in western Canada employ intermediate and potentially inferior migratory routes.

It seems possible that pairs of migrant birds originating from either side of the migratory divide in Siberia could produce offspring that take such inferior, intermediate migratory routes, that is, across the mountains and deserts of Central Asia. Such a process might contribute to the small but noteworthy proportion of Siberian populations found migrating on this route. It might also have contributed to the historical establishment of breeding populations in the Tibetan-Himalayan region of seven passerine species with breeding ranges further north in the Palearctic zoogeographic region.

The Nature and Scale of Passerine Migration through the Tibetan-Himalayan Region

A sizeable majority of migrant passerines appear to avoid the Tibetan-Himalayan region on their flights (Moreau, 1972; Irwin & Irwin, 2005), but a large number appear nevertheless to pass through or over the mountain ranges. Numbers of long-distance migrants observed at Thikse were small, but they came from a diverse range of species, and their increased appearance during spells of cloudy weather suggests that many more birds were passing over than were recorded. Based on examination of Google Earth images, the Thikse Forestry plantation represents a small proportion of the riverine scrub in the

Upper Indus Valley. Applying appropriate multipliers to compensate for birds that fly over without being recorded, and for those that use other areas of similar habitat in the region, would result in totals of passerine migrants in the tens to hundreds of thousands. Although long-distance migrants from the north represent a small proportion of these, many thousand must pass through and over the region every spring and autumn.

Disorientation of migrating birds may increase in mountainous regions (e.g. Muheim & Jenni, 1999) and it seems likely that some of the unexpected species recorded at Thikse, such as the Black-browed Reed Warbler, were genuine vagrants rather than birds following regular migration routes. The status of many migrant passerines recorded in the Tibetan-Himalayan region as regular migrants or disoriented vagrants is one of many questions that remain to be answered.

The number and variety of migrant passerines recorded at Thikse, deep in the western ranges of the Himalayas in an area with no obvious north–south-oriented valleys, suggests that passerine migration throughout the Tibetan-Himalayan region probably takes place on a previously unrecognized scale. It seems possible that sites further to the west, where the Hindu Kush represents a relatively narrow barrier, would support larger numbers of migrants. Passerine migration through the central and eastern Himalayas in Nepal, Bhutan and Tibet probably involves smaller numbers of fewer species than migration through Ladakh because of the extent of the Tibetan Plateau to the north. Much remains to be learned about the nature and extent of migration around the eastern end of the Himalayas.

Systematic studies at several sites are needed to unravel the full nature and scale of migration through the Himalayas. The extent to which different species follow valleys or migrate on broad fronts at high altitude could be studied using radar and moon-watching at a number of stations. This research could be complemented by mist-netting programmes and diurnal observations of migration. The establishment of bird observatories as focal points for these studies would provide a wealth of new knowledge. A start has been made with the establishment of the Bird Observatory at Koshi Tappu in Nepal (Bird Guides, 2014).

Acknowledgement

Grateful thanks are extended to the District forest Officers and the staff of the Jammu & Kashmir Stage Forestry Department in Leh, and J.C. Daniel at the Bombay Natural History Society. For funding we thank the Biological Council, the British Ecological Society, the British Ornithologists' Union, The Linnean Society, the Mount Everest Foundation, the Royal Geographical Society, and the Smithsonian Institute, and for encouragement and references we thank Carol and Tim Inskipp.

References

Akhtar, S.A. (1955). Bird migration and fowling in Afghanistan. *Journal of the Bombay Natural History Society* **53**, 49–53.

Alerstam, T. (1990). *Bird Migration*. Cambridge, Cambridge University Press.

Alerstam, T. (2009). Flight by night or day? Optimal daily timing of bird migration. *Journal of Theoretical Biology*, **258**, 530–536.

Alexander, H.G. (1974). *Seventy Years of Birdwatching*. Berkhamsted, Poyser.

Ali, S. & Ripley, S.D. (1987–1999). *Handbook of the Birds of India and Pakistan, Together with those of Bangladesh, Nepal, Bhutan and Sri Lanka*. Vols. **5**–10. (second ed.) New Delhi, Oxford University Press.

Beaman, M.A.S. (1973). *Report of the ornithological Cambridge expedition to the Himalayas*. Unpublished report.

Berthold, P. (2001). *Bird Migration – A General Survey*. Second edition. Oxford, Oxford University Press.

Berthold, P. & Querner, U. (1981). Genetic basis of migratory behaviour in European warblers. *Science*, **212**, 77–79.

Bird Guides. (2014). Establishment of a new bird observatory in Nepal. https://birdguides.com/webzine/article.asp?a=2430.

Bruderer, B. (1996). Vogelzugforschung im Bereich der Alpen 1980–1995. *Ornithologische Beobachter*, **93**, 119–130.

Cheng Tso-hsin (1987). *A Synopsis of the Avifauna of China*. Beijing, Science Press.

Christison, A.F.P. (1939). On the occurrence of the European Redstart (*Phoenicurus phoenicurus*) in British Baluchistan. *Journal of the Bombay Natural History Society*, **41**, 434–435.

Cramp, S., Brooks, D.J., Dunn, E., *et al.* (1992). *Handbook of the Birds of Europe, the Middle East and North Africa. The Birds of the Western Palearctic*. Volume VI, Warblers. Oxford, Oxford University Press.

del Hoyo, J., Elliott, A. & Christie, A.D. (2004–2011). *Handbook of the Birds of the World*. Volumes **9**–16. Barcelona, Lynx Edicions.

Delany, S.N., Denby, C. & Norton, J. (1982). Ornithology. In Denby, C. (ed.). *University of Southampton Ladakh Expedition 1980 Report*, pp. 5–151. Southampton, University of Southampton.

Delany, S., Garbutt, D., Williams, C., *et al.* (2014). The Southampton University Ladakh Expeditions 1976–1982: Full details of nine species previously unrecorded in India and four second records. *Indian BIRDS*, **9**, 1–13.

Delmore, K.E. & Irwin, D.E. (2014). Hybrid songbirds employ intermediate routes in a migratory divide. *Ecology Letters*, **17**, 1211–1218.

DeLuca, W.V. Woodworth, B.K. Rimmer, C.C., *et al.* (2015). Transoceanic migration by a 12 g songbird. *Biology Letters*, **11**, 4.

Denby, C. & Phillips, A. (1977). Ornithology. In: Fraser, S. (ed.) *University of Southampton Ladakh Expedition 1976 Report*. Southampton, University of Southampton.

Dokter, A.M., Liechti, F., Stark, *et al.* (2011). Bird migration flight altitudes studied by a network of operational weather radars. *Journal of the Royal Society Interface*, **8**, 30–43.

Dolnik, V.R. (1990). *Bird Migration across Arid and Mountainous Regions of Middle Asia & Kazakhstan*. In Gwinner, E. (ed.), *Bird Migration: Physiology and Ecophysiology*, pp. 368–386. Berlin, Springer-Verlag.

Grimmett, R., Roberts, T. & Inskipp, T. (2008). *Birds of Pakistan*. London, Christopher Helm.

Helbig, A. J. (1991). Inheritance of migratory direction in a bird species: a cross-breeding experiment with SE and SW migrating Blackcaps (*Sylvia atracapilla*). *Behavioural Ecology and Sociobiology*, **28**, 9–12.

Hodgson, B.H. (1833). On the migration of the Natatores and Grallatores, as observed in Kathmandu. *Asiatic Research*, **18**, 22–128.

Holmes, P.R. (1986). The avifauna of the Suru River Valley, Ladakh. *Forktail*, **2**, 21–41.

Inskipp, C. (1985). Migration across the Nepalese Himalaya. *Oriental Bird Club Bulletin*, **1**, 8–11.

Inskipp, C. & Inskipp, T. (1985). *A Guide to the Birds of Nepal*. London, Croom Helm.

Irwin, D.E., Bensch, S. & Price, T.D. (2001). Speciation in a ring. *Nature*, **409**, 333–337.

Irwin, D.E. & Irwin, J.H. (2005). *Siberian Migratory Divides: The Role of Seasonal Migration in Speciation*. Chapter 3 in Greenberg, R., & Marra, P. (eds.), *Birds of Two Worlds: The Ecology and Evolution of Migration*. Baltimore, MD, John Hopkins University Press.

Kinnear, N.B. (1922). On the birds collected by Mr. A.F.R. Wollaston during the First Mt. Everest Expedition. *Ibis*, **64**, 495–526.

Kinnear, N.B. (1931). The Barred Warbler (*Sylvia nisoria*) in Gilgit. *Ibis*, **73**, 575.

Liechti, F., & Schaller, E. (1999). The use of low-level jets by migrating birds. *Naturwissenschaffen*, **86**, 549–551.

Lislevand, T., Chutný, B., Byrkjedal, I., Pavel, V., Briedis, M., Adamik, P. & Hahn, S. (2015). Red-Spotted Bluethroats *Luscinia s. svecica* migrate along the Indo-European Flyway: a geolocator study. *Bird Study*, **62**, 508–515.

Ludlow, F. (1927). Birds of the Gyangtse neighbourhood, southern Tibet, part 1. *Ibis*, **69**, 644–659.

Ludlow, F. (1928). Birds of the Gyangtse neighbourhood, southern Tibet, parts 2 & 3. *Ibis*, **70**, 51–73, 211–232.

Ludlow, F. (1944). The birds of south-east Tibet. *Ibis*, **86**, 43–86, 176–208, 348–389.

Ludlow, F. (1950). The birds of Lhasa. *Ibis*, **92**, 34–45.

MacKinnon, J. & Phillips, K. (2000). *A Field Guide to the Birds of China*. Oxford, Oxford University Press.

Moreau, R.E. (1972). *The Palearctic-African Bird Migration Systems*. London, Academic Press.

Muheim, R. & Jenni, L. (1999). Nocturnal orientation of robins *Erithcus rubecula*: Birds caught during migratory flight are disoriented. *Acta Ethol*, **2**, 43–50.

Pfister, O. (2004). *Birds and Mammals of Ladakh*. New Delhi, Oxford University Press.

Proud, D. (1949). Notes on the birds of the Nepal Valley. *Journal of the Bombay Natural History Society.*, **48**, 695–719.

Proud, D. (1952). Further notes on the birds of the Nepal Valley. *Journal of the Bombay Natural History Society*, **50**, 667–670.

Proud, D. (1955). More notes on the birds of the Nepal Valley. *Journal of the Bombay Natural History Society*, **53**, 57–78.

Rasmussen, P.C. & Anderton, J.C. (2012). *Birds of South Asia: The Ripley Guide*. Second edition. Smithsonian Institute and Lynx Edicions.

Ripley, S.D. (1982). *A Synopsis of the Birds of India and Pakistan*. Second edition. Bombay (Mumbai), Bombay Natural History Society.

Rogacheva, H. (1992). *The Birds of Central Siberia*. Husum, Druck und Verlagsgeschellschaft.

Sharpe, R.B. (1891). *Scientific Results of the Second Yarkand Mission*. Calcutta, Government of India.

Sillem, J.A. (1934). Ornithological results of the Netherland Karakorum Expedition 1929/30. *Orgaan-Club van Nederlandse Vogelkundigen*, **7**, 20.

Spierenburg, P. (2006). *Birds in Bhutan: Status and Distribution*. Bedford, Oriental Bird Club.

Stutchbury B.J.M., Tarof S.A., Done, T., *et al.* (2009). Tracking long-distance songbird migration by using geolocators. *Science*, **323**, 896.

Svensson, L. (1987). More about Phylloscopus taxonomy. *British Birds*, **80**, 580–581.

Vaurie, C. (1972). *Tibet and Its Birds*. London, Witherby.

Williams, C.T. & Delany, S.N. (1979). Ornithology. In Fraser, S. (ed.), *University of Southampton Himalayan Expedition 1977 Report*. Southampton, University of Southampton, pp 37–66, 104–229.

Williams, C.T. & Delany, S.N. (1983). *University of Southampton Ornithological Project, Ladakh, 1981–82 Preliminary Report*. Southampton, University of Southampton.

Williams, C.T & Delany, S.N. (1985). Migration through the north-west Himalaya – some results of the Southampton University Ladakh Expeditions-1. *Oriental Bird Club Bulletin*, **2**, 10–14.

Williams, C.T. & Delany, S.N. (1986). Migration through the north-west Himalayas – some results of the Southampton University Ladakh Expedition, part 2. *Oriental Bird Club Bulletin*, **3**, 11–16.

5 Wader Migration across the Himalayas

Simon Delany, Charles Williams, Clare Sulston, John Norton and David Garbutt

The Spectacular Migrations of Waders

Studies in the twenty-first century have confirmed that waders (known in North America as shorebirds) undertake some of the most astonishing animal migrations yet discovered. The Bar-tailed Godwits *Limosa lapponica* of the subspecies *bauri* that breed in Alaska have been found to migrate across the Pacific Ocean to winter in New Zealand, a non-stop journey of more than 11,000 km that takes more than a week (Gill *et al.*, 2005, 2009). The Red-necked Phalaropes *Phalaropus lobatus* that breed in the Shetland Isles north of Scotland have been tracked to wintering grounds at sea in the Humboldt Current, off the coast of Peru, an 11,000 km journey with stops in Iceland, Greenland and the eastern seaboard of the United States (Smith *et al.*, 2014). There has long been debate about the extent to which these accomplished and powerful travellers migrate directly across inhospitable and challenging terrain such as the Himalayas, or use alternative routes avoiding elevated regions or following their frequently preferred coastal habitats (Alerstam, 1977).

Bruderer and Liechti (1999) used radar and moon-watching to compare the scale of nocturnal migration of waders across the Mediterranean Sea at its confluence with the Atlantic Ocean, with that across the Mediterranean itself. They found a reduced passage of waders away from the Atlantic seaboard, with a majority of birds preferring to follow the coastlines of Western Europe and Africa. On one hand, this might be expected, because many wader species depend on coastal and estuarine habitats outside the breeding period. On the other hand, this finding might be considered surprising, because these powerful migrants that regularly cross the world's oceans and deserts might be expected to fly directly across the Mediterranean in order to shorten their migrations.

Migration Routes of Siberian Breeding Waders

There is evidence of heavy wader passage in Central Asia, to the northwest of the Tibetan-Himalayan region, and it is clear that millions of waders that breed in the tundra and boreal zones of Siberia avoid the mountains and deserts of the Tibetan-Himalayan region by migrating in a southwesterly direction, with many of them wintering in Africa (e.g. Delany *et al.*, 2009). Schielzeth *et al.* (2008) reported the results of waterbird censuses at the Tengiz–Korgalzhyn lake system in central Kazakhstan between 1999 and 2008, describing the numbers of 37 species of Siberian breeding wader species at this key staging area. Totals of more than half a million Red-necked Phalarope, nearly 200,000 Ruff *Philomachus pugnax*, 60,000 Little Stint *Calidris minuta*, 30,000 Dunlin *Calidris alpina*, 8000 Black-tailed Godwit *Limosa limosa*, 7000 Curlew Sandpiper *Calidris ferruginea* and 2500 Spotted Redshank *Tringa erythropus* were estimated to occur on spring migration. It is clear that as is the case with passerines (Delany *et al.*, Chapter 4), a considerable majority of birds breeding to the north of the Tibetan-Himalayan region follow migration routes that avoid the mountain barrier, and that migration over these ranges is a strategy followed by a minority of individuals. The number and variety of waders recorded in the Himalayas and Tibet nevertheless point to this minority still involving considerable numbers of some species.

Golowatin (2006), cited in Lappo *et al.* (2012), estimated that 20,000–100,000 Wood Sandpipers migrate through the lower Ob river valley, due north of the Tibetan-Himalayan region each year. The Ob rises in the Altai mountains, and birds continuing upstream on southward migration would traverse the Tibetan Plateau. Only 280–545 Wood Sandpipers were estimated to use the Tengiz–Korgalzhyn lake system in spring (Schielzeth *et al.*, 2008), and the small number of birds at this key staging area for Siberian breeding waders supports the possibility that large numbers of Wood Sandpipers take a direct route across the Tibetan-Himalayan region. Wood Sandpiper is also one of the most frequently recorded wader species on migration through the Himalayas (see later in this chapter).

Millions of Siberian breeding waders also migrate in a southeasterly direction to Southeast Asia and Australasia (e.g. Bamford *et al.*, 2008), but the extent to which waders migrate around the eastern end of the Tibetan Plateau remains poorly known.

The Wader Species that Migrate over the Tibetan-Himalayan Region

Distribution maps in del Hoyo *et al.* (1996) show the breeding and wintering ranges of wader species that migrate over or around the Tibetan-Himalayan region (Table 5.1.) The scientific names of all species are included in this table. Two species, Ibisbill and Solitary Snipe, are resident, or migrate altitudinally within the Tibetan-Himalayan region, and a further five species are breeding summer visitors which winter in lowland South Asia and beyond. By far the largest category of visitors to the region are passage migrants, and a further 34 species have breeding distributions to the north of the region and wintering ranges to the south. Of these 34 species whose distributions suggest migration over or through the Tibetan-Himalayan region, at least 30 have actually been recorded in the region.

Table 5.1. The origins and destinations of wader species migrations in the Tibetan-Himalayan region summarized from distribution maps in del Hoyo *et al.*, 1996. Species in bold have been recorded on migration in the region.

1. **Resident/altitudinal migrant within Tibetan/Himalayan region**
   Ibisbill *Ibidorhyncha struthersii*, Solitary Snipe *Gallinago solitaria*
2. **Summer breeding visitor to Tibetan-Himalayan region, wintering mainly in lowland South Asia and beyond**
   Little Ringed Plover *Charadrius dubius*, Lesser Sandplover *Charadrius mongolus*, Wood Snipe *Gallinago nemoricola*, Common Redshank*Tringa totanus*, Common Sandpiper *Actitis hypoleucos*
3. **Passage migrant – breeding mainly in tundra zone, wintering mainly on tropical and temperate coasts**
   Pacific Golden Plover *Pluvialis fulva*, Grey Plover *Pluvialis squatarola*, Common Ringed Plover *Charadrius hiaticula*, Bar-tailed Godwit *Limosa lapponica*, **Whimbrel *Numenius phaeopus***, Ruddy Turnstone *Arenaria interpres*, Great Knot *Calidris tenuirostris*, Red Knot *Calidris canutus*, Sanderling *Calidris alba*, **Curlew Sandpiper *Calidris ferruginea*, Dunlin *Calidris alpina*, Little Stint *Calidris minuta***, Broad-billed Sandpiper *Limicola falcinellus*, **Red-necked Phalarope *Phalaropus lobatus***
4. **Passage migrant – breeding mainly in tundra zone, wintering mainly inland in tropical and temperate regions**
   Jack Snipe *Lymnocryptes minimus*, Pintail Snipe *Gallinago stenura*, Spotted Redshank *Tringa erythropus*, Temminck's Stint *Calidris temminckii*, Ruff *Philomachus pugnax*
5. **Passage migrant – breeding mainly in boreal zone, wintering mainly on tropical and temperate coasts**
   Eurasian Curlew *Numenius arquata*, Terek Sandpiper *Xenus cinereus*
6. **Passage migrant – breeding mainly in in boreal zone, wintering mainly inland in tropical and temperate regions**
   Eurasian Woodcock *Scolopax rusticola*, Common Snipe *Gallinago gallinago*, Marsh Sandpiper *Tringa stagnatilis*, Common Greenshank *Tringa nebularia*, Green Sandpiper *Tringa ochropus*, Wood Sandpiper *Tringa glareola*
7. **Passage migrant – breeding mainly in steppe or semi-desert zone, wintering mainly inland in tropical and temperate regions**
   Black-winged Stilt *Himantopus himantopus*, Pied Avocet *Recurvirostra avosetta*, Eurasian Oystercatcher *Haematopus ostralegus*, Stone Curlew *Burhinus oedicnemus*, **Collared Pratincole *Glareola pratincola***, Oriental Pratincole *Glareola maldivarum*, Black-winged Pratincole *Glareola nordmanni*, **Northern Lapwing *Vanellus vanellus***, Sociable Lapwing *Vanellus gregarius*, White-tailed Lapwing *Vanellus leucurus*, **Kentish Plover *Charadrius alexandrinus***, Swinhoe's Snipe *Gallinago megala*, Black-tailed Godwit *Limosa limosa*

The species sequence in this and all tables follows del Hoyo *et al.*, 1996.

Two species (Common Redshank and Common Sandpiper) have breeding distributions both within the Tibetan-Himalayan region and extensively to the north. Common Sandpiper is numerous throughout the taiga zone in Central Siberia (Rogacheva, 1992), but its distribution within the Tibetan-Himalayan region is mostly confined to the southern slopes of the Great Himalayan Range. Common Redshank, on the other hand, is found only locally in the southern parts of Central Siberia (Rogacheva, 1992), but breeds widely in the Tibetan-Himalayan region. This suggests that a high proportion of the Common Sandpipers recorded on migration in the region are migrants from the north, but that Common Redshanks are more likely to be birds which have bred within the Tibetan-Himalayan region.

Ring Recoveries of Waders Suggesting Trans-Himalayan Migration

Large numbers of waders have been ringed by the Indian and Russian national bird-ringing programmes, but only partial results have been published in English in recent years (Veen *et al.*, 2005). Publications from the past century (e.g. McClure *et al.*, 1972; Pavlov, 1985; McClure, 1998) present details of a number of recoveries to the north in Siberia, and to the south in lowland India. This strongly suggests that large numbers of waders migrate around or over the region, bearing in mind the very low recovery rates of bird rings in Asia (McClure *et al.*, 1972, 1998).

The most frequently recovered ringed species is Ruff, and 20 birds of this species ringed at Bharatpur in northern lowland India have been recovered in Siberia and Central Asia. A total of 15 of these recoveries came from east Siberia, and the overall pattern of recoveries suggests a migration route west of the Himalayas and Tibetan Plateau, heading north through Kashmir, Pakistan, Afghanistan and Kazakhstan, then turning more to the east towards Yakutia and Magadan (McClure, 1998).

Most of the ring recoveries of species other than Ruff involve movements from Bharatpur, and from Point Calimere on the southeast Indian coast, to southwest Siberia. Three Wood Sandpipers, three Marsh Sandpipers, three Common Snipe and one Common Redshank ringed in lowland India have been recovered in the Novosibirsk-Tomsk region of southwest Siberia, and single Wood Sandpiper and Curlew Sandpipers ringed in southwest Siberia have been recovered in India. One Green Sandpiper ringed in European Russia was also recovered in Kerala (McClure, 1998; Veen *et al.*, 2005).

The History of Wader Records in the Tibetan-Himalayan Region

Published observations in the Tibetan-Himalayan region between 1833 and the twenty-first century have revealed that more than 30 Siberian breeding wader species migrate directly over or through the region. Most have been recorded on only a few occasions, and fewer than half of these species occur regularly in appreciable numbers.

Hodgson (1833) observed Common Snipe, Jack Snipe and Eurasian Woodcock on migration in the Kathmandu Valley, and also the first of only two published records of Eurasian Oystercatcher migrating across the Himalayas. Proud (1949, 1952, 1955) in the same area noted that Temminck's Stints 'swarm all along the river banks in August', and also recorded regular migration of three species of sandpiper (Wood, Green and Common), Ruff and three species of snipe (Common, Pintail and Jack). More recently in Nepal, wader passage at Kosi Barrage, in the lowlands south of the Great Himalaya range, was most marked in spring, between the end of February and early May, with most birds passing through in March and April. Typical totals of about 1000 to 1500 waders of several species have been recorded in spring, and numbers in autumn are less well studied but are smaller

(Inskipp & Inskipp, 1985). Further east, in Bhutan, Spierenburg (2006) reported a similar pattern of wader migration, and noted that water levels were generally more favourable for waders in spring than in autumn, and suggested that some species might undertake loop migrations, following different routes on their spring and autumn migrations.

Also in Nepal, Kinnear (1922) reporting on observations from the first mountaineering expedition to attempt an ascent of Mount Everest, recorded that at night, the climbers heard migrating waders from their tents at high altitude. Species identified were Eurasian Curlew and probable Godwits. This expedition also found the corpses of a number of migrant birds that had died in bad weather, including a Temminck's Stint, a Pintail Snipe and a Painted Snipe, all in snow at an altitude of 5200 m a.s.l. The Great Himalaya Range forms the northern boundary of the distribution range of Painted Snipe, which is partially migratory or nomadic (del Hoyo *et al.*, 1996), and this was an unexpected species to find at such high altitude in the Himalayas.

A series of observations in the Gyangtse region of Tibet, south of Lhasa and close to the border with Bhutan and Sikkim, centred on Dochen Lake (referred to in the papers as Hram Tso) at 4470 m a.s.l., were published by Ludlow (1927, 1928, 1944, 1950) and Maclaren (1948). 'Several flocks' of Pacific Golden Plovers used lakes in southern Gyangtse in 'considerable numbers' on autumn migration, when Temminck's Stint, Ruff, Black-winged Stilt, Pied Avocet, Common Redshank, Common Snipe and Northern Lapwing were also regularly recorded. Eurasian Curlew and Common Greenshank were less common in autumn, and Ludlow drew attention to a single record of Whimbrel shot at Dochen Lake in September 1908. Northern Lapwing, Common Greenshank, Common Redshank, Temminck's Stint and Common Snipe were also recorded on spring migration, and Red-necked Phalarope was only recorded in spring, when several were seen along the banks of the Tsangpo River between Lilung and Tamuyen. Maclaren (1948) supplemented Ludlow's records with observations made in May 1946, a period of low water at Dochen Lake, when flocks of stints (presumably, Temminck's Stints), Common Redshanks and Pied Avocets (numbering 55 and 25) were gorging themselves on a super-abundance of midges (Chironomidae) washed into the water, and when three Common Greenshanks, three Dunlin and a Curlew Sandpiper were also present. The use of high-altitude lakes to refuel may indicate opportunistic and flexible behaviour by these species to enhance their chances of survival on their northern migration, and maximize the chances of arriving at the breeding grounds in good condition for breeding. It also seems possible that feeding at these lakes is a regular strategy followed by a minority of individuals of a small number of wader species (see Prins *et al.*, Chapter 18).

Ludlow used observations of Pintail Snipe at Dochen Lake to support his view that migrants at this and similar high-altitude lakes were more likely to continue their journey to South Asia by a direct route across the Himalayas in preference to using a route that would involve returning to the valley of the Tsangpo River and following it along a very indirect route. Ludlow considered the mostly clear weather conditions in southern Tibet favourable for migration in autumn and thought the prevailing southwesterly winds in spring would also be beneficial for migrants heading north (see Heise, Chapter 15).

In the western Himalayan region, Sillem (1934) recorded Temminck's Stint, Green and Wood Sandpipers and Common Greenshank in the region of the Kara-tagh Pass at 5340 m a.s.l. in the eastern Karakoram in mid-August. Knowledge of wader migration in the western Himalayan region was enhanced by the four expeditions to Ladakh by Southampton University in the period 1976 to 1982, and the rest of this chapter will consider the observations of waders by these expeditions in some detail. Altogether, 27 out of the 41 wader species listed in Table 5.1 were recorded during these systematic ornithological observations in Ladakh (Denby & Philips, 1977; Williams & Delany, 1979, 1983, 1985, 1986; Delany *et al.*, 1982).

Autumn Wader Passage through Ladakh

Suru Valley

The Southampton University expeditions to Ladakh visited the valley of the Suru River, a southern tributary of the Indus flowing north from its source in the Great Himalaya Range, in the second half of July in 1977 and from late July to early August in 1980. Waders observed in the Suru Valley included small numbers of resident Ibisbills, and summer-visiting, breeding Lesser Sandplovers, Redshanks and Common Sandpipers, but a number of early autumn passage migrant waders of seven species were also recorded.

The most important site for waders in the Suru Valley is the Rangdum Swamp, a freshwater marsh at an altitude of 4260 m a.s.l. Up to 18 Lesser Sandplovers, including two pairs each with two downy young, and seven pairs of Common Redshanks were found nesting here, and the most numerous passage migrant waders were Wood Sandpipers, with more than 40 recorded on 30 July 1980. There were 20 Temminck's Stints at Rangdum on the same date, and single Common Greenshanks in 1977 (on 16 and 18 July), and three Black-winged Stilts and a Curlew Sandpiper in 1980 (on 6 August and 29 July, respectively). On 18 July 1977, a flock of 33 Ruffs at Rangdum included a number of males in full breeding plumage, but only two Ruffs, a male and a female, were seen in 1980, on 30 July. Elsewhere in the Suru Valley, one or two Green Sandpipers were seen almost daily in both 1977 and 1980.

The best day for waders at Rangdum was 30 July 1980, a day of steady rain with the wind gusting to Beaufort force 6. These adverse weather conditions would have interrupted the migrations of waders (and other species) over the Himalayas, and accounted for the flocks of Wood Sandpipers and Temminck's Stints and two Ruffs mentioned earlier, and for the Curlew Sandpiper recorded the previous day. The appearance of flocks of waders in bad weather suggests that there is regular, high-level, but relatively small scale migration over the Suru Valley region in late July and early August, and that many birds must pass through unobserved in clear weather.

Upper Indus Valley

The methods the Southampton University expeditions used in 1977, 1980 and 1981–1982 are described in Delany *et al.* (Chapter 4). The bird-ringing programme did not include waders, although small numbers of Green Sandpiper, Common Snipe and Ibisbill were trapped (Figures 5.1, 5.2 and 5.3). Systematic observations of birds along a 13 km stretch of the Upper Indus Valley between Thikse and Choglamsar included frequent observations of waders. Visits were made every two to three days downstream of Thikse to Shey and Choglamsar in the autumn migration seasons of 1980 and 1981 and the spring of 1982. Observations were less intense in 1977 because there

Figure 5.1. Ibisbill, Thikse, Ladakh, January 1982. A sparsely distributed resident of gravel-bottomed rivers throughout the Tibetan-Himalayan region, birds from the highest altitudes migrate to lower-lying river valleys for the winter. Photo, by Clare Sulston. (A black-and-white version of this figure will appear in some formats. For the colour version, please refer to the plate section.)

Figure 5.2. Common Snipe, Thikse, Ladakh, autumn 1980. First recorded in the Himalayas by Hodgson (1833), this is a regular long-distance migrant throughout the Himalayas in small numbers. Photo, by John Norton. (A black-and-white version of this figure will appear in some formats. For the colour version, please refer to the plate section.)

Figure 5.3. Green Sandpiper, Thikse, Ladakh, autumn 1981. The most frequently recorded wader in the Tibetan-Himalayan region, this species is very regular on spring and autumn migration and infrequent in other seasons. Photo, by Clare Sulston. (A black-and-white version of this figure will appear in some formats. For the colour version, please refer to the plate section.)

were only two observers, and this slightly lower level of coverage resulted in fewer observations of species that used these habitats in this year.

Autumn Wader Migration through the Upper Indus Valley

Figure 5.4 summarizes the numbers of the most commonly observed four migratory wader species along a 13 km stretch of the Upper Indus Valley in the autumn migration seasons of 1977, 1980 and 1981. Table 5.2 summarizes the observations of passage migrant waders recorded in these seasons in smaller numbers. We appear to have missed the start of the autumn migration period, and the numbers of waders, particularly Wood Sandpipers, Temminck's Stints and Ruffs, seen in the Suru Valley in the early autumn periods of 1977 and 1980 confirmed that migration was already well under way.

The highest numbers of waders in the Upper Indus Valley were recorded in the second half of August of each year, with numbers decreasing through September, before a second peak consisting almost entirely of Northern Lapwings in late November 1981 (Table 5.2). The earlier expeditions departed Thikse before the end of November and made no observations this late in the season. The most frequently recorded autumn migrant wader species was Green Sandpiper, with 450 bird-days recorded over the three autumns, followed by Greenshank (288 bird-days) Temminck's Stint (268 bird-days) and Common Sandpiper (132 bird-days). The main passage period for both Green and Common Sandpipers was from the middle of August to

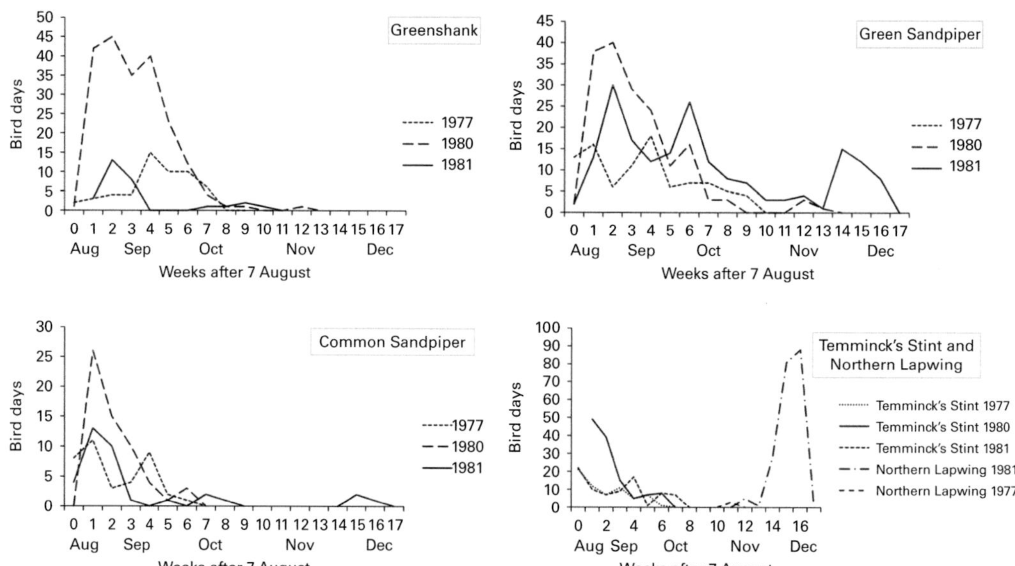

Figure 5.4. Wader species for which more than 100 bird-days were recorded along a 13 km stretch of the Upper Indus Valley, between Thikse and Choglamsar, in the autumns of 1977, 1980 and 1981. Totals are number of bird-days in each seven-day period.

the end of September, and there was a second small peak of Green Sandpipers in the second half of November in 1981. Green Sandpipers occur in Ladakh throughout the year, but breeding has not been recorded, and they are noticeably more numerous during the spring and autumn passage periods. Common Sandpipers breed scarcely in western Ladakh, but are very regular on migration in the Upper Indus Valley. Passage of Common Greenshanks and Temminck's Stints was less protracted than for Green and Common Sandpipers, peaking between mid-August and mid-September.

Table 5.2 shows that other species were more scarce on autumn passage, with 48 bird-days for Black-winged Stilt over the three autumns, 41 of Wood Sandpiper and smaller numbers, in descending order, of Pacific Golden Plover, Redshank, Kentish Plover, Lesser Sandplover and Common Snipe. There were fewer than 10 records over the three autumns of a further eight species (Table 5.2).

An observation of 30 Black-winged Stilts by mountaineers in Ladakh in August 1977 indicates that flocks of this species do migrate directly over the Himalayas. The flock was seen at a mountain lake above Nimaling at an altitude of 5200 m a.s.l. on 31 August 1977. The mountaineer Simon Fraser wrote in his account of the ascent of peaks in the Kang Yatse region: 'On the way down we were surprised to see a flock of Black-winged Stilts flying gracefully round the lake in tight formation rather like a pack of racing cyclists' (Fraser 1979). This is the largest flock recorded in the Tibetan-Himalayan region, and the location indicates that the birds were migrating directly across high passes en route to the Indian plains.

Table 5.2. Wader species for which fewer than 50 bird days were recorded along a 13 km stretch of the Upper Indus Valley, between Thikse and Choglamsar, in the autumns of 1977, 1980 and 1981. Totals are number of bird-days in each seven-day period.

Autumn 1977, 1980, 1981		7.8–13.8	14.8–20.8	21.8–27.8	28.8–3.9	4.9–10.9	11.9–17.9	18.9–24.9	25.9–1.10	2.10–8.10	9.10–15.10	16.10–22–10	23.10–29.10	30.10–5.11	6.11–12.11	13.11–19.11	20.11–26.11	27.11–3.12
Black winged Stilt	**1977**	1	2	2	2		2	2				1						
	1980		5	4	2	17												
	1981		4	1		1	1											
Grey Plover	**1980**							3										
Pacific Golden Plover	**1977**		2				1	1										
	1980			26	9													
Kentish Plover	**1977**	1	3	1		1		1	4									
	1980			9	5	3		1										
	1981		3	1				1										
Lesser Sandplover	**1980**		6	2	3													
	1981		1			4												
Spotted Redshank	**1977**		1															
Common Redshank	**1977**	1	1															
	1980		10	10	4				1									
	1981	3	1		1												2	1
Marsh Sandpiper	**1977**						2											
Wood Sandpiper	**1977**	1	1			2	2		2	4								
	1980		1	4	1	1		5										
	1981	2	1	7		3	1	3										
Terek Sandpiper	**1977**	1																
	1980			1				3										
Eurasian Curlew	**1977**							2										
Common Snipe	**1980**								19	5		1						
	1981																1	1
Pintail Snipe	**1981**							1										
Little Stint	**1977**	2					1											
	1980		3	3														
Curlew Sandpiper	**1977**			1														
	1980		1	1														
	1981				2													
Ruff	**1977**							1										
	1980		1					2										

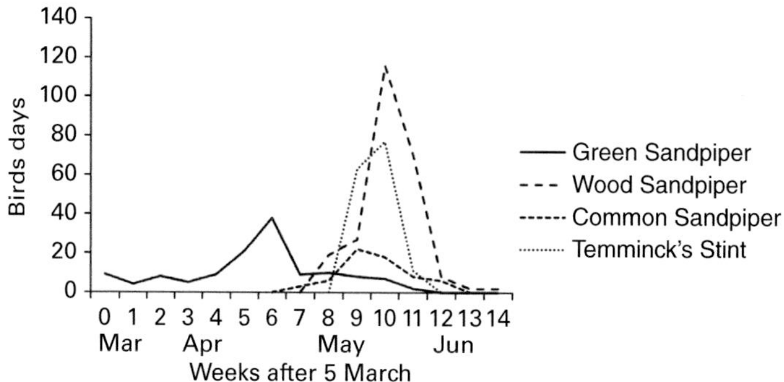

Figure 5.5. Wader species for which more than 60 bird-days were recorded along a 13 km stretch of the Upper Indus Valley, between Thikse and Choglamsar, in the spring of 1982. Totals are number of bird-days in each seven-day period.

Spring Wader Migration through the Upper Indus Valley in 1982

Figure 5.5 and Table 5.3 show that spring migration of waders through Ladakh involved fewer species over less protracted periods than autumn passage, although overall numbers of birds observed were similar. The bulk of spring passage was made up of three species: Wood Sandpiper (244 bird-days), Temminck's Stint (151) and Green Sandpiper (130). Wood Sandpiper numbers peaked around the middle of May and Temminck's Stint a little earlier, in the first half of the month. Green Sandpiper records covered a longer period, peaking between mid-April and mid-May. Overall, there was a strong mid-May peak in numbers of waders observed on spring migration in the Upper Indus Valley in 1982.

Species that occurred in higher numbers in spring than in autumn were Temminck's Stint, Common Snipe, Marsh Sandpiper and especially Wood Sandpiper. Green and Common Sandpipers occurred in comparable numbers in spring and autumn. Two species, an unidentified (probably Collared) Pratincole, and Red-necked Phalarope, occurred in spring that were not recorded in autumn. Wood Sandpipers were considerably more numerous on spring passage through the Upper Indus Valley than in autumn, but it was the most commonly observed wader species in the Suru Valley earlier in the autumn in 1997 and 1980. It seems possible that we missed the bulk of the autumn migration of this species because it took place before the start of observations in the Upper Indus Valley. It is also possible that the direct north-to-south route offered by the Suru Valley concentrated the migration of this species, so that more observations were made than later in the autumn further to the east.

Common Greenshank, one of the most numerous species on autumn passage, was only recorded once in spring. Northern Lapwing, with three bird-days, and Pacific Golden Plover with just one, were also considerably less numerous on spring migration.

Table 5.3. Wader species for which fewer than 40 bird-days were recorded along a 13 km stretch of the Upper Indus Valley, between Thikse and Choglamsar, in the spring of 1982. Totals are number of bird-days in each seven-day period. Note that Little Ringed Plovers bred on the Indus, being recorded with two young at Choglamsur on 23 June 1982, and departed each year in mid-September.

Spring 1982	5.3–11.3	12.3–18.3	19.3–25.3	26.3–1.4	2.4–8.4	9.4–15.4	16.4–22.4	23.4–29.4	30.4–6.5	7.5–13.5	14.5–20.5	21.5–27.5	28.5–3.6	4.6–10.6	11.6–17.6
Ibisbill	4	4	9	3	3	6	6	4	1	8	4	10	11	8	7
Northern Lapwing				1									1	1	
(Collared) Pratincole							1								
Pacific Golden Plover														1	
Common Redshank			1	1		1			1	1		3	3		
Marsh Sandpiper						1	1			1	8				
Common Greenshank										1					
Common Snipe						1	6	6	10	14					
Little Stint											1	1			
Curlew Sandpiper						4									
Ruff						1	2	1							
Red-necked Phalarope													1		

We recorded a further seven species in small numbers on autumn migration, but not in spring.

All of the more numerous wader species migrating through Ladakh, and most of the scarce species as well, share a preference for freshwater habitats throughout the year, and all have a tendency to occur individually or in small, dispersed flocks. We recorded a few coastal wintering species ('shorebirds') and those which habitually form large congregations, but in very small numbers. Most of these species appear to follow coastlines on their migrations and so are only occasionally recorded in the Tibetan-Himalayan region.

The Nature and Scale of Wader Migration through the Tibetan-Himalayan Region

Shortly before the use of satellite transmitters and geolocators started to confirm the epic nature of many wader migrations in the early 2000s, Kvist *et al.* (2001) suggested that previous calculations had overestimated the cost of carrying heavy fat loads to fuel migration. This revelation made subsequent discoveries less surprising, and in the context of trans-Himalayan migration, makes direct, long-distance flights across these ranges appear less arduous and so more likely.

Although at least 30 Siberian breeding wader species have been recorded migrating through the Tibetan-Himalayan region, only four appear to do so in sizeable numbers on a regular basis, in approximately the following order of abundance: Green Sandpiper, Temminck's Stint, Wood Sandpiper and Common Greenshank. Common Redshank and Common Sandpiper are also frequently recorded, and their relative distribution (see earlier in this chapter) suggests that a high proportion of Common Sandpipers and a low proportion of Common Redshanks observed in spring and autumn in Ladakh are likely to be long-distance passage migrants from Siberia. Northern Lapwing may also be a relatively numerous migrant, but there is only one season's data from their late autumn migration period. Additional species mentioned regularly in the literature and recorded in moderate or low numbers in Ladakh are Black-winged Stilt, Pied Avocet, Pacific Golden Plover, Kentish Plover, Pintail Snipe, Common Snipe, Marsh Sandpiper, Terek Sandpiper, Curlew Sandpiper and Ruff. These are all species that often spend most of their life cycles in freshwater habitats, in many cases away from coasts, and they are also mostly species that do not habitually congregate in large numbers.

There is still a marked lack of records in the Tibetan-Himalayan region of wader species that strongly favour coastal habitats outside the breeding season, for example, Common Ringed Plover, Bar-tailed Godwit, Great Knot, Red Knot, Sanderling and Ruddy Turnstone. All these species have breeding ranges in Siberia, to the north of the Tibetan-Himalayan region, and non-breeding ranges to the south. It seems likely that these species concentrate along coastlines on their migrations, and usually avoid routes that would take them deep into continental Central Asia.

Observed wader migration through Ladakh is characterized by the regular occurrence of small parties of birds, and by increased numbers of a wider variety of species during and after cloudy and rainy weather. This suggests that there is regular, broad-front migration over the Himalayas in spring and autumn, with many birds passing over the ranges without stopping, and so going unobserved. It is difficult to estimate the scale of under-recording that takes place, but observations throughout the Tibetan-Himalayan region suggest that the migrations of waders are widespread and that these migrations seem likely to involve millions of birds. To give a simplified guesstimate, assuming the Himalayan range is 2400 km in length, and the migration season lasts 30 days, with birds migrating for 10 hours per day and passing over at an average rate of two birds per kilometre per hour, a total of 1.4 million waders would pass over the range. This is a small but nevertheless important proportion of the waders that migrate out of Siberia each year.

Acknowledgements

Grateful thanks are extended to the District forest Officers and the staff of the Jammu & Kashmir Stage Forestry Department in Leh, and J.C. Daniel at the

Bombay Natural History Society. For funding we thank the Biological Council, the British Ecological Society, the British Ornithologists' Union, The Linnean Society, the Mount Everest Foundation, the Royal Geographical Society, and the Smithsonian Institute, and for encouragement and references we thank Carol and Tim Inskipp.

References

Alerstam, T. (1977). Why do migrating birds fly along coastlines? *Journal of Theoretical Biology*, **65**, 699–712.

Bamford, M., Watkins, D., Bancroft, W., *et al.* (2008). *Migratory Shorebirds of the East Asian-Australasian Flyway: Population Estimates and Internationally Important Sites*. Canberra, Wetlands International Oceania.

Bruderer, B. & Liechti, F. (1999). Bird migration across the Mediterranean. In Adams, N.J. & Slotow, R.H. (eds.), *Proc. 22 Int. Ornithol. Congr., Durban: 1983–1999.* Johannesburg: BirdLife South Africa.

Delany, S.N., Denby, C. & Norton, J. (1982). Ornithology. In Denby, C. (ed.), *University of Southampton Ladakh Expedition 1980 Report*, pp. 5–151. Southampton, University of Southampton.

del Hoyo, J., *et al.* (1996). *Handbook of the Birds of the World Volume 3: Hoatzin to Auks*. Barcelona, Lynx Edicions.

Delany, S., Scott, D.A., Dodman, T. & Stroud, D. (eds.) (2009). *An Atlas of Wader Populations in Africa and Western Eurasia*. Wetlands International, Wageningen, The Netherlands.

Denby, C. & Phillips, A. (1977). Ornithology. In Fraser, S. (ed.), *University of Southampton Ladakh Expedition 1976 Report*. Southampton, University of Southampton.

Fraser, S. (ed.) (1979). *University of Southampton Himalayan Expedition 1977: Report.* Southampton, University of Southampton.

Gill, R.E. Jr, Piersma, T., Hufford, G., *et al.* (2005). Crossing the ultimate ecological barrier: evidence for an 11 000-km-long-nonstop flight from Alaska to New Zealand and eastern Australia by Bar-tailed Godwits. *Condor*, **107**, 1–20.

Gill, R.E. Jr, Tibbitts, T.L., Douglas, D.C., *et al.* (2009). Extreme endurance flights by landbirds crossing the Pacific Ocean: ecological corridor rather than barrier? *Proceedings of the Royal Society B*, **276**, 447–457.

Hodgson, B.H. (1833). On the migration of the Natatores and Grallatores, as observed in Kathmandu. *Asiatic Research*, **18**, 22–128.

Inskipp, C. (1985). Migration across the Nepalese Himalayas. *Oriental Bird Club Bulletin*, **1**, 8–11.

Inskipp, C. & Inskipp, T. (1985). *A Guide to the Birds of Nepal*. London, Croom Helm.

Kinnear, N.B. (1922). On the birds collected by Mr. A.F.R. Wollaston during the First Mt. Everest Expedition. *Ibis*, **4**, 495–526.

Kvist, A., Lindström, Å., Green, M., Piersma, T. & Visser, H. (2001) Carrying large fuel loads during sustained bird flight is cheaper than expected. *Nature*, **413**, 730–732.

Lappo, E.G., Tomkovich, P.S. & Syroeckovskiy, E.E. (2012). *Atlas of Breeding Waders in the Russian Arctic*. Moscow, Publishing House (in Russian with English summaries).

Ludlow, F. (1927). Birds of the Gyangtse neighbourhood, southern Tibet, part 1. *Ibis*, **69**, 644–659.

Ludlow, F. (1928). Birds of the Gyangtse neighbourhood, southern Tibet, parts 2 & 3. *Ibis*, **70**, 51–73, 211–232.

Ludlow, F. (1944). The birds of south-east Tibet. *Ibis*, **86**, 43–86, 176–208, 348–389.

Ludlow, F. (1950) The birds of Lhasa. *Ibis*, **92**, 34–45.

Maclaren, P.I.R. (1948). Notes on the birds of the Gyantse Road, southern Tibet. *Ibis*, **90**, 199–205.

McClure, H.E (1998). *Migration and Survival of the Birds of Asia*. Bangkok, White Lotus.

McClure, H.E. & Leelavit, P. (1972). *Birds Banded in Asia during the MAPS Program, by Locality, from 1963 through 1971. 1–478*. Bangkok, U.S. Army Research and Development Group.

Pavlov, D.S. (ed.) (1985). *Migrations of Birds of Eastern Europe and Northern Asia. Gruiformes and Charadriiformes*. Academy of Sciences of the USSR, Moscow (in Russian).

Proud, D. (1949). Notes on the birds of the Nepal Valley. *Journal of the Bombay Natural History Society*, **48**, 695–719.

Proud, D. (1952). Further notes on the birds of the Nepal Valley. *Journal of the Bombay Natural History Society*, **50**, 667–670.

Proud, D. (1955). More notes on the birds of the Nepal Valley. *Journal of the Bombay Natural History Society.*, **53**, 57–78.

Rogacheva, H. (1992). *The Birds of Central Siberia*. Husum, Druck und Verlagsgeschellschaft.

Schielzeth, H., Eichorn, G., Heinecke, *et al.* (2008). Waterbird population estimates for a key staging site in Kazakhstan: a contribution to wetland conservation on the Central Asian Flyway. *Bird Conservation International*, **18**, 71–86.

Sillem, J.A. (1934). Ornithological results of the Netherland Karakorum Expedition 1929/30. *Orgaan-Club van Nederlandse Vogelkundigen* **7**, 20.

Smith, M., Bolton, M., Okill, D.J., *et al.* (2014). Geolocator tagging reveals Pacific migration of Red-necked Phalarope Phalaropus lobatus breeding in Scotland. *Ibis*, **156**, 870–873.

Spierenburg, P. (2006). *Birds in Bhutan: Status and Distribution*. Bedford, Oriental Bird Club.

Veen, J., Yurlov, A.K., Delany, S.N., *et al.* (2005). *An Atlas of Movements of Southwest Siberian Waterbirds*. Wetlands International, Wageningen, The Netherlands.

Williams, C.T. & Delany, S.N. (1979). Ornithology. In Fraser, S. (ed.), *University of Southampton Himalayan Expedition 1977 Report*, pp. 37–66, 104–229. Southampton, University of Southampton.

Williams, C.T. & Delany, S.N. (1983). *University of Southampton Ornithological Project, Ladakh, 1981–82 Preliminary Report*. Southampton, University of Southampton.

Williams, C.T & Delany, S.N. (1985). Migration through the north-west Himalaya – some results of the Southampton University Ladakh Expeditions-1. *Oriental Bird Club Bulletin*, **2**, 10–14.

Williams, C.T. & Delany, S.N. (1986). Migration through the north-west Himalayas – some results of the Southampton University Ladakh Expedition, part 2. *Oriental Bird Club Bulletin*, **3**, 11–16.

6 Raptor Migration across and around the Himalayas

Matías A. Juhant and Keith L. Bildstein

History of Research

Given the remote and rugged terrain of the Himalayas, it is not surprising that published observations of raptor migration in the region have, until recently, been sporadic.

The earliest useful record we could find is Scully's descriptions of raptors collected during 19 months in the Gilgit region in 1879–1880 that included seasonal occurrences and status of 15 species: Western Osprey *Pandion haliaetus* (English and scientific names follow Gill & Donsker, 2015), Black Kite *Milvus migrans*, Griffon Vulture *Gyps fulvus*, three harriers *Circus* spp., three hawks *Accipiter* spp., Common Buzzard *Buteo buteo*, Booted Eagle *Hieraaetus pennatus* and four falcons *Falco* spp. (Scully, 1881). Other early reports include those of Donald (1905, 1923), who in February 1905 observed a large concentration of Western Marsh *Circus aeruginosus*, Pallid *C. macrourus* and Pied harriers *C. melanoleucos* in eastern India south of the Himalayas (Donald, 1905), which, presumably, had travelled across or around the Himalayas from their breeding grounds in central and eastern Asia to overwintering sites along the Indo-Gangetic Plain south of the Himalayas (Naoroji, 2006). Eighteen years later, Donald recorded approximately 40 *Aquila* eagles flying east-north-east to west-south-west at intervals of one to five minutes at Jhatingri (Donald, 1923). These were the first of numerous reports of a substantial movement of raptors along what we call the East-to-West Southern Corridor (see later in this chapter).

Following these early observations, there were few further records until the 1970s, when several regional and international authorities began reporting their work in the area, most of which was concentrated in the central Himalayas of Nepal.

Koning (1976) described overwintering concentrations of raptors in the Indo-Gangetic Plain of Pakistan during five consecutive winters in 1970–1974 that included close to 10,000 sightings representing 33 species. Thiollay (1979) recorded 151 migrants of 15 species during several days of observation in Mustang, central Nepal in late September–early October 1978. Based on a series of observations between 1971 and 1980, that built on Donald's earlier work (Donald, 1923), Fleming reported an approximately 1000-km-long east-to-west autumn migration corridor along the southern slopes of the Himalayas stretching from northeastern India through Nepal and into northwestern India. The flights were dominated by large numbers of Steppe Eagles *Aquila nipalensis* (Figure 6.1), together with smaller numbers of Greater Spotted Eagles *Clanga clanga* and Eastern Imperial Eagles

Figure 6.1. Adult Steppe Eagle on 11 November 2014 in Thulakharka (Nepal). Photo credit Hawk Mountain Sanctuary Archives (A black-and-white version of this figure will appear in some formats. For the colour version, please refer to the plate section.)

Aquila heliaca. Single-day counts ranged from 1 to 630 birds, with migrants passing at altitudes of between 1700°m to 3800 m above sea level (a.s.l.) between 9:00 and 16:00 local time, in flocks of up to 20 individuals (Fleming, 1983). Fleming concluded that the approximately 45,000 migrants, which passed mainly from late October to late November, presumably travelled from breeding areas in central Asia towards wintering grounds presumably in India (Naoroji, 2006; Pande *et al.*, 2013).

De Roder (1989) conducted the first systematic counts of visible raptor migrants in the Himalayas near Annapurna in central Nepal between 20 October and 7 November 1985 when he recorded close to 8500 raptors representing 19 species; more than 90% were identified as Steppe Eagles (28% juveniles birds, 28% subadults and 44% adults). De Roder concluded that between 10,000 and 20,000 Steppe Eagles migrated through the region in autumn (De Roder 1989), an estimate Bijlsma (1991) and Inskipp and Inskipp (1985) confirmed.

In a global directory of raptor migration 'watch-sites' (Zalles & Bildstein, 2000), the major raptor migration routes and bottlenecks described included two in Nepal and one in western China. On the heels of that publication, additional attention began to focus on raptor migration in the region. Finally, Den Besten conducted the first systematic counts of visible raptor migrants in northwest India in the autumn of 2001 and spring of 2002. Counting at a site approximately 50 km west of Donald's (1923) observation points, Den Besten (2004) reported nearly 13,850 raptors between 23 October and 30 November in autumn, and from 19 February through 29 March in spring, including 16 species, 97% of which were Steppe Eagles.

In addition, systematic raptor migration counts have been carried out since 2000 at several sites between northwest India and Nepal. DeCandido *et al.* (2001) counted autumn migrants near the Annapurna range for eight days in late October–early November in 1999, recording a total of 950 migratory raptors of eight species, 86% of which were identified as Steppe Eagles. Of these, 15% were juveniles, 65% subadults and 20% adults. During this count, a local teacher, Gurung, became involved in the fieldwork. Although military conflict affected the migration count for several seasons, Gurung and his family conducted an autumn migration count from late October through early December 2003 (Gurung *et al.*, 2004), during which they recorded close to 7000 migratory raptors representing 10 species, 94% of which were identified as Steppe Eagles. Since 2012, Subedi and DeCandido have conducted counts from mid-September to early December in Thula Kharka, central Nepal (Subedi & DeCandido, 2013, 2014; Subedi, 2015), and nowadays, incipient long-term counts of visible migrants are well under way, at least in this part of the region.

The Biogeographic Context

Approximately 40% of all raptor species migrate, with the tendency to undertake migration increasing with increasing latitude and seasonality, as the latter often acts to reduce prey availability for raptors in winter (Bildstein, 2006). Asia, the world's largest continent, is host to at least 66 of the world's 130 or so species of migratory raptors, and most of these occur in the Palearctic region north of the Himalayas. Of these 66 species, 45 are known to migrate from and through the Himalayas (Table 6.1). That said, many raptors circumnavigate the region en route to wintering areas principally in Southeast Asia, the Indian subcontinent, the Middle East and eastern and southern Africa. Although many researchers (e.g. Moreau, 1972; Newton, 2008; Hawkes *et al.*, 2012) have suggested that avian migrants follow these longer routes because of the lower temperatures, higher altitudes and lower oxygen levels they would face if they took a more direct route over the mountains, circumstances specific to migratory raptors also may play a role.

Although raptors are relatively large-bodied, they have oversized wings and, as a result, are lightly wing-loaded birds, many of which use low-cost soaring flight to complete their diurnal migrations (Bildstein, 2006). 'Soaring,' or updraught-assisted non-flapping flight (*sensu* Kerlinger, 1989), is an energetically efficient form of flight, and most species of migratory raptor employ it, at least episodically. In fact many long-distance migrants are so-called obligate soaring migrants (*sensu* Bildstein, 2006) that depend on soaring flight to complete their migratory journeys. Updraughts necessary for soaring flight include both thermals, or pockets of warm rising air, and deflection or 'orographic' updraughts that occur when horizontal winds strike surface discontinuities, including mountains (Bildstein, 2006; see Ohlmann, Chapter 14, and Heise, Chapter 15). In many parts of the world, particularly in the tropics, raptors congregate in large flocks along 'thermal pathways' (*sensu* Berthold, 2001) while migrating. The high-altitude, rugged terrain of the Himalayas precludes this type of pathway and hence its use by migrating birds of prey. Orographic updraughts do occur, particularly along the southernmost slopes of the Himalayas, and it is not surprising that this creates the region's only well-known raptor migration pathway and that this corridor is dominated numerically by Steppe Eagles, a raptor known to be an obligate soaring migrant (Spaar & Bruderer, 1996; for details, see Heise, Chapter 15). Raptors, including falcons and, to some extent, accipiters and harriers,

Table 6.1. Raptors that migrate across the Himalayan region, including the western Himalayas (Pakistan and northwestern India), central Himalayas (Nepal) and eastern Himalayas (northeastern India and Bhutan). (RM = Regular migrant, V = Vagrant, ? = status uncertain, ND = no data). [1–5] Conservation status BirdLife International (2015).

	Himalayan region		
Migratory species	Western	Central	Eastern
Complete migrants[a]:			
Western Osprey *Pandion haliaetus*[1, A]	RM	RM	RM
Short-toed Snake Eagle *Circaetus gallicus*[1, A]	RM	RM	ND
Pallid Harrier *Circus macrourus*[3, A]	RM	RM	ND
Pied Harrier *Circus melanoleucos*[1, A]	ND	RM	?
Montagu's Harrier *Circus pygargus*[1, A]	RM	RM	ND
Greater Spotted Eagle *Clanga clanga*[2, A]	RM	RM	RM
Steppe Eagle *Aquila nipalensis*[1, A]	RM	RM	RM
Lesser Kestrel *Falco naumanni*[1, A]	RM	RM	RM
Amur Falcon *Falco amurensis*[1, A]	?	RM	RM
Eurasian Hobby *Falco subbuteo*[1, A]	RM	RM	?
Partial migrants[b]:			
Black-winged Kite *Elanus caeruleus*[1, B]	?	?	?
Black Baza *Aviceda leuphotes*[1, B]	V	RM	RM
Crested Honey-Buzzard *Pernis ptilorhynchus*[1, A]	RM	RM	RM
Red Kite *Milvus milvus*[3, A]	V	V	ND
Black Kite *Milvus migrans*[1,C]	RM	RM	RM
Pallas's Fish Eagle *Haliaeetus leucoryphus*[2, B]	RM	?	?
White-tailed Eagle *Haliaeetus albicilla*[1, A]	?	RM	?
Lammergeier (Bearded Vulture) *Gypaetus barbatus*[3, B]	?	ND	ND
Egyptian Vulture *Neophron percnopterus*[4, A]	RM	RM	ND
White-rumped Vulture *Gyps bengalensis*[5, B]	ND	?	ND
Himalayan Vulture *Gyps himalayensis*[3, B]	RM	RM	?
Griffon Vulture *Gyps fulvus*[1, C]	?	?	?
Cinereous Vulture *Aegypius monachus*[3, A]	RM	RM	ND
Western Marsh Harrier *Circus aeruginosus*[1, A]	RM	RM	?
Eastern Marsh Harrier *Circus spilonatus*[1, A]	V	V	ND
Hen Harrier *Circus cyaneus*[1, A]	RM	RM	RM
Shikra *Accipiter badius*[1, B]	ND	?	ND
Besra *Accipiter virgatus*[1, B]	?	?	RM
Eurasian Sparrowhawk *Accipiter nisus*[1, C]?	RM	RM	RM
Northern Goshawk *Accipiter gentilis*[1, A]	?	?	?
Common Buzzard *Buteo buteo*[1, A]	RM	RM	RM
Long-legged Buzzard *Buteo rufinus*[1, A]	RM	RM	RM
Upland Buzzard *Buteo hemilasius*[1, B]	?	?	?
Eastern Imperial Eagle *Aquila heliaca*[2, A]	RM	RM	ND
Golden Eagle *Aquila chrysaetos*[1, B]	?	?	ND
Bonelli's Eagle *Aquila fasciata*[1, A]	ND	?	?
Booted Eagle *Hieraaetus pennatus*[1, A]	RM	RM	RM
Common Kestrel *Falco tinnunculus*[1, B?]	ND	RM	RM
Merlin *Falco columbarius*[1, A]	?	?	?
Oriental Hobby *Falco severus*[1, B]	RM	RM	RM

Table 6.1. (cont.)

Migratory species	Himalayan region		
	Western	Central	Eastern
Saker Falcon *Falco cherrug*[4, A]	RM	RM	ND
Peregrine Falcon *Falco peregrines*[1, A]	RM	RM	?
Irruptive migrants[c]:			
Slender-billed Vulture *Gyps tenuirostris*[5, B]	?	ND	ND
Crested Serpent Eagle *Spilornis cheela*[1, B]	?	?	ND
Mountain Hawk-Eagle *Nisaetus nipalensis*[1, B]	ND	?	?
Total	**39**	**43**	**29**

[1] Least concern
[2] Vulnerable
[3] Near threatened
[4] Endangered
[5] Critically endangered
[a] Species or regional populations in which at least 90% of all individuals regularly migrate.
[b] Species or regional populations in which fewer than 90% of all individuals regularly migrate.
[c] Species or regional populations in which the extent of movement varies annually, typically due to between year shifts in prey abundance.
[A-C:] Type of migrant with:
[A] Palearctic migrant,
[B] Indomalayan migrant, and
[C] Indomalayan and Palearctic migrant. English and scientific names follow Gill and Donsker (2015).

Notes: 1. Eastern Buzzards *Buteo japonicus* and Himalayan Buzzards *B. burmanicus* are included as Common Buzzards *B. buteo*, not as separate species as proposed by Gill and Donsker (2015) and Riesing *et al.* (2003), because of identification concerns.
2. Because of the lack of information on the migration status of Red-headed Vulture *Sarcogyps calvus*, Lesser-spotted Eagle *Clanga pomarina* and Barbary Falcon *Falco pelegrinoides* in the Himalayan region, these species are not included in the migration list. That said, all three have been recorded as migrants at the Khare watch-site by de Roder (1989), Gurug *et al.* (2004) and Thiollay (1979).

that migrate principally by using flapping flight do apparently migrate across the Himalayas, but only across broad fronts, making the extent of their movements far more difficult to assess and quantify at visible migration watch-sites (see Dixon *et al.*, Chapter 8).

The emerging development of satellite-tracking technology promises to increase our knowledge regarding the extent of such movements (Dodge *et al.*, 2014). It thus appears that the Himalayas present a formidable barrier to raptor migration and that these mountains, together with the Tibetan Plateau, present a significant barrier to migrating raptors that limits the existence of concentrated migratory movements.

Patterns of Raptor Migration in the Tibetan-Himalayan Region

As is true in most regions in the world, the visible migrations of raptors are better known for autumn (i.e. post-nuptial or outbound) movements than for spring (i.e. pre-nuptial or return) movements, largely because the former follow the breeding season and occur when populations are at their annual peak, whereas the latter occur after a period of winter mortality (Bildstein, 2006). Because of this, much of what follows focuses on the geography of autumn movements of raptor migrants through the region. Fortunately, the movements of a small but growing number of raptors are now being tracked by satellite in central and eastern Asia (Dixon *et al.*, 2011; Dixon *et al.*, 2012; Dixon *et al.*, Chapter 8; Batbayar & Lee, Chapter 7), following similar work in North America and Western Europe (Bildstein, 2006; Dodge *et al.*, 2014), giving avian movement ecologists the opportunity to follow individual raptors on both their outbound and return migrations. Information from these studies, albeit largely anecdotal to date, is summarized later in this chapter, along with findings from raptor migration watch-sites.

The Himalayas, which span approximately 2400 km east to west through parts of China, Bhutan, Nepal, India and Pakistan, are traversed or partially circumnavigated by at least 45 species of migratory raptors each autumn (Table 6.1). Presumably, most if not all of these migrants also migrate through the region in slightly smaller numbers in spring. It is believed that in autumn, most raptors migrate across the Himalayas on a broad front, using valley or river systems as short-distance passageways and navigational aids, and thereafter concentrate south of it along a major east-to-west flyway (Roberts, 1991; Chettri *et al.*, 2006; DeCandido *et al.*, 2013). These raptors include populations of 30 Palearctic migrant species that breed in central and eastern Asia and that migrate through and around the Himalayas to overwinter in the Indian subcontinent, the surrounding areas, the Middle East and eastern and southern Africa. A further 18 populations of Indo-Malayan migrant raptor species breed in the Himalayas and migrate shorter distances altitudinally, longitudinally or latitudinally within the region. Of the 45, 10 are complete migrants, 32 are partial migrants and 3 are classified as irruptive migrants (*sensu* Bildstein, 2006) (see Table 6.1 for details and definitions of types of migrants). Based on satellite-tracking data, a small number of banding recoveries and wing-tag re-sightings, raptor migration watch-site counts and sporadic observations, we characterize the movements as occurring along four corridors of uncertain magnitude: the (1) *Western Circum-Himalayan Corridor*, (2) *Eastern Circum-Himalayan Corridor*, (3) *East-to-West Southern Corridor* and (4) more broad-frontal *Trans-Himalayan Corridor* (Figure 6.1).

Circum-Himalayan corridors. Moreau (1972) suggested that the Himalayas and the Tibetan Plateau formed an effective barrier for many avian migrants. This appears to be true at least for raptors, with individuals of many species avoiding a direct crossing by circumnavigating this high-altitude barrier either to the west or east, or, in a few instances, by wintering to the north.

In autumn, the western edge of the Himalayas appears to be a concentrating point for several streams of migrants that avoid the Himalayas by migrating south through the Indus River Valley (Roberts, 1991). In spring, a first-year male Pallid Harrier fitted with a satellite tracking device on its wintering grounds on the western edge of the Indo-Gangetic Plain in northwestern India avoided the Himalayas completely by migrating north from India through western Pakistan and Afghanistan, and then north to its presumed breeding grounds in central Kazakhstan (Terrauble *et al.*, 2012). Additional satellite-tracking and ringing data suggest that Pallid Harriers breeding in

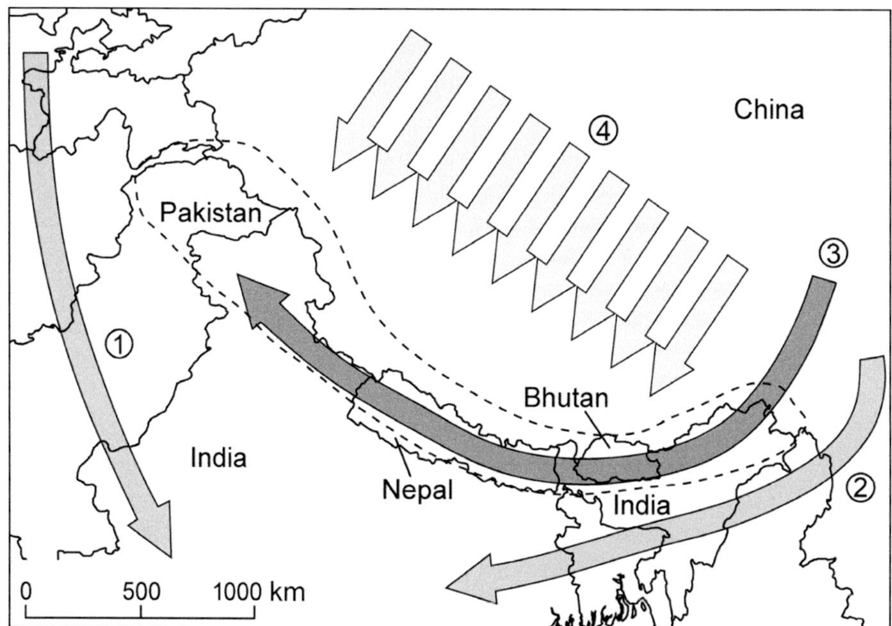

Figure 6.2. Autumn raptor migration corridors across and around the Himalayas: (1) Western Circum-Himalayan Corridor, (2) Eastern Circum-Himalayan Corridor (3) East-to-West Southern Corridor, (4) Broad-Frontal Trans-Himalayan Corridor (see text for details).

Kazakhstan overwinter both in sub-Saharan Africa and the Indian subcontinent (Terrauble *et al.*, 2012), indicating that movements of this tracked bird are probably representative of this population of the species. As many as 3000 harriers, mainly Montagu's Harriers *C. pygargus* together with lesser numbers of Pallid and Western Marsh Harriers, have been recorded in the wintering grounds in the Indo-Gangetic Plain of western India (Clarke *et al.*, 1998). These observations, together with the fact that relatively few harriers (< 70 individuals) have been seen on passage at migration watch-sites in the Himalayas (e.g. de Roder, 1989; Subedi & DeCandido, 2013; 2014; Subedi, 2015) (Figure 6.2), suggests that many, if not most, members of this genus avoid crossing the Himalayas via the western Circum-Himalayan Corridor.

On the eastern edge of the Himalayas, Amur Falcons *Falco amurensis* congregate temporarily in large numbers in autumn in northeastern India and Bangladesh to feed and fatten prior to embarking on the second stage of migration through the Indian subcontinent and across the Indian Ocean to equatorial eastern and southern Africa (Clement & Holman, 2001; Bildstein, 2006; Dixon *et al.*, 2011). Although fewer than 150 of these kestrel-sized falcons have been recorded migrating through Nepal in autumn (e.g. de Roder 1989; Subedi & DeCandido, 2013; 2014; Subedi, 2015), it is not clear whether they move along north–south or east–west axes, or both (Inskipp & Inskipp, 1985). Presumably, most individuals of this species follow an elliptical course around the Himalayas that has them migrating west and north of the mountains during their return migration in spring (Bildstein, 2006; Anderson, 2009).

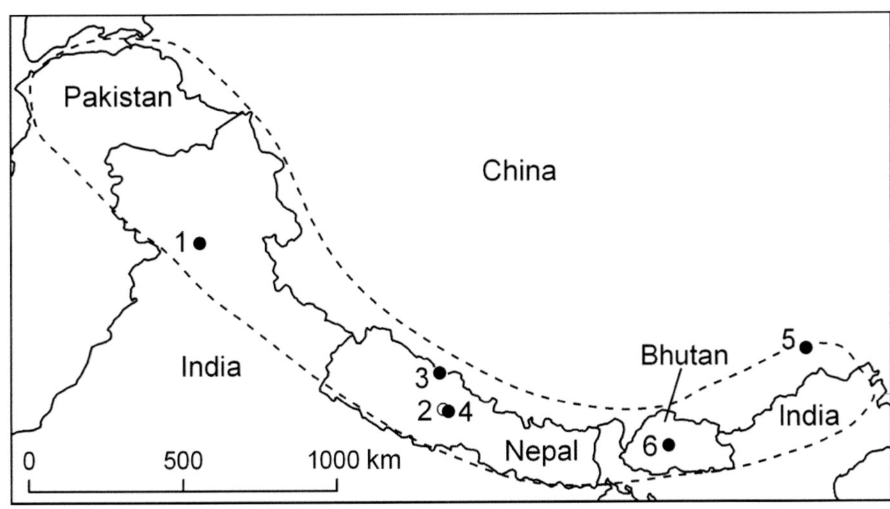

Figure 6.3. Locations of the six raptor-migration watch-sites in the Himalayas: (1) McLeod Ganj, northwestern India, (2) Khare, central Nepal, (3) Upper Kali Gandaki, northwestern Nepal, (4) Thula Khara, central Nepal, (5) Güncang, Tibetan Autonomous Region, western China; (6) Lawala Pass, western Bhutan (see text for details of species and counts).

The northern edge of the Himalayas is used both as a migration corridor and a wintering grounds for at least three migratory raptor species. A Steppe Eagle and two Saker Falcons *F. cherrug* fitted with satellite-tracking units in Mongolia and the Russian Altai, respectively, left their breeding grounds in early autumn and overwintered on the Tibetan Plateau north of the Himalayas (Eastham, 1998; Ellis *et al.*, 2001; Potapov *et al.*, 2002). Although individuals of both these species also overwinter in Africa and in the Indian subcontinent (Koning, 1976; Ferguson-Lees & Christie, 2001; Meyburg *et al.*, 2012; Pande *et al.*, 2013), it remains unclear how they move through or around the Himalayas. Also, a Peregrine Falcon *F. peregrinus calidus* fitted with a satellite-tracking device in the eastern Taymyr Peninsula, northern Russia, overwintered on the western Tibetan Plateau (Dixon *et al.*, 2012), although most individuals from this population overwinter further south, in the Indian subcontinent (Dixon *et al.*, 2012; Dixon *et al.*, Chapter 8).

Much remains to be learned about the extent to which raptors circumnavigate the Himalayas both during outbound migration in autumn and return migration in spring. The growing use of satellite-tracking devices on medium-sized and larger birds of prey promises to reveal much about the scale and nature of such movements (cf. Dodge *et al.*, 2014).

The East-to-West Southern Corridor. This well established raptor-migration corridor (de Roder, 1989; DeCandido *et al.*, 2013; Subedi & DeCandido, 2013, 2014; Subedi, 2015) may be the region's most significant, and many individuals avoid traversing the Tibetan Plateau by travelling east to west along the southern Himalayas along a mid-altitude corridor roughly parallel to the most southerly southern slopes of the range. Migrants presumably follow an initial route to the east of the Himalayas (i.e. circa 100°E), from north to south and then turn west once they reach a point that is aligned with its southern slopes (i.e. circa 27°N). This 'geographic detour' enables the birds that take it to soar at low cost in orographic updraughts (Bildstein, 2006) while migrating at lower altitudes than they would otherwise use had they crossed the Tibetan Plateau east-to-west (see Ohlmann, Chapter 14; Heise, Chapter 15).

Steppe Eagles are the most studied, as well as the most common migrants in this corridor (DeCandido *et al.*, 2013). A recent estimate suggests that as many as 50,000 (DeCandido *et al.*, 2013) *Aquila* and *Clanga* spp., mainly Steppe Eagles, migrate along this corridor each autumn, concentrating at lower-altitude (i.e. 800–4500 m) bottlenecks near Annapurna, Kali Gandaki, Arun and Dudh Kosi in Nepal, and at Kangra and Dharamsala in northwestern India (Pande *et al.*, 2013). An average of 7150 Steppe Eagles has been recorded in three consecutive autumn migration counts at Thula Kharka, central Nepal, with juveniles (i.e. less than one year old) greatly outnumbered by sub-adult (two to four years old) and adult eagles (i.e. five years old or older) (Subedi, 2015).

The Trans-Himalayan Corridor. Although little information has been collected based on visible migration along this most likely diffuse corridor (Inskipp & Inskipp, 1985), satellite tracking suggests that it may be significant for several species, including Peregrine Falcon (Dixon *et al.*, 2012; Dixon *et al.*, Chapter 8), or even SteppeEagle (Batbayar & Lee, Chapter 7), and possibly Cinereous Vultures (*Aegypius monachus*), with migrants travelling along major valleys and gorges of up to approximately 5000 m, including the Kali Gandaki River Valley and the Tsangpo-Brahmaputra River Valley (Chettri *et al.*, 2006).

Five of the 10 Peregrine Falcons fitted with satellite-tracking units on their breeding grounds in the eastern Taymyr Peninsula in northern Russia in 2011 migrated through the Himalayas on a broad front between northwestern Pakistan and Bhutan (Dixon *et al.*, 2012; Chapter 8). Although the number of tracked individuals is small, it appears likely that Arctic Peregrine Falcons *F. p. calidus* migrating through the Himalayas breed in northern Russia between the Gydan Peninsula and the Lena Delta (Dixon *et al.*, 2012; Chapter 8). In addition, an average of 21 Peregrine Falcons has been recorded at Thula Kharka in central Nepal (Subedi, 2015).

Re-sightings of Cinereous Vultures wing-tagged on the breeding grounds in central Mongolia suggest that although most of this population overwinters in South Korea, a few individuals overwinter in the lowlands of central Nepal in the Himalayas, presumably having travelled either along the eastern edge of the range or through it (Batbayar *et al.*, 2008; Kenny *et al.*, 2008). An average of 59 Cinereous Vultures has also been recorded at Thula Kharka in central Nepal (Subedi, 2015).

Finally, travelling across Himalayas may be risky for migrating raptors. A Steppe Eagle was found dead at 7925 m on the South Col of Mount Everest in 1960 (Singh, 1961).

Altitudinal movements. Altitudinal migration consists of relatively short-distance seasonal movements between high-altitude breeding sites and lower-altitude overwintering areas (Newton, 2008). Such migration almost certainly occurs in the Himalayas. Unfortunately, the relatively short distances involved, coupled with the fact that such movements often occur coincidentally with longer-distance latitudinal movements, means that altitudinal migration is far less well understood than its latitudinal counterpart (Bildstein, 2006).

What is known for the Himalayas is that the degree to which raptors migrate altitudinally in winter can depend on weather conditions (Inskipp & Inskipp, 1985; Grimmentt *et al.*, 1999, 2000; Inskipp *et al.*, 1999). Pallas's Fish Eagle *Haliaeetus leucoryphus*, a species that migrates latitudinally between the Tibetan Plateau and the lowlands of the eastern Himalayas, and as such is something of a north–south migrant, does so in part because its ability to secure sufficient aquatic prey in winter is affected by the icing over of high-altitude streams and rivers (Ali, 1977; Ali & Ripley, 1978). Observations of small numbers of Crested Serpent Eagles *Spilornis cheela* at migration watch-sites at Thula Kharka in central Nepal indicate that this reptile-eating species may also migrate altitudinally in the region (Subedi & DeCandido, 2013, 2014). Other species that are known to migrate altitudinally elsewhere in their ranges have been recorded infrequently at one or more watch-sites in the Himalayas, including White-rumped Vulture *G. bengalensis*, Himalayan Vulture *G. himalayensis*, Shikra *Accipiter badius*, Besra *A. virgatus*, Upland Buzzard *B. hemilasius* and Mountain Hawk-Eagle *Nisaetus nipalensis* (De Roder, 1989; Den Besten, 2004; Gurung *et al.*, 2004; Feijen *et al.*, 2005; Subedi & DeCandido, 2013, 2014).

Migration Strategies Demonstrated by Occurrences during the Non-breeding Season in and around the Himalayas

Based on satellite-tracking data, we know that Palearctic populations of Steppe Eagle, Saker Falcon and Peregrine Falcon overwinter on the Tibetan Plateau, often at elevations in excess of 4500 m. a.s.l. (Eastham, 1998; Ellis *et al.*, 2001; Potapov *et al.*, 2002; Dixon *et al.*, 2012).

Larger raptors, including obligate and facultative scavengers such as vultures and Steppe Eagles, also congregate in winter in large numbers around rubbish and carrion dumps in the southern Himalayas and immediately to the south (Chhangani & Mohnot, 2008; Pande *et al.*, 2013). Observations during 10 winters between 1995 and 2004 in Rajasthan, northwestern India showed that Griffon, Himalayan and Cinereous vultures overwinter regularly in the area, with numbers peaking in January–February (Chhangani & Mohnot, 2008).

Six species of harriers also overwinter in large numbers in the Indian subcontinent, mainly in open habitats (Clarke *et al.*, 1998; Verma, 2007; Verma & Prakash, 2007). Montagu's and Western Marsh Harriers are the most common, whereas Pied, Pallid, Hen *C. cyaneus* and Eastern Marsh *C. spilonatus* harriers are less common (Naoroji, 2006).

Steppe Eagle, a species that overwinters in Africa (Meyburg *et al.*, 2012), also overwinters in both the Indian subcontinent (Naoroji, 2006; Sharma & Sundar, 2009; Pande *et al.*, 2010, 2013) and the Tibetan Plateau (Ellis *et al.*, 2001), with many individuals congregating in large numbers around rubbish dumps in Rajasthan, Gujarat, Maharashtra and Karnataka in western and northwestern India (Sharma & Sundar, 2009; Pande *et al.*, 2010, 2013).

Several Peregrine Falcons fitted with satellite units in the Eastern Taimyr Peninsula, northern Russia overwintered in the Indian subcontinent along the Indo-Gangetic Plain (Dixon *et al.*, 2012; Chapter 8).

Recently and Currently Active Raptor Migration Watch-Sites

Visible raptor migration has been observed at six raptor migration watch-sites (*sensu* Zalles & Bildstein, 2000) in and around the Himalayas (Figure 6.1). At McLeod Ganj (900 m; Figure 6.2), in Himachal Pradesh, northwestern India, counts were conducted on 38 days in autumn 2001 (between 23 October and 30 November) and 29 days in spring 2002 (between 19 February and 29 March). Sixteen species were seen either on autumn or spring migration, or in both seasons. In autumn, a total of 8228 individuals was recorded, including 1 Crested Honey-buzzard *Pernis ptilorhynchus*, 11 Greater Spotted Eagles, 3 Booted Eagles, 8194 Steppe Eagles, 2 Eastern Imperial Eagles, 11 Eurasian Sparrowhawks *A. nisus*, 1 Northern Goshawk *A. gentilis*, 2 Hen Harriers, 1 Long-legged Buzzard *B. rufinus*, 1 Common Buzzard, and 1 Common Kestrel *F. tinnunculus*. In spring a total of 5618 individuals was recorded, including 4 Crested Honey-buzzards, 15 Greater Spotted Eagles, 9 Booted Eagles, 5204 Steppe Eagles, 1 Eastern Imperial Eagle, 1 Shikra, 35 Eurasian Sparrowhawks, 4 Northern Goshawks, 1 Western Marsh Harrier, 3 Hen Harriers, 1 Pallid Harrier, 80 Black Kites, 1 Upland Buzzard, 180 Long-legged Buzzards, 2 Common Buzzards, 64 Common Kestrels, and 13 unidentified raptors (Den Besten, 2004).

At Khare (1650 m; Figure 6.2), in the Janakpur Zone, central Nepal, counts were conducted in 1984, 1985, 1999 and 2003. Twenty-five species were seen in autumn between late October and early December. The counts that follow are the highest recorded numbers in any year: 10 Lammergeiers (a.k.a. Bearded Vulture *Gypaetus barbatus* (Figures 6.4, 6.5), 74 Egyptian Vultures *Neophron percnopterus*, 2 Crested Honey-Buzzards, 3 White-rumped Vulture, 233 Himalayan Vulture, 14 Red-headed Vultures *Sarcogyps calvus*, 8 Booted Eagles, 7852 Steppe Eagles, 9 Eastern Imperial Eagles, 3 Shikras, 13 Besras, 7 Eurasian Sparrowhawks, 1 Western Marsh Harrier, 66 Hen Harriers, 2 Pallid Harriers, 3 Montagu's Harriers, 97 Black Kites, 30 Long-legged Buzzards, 32 Common Buzzards, 77 Lesser Kestrels *F. naumanni*, 138 Amur Falcons, 19 Eurasian Hobbys *F. subbuteo*, 3 Saker Falcons, 2 Peregrine Falcons and 2 Barbary Falcons *F. pelegrinoides* (De Roder, 1989; Bijlsma, 1991; Zalles & Bildstein, 2000; DeCandido *et al.*, 2001; Gurung *et al.*, 2004).

At Upper Kali Gandaki (800–4500 m; Figure 6.2) in Dhaulagiri Zone, northwestern Nepal, counts were conducted in 1978 and 1984. Seventeen species were seen in autumn between late September and early November. The counts that follow are the highest recorded in either year: 1 Western Osprey, 1 Crested Honey-buzzard, 16 Booted Eagles, 63 Steppe Eagles, 4 Eastern Imperial Eagles, 1 Lesser-spotted Eagle, 23 Eurasian Sparrowhawks, 2 Northern Goshawks, 3 Western Marsh Harriers, 3 Hen Harriers, 1 Pallid Harrier, 32 Black Kites, 5 Long-legged Buzzard, 20 Common Buzzards,

Figure 6.4. Soaring adult Lammergeyer on 23 November 2014 in Thoolakharka (Nepal; on maps referred to as Thuli Kharka: 28°22 N, 82°19'E), which is a well-known raptor migration site. Photo credit Hawk Mountain Sanctuary Archives. (A black-and-white version of this figure will appear in some formats. For the colour version, please refer to the plate section.)

Figure 6.5. Adult Lammergeyer on 26 November 2014 in Thoolakharka (Nepal). For more information on this spot, see http://raptorsofnepal.blogspot.nl/ Photo credit Hawk Mountain Sanctuary Archives. (A black-and-white version of this figure will appear in some formats. For the colour version, please refer to the plate section.)

11 Lesser Kestrels, 39 Common Kestrels and 7 Eurasian Hobbys (Thiollay, 1979; Bijlsma, 1991; Zalles & Bildstein, 2000).

At Thula Kharka (2050 m; Figure 6.2), in Gandaki Zone, central Nepal, counts were conducted during 80 days in autumn 2012 (15 September to 4 December) and during 85 days in autumn 2013 (15 September to 8 December). Thirty-four species were seen in one or both years. The counts that follow are the means for both years: 4 Western Ospreys, 32 Egyptian Vultures, 511 Crested Honey Buzzard, 63 White-rumped Vultures, 1742 Himalayan Vultures, 64 Griffon Vultures, 65 Cinereous Vultures, 15 Crested Serpent Eagles, 2 Short-toed Snake Eagles *Circaetus gallicus*, 3 Mountain Hawk-Eagles, 2 Greater Spotted Eagles, 132 Booted Eagles, 7640 Steppe Eagles, 0.5 Eastern Imperial Eagle, 5 Golden Eagles *A. chrysaetus*, 21 Bonelli's Eagles *A. fasciata*, 108 Eurasian Sparrowhawks, 6 Northern Goshawks, 0.5 Eastern Marsh Harrier, 14 Hen Harriers, 1 Pallid Harrier, 0.5 Pied Harrier, 416 Black Kites, 1 Pallas's Fish Eagle, 1 White-tailed Eagle *H. albicilla*, 6 Upland Buzzards, 7 Long-legged Buzzards, 141 Common Buzzards, 67 Lesser Kestrels, 200 Common Kestrels, 84 Amur Falcons, 145 Eurasian Hobbys, 3 Saker Falcons, 23 Peregrine Falcons and 82 unidentified raptors (Subedi & DeCandido, 2013, 2014).

At Güncang (3170 m; Figure 6.2), in the Tibetan Autonomous Region, western China, counts were conducted sporadically in 1986. Seven species were seen in spring between 10 and 16 March, including 48 Steppe Eagles, 2 Eurasian Sparrowhawks, 1 Northern Goshawk, 3 Hen Harriers, 167 Black Kites, 2 White-tailed Eagles and 2 Common Buzzards (Robson, 1986; Zalles & Bildstein, 2000).

At Lawala Pass (3400 m; Figure 6.2), in Thimphu Province, in western Bhutan, counts were conducted sporadically in 1999. Seven species were seen in autumn between 13 and 14 November, including 1 Egyptian Vulture, 10 Himalayan Vultures, 2 Mountain Hawk-Eagles, 6 Steppe Eagles, 1 Golden Eagle, 4 Hen Harriers, 1 Peregrine Falcon and 22 unidentified raptors (Feijen *et al.*, 2005).

The Scale and Diversity of Raptor Migration across and around the Tibetan-Himalayan Region

Fieldwork involving counts of visible migrants at traditional raptor migration watch-sites suggests, that with the possible exception of Steppe Eagle, relatively few species of raptors migrate in large numbers across the Himalayas, and those that do migrate do so across a relatively broad front. The small number of migration watch-sites in the region, together with little if any information regarding the representative nature of the sites both in terms of number of individuals and number of species, make it all but impossible to gauge the total number of raptors that migrate through the region. That said, with a minimum of 45 migratory species, including 30 regular migrants, the region hosts one of the greatest assemblages of migratory raptors anywhere, in part because it occurs at the junction of the Palearctic and Indo-Malayan Biogeographic Realms, two of the world's most raptor-diverse realms (Ferguson-Lees & Christie, 2001). Much remains to be learned about raptor migration in this still 'ornithologically' remote region. Counts at raptor migration watch-sites, together with the emerging technology of satellite tracking, promise to close many gaps in our current knowledge, particularly regarding how weather and terrain combine to shape the movements of birds of prey in truly mountainous areas.

Conservation Considerations

Twelve of the 45 raptor species that migrate across and around the Himalayas are of global conservation concern, including seven Palearctic and five Indo-Malayan species (Table 6.1). According to BirdLife International (2015), the following species are classified as Globally Threatened: White-rumped Vulture and Slender-billed Vulture (*Gyps tenuirostris*) in the category Critically Endangered; Egyptian Vulture and Saker Falcon in the category Endangered; Greater Spotted Eagle, Pallas's Fish Eagle and Eastern Imperial Eagle in the category Vulnerable; and Pallid Harrier, Red Kite, Lammergeier, Himalayan Vulture and Cinereous Vulture in the category Near Threatened. Conservation concerns vary among the species, although the main threats include unintentional poisoning via veterinary drugs (Naoroji, 2006; Ogada *et al.*, 2012), pesticide misuse, human persecution, the live-bird trade and land-use change (Naoroji, 2006). Indeed, since the early 1990s, several populations of Asian vultures have declined by more than 95 percent (Ogada *et al.*, 2012). Several non-steroidal anti-inflammatory drugs (NSAIDs), including most notably Diclofenac, have been shown to be the major cause of the collapse of populations of vultures. Vultures are exposed to Diclofenac and other NSAIDs when they feed on drug-treated carcasses of livestock and die from kidney failure, with clinical signs of extensive visceral gout and renal damage (Das *et al.*, 2011). Fortunately, veterinary use of Diclofenac has been banned in India, Nepal and Pakistan, and in some regions vulture population declines have slowed (Galligan *et al.*, 2014).

In addition, the draining of swamps, marshes and other wetlands, together with chemical contamination of the system, poses a direct threat to fish eaters, including Pallas's Fish Eagles (Naoroji, 2006).

The six species of harriers that overwinter in the Indian subcontinent depend on natural grasslands both as roosting and foraging areas that are currently being reduced and fragmented for agricultural and other uses and that have been planted with invasive alien species (Naoroji, 2006; Verma, 2007).

Saker Falcons are potentially under threat from over-trapping for falconry (Eastham *et al.*, 2000). Most of the legal and illegal trapping takes place in autumn and early winter when the falcons are migrating (Potapov *et al.*, 2002).

Suggestions for Future Studies

Previous investigations of raptor migration across the Himalayas have focused on opportunistic (e.g. Bijlsma, 1991) and planned counts of visible migrants (e.g. Subedi & DeCandido, 2013; 2014) at migration watch-sites and other likely and accessible locations (Bildstein *et al.*, 2008). With this in mind, sampling the numbers of migrating raptors across a geographically expanded network of migration sites, coupled with greatly increased coverage in spring, would offer a cost-effective and efficient method for long-term monitoring of these birds (Zalles & Bildstein, 2000;

Bildstein, 2006; Bildstein *et al.*, 2008), particularly when the species in question concentrate along well-defined narrow fronts while moving through the region (Farmer *et al.*, 2010). Unfortunately, most practitioners in the field still view the migratory movements of birds of prey as 'fixed, and all-but immutable in place despite a growing body of field and experimental evidence that suggests otherwise. In fact, the picture that is now emerging from the literature indicates that migration behaviour in general, and migration geography in particular, are dynamic and flexible attributes of many species, and that both can shift quickly in response to changing ecological conditions' (Bildstein *et al.*, 2008). Important cases in point are the roles of global climate change and land-use change on bird migration. Elsewhere, one sees the effects of these factors, but in the Himalayas, knowledge regarding potential impacts is still in its infancy. Conservationists and governments urgently need clear baseline information to assess this, and our chapter does exactly that.

In addition, we also believe that counts at migration watch-sites alone are insufficient for monitoring and understanding raptor migration in the region and that the new investigative technology of satellite tracking will play an essential role in furthering knowledge in the field. The implementation of new systems employing GPS-logging, solar-powered 'mini' units on small as well as large birds of prey, coupled with cellular-telephone-based tracking units, will play a major role in helping raptor migration scientists better understand the geographical and temporal nuances in migration behaviour, and thereby help interpret watch-site count data more accurately. This is especially likely if new efforts follow the lead offered by recent studies of the physiological costs of migratory flights across the Himalayas by researchers studying the movements of Bar-headed Geese (*Anser indicus*) travelling through the region (Hawkes *et al.*, 2012, Chapter 16; Bishop *et al.*, 2015). Specific questions that need to be addressed include the extent to which raptors in general are capable of high-altitude flight such as that exhibited by raptors such as Rüppell's *Gyps rueppellii* and Griffon Vultures (Laybourne, 1974; Bögel, 1990).

Simply put, as we move forward, it is essential that we work together with volunteer raptor-watchers and the greater conservation and scientific communities, and that we use adaptive data management, data sharing and effective collaborations in our efforts (Dodge *et al.*, 2014).

References

Ali, S. (1977). *Field Guide to the Birds of the Eastern Himalayas*. London: Oxford University Press.

Ali, S. & Ripley, S.D. (1978). *Handbook of the Birds of India and Pakistan*. Vol. **1**. Bombay: Oxford University Press.

Anderson, R.C. (2009). Do dragonflies migrate across the western Indian Ocean? *Journal of Tropical Ecology*, **25**, 347–358.

Batbayar, N., Reading, R., Kenny, D., Natsagdorj, T. & Won Kee, P. (2008). Migration and movement patterns of Cinereous Vultures in Mongolia. *Falco*, **32**(2), 5–7.

Berthold, P. (2001). *Bird Migration: A General Survey*. Oxford: Oxford University Press.

Bildstein, K.L. (2006). *Migrating Raptors of the World: Their Ecology and Conservation*. New York: Cornell University Press.

Bildstein, K.L., Smith, J.P. & Ruelas Inzunza, E. (2008). The future of raptor-migration monitoring. In K.L. Bildstein, J.P. Smith, E. Ruelas Inzunza & R.R. Veit, eds., *State of North America's Birds of Prey*. Cambridge, MA, and Washington, DC: Nuttall Ornithological Club & American Ornithologists' Union, pp. 435–446.

Bijlsma, R.G. (1991). Migration of raptors and Demoiselle Cranes over central Nepal. In R.D. Chancellor & B.U. Meyburg, eds., *Birds of Prey Bulletin*. Berlin, London and Paris: The World Working Group on Birds of Prey and Owls, pp. 73–80.

BirdLife International, 2015, *IUCN Red List for Birds. Available from:* www.birdlife.org. [6 March 2015].

Bishop, C.M., Spivey, R.J., Hawkes, L.A., *et al*. (2015). The roller coaster flight strategy of Bar-headed Geese conserves energy during Himalayan migrations. *Science*, **347**, 250–254.

Bögel, R. (1990). Measuring flight altitudes of Griffon Vultures by radio telemetry. *International Ornithological Congress*, **20**, 489–490.

Chettri, N., Rastogi, A. & Singh, O.P. (2006). Assessment of raptor migration and status along the Tsango-Brahmaputra corridor (India) by a local communities participatory survey. *Avocetta*, **30**, 61–68.

Chhangani, A.K. & Mohnot, S.M. (2008). Demography of migratory vultures in and around Jodhpur, India. *Vulture News*, **58**, 23–34.

Clarke, R., Prakash, V., Clark, W.S., Ramesh, N. & Scott, D. (1998). World record count of roosting harriers Circus in Blackbuck National, Velavadar, Gujarat, north-west India. *Forktail*, **14**, 70–71.

Clement, P. & Holman, D. (2001). Passage records of Amur Falcon Falco amurensis from SE Asia and southern Africa including first records from Ethiopia. *Bulletin British Ornithology Club*, **121**, 222–230.

Das, D., Cuthbert, R.J., Jakati, R.D. & Prakashm, V. (2011). Diclofenac is toxic to the Himalayan Vulture Gyps himalayensis. *Bird Conservation International*, **21**, 72–75.

DeCandido, R., Allen, D. & Bildstein, K.L. (2001). The migration of Steppe Eagle (Aquila nipalensis) and other raptors in central Nepal, autumn 1999. *Journal of Raptor Research*, **35**, 35–39.

DeCandido, R., Gurung, S., Subedi, T. & Allen, D. (2013). The east-west migration of Steppe Eagle *Aquila nipalensis* and other raptors in Nepal and India. *BirdingASIA*, **19**, 18–25.

Den Besten, J.W. (2004). Migration of Steppe Eagle Aquila nipalensis and other raptors along the Himalayas past Dharamsala, India, in autumn 2001 and spring 2002. *Forktail*, **20**, 9–13.

de Roder, F.E. (1989). The migration of raptors south of Annapurna, Nepal, autumn 1985. *Forktail*, **4**, 9–17.

Dixon, A., Batbayar, N. & Purev-Ochir, G. (2011). Autumn migration of an Amur Falcon Falco amurensis from Mongolia to the Indian Ocean tracked by satellite. *Forktail*, **27**, 86–89.

Dixon, A., Sokolov, A. & Sokolov, V. (2012). The subspecies and migration of breeding Peregrines in northern Eurasia. *Falco*, **39**, 4–9.

Dodge, S., Bohrer, G., Bildstein, K., *et al.* (2014). Environmental drivers of variability in the movement ecology of turkey vultures (Cathartes aura) in North and South America. *Philosophical Transactions of the Royal Society B*: 20130195.

Donald, C.H. (1905). A congregation of harriers. *Journal of the Bombay Natural History Society*, **16**, 504–505.

Donald, C.H. (1923). Migration of eagles. *Journal of the Bombay Natural History Society*, **29**, 1054–1055.

Eastham, C. (1998). Satellite tagging of Sakers in the Russian Altai. *Falco*, **12**, 10–12.

Eastham, C.P., Quinn, J.L. & Fox, N.C. (2000). Saker Falco cherrug and Peregrine Falco peregrinus falcons in Asia: determining migration routes and trapping pressure. In R.D. Chancellor & B.U. Meyburg, eds., *Raptors at Risk*. British Columbia: Hancock House, pp. 247–258.

Ellis, D.H., Moon, S.L. & Robinson, J.W. (2001). Annual movements of a Steppe Eagle (Aquila nipalensis) summering in Mongolia and wintering in Tibet. *Journal of the Bombay Natural History Society*, **98**, 335–340.

Farmer, C.J., Safi, K., Barber, D.R., Newton, I., Martell, M. & Bildstein, K.L. (2010). Efficacy of migration counts for monitoring continental populations of raptors: an example using the Osprey (Pandion haliaetus). *Auk*, **127**, 863–870.

Feijen, C., Feijen, H.R. & Shulten, G.G.M. (2005). Raptor migration in Bhutan: incidental observations. *BirdingASIA*, **3**, 61–62.

Ferguson-Lees, J. & Christie, D.A. (2001). *Raptors of the world*. Boston, MA: Houghton Mifflin.

Fleming, R.L. (1983). An east-west *Aquila* eagle migration in the Himalayas. *Journal of Bombay Natural History Society*, **80**, 58–62.

Galligan, T.H., Amano, T., Prakash, V.M., *et al.* (2014). Have population declines in Egyptian Vulture and Red-Headed Vulture in India slowed since the 2006 ban on veterinary diclofenac? *Bird Conservation International*, **24**, 272–281.

Gill, F. & Donsker, D. (2015). *IOC World Bird List (v. 5.1)*. Available from: www.worldbirdnames.org. [27 March 2015].

Grimmett, R., Inskipp, C. & Inskipp, T. (1999). *Birds of India, Pakistan, Nepal, Bangladesh, Bhutan, Sri Lanka, and the Maldives*. Princeton, NJ: Princeton University Press.

Grimmett, R., Inskipp, C. & Inskipp, T. (2000). *Birds of Nepal*. Princeton, NJ: Princeton University Press.

Gurung, S.B., Gurung, S., Gurung, S. & McCarty, K. (2004). Autumn 2003 raptor migration in Nepal. *International Hawkwatcher*, **9**, 12–15.

Hawkes, L.A., Balachandran, S., Batbayar, N., *et al.* (2012). The paradox of extreme high-altitude migration in bar-headed geese *Anser indicus*. *Proceedings of the Royal Society B*: 2012.2114.

Inskipp, C. & Inskipp, T. (1985). *A Guide to the Birds of Nepal*. 2nd ed. London: Christopher Helm.

Inskipp, C., Inskipp, T. & Grimmett, R. (1999). *Birds of Bhutan*. London: Christopher Helm.

Kenny, D., Batbayar, N., Tsolmonjav, P., Jo Willis, M., Azua, J. & Reading, R. (2008). Dispersal of Eurasian Black Vulture Aegypius monachus fledglings from the Ikh Nart Nature Reserve, Mongolia. *Vulture News*, **59**, 13–19.

Kerlinger, P. (1989). *Flight Strategies of Migrating Hawks*. Chicago: University of Chicago Press.

Koning, F.J. (1976). Notes on the birds of prey in the Indus Valley. *Journal of the Bombay Natural History Society*, **73**, 448–455.

Laybourne, R.C. (1974). Collision between a vulture and an aircraft at an altitude of 37,000 feet. *Wilson Bulletin*, **86**, 461–462.

Meyburg, B.U., Meyburg, C. & Paillat, P. (2012). Steppe Eagle migration strategies – revealed by satellite telemetry. *British Birds*, **105**, 506–519.

Moreau, R.E. (1972). *The Palaearctic-African Bird Migration Systems*. London: Academic Press.

Naoroji, R. (2006). *Birds of Prey of the Indian Subcontinent*. London: Christopher Helm.

Newton, I. (2008). *The Migration Ecology of Birds*. London: Academic Press.

Ogada, D.L., Keesing, F. & Virani, M.Z. (2012). Dropping dead: causes and consequences of vulture population declines worldwide. *Annals of the New York Academy of Sciences*, **1249**, 57–71.

Pande, S.A., Deshpande, P., Mahabal, A.S. & Sharma, R.M. (2013). Distribution of the Steppe Eagle in the Indian subcontinent: review of records from 1882 to 2013 AD. *Raptors Conservation*, **27**, 180–190.

Pande, S., Pawashe, A., Sant, N., Mahabal, A. & Dahanukar, N. (2010). Metropolitan garbage dumps: possible winter migratory raptor monitoring stations in peninsular India. *Journal of Threatened Taxa*, **2**, 1214–1218.

Potapov, E., Fox, N.C., Sumya, D. & Gombobaatar, B. (2002). Migration studies of the Saker Falcon. *Falco*, **19**, 3–4.

Riesing, M.J., Kruckenhauser, L., Gamauf, A. & Haring, E. (2003). Molecular phylogeny of the genus Buteo (Aves: Accipitridae) based on mitochondrial marker sequences. *Molecular Phylogenetics and Evolution*, **27**, 328–342.

Roberts, T.J. (1991). *The Birds of Pakistan*. Karachi: Oxford University Press.

Robson, C.R. (1986). Recent observations of birds in Xizang and Quinghai provinces, China. *Forktail*, **2**, 67–82.

Scully, L. (1881). A contribution to the ornithology of Gilgit. *Ibis*, **23**, 415–453.

Sharma, P. & Sundar, K.S.G. (2009). Counts of Steppe Eagles *Aquila nipalensis* at a carcass dump in Jorbeer, Rajasthan, India. *Forktail*, **25**, 161–163.

Singh, G. (1961). The eastern steppe eagle [Aquila nipalensis (Hodgson)] on the south col of Everest. *Journal of the Bombay Natural History Society*, **58**, 270.

Spaar, R. & Bruderer, B. (1996). Soaring migration of Steppe Eagles Aquila nipalensis in southern Israel: flight behavior under various wind and thermal conditions. *Journal of Avian Biology*, **27**, 289–301.

Subedi, T. (2015). *East to west migration of Steppe Eagle Aquila nipalensis and other raptors in Nepal: abundance, timing and age class determination*. Available from: www.nationalbirdsofpreytrust.net/reports/Subedi.pdf [25 April 2015].

Subedi, T. & DeCandido, R. (2013). *Raptor migration summary for autumn 2012, central Nepal*. Available from: www.dvoc.org/MeetingsPrograms/MeetingsPrograms2014/Resources/2012_FinalReport_Nepal.pdf. [25 April 2015].

Subedi, T. & DeCandido, R. (2014). *Raptor migration summary for autumn 2013, central Nepal*. Available from: www.birdsofnepal.org/2013.FinalReport.Nepal.pdf [25 April 2015].

Terrauble, J., Mougeot, F., Cornulier, T., Verma, A., Gavrilov, A. & Arroyo, B. (2012). Broad wintering range and intercontinental migratory divide within a core population of the near-threatened pallid harrier. *Diversity and Distribution*, **18**, 401–409.

Thiollay, J. M. (1979). La migation des grues a travers l'Himalaya et la prédation par les aigles royaux. *Alauda*, **47**, 83–92.

Verma, A. (2007). *Harriers in India: A Field Guide*. Dehradun: Wildlife Institute of India.

Verma, A. & Prakash, V. (2007).Winter roost habitat use by Eurasian Marsh Harriers Circus aeruginosus in and around Keoladeo National Park, Bharatpur, Rajasthan, India. *Forktail*, **23**, 17–21.

Zalles, J.I. & Bildstein, K.L. (2000). *Raptor Watch: A Global Directory of Raptor Migration Sites*. Cambridge and Kempton, PA: BirdLife International.

7 Steppe Eagle Migration from Mongolia to India

Nyambayar Batbayar and Hansoo Lee

Background

Steppe Eagle *Aquila nipalensis* is one of the complete long-distance migratory large eagles in Eurasia. It breeds in southwestern Russia, across Kazakhstan and Kyrgyzstan and into Mongolia and northeast China. It mainly breeds in steppe, forest steppe and semi-desert areas and is recorded nesting up to 2300 m above sea level (a.s.l.) in mountainous areas (Del Hoyo *et al.*, 1994; Ferguson-Lees & Christie, 2001). A nest of a Steppe Eagle pair was recorded at 2600 m a.s.l. in central Mongolia (pers. obs.). Steppe Eagles winter in eastern and southern Africa, in northern India and on the Tibetan Plateau in China (Ferguson-Lees & Christie, 2001). The species is declining globally due to habitat destruction, persecution, collisions with power lines and non-targeted poisoning, and was recently classified as Endangered under IUCN's Threatened Species criteria (BirdLife International 2016).

Until recently, understanding of the migrations of Steppe Eagles was limited to bird-banding projects and visual observations made at several prominent raptor migration watch-sites in the Middle East and Asia. Thousands of man-hours were devoted to monitor Steppe Eagles passing through those sites every spring and autumn (Leshem, 1985; Christensen & Sorensen, 1989; Zalles & Bildstein, 2000). However, tracking Steppe Eagles year round to document final destinations and stopover locations, and to learn about migration timing, route and behaviour and so forth was made possible by the advancement of satellite and cellular-tracking technologies in wildlife research.

Two races of Steppe Eagles are considered to intergrade around the west of the Russian Altai: western – *A. n. orientalis* and eastern – *A. n. nipalensis* (Ferguson-Lees & Christie, 2001). Migrations of western Steppe Eagles that breed in southern Russia and Kazakhstan have been documented through satellite-tracking studies (Meyburg *et al.*, 2003, Meyburg *et al.*, 2012, Javed *et al.*, 2014). All these birds wintered in the Arabian Peninsula and Africa.

Much needs to be learnt about the complete annual cycle and migration ecology and behaviour of Steppe Eagles from the eastern part of the range. It has long been known that eastern Steppe Eagles migrate across the Himalayas. Observations of Steppe Eagles passing through raptor migration watch-sites in central Nepal and India revealed that Steppe Eagles migrated through mountain passes in the Himalayas, arriving from the north, then continued moving in an east-to-west direction (Fleming Jr, 1983;

Figure 7.1. The juvenile Steppe captured on 3 August 2015. Photo: Nyambayar Batbayar.

Bijlsma, 1991; Welch & Welch, 1991; DeCandido *et al.*, 2001; Den Besten, 2004; DeCandido *et al.*, 2013; Juhant & Bildstein, Chapter 6.). Exact migration routes through the Himalayas and over the Tibetan Plateau were not, however, known before our observations.

Ellis *et al.* (2001) were the first to use satellite technology in East Asia to study migratory movements of eastern Steppe Eagles that breed in Mongolia. They tracked an adult female Steppe Eagle from southeastern Mongolia. This female eagle wintered in southeastern Tibet and returned to the same area in Mongolia where it was captured. Here, we present the latest results of our tracking study in Mongolia. For the first time we documented the full autumn migration of an eastern Steppe Eagle, including details of its crossing of the Himalayas.

Tracking Steppe Eagles on Their Migrations

A near fledging Steppe Eagle was captured by hand at the nest site in the Khurkh River Valley (elevation 1170 m a.s.l.) on 3 August 2015 (Figure 7.1). The sex of the bird was unknown and its weight was 3500 g. Prior to deploying the transmitter, we thoroughly checked the bird for stress and physical injury. We fitted a 50g solar-powered WT-300™ GPS Mobile transmitter using the backpack harness method (Figure 7.2). The transmitter was programmed to record one GPS location every two hours, collecting up to 12 GPS locations per day. The bird was returned to the

Figure 7.2. Fixing a satellite transmitter to a juvenile Steppe Eagle in Mongolia in 2014. Photo credit: In Kyu Kim.

nest after the transmitter had been fitted. The transmitter was manufactured by KoEco (Daejeon, South Korea). This type of transmitter has worldwide coverage using the Wide Band Code Division Multiple Access (WCDMA) system with global roaming, which makes it very suitable for tracking long-distance migrant birds. The transmitter communicated with nearby cellular towers once every day around 22:00 UTC time, and data were received through a web portal, www.wi-tracker.com. When these transmitters fail to communicate with cell towers or if they are in an area without cellular networks, they can store geographic coordinates acquired by GPS in on-board memory and send them during the next successful communication. The transmitter was programmed to record data on up to 200,000 GPS locations, registered time, flight altitude, instantaneous speed, heading direction, dilution of precision and voltage level. Date and time of GPS fixes were recorded in Universal Coordinated Time (UTC).

The GPS Mobile transmitter on our single Steppe Eagle transmitted a signal for 176 days after its deployment, until 26 January 2016. From this single bird, we obtained high-quality data on 2044 GPS locations, mostly at a rate of 12 fixes per day (75% of all successful fixes). Overall, 99% of the GPS fixes were of high quality.

We used the ESRI ArcMAP program (ESRI, 2011) and R programing language (R Core Team, 2013) to plot locations, measure distances, trace migration routes and handle the data. Calculation of distances between daily roosting stops and total distance tracked excluded local movements. We manually checked roost locations and land marks. We calculated mean, minimum and maximum hours spent at night roosts, and travel speeds. We used the middle value (the median) in the daily normal flight distance and speed to describe the flights. In order to reflect natural daytime hours at particular locations, we converted the recorded time to four local time zones adjusting the bird's proximity to local area, for example, Mongolia (UTC+8 hours), China (UTC+8 hours), Nepal (UTC+5:45 hours) and Indian Time Zone (UTC+5:30 hours). When the bird arrived in the southern Tibetan Plateau, the Nepal Time Zone was used to adjust local time.

Distance Travelled

The Steppe Eagle flew in a southwesterly direction across central Mongolia and continued flying through the west of the Turpan Depression and Tarim Basin, and across the eastern part of the Tibetan Plateau. The total distance tracked, excluding local movements, was 5278 km. Daily distance travelled varied during the migration and ranged from 40 km to 365 km per day. The median daily distance covered was 193 km.

Timing and Duration of Migration

Our eagle started its autumn migration on the morning of 8 October 2015 and arrived at its wintering area around midday on 15 December 2015. It took 65 days during autumn migration to reach the wintering site.

Stopovers

During the entire trip the bird made only two long stops for resting and refuelling (Table 7.1). The first stop was made 165 km south of the breeding site in an open area

Table 7.1. Major stopover locations, arrival and departure timing and stopover duration of an immature Steppe Eagle tracked from northeast Mongolia

Major stopover site(s)	Location	Coordinates	Elevation	Arrival date	Departure date	Stopover duration (days)
North of Ondorkhaan City	Central Mongolia	47°22'N, 110°40'E	1160	8-Oct-15	19-Oct-15	11
Tsangpo River Valley	Southern Tibet, China	30°00'N, 90°24'E	1030	5-Nov-15	2-Dec-15	27

with low, rolling hills located to the north of Ondorkhaan City, Khentii Province, Mongolia. Here the bird arrived on the same day that it left the breeding area and stayed for 11 days before continuing its autumn migration. The second stopover site was in the Tsangpo River Valley in Tibet. This site is located in the southern Tibetan Plateau about 85 km northwest of Lhasa and approximately 345 km north of the crossing point of the Great Himalaya Range. The bird arrived at this location in the afternoon of 5 November and stayed for 27 days until 2 December. This stopover site must have been important for refuelling before the main journey across the Great Himalaya Range.

Between the two major stopovers and the final destination, the bird made regular resting stops every evening. No night-time flight was observed. The bird often arrived at daily roost sites approximately one hour before sunset (on average 5:25 p.m., earliest 3:45 p.m. and latest 8:00 p.m.) and left in the morning when the air temperature became warmer (on average 9:10 a.m., earliest 7:45 a.m. and latest 11:00 a.m.). Overall, daily night-time rest lasted between 14 hours (42% of a day) and 18 hours (75%) with an average of 16 hours (65%). This pattern was kept fairly consistently throughout the migration. No large gaps or big changes in the

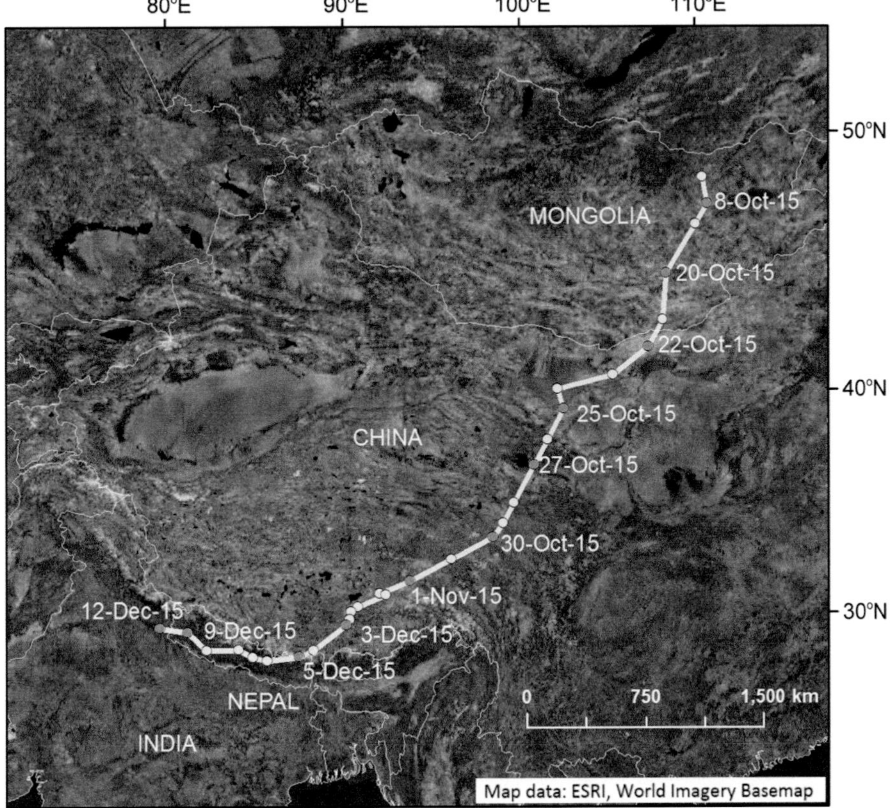

Figure 7.3. Migration route, stopover sites (red arrows) and daily roost locations (yellow circles) of an immature Steppe Eagle from northeast Mongolia.

number of hours rested and distance travelled were observed along the migration route or through the change of season (Figure 7.3). This particular Steppe Eagle made relatively short daily movements only twice when it was in southern Tibet, perhaps due to disruption by strong mountain winds or bad weather conditions, but we did not collect weather-related information.

Flight Altitude

During migration the Steppe Eagle mostly flew up to 200 m (median = 215 m) relative to the ground and occasionally reached more than 1600 m (Figure 7.5a). The highest altitude recorded was 7200 m a.s.l.. The altitude of the roost site on the day before the crossing of the Great Himalaya range was 4630 m a.s.l. (about 45 km from the pass), and, after the crossing, the eagle used a roost at 4530 m.

Through this tracking we discovered one mountain pass where Steppe Eagles cross the Himalayas. The eagle started climbing in the morning of 5 December after 9:45 a.m. Note that the precise timing of departure from the roost (or arrival at it) was not possible to determine because the transmitter was programmed to obtain a GPS location every two hours. So the precise timing could have been sometime after 9:45 a.m. The bird made its first stop at around 3:45 p.m., just after crossing the highest part of the route (at 7200 m a.s.l.) on 5 December 2015. This roost location was approximately 62 km southeast of Mount Everest (Figure 7.4).

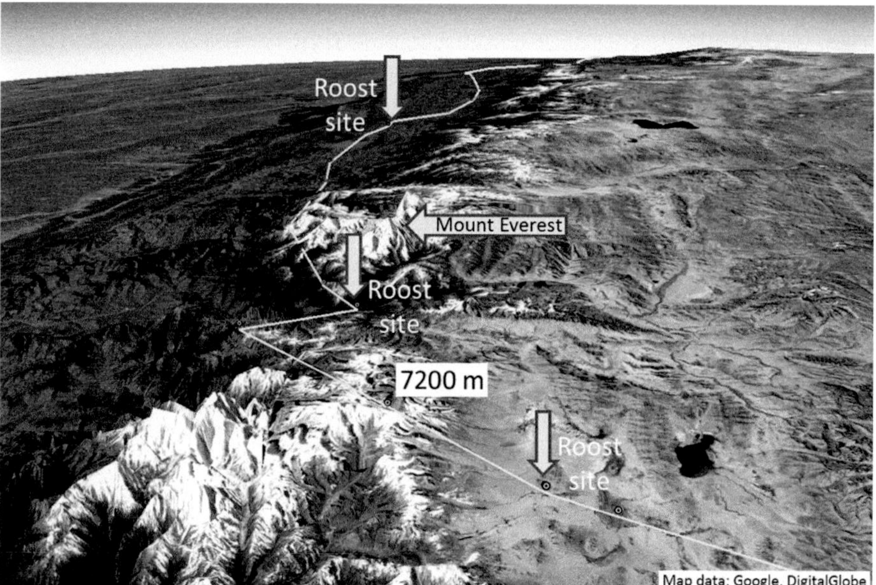

Figure 7.4. Migration path and roost locations of an immature Steppe Eagle before, during and after the crossing of the Himalayan mountain ridge. (A black-and-white version of this figure will appear in some formats. For the colour version, please refer to the plate section.)

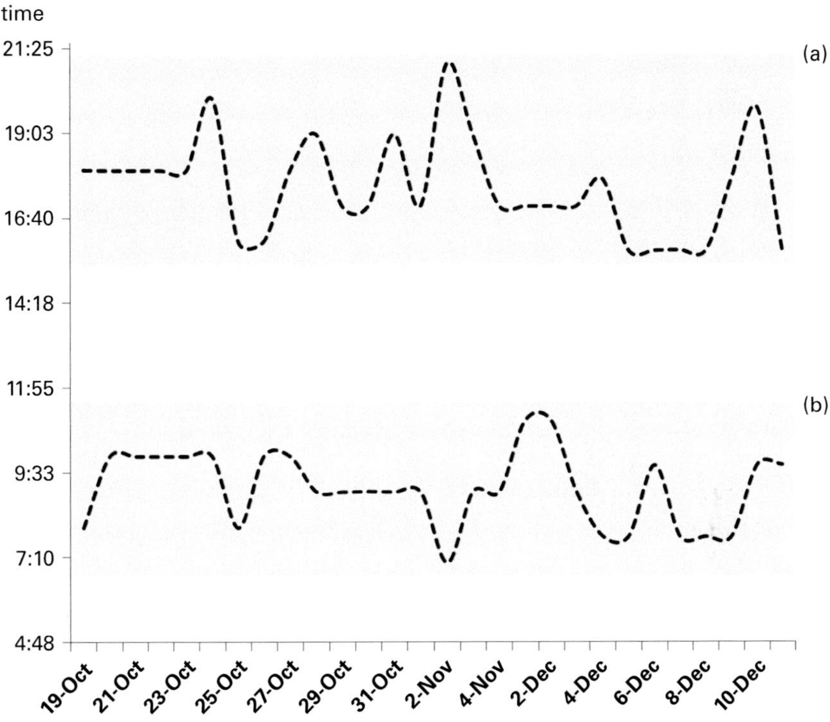

Figure 7.5. Timing of arrival at (a) and departure from (b) the daily roost sites by an immature Steppe Eagle during autumn migration.

The crossing point of the Great Himalaya range was between Omi Kangri Peak (6840 m) and Jongsong Peak (7460 m). Four GPS locations with a two-hour interval were located from the start of the climb to the landing after the crossing, covering about 117 km distance in total. Unfortunately, present-day techniques do not yet allow measurement of detailed flight trajectories during the period when the bird flew through the pass over the Himalayas.

Flight Speed

The average flight speed during migration was 44 km.h^{-1} (median was 41 km.h^{-1}) and the maximum speed was 119 km.h^{-1} (Figure 7.6b). The flight speed ranged from about 14 km.h^{-1} to about 28 km.h^{-1} between points during the Himalaya crossing (with an average speed of about 19 km.h^{-1}).

Wintering Site

After the initial arrival in the wintering area, the bird moved twice over one month until it settled down to winter in a mountainous area near Nainital, Uttarakhand State, in India.

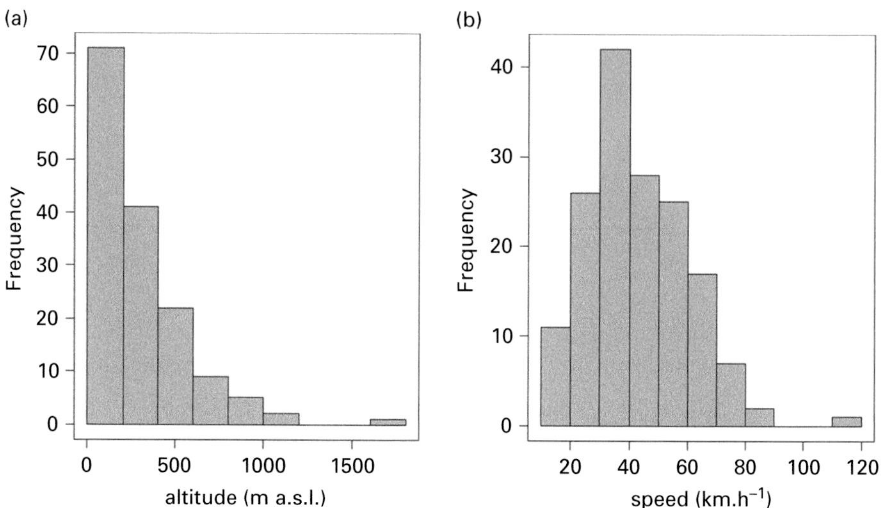

Figure 7.6. Steppe Eagle flight altitude relative to ground (a) and flight speed (b) during migration

At both locations, foraging range was limited to particular mountain valleys. Local movement at the wintering ground was not complex, because every day the bird was foraging between gullies in the same valley and returned to the same roost site, or sites very close to it. Its regular roost site was usually in a tree located at the bottom of the valley.

Lessons Learnt

The exact location of crossing points Steppe Eagles use to cross the Himalayas had not previously been described in detail. The area where our bird crossed the Himalayas was the same as the area where Bar-headed Geese crossed (Takekawa *et al.*, 2009; Hawkes *et al.*, 2011; Bishop *et al.*, 2014) (Takekawa *et al.*, Chapter 1; Hawkes *et al.*, Chapter 16).

Fleming (1983) first suggested that Steppe Eagles may be crossing the Himalayas on a broad front from Tibet. The flight path of the Steppe Eagle that we tracked did not seem to follow any particular mountain valley during the southward crossing. Perhaps Steppe Eagles do not necessarily rely on specific narrow valleys to cross the Himalayas, which could be a way to avoid turbulent winds in mountain valleys at high altitudes (see Ohlmann, Chapter 14; Heise, Chapter 15).

In our study we were the first to record the flight altitude of Steppe Eagles when crossing the Himalayas. Previously, several papers presented flight altitudes of Bar-headed Geese during trans-Himalayan migration. The tolerance of high-altitude and low-oxygen conditions by Bar-headed Geese is well documented (Scott *et al.*, 2015). It was shown that Bar-headed Geese reached a maximum altitude of 7290 m a.s.l. during their migration (Hawkes *et al.*, 2011; Hawkes *et al.*, 2012;

Bishop *et al.*, 2015; Hawkes *et al.*, Chapter 16). Although the Steppe Eagle that we followed did not reach that exact altitude, it came very close. Thus, Steppe Eagles must also experience similar challenges (low oxygen levels, lower air temperature etc.) when they are at high altitude during the crossing of the Himalayas.

The distance and duration of migrations of Steppe Eagles seem to vary considerably among geographical populations across the species global range. An adult male Steppe Eagle that migrated from Kazakhstan (west of Mongolia) to wintering grounds in Ethiopia and Sudan spent 53 days on autumn migration to cover a distance of 7315 km (Meyburg *et al.*, 2012). An adult female eagle that migrated from southeastern Mongolia spent 20 days on autumn migration, covering a distance of 1800 km to reach her wintering grounds in the southeastern Tibetan Plateau (Ellis *et al.*, 2001). Our bird, which was newly fledged when it set off, spent 65 days on autumn migration, covering a distance of 5278 km to arrive in northern India for the winter. These three studies showed markedly different pictures of the migratory behaviour of Steppe Eagles, and birds from different geographical regions may use different migration strategies (cf. Dixon *et al.*, Chapter 8). We do not, however, have enough tracking data to describe migration strategies for eastern Steppe Eagles in as much detail as the well-documented examples for western Steppe Eagles (Meyburg *et al.*, 2012).

Many previous observations made at raptor migration watch-sites in India and Nepal reported east-to-west flights of Steppe Eagles along the southern foothills of the Himalayas after they cross the main range (Juhant & Bildstein, Chapter 6). Most of those migration counts were also made in October and November. De Roder (1989) reported that peak Steppe Eagle passage was in early November in 1985, DeCandido *et al.* (2001) described peak counts at the end of October and early November, but Den Besten (2004) reported that the peak was in the third week of November in 2001. The timing of the juvenile Steppe Eagle that we tracked passing through these sites was slightly later, at the end of the first week of December. It is possible that our single bird was a late migrant.

The foraging range of our Steppe Eagle at its wintering ground was fairly restricted to particular valleys. Ellis *et al.* (2001) reported that Steppe Eagles wintering in southern Tibet also have a narrow foraging range in winter. We do not know what constitutes their main source of food at their wintering ground in northern India. Nevertheless, the foraging range of Steppe Eagles wintering in northern India and southern Tibet seem to be different from that in Africa, where they have a much larger winter range, perhaps due to differing types and availability of their main prey species (Meyburg *et al.*, 2012). Steppe Eagles are large *Aquila* eagles and they prey mostly on small mammals, up to hare and marmot size, and carrion as well (Forsman, 1998). It seems that the Himalayan region holds enough of these prey species to regularly support thousands of migrating and wintering Steppe Eagles.

Because tracking data from our single bird factually showed that the Steppe Eagles that breed in Mongolia do cross the Himalayas, it may well be deduced that some Steppe Eagles breeding in eastern Russia and China do the same. But it is important to gain an understanding of what proportion of the population of this Globally Threatened eagle species in East Asia winters in southern Tibet and the Indian subcontinent. More satellite-tracking studies with bigger sample sizes would be very useful in this regard. A recent study has shown that the non-steroidal anti-inflammatory veterinary drug diclofenac that has caused three species of *Gyps* vultures to come close to extinction may also be affecting Steppe Eagles in India (Sharma *et al.*, 2014). Thus it is important to identify such threats to Steppe Eagles along the flyway of this Endangered species.

Acknowledgements

We thank our Mongolian and Korean field assistants for their help in capturing the Steppe Eagle and deploying the transmitter. The Korean Institute of Environmental Ecology supported the funding for the transmitter.

References

Bijlsma, R.G. (1991). Migration of raptors and Demoiselle Cranes over central Nepal. *Birds of Prey Bulletin*, **4**, 73–80.

BirdLife International. (2016). Species factsheet: Aquila nipalensis. Downloaded from http://www.birdlife.org on 12/03/2016.

Bishop, C.M., Spivey, R.J., Hawkes, L.A., *et al.* (2015). The roller coaster flight strategy of Bar-headed Geese conserves energy during Himalayan migrations. *Science*, **347**, 250–254.

Christensen, S. & Sorensen, U. (1989). A review of the migration and wintering of Aquila Pomarina and Aquila Nipalensis Orientalis. *Raptors in the Modern World.* Berlin, World Working Group on Birds of Prey, 139–150.

De Roder, F.E. (1989). The migration of raptors south of Annapurna, Nepal, autumn 1985. *Forktail*, **4**, 9–17.

DeCandido, R., Allen, D. & Bildstein, K.L. (2001). The migration of Steppe Eagles (Aquila Nipalensis) and other raptors in central Nepal, autumn 1999. *Journal of Raptor Research*, **35**, 35–39.

DeCandido, R., Gurung, S., Subedi, T. & Allen, D. (2013). The east–west migration of Steppe Eagle Aquila nipalensis and other raptors in Nepal and India. *Birding Asia*, **19**, 18–25.

Del Hoyo, J., Elliott, A. & Christie, D. (1994). *Handbook of the Birds of the World, Vol. 2. New World Vultures to Guineafowl*, Barcelona, Lynx Edicions.

Den Besten, J. (2004). Migration of Steppe Eagles Aquila nipalensis and other raptors along the Himalayas past Dharamsala, India, in autumn 2001 and apring 2002. *Forktail*, **20**, 9–13.

Ellis, D.H., Moon, S.L. & Robinson, J.W. (2001). Annual movements of a Steppe Eagle (Aquila nipalensis) summering in Mongolia and wintering in Tibet. *Journal of the Bombay Natural History Society*, **98**, 335–340.

ESRI (2011). *Arcgis Desktop: Release 10*, Redlands, CA, Environmental Systems Research Institute.

Ferguson-Lees, J. & Christie, D. (2001). *Raptors of the World*, New York, Houghton Mifflin Company.

Fleming Jr, R.L. (1983). An east–west Aquila eagle migration in the Himalayas. *Journal of the Bombay Natural History Society*, **80**, 58–62.

Forsman, D. (1998). *The Raptors of Europe and the Middle East: A Handbook of Field Identification*, Carlon, T. & A.D. Poyser.

Hawkes, L.A., Balachandran, S., Batbayar, N., *et al.* (2011). The trans-Himalayan flights of Bar-headed Geese Anser indicus. *Proceedings of the National Academy of Sciences of the United States of America*, **108**, 9516–9519.

Javed, S., Khan, S., Nazeer, J., *et al.* (2014). Satellite tracking of a young Steppe Eagle from the United Arab Emirates during two spring and autumn migrations. *Ostrich*, **85**, 131–138.

Leshem, Y. (1985). Israel: an international axis of raptor migration. *ICBP Technical Publication*, **5**, 243–250.

Meyburg, B.-U., Meyburg, C. & Paillat, P. (2012). Steppe Eagle migration strategies – revealed by satellite telemetry. *British Birds*, **105**, 506.

Meyburg, B.-U., Paillat, P. & Meyburg, C. (2003). Migration routes of Steppe Eagles between Asia and Africa: a study by means of satellite telemetry. *The Condor*, **105**, 219–227.

R Core Team (2013). R: a language and environment for statistical computing. Vienna, Austria: R Foundation for Statistical Computing, www.R-Project.Org/.

Scott, G.R., Hawkes, L.A., Frappell, *et al.* (2015). Review: how Bar-Headed Geese fly over the Himalayas. Physiology, **30**, 107–115.

Sharma, A.K., Saini, M., Singh, S.D., *et al.* (2014). Diclofenac is toxic to the Steppe Eagle Aquila nipalensis: widening the diversity of raptors threatened by NSAID misuse in South Asia. *Bird Conservation International*, **24**, 282–286.

Takekawa, J.Y., Heath, S.R., Douglas, D.C., *et al.* (2009). Geographic variation in Bar-Headed Geese Anser indicus: connectivity of wintering areas and breeding grounds across a broad front. *Wildfowl*, **59**, 100–123.

Welch, G. & Welch, H. (1991). The autumn migration of the Steppe Eagle Aquila nipalensis. *Sandgrouse*, **13**, 24–33.

Zalles, J.I. & Bildstein, K.L. (2000). *Raptor Watch: A Global Directory of Raptor Migration Sites*, Birdlife International, Cambridge.

8 Peregrine Falcons Crossing the 'Roof of the World'

Andrew Dixon, Lutfor Rahman, Aleksandr Sokolov and Vasiliy Sokolov

Peregrines on the Central Asian Flyway

The polar regions exhibit extreme variation in environmental conditions associated with annual seasonal changes, particularly with regard to daylight and temperature. In the boreal summer, the Arctic experiences continuous daylight and relatively mild temperatures, whereas in winter the opposite is true. This seasonal periodicity in environmental conditions means that many birds inhabiting the Arctic are migratory, arriving when conditions are suitable in summer to breed and then subsequently departing south to avoid the challenges that the cold and darkness of winter bring for their foraging and survival.

On the Eurasian continent, the Arctic Circle stretches more than 7800 km from ca. 13°E on the coast of Norway to ca. 171°E in Russia on the Bering Strait, and includes the breeding distribution ranges of birds utilizing three major avian migratory flyways, that is, the African-Eurasian, Central Asian and East Asian–Australian Flyways. The Central Asian Flyway connects wintering areas in the Indian subcontinent and breeding regions in the Central Siberian Arctic, and although it includes no major sea or water crossings, it does encompass a series of potential ecological barriers for migratory birds (Figure 8.1; Irwin & Irwin, 2005). Ecological barriers are areas of unsuitable habitat for foraging or resting that birds must negotiate on their migratory journeys, either by traversing them rapidly or avoiding them via a detour (Newton, 2008). To reach the Indian subcontinent, an Arctic migrant on the Central Asian Flyway has to first cross the vast Central Siberian Plateau, which extends more than 3.5 million km^2 between the Yenisei and Lena and is dominated by the boreal forest (taiga). At the southern edge of the taiga, from west to east, there are a series of mountain ranges, including the Altai, the Sayan, the Khangai and the Khentii Mountains, while to the southwest of these South Siberian Mountains are the mountain ranges of the Tien Shan, stretching from Kyrgyzstan to Inner Mongolia. Beyond the South Siberian Mountains is a broad expanse of steppe and desert, the Gobi, extending from Dzungaria in northeast Kazakhstan and northwest China, through to southeast Mongolia. At the southern edge of the gravelly Gobi Desert lie the sandy expanses of the Taklamakan, Badain Jaran and Ordos Deserts of China. Finally, the connecting chain of the Kunlun, Altun and Qilian Mountains stands on the northern rim of the Qinghai-Tibetan Plateau. Also known as the 'Roof of the World', the Qinghai-Tibetan Plateau extends more than 2.5 million km^2 with an average altitude of 4500 m. To the south of the Qinghai-Tibetan Plateau, the Himalayan range, extending from the

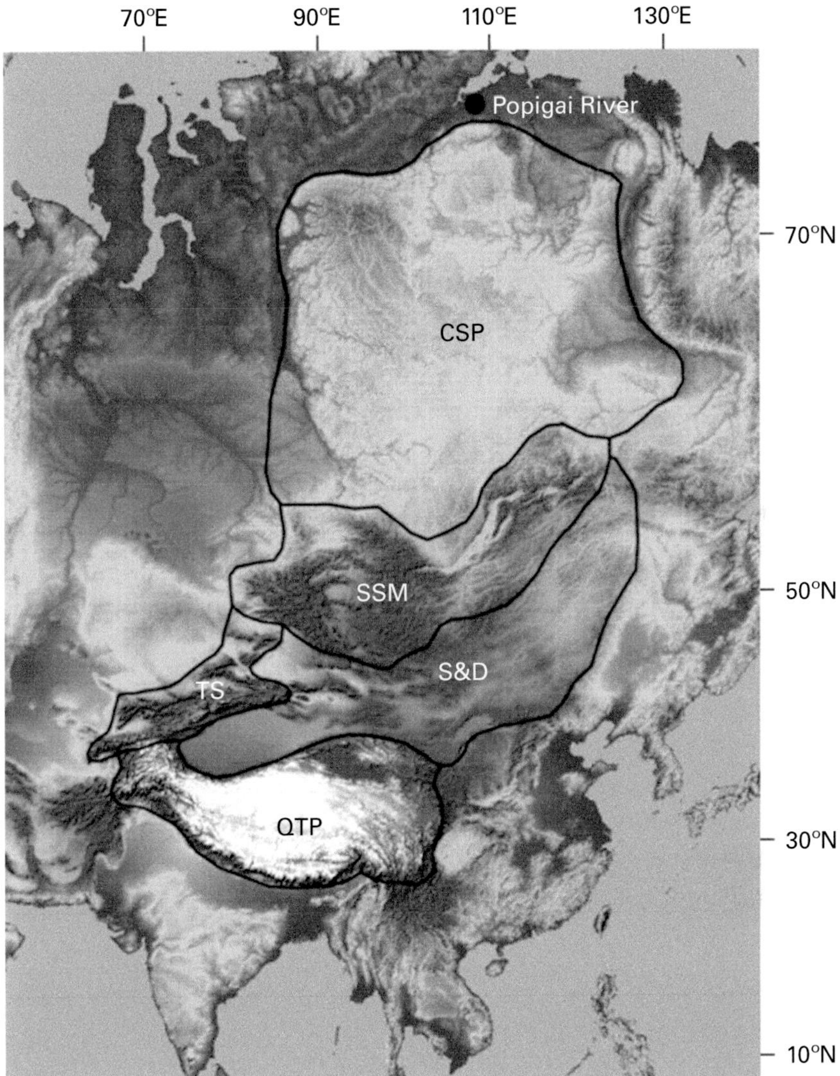

Figure 8.1. Ecological barriers for migratory Peregrines on the Central Asian Flyway. CSP: Central Siberian Plateau; SSM: South Siberian Mountains, TS: Tien Shan; S&D: Steppe and Desert; QTP: Qinghai-Tibetan Plateau and Himalayas. Popigai River where Peregrines were fitted with satellite transmitters is shown.

Hindu Kush and Karakoram in the west to the Eastern Himalayas in southeast Tibet, represents a formidable, final ecological barrier for any migrant heading for the fertile lowlands of the Indian subcontinent.

Peregrines are considered low-altitude migrants, mostly flying < 100 m above the ground (Kerlinger, 1989), with an average travel speed in flapping flight of ca. 50 km.h^{-1} (Cochran & Applegate, 1986). Data from radio transmitters indicate that during a typical day on migration, a Peregrine spends 17 h perched/roosting, 6 h in migratory flight, generally from mid-morning, and 1 h hunting (White *et al.*, 2002). Over the whole migration period, Peregrines probably spend around 20% of their time moving and the rest perched or hunting at stopover sites (Fuller *et al.*, 1998). Individuals are solitary on migration and the species exhibits a broad-front pattern of migration, although aggregations may occur along coastlines or along the edge of mountain chains, particularly in areas of prime habitat (White *et al.*, 2002).

In this chapter, we examine the extent to which the succession of potential ecological barriers in the Central Asian Flyway influences the migratory routes and behaviour of Peregrines, paying particular attention to the role played by the Qinghai-Tibetan Plateau and the Himalayas.

Peregrines Breeding in Arctic Eurasia

Peregrine Falcons have a very wide global distribution and as many as 19 subspecies have been described (del Hoyo *et al.* 1994). The species breeds throughout Arctic Eurasia, where three subspecies occur, reaching a northernmost latitude of 76.5°N on Novaya Zemlya. The nominate *F. p. peregrinus* breeds in Arctic Lapland, *F. p. calidus* across much of northern Russia and *F. p. harterti* in the Russian Far East. These subtly different forms overlap in their use of the major Eurasian migratory flyways, with *peregrinus* and *calidus* using the African-Eurasian Flyway, *calidus* using the Central Asian Flyway and

Figure 8.2. Female Peregrine deployed with a satellite-received transmitter, Popigai River (Photo: Andrew Dixon).

calidus and *harterti* using the East Asian–Australian Flyway (Dixon *et al.*, 2012). Whilst it has long been known that Peregrines from Arctic Eurasia spend the winter in the Indian subcontinent (Naoroji, 2006), their precise region of origin in the Siberian Arctic was not known before the advent of satellite telemetry (Dixon *et al.*, 2012).

In June 2011, in the north of Krasnoyarsk Krai, Russia, we trapped adult Peregrines at their nesting sites along a 280 km stretch of the Popigai River from its mouth near the Khatanga Gulf of the Laptev Sea. We caught 10 birds during incubation using noose-carpet traps, and fitted them with satellite-received transmitters (18 g Argos Solar PTT100, Microwave Telemetry Inc., Columbia, MD, USA) using a Teflon ribbon chest harness (weighing ca. 5 g; Figure 8.2). The average weight of nine females was 1130 g (range 1070–1260 g), and a single male weighed 730 g, so the combined mass of the transmitter and harness represented 2% and 3% of the body weight of females and males, respectively. The male (52474) and female (52523) of one breeding pair were both deployed with a satellite-received transmitter.

Autumn Migration

In 2011, the median distance travelled by the 10 Peregrines tracked via satellite was 6330 km (range 5490–7555 km), with these migrations taking 15–50 days (mean = 32 days) to complete. Individuals exhibited fidelity to their wintering ranges in each successive autumn migration. Over the period 2011–2014, the median departure date from the breeding range was 18 September (range 2–24 September; N = 25), and the median arrival date in the wintering range was 17 October (range 23 September–11 November; N = 24).

We plotted the first autumn migration of 10 adult Peregrines using dynamic Brownian bridge movement modelling (Kranstauber *et al.*, 2012), which provided a probable utilization distribution map showing space use that indicated stopover areas and regions of relatively slow migration (Figure 8.3). In comparison to other regions crossed during autumn migration, Peregrines travelled slowly across the Central Siberian Plateau, utilizing stopover areas in more open habitats such as marshes and river valleys. Another region indicating the use of stopover sites was the steppe and desert of southern Mongolia and northwest China.

Crossing the Central Siberian Plateau

On leaving their breeding region, all birds headed south. After they had moved ca. 2000 km (i.e. reached ca. 55°N), all had travelled on a bearing of 170°–225°, such that by the time they had traversed the Central Siberian Plateau, the passage Peregrines from the Popigai River were located mainly in the region between Novosibirsk and Lake Baikal. Peregrines completed their journey across the Central Siberian Plateau in around two weeks, though individuals varied in their behaviour at this stage of their autumn migration, with some making prolonged stops in open boggy areas and river valleys, taking more than

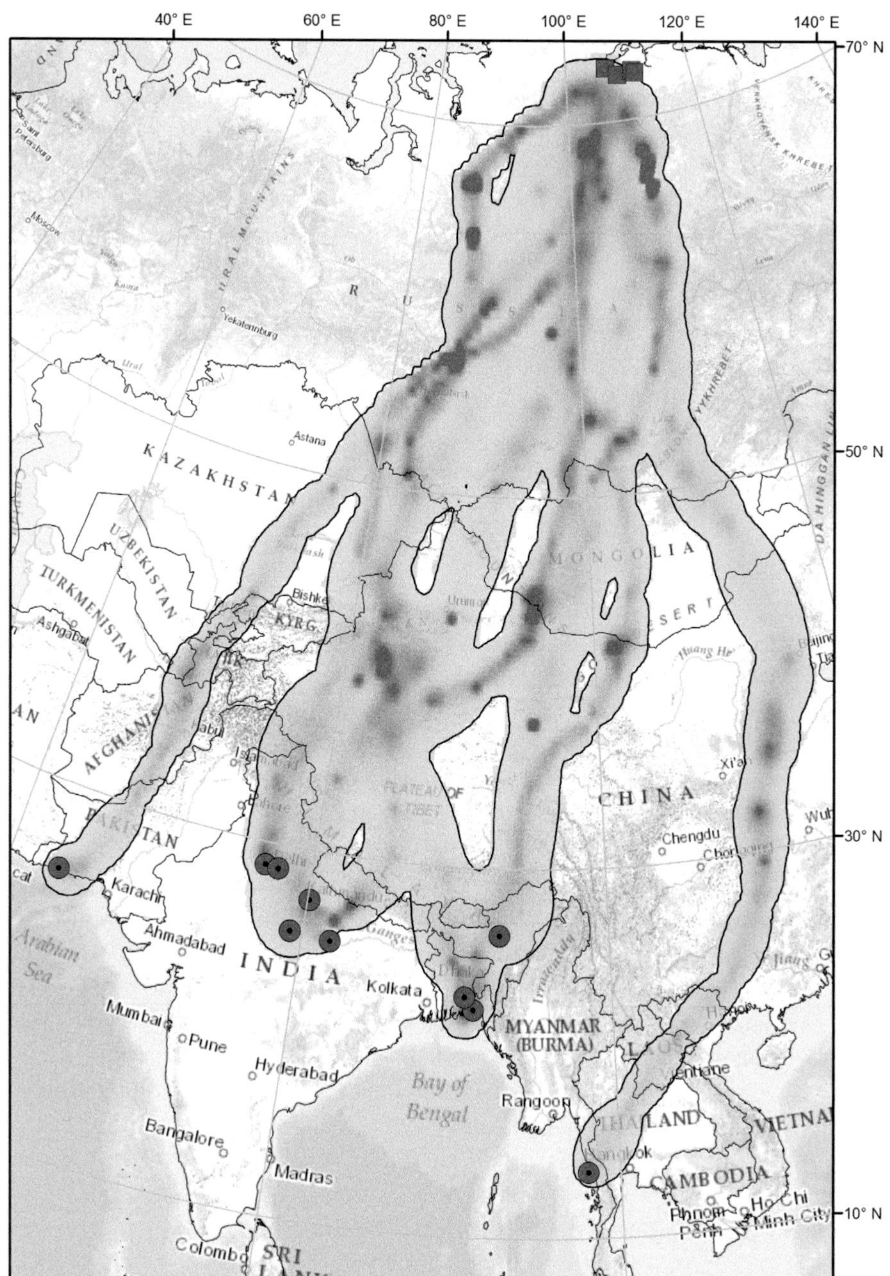

Figure 8.3. Autumn migration pathways of 10 Peregrines in 2011 from the Popigai River, Russia, plotted using dynamic Brownian bridge movement modelling based on best of day satellite locations. We calculated utilization distribution (UD) output grids for all birds at 50 km^2 resolution. We used a window length of approximately 7 days. The mean number of locations during migration period was 29 (range: 13–48). We weighted each individual UD by multiplying all pixel values by the total number of days for its migration event, and then summed the pixel values for all their weighted UDs and then re-scaled their cumulative pixel values to the sum of 1. The resulting UD represented the proportional amount of time each pixel was occupied across an individual's migration route. (A black-and-white version of this figure will appear in some formats. For the colour version, please refer to the plate section.)

a month to make the journey, whereas others completed it in just five days. Birds were not necessarily consistent over successive years in the speed of their passage across the Central Siberian Plateau. For example, one individual (52480) spent extended periods at wetlands in the catchments of the Olenyok and Lena Rivers during autumn migration in 2011 and 2012, taking, respectively, 28 and 34 days to cross the taiga zone, whereas the same bird, following a similar route, completed the journey in 14 days in autumn 2013. Another bird (52482) took 26 days to cross the Central Siberian Plateau in autumn 2011, stopping for extended periods along the Yenisei River, but only 12 days in 2012, despite taking a longer, more westerly route to reach the region of Novosibirsk. Examples of faster migration included one bird (53081) that took 9, 15, 14 and 5 days, respectively, to reach the Sayan Mountains during each autumn migration from 2011 to 2014. Other individuals showed greater consistency in their journey time. For example, one bird (52483) took 8, 7, 9 and 11 days, respectively, to reach the Mongolian border during four autumn migrations from 2011–2014. This consistently rapid traverse of the Central Siberian Plateau was achieved despite the bird varying its migration pathway by travelling both west and east of Lake Baikal in different years.

Two birds, travelling to the most easterly and westerly wintering ranges, after crossing the Central Siberian Plateau, flew along migration routes that bypassed the subsequent ecological barriers identified in this study. One bird (52523) took a route east of Lake Baikal, travelling through eastern Mongolia, the North China Plain south to Guangxi, China and then onwards through Viet Nam, Laos and Thailand to its wintering site on the Andaman coast of Myanmar (Figure 8.4a). The other bird (52624) crossed steppe and desert steppe travelling from Novosibirsk, Russia, through Kazakhstan, and south through Afghanistan to winter on the coast of Balochistan, Pakistan (Figure 8.4d).

The Peregrine Falcon is one of the most adaptable avian species, with its 20 or so generally recognized subspecies occupying every continent on earth except Antarctica (White *et al.*, 2013). Yet, despite this ubiquity, there are large areas of Eurasia where Peregrines do not breed but rather occur as passage migrants. This includes many of the regions outlined previously; in the Central Siberian Plateau, breeding Peregrines eschew the dark coniferous forest and occur mainly along major rivers, and they breed rather sparingly and locally in the South Siberian Mountains (Rogacheva, 1992), Gobi and Tien Shan, but are generally absent from the sand deserts and Qinghai-Tibetan Plateau of China (White *et al.*, 2013). The extent to which unsuitable habitats represent ecological barriers for migratory Peregrines is unknown. Satellite telemetry in North America indicated that the migratory routes Peregrines take were influenced by coastlines acting as ecological barriers, and that the Appalachian Mountains seemed to create a migratory divide between birds moving along the coast and those over the mid-continent (Fuller *et al.*, 1998; Worcester & Ydenberg, 2008).

Crossing the South Siberian Mountains

The Altai, Abakan, Sayan, Khangai and Khentii mountain ranges of southern Siberia and northern Mongolia extend from 48°N to 54°N, a north–south distance of more than

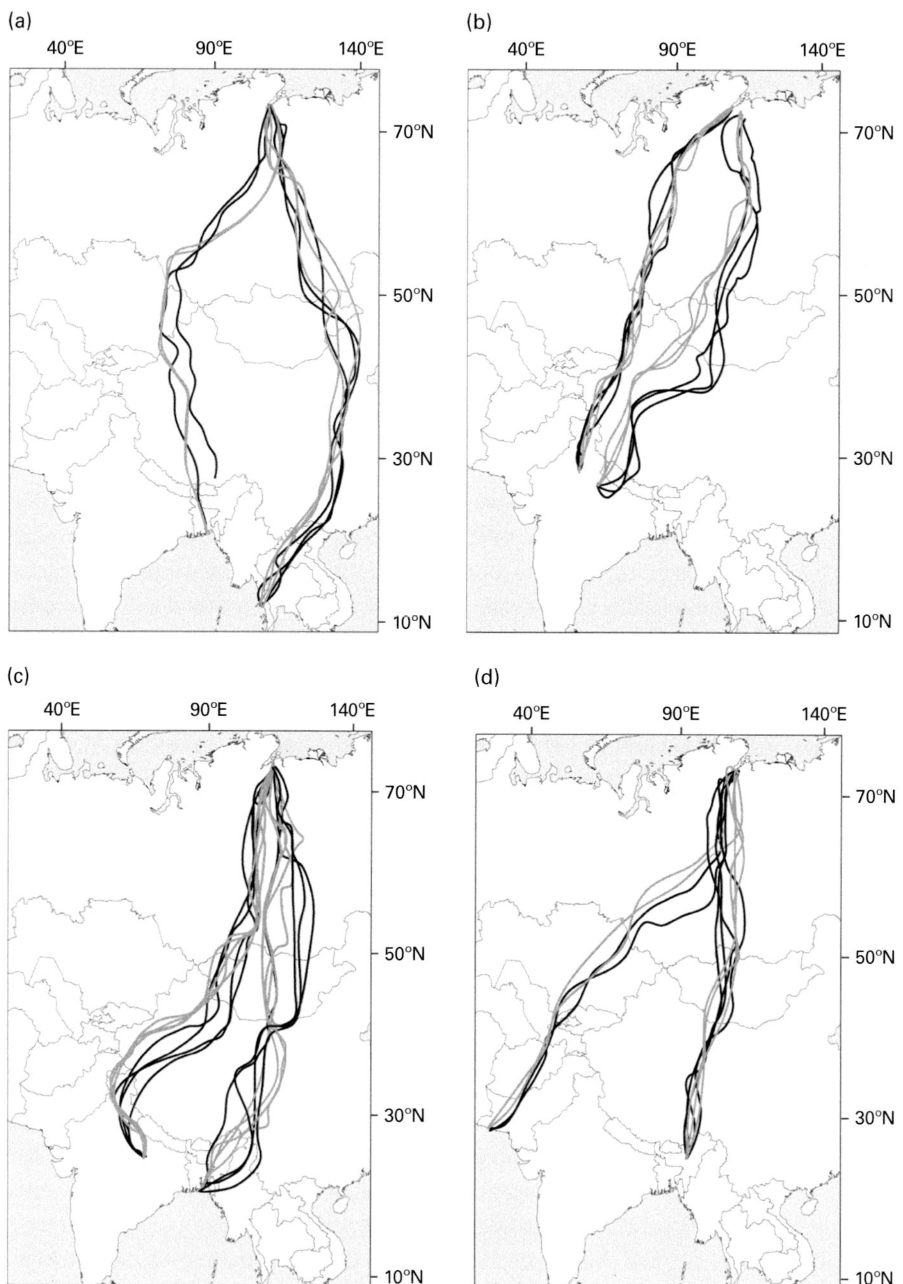

Figure 8.4 a–d. Migration pathways of eight Peregrines: black line = autumn migration; grey line = spring migration. (a) Migration pathway of a breeding pair, male (52474; left) showing one complete migration cycle and partial autumn migration where bird died crossing the Qinghai-Tibetan Plateau and female (52523; right) showing three complete migration cycles. (b) Three complete migration cycles of two females (52482; left and 52480; right). (c) Four complete migration cycles of two females (52475; left and 52483; right). (d) Two complete migration cycles of female (52624; left) and four complete migration cycles of female (53081; right).

650 km, which presents a potential barrier to Peregrine migration at the southern edge of the boreal forest. None of the migratory Peregrines crossed the Russian Altai and Abakan Mountains on autumn migration. Three birds took a route to the west, taking them down through eastern Kazakhstan, with five taking a route to the east through the Sayan, Khangai or Khentii Mountains. These mountain ranges were crossed rapidly, typically within ca. three days, and none of the birds made prolonged stopovers within the mountains.

Crossing the Deserts

For a southbound Peregrine that has traversed the South Siberian Mountains, the arid, rocky deserts of the Dzungarian Basin and Gobi, and the sands of the Taklamakan, Badain Jaran and Ordos Deserts form a band of inhospitable habitat some 700 to 800 km wide. One bird (53081) crossed the region rapidly in two to four days with little deviation over a distance of ca. 700 km during each of its autumn migrations from 2011–2014. However, this was the exception, as the other four birds made considerable deviations while crossing the deserts. One bird (52483; Figure 8.4c) consistently, over four successive autumn migrations, made a westward deviation at the Mongolia–China border in Omnogovi, to circumnavigate the sand dunes of the Badain Jaran Desert, each time passing to the northwest. Another (52475; Figure 8.4c), on crossing the eastern fringe of the Tien Shan, made a long westward deviation taking it around the sandy Taklamakan Desert over four successive migrations to reach the Karakoram Mountain range. A third bird (52480: Figure 8.4b), in three successive migrations from 2011 to 2013, also made large westward deviations on its flight path before reaching the Qilian Mountains in China, that resulted in it skirting along the southern edge of the Taklamakan Desert to reach a point on the west of the Kunlun Mountains. Further evidence that the Taklamakan represents a formidable barrier was that one bird (52474), after crossing the Tien Shan range, took three attempts before it was able to successfully make its way over the 450 km wide sandy expanse. Having arrived at the northern edge of the desert on the 4 October 2011, it remained in agricultural land before making an abortive attempt to cross the desert on 6 October, turning back after travelling some 250 km. The bird remained in fertile farmland at the northern edge of the desert until 10 October, when it made another attempt to cross, turning back after ca. 120 km and returning to farmland; the bird eventually crossed on its third attempt on 13 October 2011. Passage by Peregrines over the deserts was generally rapid, but some birds made occasional stopovers at wetland and agricultural oases en route.

Crossing the 'Roof of the World'

Eight of the 10 satellite-tagged Peregrines headed for wintering sites in India and Bangladesh, which involved journeys across the high-altitude Qinghai-Tibetan Plateau and Himalayan mountain range. The distance travelled across this geographical region by individuals ranged from 485 to 1755 km depending on where they crossed, with an average distance travelled of 1080 km over a period of 3.4 days (Table 8.2). Movement across the plateau and mountains was rapid and none of the Peregrines used diurnal stopover sites on this section of their autumn migration.

The specific migration pathways of eight individuals across the Qinghai-Tibetan Plateau and Himalayas during autumn migration in 2011 were as follows: one bird (52474), spent the night of 14 October in the Kunlun Mountains in the north of the Qinghai-Tibetan Plateau and crossed the Himalayas on 17 October in western Bhutan, to reach its wintering destination in the Sundarbans of Bangladesh. Another individual (52483) wintering in the Sundarbans crossed the Himalayas in southeast Tibet on 10 October, having started its journey across the Qinghai-Tibetan Plateau in the Qilian Mountains of northern Qinghai on 5 October. Another (53081) took a similar route from the Qilian Mountains on 2 October before crossing the Himalayas in southeast Tibet on 5 October to reach its wintering destination in Assam, India. A further bird (52475) spent the night of 3 November in the Kunlun Mountains, before crossing the western region of the Qinghai-Tibetan Plateau, roosting overnight mid-journey and eventually crossing the Himalayas in the region of Uttarakhand on 5 November to reach its destination near Allahabad, India. Two birds (52476 and 52482) wintering in northern India, both crossed the Himalayas in the Jammu and Kashmir region, travelling via the Karakoram Mountain range. A bird (52705) wintering further east in southern Uttar Pradesh took a more easterly route, heading from the Kunlun Mountains on 15 October, across the far western flank of the Qinghai-Tibetan Plateau to cross the Himalayas in the region of Uttarakhand the following day. A female (52580) spent the night of 2 November in the Kunlun Mountains and began its journey over the Qinghai-Tibetan Plateau the following morning, spending the next night in the central Changthang and eventually crossing the Himalayas in the Western region of Nepal on 5 November to reach its wintering range near Hardoi in Uttar Pradesh, India.

We tracked three birds over four successive autumn migrations (2011 to 2014), three over three migrations (2011 to 2013) and one over two full southbound migrations (2011 to 2012). Of those that we tracked for four successive years, two were extremely consistent in their passage over the Qinghai-Tibetan Plateau and Himalayas, with one (53081) entering the plateau on the western edge of the Qilian range in Qinghai, crossing rapidly and entering India over the Himalayas in southeast Tibet, to winter at a range in Assam (Figure 8.4d), and the other (52475) crossed the Himalaya in the Jammu and Kashmir region via the Karakoram Mountain range before heading southeast to its wintering range in Allahabad, India (Figure 8.4c). The third bird (52483) entered the Qinghai-Tibetan Plateau at the Qilian Mountains in the same place each year, but in 2012 to 2014 it travelled 300–500 km further east along the Qinghai–Sichuan border to cross the Himalayas in the east and reach its wintering range in the Sundarbans of Bangladesh via Myanmar (Figure 8.4c). One of the three birds

(52480) tracked over three migrations travelled a similar path in each year, starting its crossing from the Kunlun Mountains, passing through the Changthang on the Qinghai-Tibetan Plateau and crossing the Himalayas in the western region of Nepal to reach its wintering range in Uttar Pradesh (Figure 8.4b). Another (52482) followed a similar path each autumn, crossing the Qinghai-Tibetan Plateau and the Himalayas in the Jammu and Kashmir region via the Karakoram Mountain range before heading southeast to its wintering range in Haryana, India (Figure 8.4b).

Notably, one bird (52474) that made an extremely long (1100 km) journey across the Qinghai-Tibetan Plateau in just two days in 2011 struggled to make the same journey the following year and eventually succumbed in a river valley 75 km southeast of Lhasa, Tibet after travelling 730 km in 11 days, with the last 300 km of this journey taking a week to complete (Figure 8.4a).

Wintering Areas in South Asia

The eight Peregrines we tracked via satellite on the Central Asian Flyway occupied wintering ranges on the Indian subcontinent from 91°E, 22°N in the Sundarbans, Bangladesh to 77°E, 29°N in northern India.

The land cover in the wintering ranges was varied. Two birds with ranges ca. 60 km apart in the eastern Sundarbans, Bangladesh occupied estuarine mangroves. In Assam, northeastern India, a bird wintered in the agricultural landscape of the Brahmaputra Valley. A further three birds occupied agricultural land in Uttar Pradesh, India, with their wintering ranges being spaced 220–285 km apart, with another occupying agricultural land in the neighbouring state of Haryana. This latter bird established its wintering range just 45 km from another, which occupied a range in the urban centre of New Delhi. In Balochistan Province of Pakistan, a bird wintered at the rocky, desert coast of the Arabian Sea, and another coastal winter range on the Andaman Sea was established by a bird wintering on a wooded peninsula at the Dawei River mouth in the Tanintharyi region of Myanmar. All birds that were tracked during successive migrations returned to the same wintering ranges.

Spring Migration

Over the period 2012 to 2015, the median departure date of Peregrines from their wintering ranges was 27 April (range 19 April–19 May; N = 23) and the median arrival date in their breeding ranges was 26 May (range 8 May–28 June; N = 23).

The migration pathways used when returning to breeding ranges on the Popigai River were broadly similar to those used on autumn migration, especially when crossing the Central Siberian Plateau, although some individuals exhibited consistent differences in their autumn and spring migration pathways along sections of their routes (Figure 8.4a–d). One female (52480) consistently used a more westerly return pathway in spring, crossing the Himalayas ca. 400 km further west, in the Uttarakhand region of India, than in autumn. This pathway involved a rapid crossing of the Qinghai-Tibetan Plateau and Taklamakan Desert, resulting in a clockwise loop-migration pattern until the bird re-joined its autumn pathway at the headwaters of the River Lena in the Central Siberian Plateau (Figure 8.4b). A second female (52475) also made a more westerly crossing of the Himalayas on its return spring migration, travelling along the Himalayan foothills some 600 km further west before crossing in the Gilgit-Baltistan region of Pakistan. Its clockwise migratory loop took it along the northern edge of the Taklamakan to the Gobi of western Mongolia where it re-joined its autumn pathway (Figure 8.4c). A female (52483) that skirted the eastern edge of the Qinghai-Tibetan Plateau during autumn migration crossed the Himalayas via the Tsangpo Gorge above the Brahmaputra Valley on spring migration, an important area for migratory raptors (Chettri *et al.*, 2006). On crossing the eastern Himalayas the bird headed northeast to Sichuan, passing east of the higher alpine grasslands of the Qinghai-Tibetan Plateau (Figure 8.4c). This bird was remarkably consistent in its use of a wetland site near Ejin, Inner Mongolia, in the northwest of the Baidan Jaran Desert.

Spring migration from wintering sites in the lowlands of India and Bangladesh involves a rapid altitudinal climb to cross the Himalayas. Most birds travelled from the lowlands (< 300 m a.s.l.) south of the Himalayan foothills to the high-altitude mountains and plateau (> 5000 m a.s.l.) within two days. For example, one bird (52480) was at an altitude of ca. 135 m a.s.l. near Hardoi, Uttar Pradesh at 05:14 on 23 April and 27 h later it was 500 km away at an altitude ca. 5600 m in Tibet having crossed the Himalayas in the region of far western Nepal.

Negotiating Ecological Barriers

An obvious way to negotiate a barrier to migration is to travel around it. The two Peregrines that avoided all barriers after crossing the Central Siberian Plateau occupied winter ranges in coastal regions of Pakistan at 63°E and Myanmar at 98°E. In contrast to the birds using the Central Asian Flyway and wintering in the Indian subcontinent, the migration routes taken by these two birds were within the African-Eurasian and East Asian–Australian Flyways, respectively.

For those Peregrines on the Central Asian Flyway that crossed successive ecological barriers, the most notable deviations occurred prior to crossing the South Siberian Mountains, the deserts, the Qinghai-Tibetan Plateau and the Himalayas (Figure 8.4 a–d). Deviations made by some of the migratory Peregrines along their flight path to avoid crossing the sandy Taklamakan and Badain Jaran Deserts in China indicate that these habitats represent significant ecological barriers for migratory Peregrines. Similarly, it is notable that the Peregrines we tracked across the Qinghai-Tibetan Plateau and Himalayas made deviations that increased the total distance travelled to their wintering ranges but significantly reduced the distance travelled across the 'Roof of the World'. One bird (52474; Figure 8.4a) stopped moving (presumed dead) during migration across the plateau, with its migration speed slowing greatly in the week before it finally died.

In order to explore how geographic barriers can influence migration, we examined the speed of migration across specific barriers (Figure 8.1; the Central Siberian Plateau, the Tien Shan and South Siberian Mountains, the Steppe and Deserts and the Qinghai-Tibetan Plateau and Himalayas), using a generalized linear mixed-effect model (GLMM). The GLMM incorporated individual birds and year as a random variable, and we ranked birds to indicate intrinsic variation among individuals using a mean rank score for four biometric measures taken when the birds were first trapped: body mass, wing chord (from carpal joint to the tip of the longest primary), body girth at chest and full body length. Mean rank score, barrier type, distance travelled across each barrier and migration season (autumn or spring) were included as fixed variables. The most parsimonious models explaining variation in migration speed incorporated barrier type and distance travelled across the barrier (Table 8.1), with Peregrines migrating more quickly across the mountainous barriers of the Qinghai-Tibetan Plateau and the Himalayas and the South Siberian Mountains (Table 8.2).

Table 8.1. Model selection results for migration speed (km.h^{-1}) of Peregrines in relation to: the type of ecological barriers they encounter, distance crossed over the barriers, mean rank score of four biometric parameters of individual birds, and the migration season.

Models	K	AICc	ΔAICc	w_i
Barrier	7	670.8	0.00	0.39
Barrier + Distance	8	671.9	1.04	0.23
Barrier + Rank	8	673.3	2.52	0.11
Barrier + Migration	8	673.5	2.68	0.10
Barrier + Distance + Rank	9	674.3	3.47	0.07

K = the number of parameters, AICc = Akaike's Information Criterion corrected for small sample sizes, ΔAICc = difference in AICc values from the top ranked model, wi = Akaike weight, to show the weight of evidence in favour of a model relative to all the models tested.

Table 8.2. Mean distance, duration and speed of migration (95% Confidence Limits) of Peregrines across migratory barriers. Note that speed of migration refers to time taken to cross migratory barriers, not actual flight speed. N = the number of crossings measured during autumn and spring migration. The speed at which a bird crosses a geographical barrier can be achieved by flying faster or flying for longer periods of time (Adamík *et al.*, 2016).

Barrier	N	Distance (km)	Duration (days)	Speed (km.h^{-1})
Central Siberian Plateau	47	1980 (1887–2073)	10.5 (8.7–12.3)	9.4 (8.4–10.4)
Tien Shan & South Siberian Mountains	37	784 (681–887)	3.3 (2.7–3.9)	11.5 (9.9–13.1)
Steppe & Desert	38	1249 (1037–1461)	6.9 (5.5–8.3)	9.8 (7.6–12.0)
Qinghai-Tibetan Plateau & Himalaya	36	1081 (948–1214)	3.4 (2.6–4.2)	15.5 (13.5–17.5)

In this respect, Peregrines behaved similarly to Demoiselle Cranes *Grus virgo*, which also cross the Central Asian deserts and Himalayas relatively quickly (Kanai *et al.*, 2000, Higuchi & Minton, Chapter 3).

Peregrines are infrequently sighted at raptor migration watch points in the Himalayas (De Roder, 1989; DeCandido *et al.*, 2001), but such sites often tend to target aggregations of soaring species and do not necessarily reflect the movements of raptors migrating on a broad front (Bildstein, 2006; Juhant & Bildstein, Chapter 6). Furthermore, falcons which migrate by flapping flight are likely to cross mountain ranges at higher altitudes than raptors which depend on soaring flight (Newton, 2008; see also Groen & Prins, Chapter 17). Unlike GPS data, the Argos satellite location system does not enable us to accurately determine exactly where individual Peregrines crossed the Himalayas, nor were we able to determine the flight altitude of the birds we tracked. While some of the migratory Peregrines we tracked avoided crossing the 'Roof of the World', others did cross this high-altitude region, although they made efforts to minimize the time spent on this crossing by migrating rapidly and deviating to take shorter routes. The Bar-headed Goose, another trans-Himalayan migratory species, has physiological adaptations for high-altitude migration, and in comparison to other geese, has proportionally bigger lungs, improved oxygen supply to the muscles and heart, a denser capillary network and haemoglobin that carries more oxygen (e.g. Butler, 2010, Hawkes *et al.*, 2011; Hawkes *et al.*, chapter 16). It would be interesting to determine if similar adaptations are exhibited in Peregrine populations on the Central Asian Flyway that also undertake high-altitude migratory journeys.

References

Adamík, P., Emmenegger, T., Briedis, M., *et al.* (2016). Barrier crossing in small avian migrants: individual tracking reveals prolonged nocturnal flights into the day as a common migratory strategy. *Science Report* **6**, 21560.

Bildstein, K. (2006). *Migrating Raptors of the World: Their Ecology and Conservation*. Ithaca, NY: Cornell University Press.

Butler, P.J. (2010). High fliers: the physiology of bar-headed geese. *Comparative Biochemistry and Physiology Part A: Molecular & Integrative Physiology*, **156**, 325–329.

Chettri, N., Rastogi, A. & Singh, O.P. (2006). Assessment of raptor migration and status along the Tsangpo-Brahmaputra corridor (India) by a local communities participatory survey. *Avocetta*, **30**, 61–68.

Cochran, W.W. & Applegate, R.D. (1986). Speed of flapping flight of Merlins and Peregrine Falcons. *Condor*, **88**, 397–398.

DeCandido, R., Allen, D. & Bildstein, K.L. (2001). The migration of Steppe Eagles (Aquila nipalensis) and other raptors in central Nepal, autumn 1999. *Journal of Raptor Research*, **35**, 35–39.

del Hoyo, J., Elliott, A. & Sargatal, J. (1994). *Handbook of the Birds of the World,* vol. 2: *New World Vultures to Guinea Fowl*. Lynx Edicions, Barcelona, Spain.

DeRoder, F.E. (1989). The migration of raptors south of Annapurna, Nepal, autumn 1985. *Forktail*, **4**, 9–17.

Dixon, A., Sokolov, A. & Sokolov, V. (2012). The subspecies and migration of breeding Peregrines in northern Eurasia. *Falco*, **39**, 4–9.

Fuller, M.R., Seegarm W.S. & Schueck, L.S. (1998). Routes and travel ranges of migrating Peregrine Falcons Falco peregrinus and Swainson's Hawks Buteo swainsoni in the western hemisphere. *Journal of Avian Biology*, **29**, 433–440.

Hawkes, L.A., Balachandran, S., Batbayar, N., *et al.* (2011). The trans-Himalayan flights of Bar-headed Geese (Anser indicus). *Proceedings of the National Academy of Science*, **108**, 9516–9519.

Irwin, D.E. & Irwin, J.H. (2005). Siberian migratory divides: the role of seasonal migration in speciation. In R. Greenberg and P.P. Marra, eds., *Birds of Two Worlds: The Ecology and Evolution of Migration*. Baltimore, MD: John Hopkins University Press, pp. 27–40.

Kanai, Y., Minton, J., Nagendran, M., *et al.* (2000). Migration of Demoiselle Cranes in Asia based on satellite tracking and fieldwork. *Global Environment Research*, **2**, 143–153.

Kerlinger, P. (1989). *Flight Strategies of Migrating Hawks*. Chicago, IL: University of Chicago Press.

Kranstauber, B., Kays, R., LaPoint, S.D., Wikelski, M. & Safi, K. (2012). A dynamic Brownian bridge movement model to estimate utilization distributions for heterogeneous animal movement. *Journal of Animal Ecology*, **81**, 738–746.

Naoroji, R. (2006). *Birds of Prey of the Indian Subcontinent*. London: Christopher Helm.

Newton, I. (2008). *The Migration Ecology of Birds*. Oxford: Academic Press.

Rogacheva, H. (1992). *The Birds of Central Siberia*. Husum: Husum Druck-und Verlaggesellschaft.

White, C.M., Cade, T.J. & Enderson, J.H. (2013). *Peregrine Falcons of the World*. Barcelona: Lynx Edicions.

Worcester, R. & Ydenberg, R. (2008). Cross-continental patterns in the timing of southward Peregrine falcon migration in North America. *Journal of Raptor Research*, **42**, 13–19

Part II

Physiography of the Highest Barrier on Earth

9 Geological Origin and Evolution of the Himalayas

Michael Searle

Plate Tectonics Created the Himalayas

The Himalayan mountain ranges were formed as a result of the collision of the Indian tectonic plate with Asia approximately 50 million years ago. India was part of the great southern supercontinent called Gondwana which rifted apart from the northern supercontinent called Laurasia during the Permian approximately 250 million years ago. The intervening ocean, Tethys, was expanding by continental drift, the plates being pushed apart by seafloor spreading. Tethys reached its widest extent during the Jurassic and Cretaceous periods of the Late Mesozoic era. Approximately 130 million years ago, India rifted away from its neighbouring continental masses, Madagascar, southern Africa and Antarctica, and became an island surrounded by newly formed oceanic spreading ridges in the Indian Ocean. India moved at rapid plate tectonic rates, northwards across the Indian Ocean until 50 million years ago when the Indian continental plate first came into contact with Asia, closing the Tethys Ocean that lay in between. Prior to this collision, great thrust sheets of oceanic rocks (deep-sea sedimentary rocks, basalts, gabbros) were thrust onto the northern continental margin of India from the Tethys Ocean. Remnants of these oceanic rocks can be seen today along the Indus–Yarlung Tsangpo suture zone in Ladakh and south Tibet (Figure 9.1), the zone of collision between India and Asia and also in the high mountains of the Spontang area in Zanskar (Gansser, 1964; Searle, 2013).

Himalayan Geology

After the continental collision 50 million years ago, the northward motion of India slowed to about 5.5 $cm.y^{-1}$ but the Indian plate continued to penetrate northwards, indenting into Asia. The collision of India with Asia resulted in large-scale folding and thrusting, which effectively resulted in crustal shortening and thickening along the proto-Himalayan Mountains (Figure 9.1). The Asian plate margin in northern Ladakh and south Tibet was an Andean-type continental margin and therefore probably topographically elevated long before the Indian plate collision, and has remained topographically elevated since then. The Indian plate Himalaya was

Figure 9.1. The Tibetan Plateau (to the right) and the Gangetic Plains of India (to the left) and in between the snow-clad Himalayas as photographed by the NASA Space Shuttle (Photograph courtesy of NASA). (A black-and-white version of this figure will appear in some formats. For the colour version, please refer to the plate section.)

a low-lying continental margin, but started to shorten and uplift after the emplacement of ophiolites (slabs of oceanic rocks emplaced onto continental margins) during the Late Cretaceous (65 million years ago). Before the collision, the Tethys oceanic lithosphere was subducting northwards beneath the Asian continental margin. During this time before the Indian plate collision, the Gangdese range in Tibet and the Ladakh range in northern India formed an Andean-type margin dominated by granite batholiths and explosive volcanism. Both granite intrusion and andesitic volcanism ended about 50 million years ago when the Indian and Asian plates collided, ending the oceanic subduction beneath Asia.

The final marine incursion along the suture zone has been dated at 50.5 million years old from small marine foraminifera (*Nummulites* sp.) in the shallow marine limestones from Pakistan, Ladakh and south Tibet (Green *et al.*, 2008). The actual zone of plate collision, called the Indus–Yarlung Tsangpo suture zone, after the two

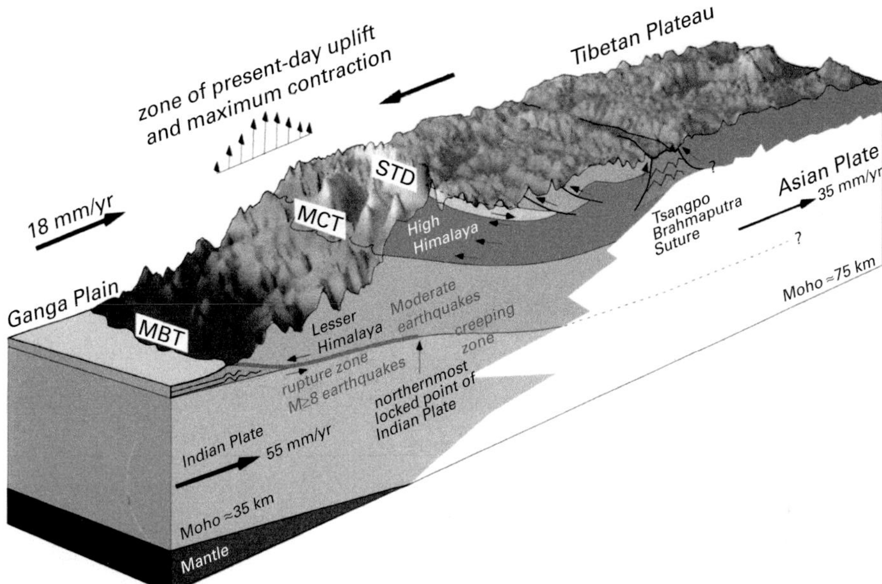

Figure 9.2. Block diagram of the Indian Plate dipping under the Asian Plate. MBT stands for 'Main Boundary Thrust', MCT for 'Main Central Thrust' and STD for 'South Tibetan Detachment'. (A black-and-white version of this figure will appear in some formats. For the colour version, please refer to the plate section.)

great rivers that flow along it, contains highly deformed remnant ophiolites and marine sedimentary rocks that were deposited in the Tethyan oceanic realm. These marine suture zone rocks are overlain by continental 'molasse-type' sediments (conglomerates, red sandstones, mudstones, etc.) that were deposited in lakes and braided river systems. The youngest subduction-related Andean-type granites along the Ladakh range are 47 million years old, dated by uranium-lead isotopes in zircon crystals. This dates the ending of oceanic lithospheric subduction beneath Tibet. Following the India–Asia collision, the Indian plate continued to break apart, the upper crustal levels crumpling and folding to form the Himalayas, the lower crustal levels underthrusting Tibet towards the north, 'jacking up' the Tibetan plateau progressively from south to north (Figures 9.2, 9.3).

The continental collision resulted in further crustal shortening and thickening causing uplift of the Himalayas. The northern ranges of the Himalayas in Zanskar, Lahul-Spiti and southernmost Tibet show spectacular folds in the Permian to Cretaceous limestones that were deposited along the northern continental margin of India. Pressure and temperature increases with depth during crustal thickening, resulting in the transformation of upper crustal sedimentary rocks (limestones, sandstones, shales etc.) to metamorphic rocks (gneisses, schists) and at the highest temperatures, by partial melting, resulting in firstly migmatites and eventually granites. Metamorphic rocks and granites have key index minerals that can reveal the temperature, pressure and depth of formation. These minerals include andalusite, kyanite and sillimanite (Searle *et al.*, 2010).

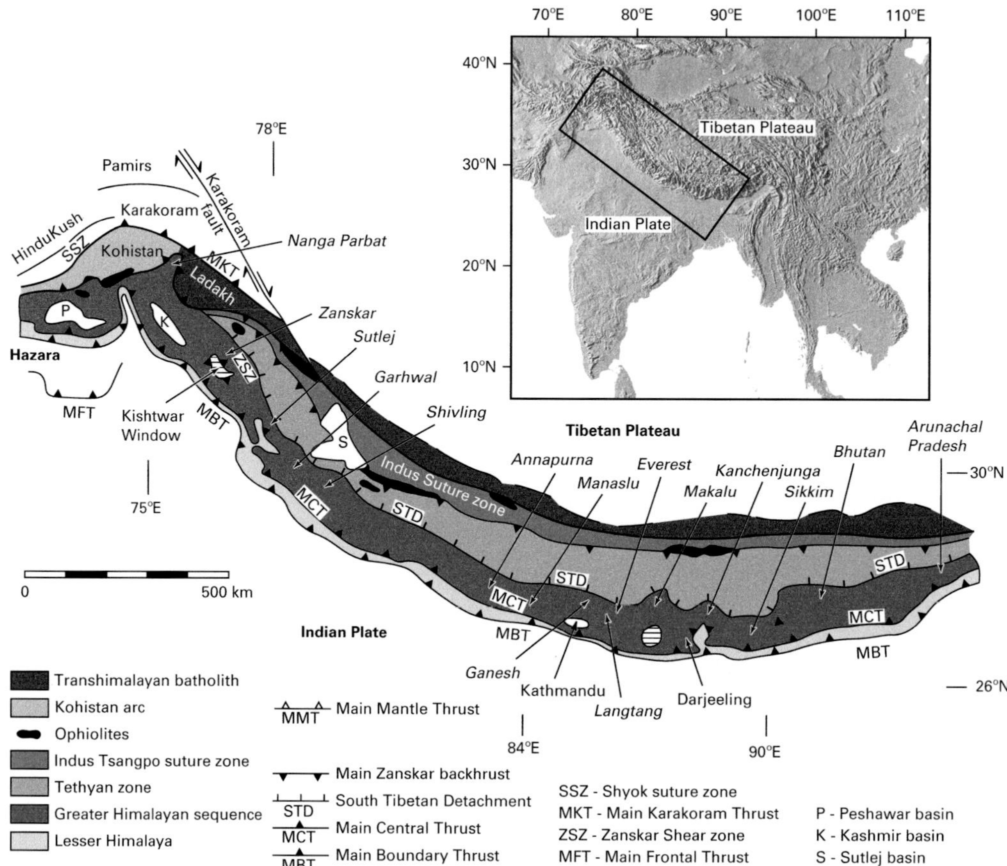

Figure 9.3. Simplified geology of the Himalayas between the Asian Plate (with the Tibetan Plateau) and the Indian Plate (with the Gangetic Plain). (A black-and-white version of this figure will appear in some formats. For the colour version, please refer to the plate section.)

Tiny uranium-bearing radioactive minerals such as monazite and zircon grow in these metamorphic and granitic rocks and can be dated using uranium-lead isotopes. Ages of Himalayan metamorphic minerals range from around 35–28 million years (kyanite gneisses) through to 24–14 million years (sillimanite grade gneisses and granites). These ages show that the Himalayan crust was thick and topographically elevated during this period, at least from 35 million years ago.

Geologically, the climax of Himalayan mountain building occurred during the Miocene. Between about 21 and 12 million years ago, metamorphic processes became so enhanced that the rocks actually started to melt, forming the famous Himalayan leucogranites. These granites formed, surprisingly, in the middle part of the crust where temperatures reached more than 750–800°C. The metamorphic backbone of the Himalayas forms most of the high peaks and this mid-crustal layer is bounded along the base by a crustal-scale south-directed thrust fault and shear zone, termed the Main Central Thrust Zone. Along the entire 2000 km length of the Himalayas, the southern part of the Greater Himalaya shows inverted metamorphic isograds (lines of equal temperature and pressure); in other words, the higher up the crust, the deeper and more metamorphosed rocks are exposed. Whereas the entire southern part of the Himalayas shows this up-side-down or inverted metamorphism, the northern part shows a right-way-up metamorphic isograd sequence along the South Tibetan Detachment, a north-dipping, low-angle normal fault zone. Thus the metamorphic rocks of the Greater Himalayan ranges have been thrust towards the south, bounded by the Main Central Thrust below (south) and the South Tibetan Detachment above (north). This process, termed *channel flow*, was initiated by partial melting of the crust to form granites, probably around 23 million years ago (Figure 9.4).

Many of the highest peaks of the Himalayas are formed of Miocene leucogranite that crystallized between 21 and 12 million years ago. These include Shivling (6543 m a.s.l.), Meru (6672 m) and the Bhagirathi (6856 m) peaks of the Indian Himalayas, Manaslu (8163 m), Langtang Lirung (7234 m), Nuptse (7861 m), Makalu (8463 m) and Kangchenjunga (8598 m) in Nepal and Shisha Pangma (8027 m) in south Tibet. Everest (8850 m) is composed of granite in the lower parts around the Nepalese side basecamp, but the summit area is composed of Ordovician limestone, approximately 450 million years old. These limestone fragments collected from the summit of Everest still show crinoid fragments, remnants of sea lilies that once lived in a shallow tropical sea. The South Tibetan Detachment normal fault cuts right through the summit pyramid of Everest, where sedimentary rocks outcrop above metamorphic rocks (Everest series gneisses). A second low-angle normal fault places the Everest series gneisses exposed along the southwest face of Everest above the sillimanite gneisses and granites exposed around the Khumbu icefall and lower Khumbu Valley (Searle *et al.*, 2006). The two exceptions amongst the highest peaks are Dhaulagiri (8167 m) and Annapurna (8091 m), both massifs composed of Cambrian-Ordovician limestone and dolomite above the South Tibetan Detachment (Figure 9.5).

Most of the deformation and metamorphism along the Great Himalaya range occurred between ~34–10 million years ago, during the Oligocene and Miocene when the Himalayas probably attained their elevated topography. The youngest ages of metamorphic rocks and granites along the Greater Himalaya are about 12–10 million years old. Since then, deformation has propagated southwards to the Lesser Himalaya and the southernmost ranges of the Himalayas. The active plate boundary today is the seismically active Main Boundary Thrust, which forms the southern margin of the elevated topography of the Himalayas. Most of the historic earthquakes along the Himalayas, including the 25 April 2015 Gorkha earthquake in Nepal, have occurred along this thrust fault.

Himalayan Rivers and Glaciers

The rise of the Himalayan Mountains during the Oligocene–Early Miocene period (34–15 million years ago) would have created a significant land barrier to bird

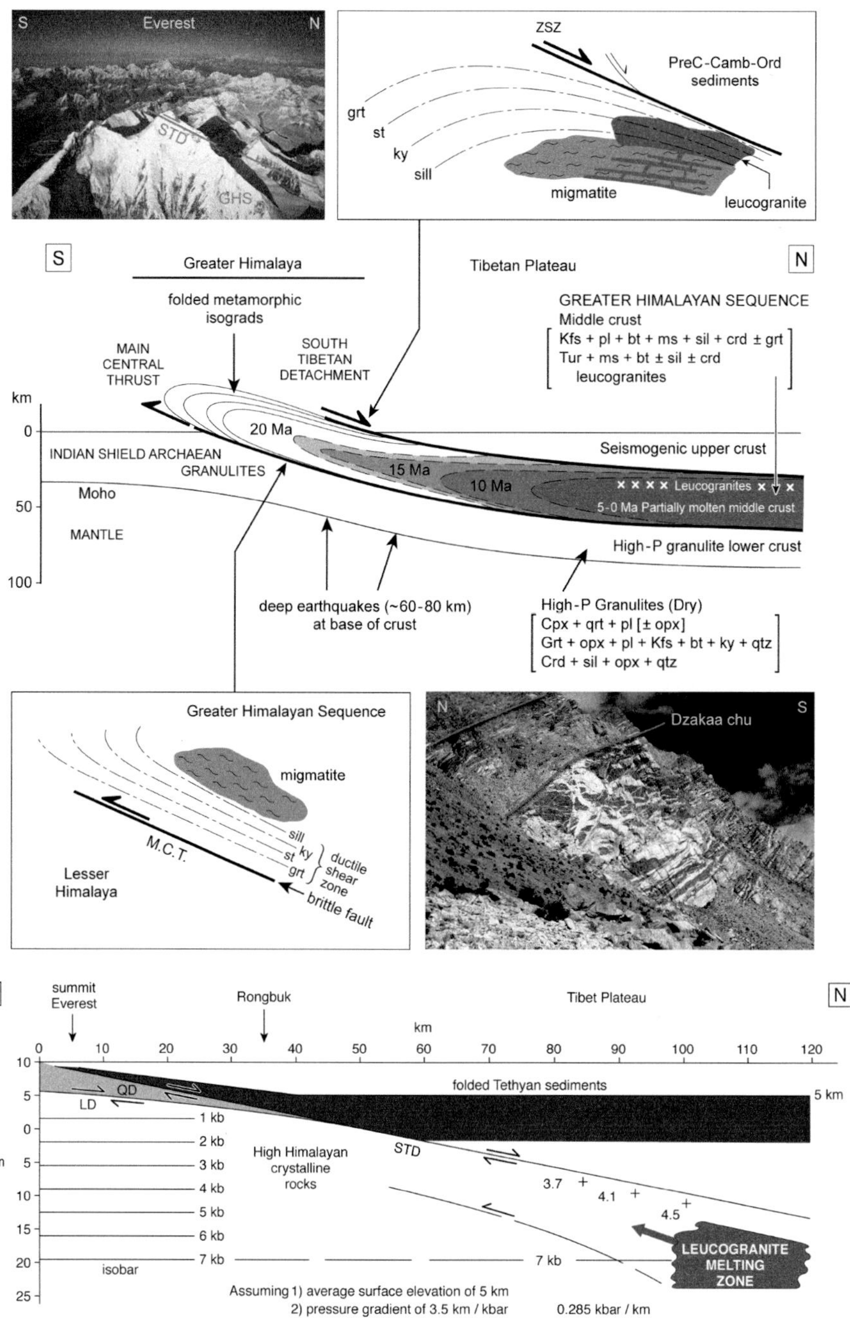

Figure 9.4. The Himalayan channel flow after Searle *et al.* (2006) showing right way up metamorphic isograds along the top of the Greater Himalayan slab and inverted isograds along the base wrapping around a southward extruding partially molten layer of middle crust. The upper fault of the South Tibetan Detachment clips the summit region of Everest and dips north beneath Tibet. (A black-and-white version of this figure will appear in some formats. For the colour version, please refer to the plate section.)

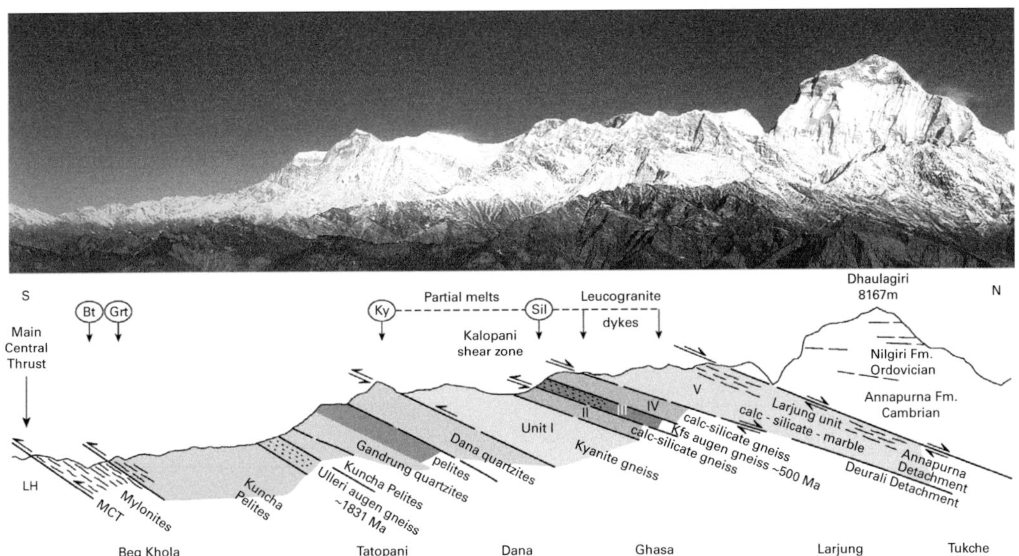

Figure 9.5. Panorama of the Dhaulagiri Himalaya above the Kali Gandaki River showing the tectonic-stratigraphic units of the Greater Himalaya, location of the metamorphic isograds (Bt-biotite; Grt-garnet; Ky-kyanite; Sil-sillimanite), the Main Central Thrust (MCT) and the low-angle normal faults beneath the unmetamorphosed sedimentary rocks of Dhaulagiri, after Searle (2010). (A black-and-white version of this figure will appear in some formats. For the colour version, please refer to the plate section.)

migration that continues to this day. As the Himalayan mountains rose in elevation, river systems were initiated, carrying rainfall and run-off southwards towards the plains of India. However, three major antecedent rivers were initiated prior to the uplift of the Himalayas – the Indus, Sutlej and Yarlung Tsango (Brahmaputra) Rivers. All three rivers rise around Mount Kailash in southwest Tibet, and flow for some distance across Tibet before cutting deep gorges through the Himalayas. The Indus flows west into Ladakh, cutting an impressive 6 km deep canyon between the Nanga Parbat (8125 m) and Haramosh massifs before turning south across the Kohistan and Pakistan Himalayas to drain into the Arabian Sea. The Sutlej cuts across the Garhwal–Kumaon Himalaya of northern India before joining the Indus. The Yarlung Tsangpo River flows for more than 1000 km eastwards across southern Tibet before plunging through some deep canyons between the Namche Barwa (7755 m) and Gyala Peri (7150 m) massifs of southeast Tibet. The river winds through the great gorge country before disgorging onto the plains of Assam. The Brahmaputra then flows west around the uplifting Shillong plateau before joining with the great Ganges to flow out into the Bay of Bengal.

Most Himalayan rivers initiated during uplift of the Himalayas and progressively cut back northwards through headwall erosion. The Himalayan Mountains seen today have been shaped by fluvial erosion and more recently by glacial erosion in the past 2 million years. The upper valleys of the Himalayas show classic U-shaped profiles carved by glaciers, whereas the lower parts show deeply incised V-shaped valleys. Almost all glaciers of the higher Himalayas have been rapidly retreating since the last glacial maxima as a result of global warming (see Owen, Chapter 10 and Bookhagen, Chapter 11).

Along the southern margin of the Himalayas in the Terai region, young and active folding in the Siwalik Hills has diverted many river courses. The erosional detritus from the Himalayas has been carried south by the river systems and deposited in a foreland basin south of the Himalayas. The Ganges takes discharge and debris eroded from the Himalayas out to the ocean, where it is deposited in the Bengal Fan. Along the western Himalayas, the Indus and Sutlej collect rainfall and erosional debris and transport it south into the Indus Fan in the Arabian Sea.

Himalayan Uplift and Climate Change

The Earth's climate has continually changed with time. The India–Asia Plate collision 50 million years ago was one of the largest continental collisions the Earth has witnessed in the past 400 million years. Since the beginning of plate tectonic processes some 2.5 billion years ago, other major collisional mountain ranges include the Caledonian–Appalachian range in Scotland and eastern North America, the subduction-related Andean mountain range in South America, the Hercynian or Variscan orogeny of central Europe and Paleo-Tethyan collisions across northern and central Asia. The closing of the Tethys Ocean that lay in between India and Asia resulted in major tectonic and climatic changes. A sudden rise of strontium isotope composition ($^{87}Sr/^{86}Sr$) of seawater indicates an increase in weathering and erosion that is interpreted to follow an increase of topographic elevation along the Himalayas (Searle *et al.*, 2011). Peak kyanite grade metamorphism along the Greater Himalaya at 35–30 million years ago indicates crustal thickening, elevated topography and increased erosion and exhumation rates. High mountains must have existed along the Himalayas at least as far back as 35 million years ago. Most Himalayan granites crystallized between about 21 and 18 million years ago, and this period marks an extremely rapid increase in $^{87}Sr/^{86}Sr$ ratios in the oceans known from deep sea drilling cores. The fossil record shows that major faunal changes in northern India and Pakistan occurred at this time. The high peaks of the Himalayas would have formed a barrier to movements of animals and birds.

The Asian monsoon system is thought to have originated between 25 and 22 million years ago, although older wet periods such as those in the Eocene (55–34 million years ago) are also thought to have resulted from enhanced greenhouse conditions (Licht *et al.*, 2014). Approximately 8 million years ago, increased aridity resulted in major changes in the vegetation of northern India and Pakistan, which changed from C_3 dominantly forests to C_4 dominantly grassland vegetation. Around 7 million years ago, there is evidence that the Indian summer monsoon strengthened and the climate became increasingly stormier. The monsoon was probably initiated much earlier if it was linked to the uplift of the high plateau of Tibet. The start of the Quaternary period, 2.5 million years ago, coincided with global cooling, intensified glaciation along the high Himalayas, increased CO_2 in the atmosphere, increased variability of the monsoon and the aridification of Tibet. Tibet lies in the rain shadow of the southwestern monsoon, so it became increasingly dry and cold with the uplift of the Himalayas. The widespread loess deposition in central China also started around this time.

The increased glacial ice scoured deep valleys in the Greater Himalaya, leading to the maximum advance of glaciers in the Everest region at 20,000 years ago. Since then, the Earth has been warming again, the ice sheets melting and glaciers retreating. Extremely young uranium-lead ages of zircons and monazite crystals from the western (Nanga Parbat–Haramosh) and eastern (Namche Barwa; Figure 9.1) syntaxes of the Himalayas suggests ongoing collision tectonics, crustal thickening and extreme topography, coupled with high erosion and exhumation rates. The Karakoram ranges of northern Pakistan and northern Ladakh have been topographically elevated at least since 54 million years ago, and possibly much longer, but these mountains also show extremely young metamorphism, rapid uplift and high exhumation – erosion rates that continue to this day.

In conclusion, by using data from geology and geochronology we can speculate on the evolution of topographic uplift caused by the influence of colliding continental plates along the Alpine Himalayan chain. There had almost certainly been a high topographic barrier along the Pamir, Karakoram and Hindu Kush ranges in Afghanistan, north Pakistan and westernmost Tibet since the mid-Cretaceous, approximately 120 million years ago. This uplift was associated with successive plate collisions with the stable Tarim Block. The Tibetan plateau region was probably as high as the present-day Andes along the southern margin of Asia (Ladakh–south Tibet) before the collision of India 50 million years ago. Since the collision, Tibet has continued to rise as a result of the underthrusting of India beneath Tibet. The Indian plate Himalaya was low-lying until about 65 million years ago, when slabs of oceanic rocks were thrust southwards from Tethys onto India, and uplifted more rapidly following the collision with Asia and the closing of the Tethys Ocean. The Himalayas probably reached peak elevation some time from 35–20 million years ago, when the thick crust and regional metamorphism eventually led to widespread partial melting to form granites. Major climate and vegetation changes were affected directly by tectonics and mountain building. During the past 25 million years, the monsoon effects resulted in increasing rainfall and erosion along the Himalayas and increased aridity in Tibet. The major rivers rising in Tibet and cutting through the Himalayas provide major bird migration routes to and from the Indian subcontinent.

References

Gansser, A. (1964). *Geology of the Himalayas*. Chichester: John Wiley & Sons.

Green, O.R., Searle, M.P., Corfield, R.I. & Corfield, R.M. (2008). Cretaceous-Tertiary carbonate platform evolution and the age of the India-Asia collision along the Ladakh Himalaya (Northwest India). *Journal of Geology*, **116**, 331–353.

Licht, A., Van Cappelle, M., Abels, H.A., *et al.* (2014). Asian monsoons in a late Eocene greenhouse world. *Nature Geosciences*, **513**, 501–506.

Searle, M.P. (2010). Low-angle normal faults in the compressional Himalayan orogeny: evidence from the Annapurna-Dhaulagiri Himalaya, Nepal. *Geosphere*, **6**, 296–315.

Searle, M. (2013). *Colliding Continents: A Geological Exploration of the Himalaya, Karakoram and Tibet.* Oxford: Oxford University Press.

Searle, M.P., Cottle, J.M., Streule, M.J. & Waters, D.J. (2010). Crustal melt granites and migmatites along the Himalaya: melt source, segregation, transport and granite emplacement mechanisms. *Transactions of the Royal Society of Edinburgh*, **100**, 219–233.

Searle, M.P., Elliott, J.R., Phillips, R.J. & Chung, S.L. (2011). Crustal-lithospheric structure and continental extrusion of Tibet. *Journal of the Geological Society, London*, **168**, 633–672.

Searle, M.P., Law, R.D. & Jessup, M.J. (2006). Crustal structure, restoration and evolution of the Greater Himalaya in Nepal–South Tibet: implications for channel flow and ductile extrusion of the middle crust. In R.D. Law, M.P. Searle & L. Godin, eds., *Channel Flow, Ductile Extrusion and Exhumation in Continental Collision Zones.* London: Geological Society, Special Publication **268**, pp.355–378.

10 Late Quaternary Glacier Fluctuations in the Himalayas and Adjacent Mountains

Lewis A. Owen

Glaciers in the Himalayas and Tibet

The mountains of the Himalayas and Tibetan Plateau constitute the most glaciated region outside of the polar realms, and contain some of the world's longest valley glaciers. The extent of these glaciers has varied significantly over geologic time in response to natural climate change. Recently, there is much evidence to suggest that glaciers have been retreating due to human-induced climate change, although in some regions they have been advancing (National Academy, 2012). Glaciers throughout the Himalayan-Tibetan region have a strong influence on the hydrology of the rivers, lakes and wetlands (Bookhagen, Chapter 11), and, consequently, fluctuating glaciers lead to significant environmental and biotic changes. The extent of glaciers has varied considerably throughout the Quaternary (~ past 2.6 million years), which has had a profound influence on the evolution of the mountain landscapes and environments, and the biota that inhabit or migrate over these mountains.

Defining the former extent of glaciation requires detailed geologic studies involving remote sensing, field mapping, analysis of landforms and sediments and geochronology. During the past few decades, significant advances in geologic methods have led to a plethora of studies on the glaciation of the mountains of the Himalayas and Tibet. Many of these studies were reviewed in Owen and Dortch (2014), and regional syntheses have been provided in Elhers *et al.* (2004, 2011), Dortch *et al.* (2013), Murari *et al.* (2014), Solomina *et al.* (2015) and, notably, Kamp and Owen (2011) and Zhou *et al.* (2011). These studies examine the extent and timing of glaciation over various time spans throughout the Quaternary. This chapter builds on this work and focus on the evidence for and the nature of glacier fluctuations during the Quaternary, specifically for the last glacial through to the present interglacial (the Holocene), spanning approximately the past 100,000 years (100 ka). This will help define the nature of glaciation in the Himalayas and Tibet, and examine the possible climate change forcing factors for glaciation. Focus is placed on the extent of glaciation during the past 20 ka, which has had the most recent imprint on Himalayan-Tibetan environments. There has been much variation in glaciation during the past 20 ka, which, of course, had significant consequences for the biota of the Himalaya and Tibet. Geographically, focus is on the Greater Himalaya, Trans-Himalaya and the mountains of southernmost part of Tibet, which are the dominant topographic barriers for bird migration. However, from a geological perspective, the Himalayas and Tibet should be considered as a whole, since they are essentially a coherent tectonic-geographic realm, where glaciation influences the great hydrological systems of central and southern Asia, many of which drain the region and include such rivers as the Tsangpo-Brahmaputra, Yangtze, Yellow, Indus, Ganges and Irrawaddy.

The Quaternary has been characterized by glacial and interglacial periods. During the past ~1 million year (1 Ma), glacials have lasted about 100 ka and have been characterized by periods of extensive glaciation that produce large, Northern Hemisphere ice sheets. Interglacials, which have only lasted about 10–20 ka, are warmer times when glaciers were much less extensive, and the Northern Hemisphere ice sheets, with the exception of the Greenland ice sheet (Lowe & Walker, 2015), essentially disappeared. Glaciers have fluctuated considerably within individual glacials and interglacials, on varied timespans, which were quasi-cyclic or rhythmic, ranging from annual, decadal, centennial, millennial, to tens of thousands of years, in response to natural climate change. The processes forcing natural climate change are debated, but they are probably a combination of changes in the Earth's orbital parameters (on 10^{3-4} year timescales), autocyclic processes such as changes in ice sheet movement and oceanic circulation (10^{1-4} years), solar variability (10^{1-3} years) and volcanic activity (10^{-2} years) (see Lowe & Walker, 2015, for fuller discussion of the nature of Quaternary climate change).

Regional Setting

To fully appreciate the nature of late Quaternary glaciation in the Himalayas and Tibet, it is important to examine the regional setting, including its geologic, topographic and climatic context. The mountains and plateau areas of the Himalaya and Tibet were formed by the collision of island arcs, a curved chain of volcanic islands, with the Eurasian continental plate, and the ultimate collision of the Indian and Eurasian continental plates at ~50 Ma, and the subsequent continued northward movement of the Indian plate into Asia (see Searle, Chapter 9). This continental–continental collision resulted in the formation of the Tibetan Plateau and a series of approximately east–west-trending ranges within and along its margins, stretching ~2000 km and ~1500 km in an east–west and north–south direction, respectively (Figures 10.1, 10.2). The combined average elevation of the region is ~5000 m above sea level (a.s.l.; Fielding *et al.*, 1994). The Greater Himalaya and Trans-Himalaya, along the southern margin of this region, are the most impressive and imposing, containing all the world's 8000 m-high peaks. Significant mountain ranges that traverse Tibet include the Nyaingentanglha Shan, Tanggula Shan, Bayan Har Shan, Kunlun Shan, Altun Shan and Qilian Shan. All these mountain ranges are glaciated. However, vast tracks of the interior of the Tibetan Plateau lack any evidence of glaciations, and were probably never glaciated (Owen & Dortch, 2014).

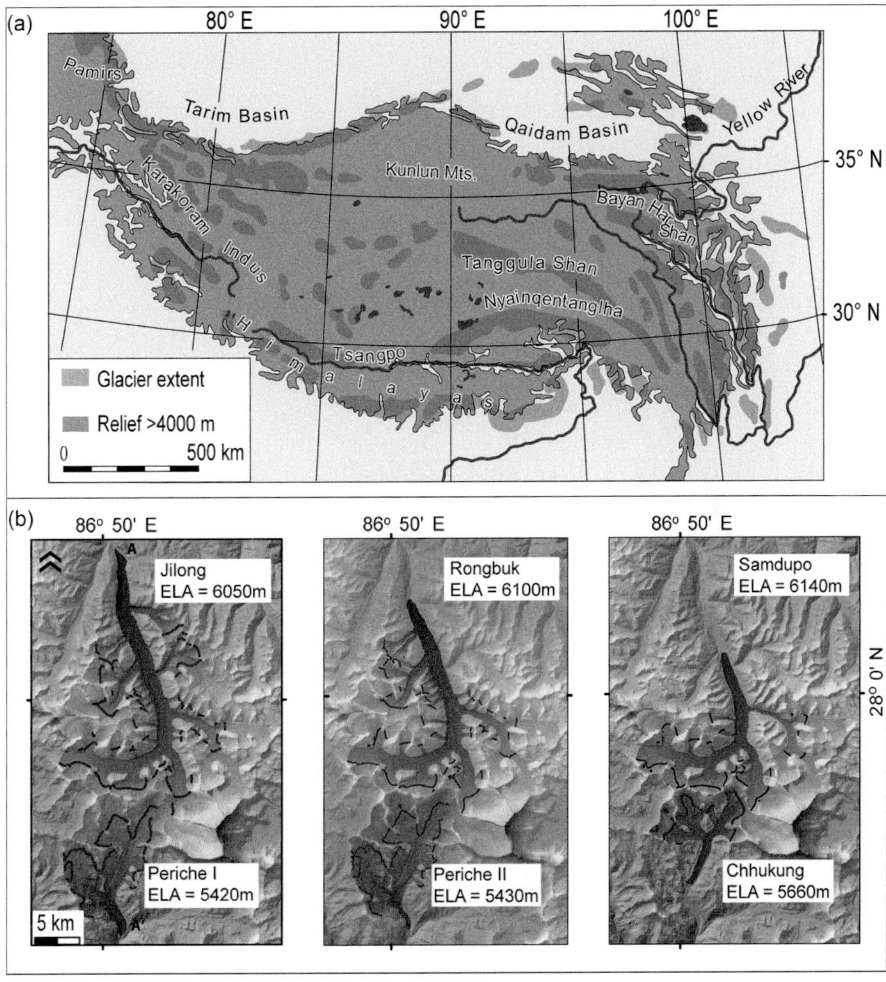

Figure 10.1. a) Reconstruction of the maximum extent of glaciation across the Himalayas and Tibet and the bordering mountains based on detailed field mapping of glacial and associated landforms and sediments (from Owen *et al.*, 2008 and Owen, 2010, after Shi, 1992 and Li *et al.*, 1991). Light grey areas represent relief over 4000 m a.s.l. and dark blue grey areas represent areas that were glaciated. Note the maximum extent of glaciation on this map probably did not occur at the same time across the region. b) Example of detailed reconstruction for the Late Quaternary glaciation of a selected area of the Himalayas (after Owen *et al.*, 2009). The reconstructed glaciers in this area advanced from the northern and southern slopes of Mount Everest (peak shown by the red star in the left image) plotted on a digital elevation model (from ASTER imagery). From left to right are the Jilong and Periche I glacial stages dated to early Late Glacial Maximum (LGM), the Rongbuk and Periche II glacial stages dated to late global LGM and the Samdupo and Chhukung glacial stages dated to the Early Holocene. The equilibrium-line altitudes (ELAs) for glaciers for each glacial stage are shown in each panel and the present ELAs for the northern and southern sides of Mount Everest are ~6200 and 5700 m a.s.l., respectively. (A black-and-white version of this figure will appear in some formats. For the colour version, please refer to the plate section.)

Figure 10.2. a) Low-oblique photograph looking southwest across a stretch of the Himalayas and southern Tibet, with Mount Everest in the centre of the frame, taken from the Space Shuttle in March 1996 (NASA image STS076-727-080). This illustrates typical topography, and glacier and snow cover across a stretch of the mountain belt and plateau. Note the green humid Indo-Gangetic Plain in the top left corner and the brown semiarid interior of Tibet essentially devoid of snow and glacier cover in the lower right corner. b) View of the north face of Mount Everest rising to 8848 m a.s.l. and Rongbuk Glacier and its moraines in the foreground. (A black-and-white version of this figure will appear in some formats. For the colour version, please refer to the plate section.)

Two major climatic systems dominate the region: mid-latitude westerlies and the Asian monsoon (Benn & Owen, 1998; Figure 10.4). The influence of these two climatic systems varies across the region (see also Bookhagen, Chapter 11). The majority of the southern and eastern regions experience a pronounced summer precipitation maximum, reflecting moisture advected northwards from the Indian Ocean by the southwest monsoon. Summer precipitation declines sharply northwards across the main Himalayan chain and little falls over western and central Tibet (Benn & Owen, 1998; Owen *et al.*, 2009; Owen & Dortch, 2014). In contrast, the mid-latitude westerlies produce a winter precipitation maximum at the western end of the Himalayas, Trans-Himalaya and western Tibet, as a consequence of moisture advected from the Mediterranean, Black and Caspian Seas (Benn & Owen, 1998; Owen *et al.*, 2009; Owen & Dortch, 2014). This results in strong north–south and west–east precipitation gradients, with some of the wettest places on Earth along the southern and southeastern margins of the Himalaya and Tibet, to the high mountains deserts in the interior of Tibet (Figure 10.2) (see also Rawat, Chapter 12). There are also strong microclimatic variations within individual mountain ranges and valleys (Owen & Dortch, 2014). The distribution and timing of precipitation has a strong influence on glaciation. Moreover, the relative roles of these climate systems have fluctuated over time, responding to natural climate change over durations ranging from decades to hundreds of millennia (see the previous section).

Benn and Owen (1998) hypothesized that the relative roles of the dominant climate systems, the South Asian monsoon and the mid-latitude westerlies varied significantly throughout the Quaternary. They suggested that this probably resulted in asynchronous glaciation throughout the Himalayas and Tibet. Numerous researchers have attempted to address this hypothesis since its publication (see details later in this chapter).

Vegetation throughout the region reflects climate and altitude (see Rawat, Chapter 12). The vegetation varies from thick subtropical forests in the foothills of the Himalayas and southeastern Tibet, to deciduous forests at higher altitude and eventually alpine meadow and mountain tundra at the highest elevations, and xerophytic vegetation persists in the arid interiors of the Tibetan Plateau (Owen & Dortch, 2014).

The strong climate gradients, microclimatic variability, varied geology, topography and vegetation result in a great diversity of environments and habitats for migratory birds. To simplify, the Himalayan-Tibetan region can be characterized as one of immense geomorphic diversity, dominated by glacial, fluvial and mass movement processes. As will be apparent later, this diversity makes it very challenging to formulate simple statements about the timing and extent of Late Quaternary glaciation throughout the Himalayan-Tibetan region.

Present-Day Glaciers

Accurately defining the extent of contemporary glaciers throughout the Himalaya and Tibet is challenging despite our access to sophisticated remote-sensing technologies. This is because of the extremely large area to be covered, and because many (but not all) of the glaciers are debris mantled, making it difficult to define their limits from morainic

Figure 10.3. Examples of glaciers and associated terrain of the Himalayas and southern Tibet illustrating the wide variety of forms. a) Relatively debris-free continental glacier advancing from the Gurla Mandhata (Naimon'anyi; the peak (left) is at 7694 m a.s.l.) down the Namarodi Valley in southern Tibet. Near its snout, the glacier is expanding to become almost a piedmont-type glacier. b) Small, cold, continental v-shaped valley glaciers in Zanskar, northern India. c) View across the debris-covered Baltoro Glacier at Concordia in northern Pakistan with K2 at 8611 m a.s.l. (left) and Broad Peak at 8047 m a.s.l. (right) in the background. At > 60 km long, the Baltoro Glacier is one of the longest mountain glaciers outside of the polar regions and at Concordia the glacier is ~3 km wide. d) View north at Kedarnath town (at ~3600 m a.s.l.) with Chorabari Glacier in the background somewhat obscured by the large latero-frontal moraines that define the margins of this maritime debris-mantled glacier (photograph taken in September 2010). The ridges on the left and right sides of the view are moraines that formed during the early Holocene. Kedarnath was tragically destroyed during a flood in June 2013 during a multi-day cloudburst and drainage of a lake associated with the Chorabari Glacier. (A black-and-white version of this figure will appear in some formats. For the colour version, please refer to the plate section.)

debris along their margin left by recent and historical retreats. The lowest elevations to which glaciers advance vary between climate regions. Generally, glaciers advance to the lowest elevations in the wettest regions, and in some cases as low as 3000 m a.s.l. Even within the same climatic region, the snout elevation of glaciers can vary considerably. This is because the extent of a glacier is controlled by numerous factors such as the shape and aspect of the valley, and the amount of snow blow and/or avalanche contribution to a glacier (Benn *et al.*, 2005).

A common way to compare the degree and extent of glaciation across a region is to define the altitude of the equilibrium lines on glaciers. The equilibrium line is the line on a glacier where accumulation and ablation of snow and ice are in balance over several or more years. Clearly, the equilibrium-line altitudes (ELAs) for glaciers across the Himalayas and Tibet vary considerably as a consequence of the different climatic zones. Owen and Benn (2005) provided a summary and review of ELAs for the Himalayas and Tibet showing that ELAs can vary from < 4300 m to > 6200 m a.s.l. Defining the ELAs is particularly useful because ELAs for former glaciers during past glaciations can be reconstructed using several geomorphic methods, which in turn provide a means to quantify and compare the degree of past glaciation. The difference in altitude between the present and past ELA is referred to as the *ELA depression* (ΔELA). ΔELAs can range from a few metres from one year to the next to more than 1000 meters during a very extensive glaciation.

Derbyshire (1981), Benn and Owen (2002) and Owen and Dortch (2014) highlighted the strong influence of climate and topography on the nature of glacial systems throughout the Himalayas and Tibetan Plateau. Three main glacier types are present in the Himalayan-Tibetan region (Derbyshire, 1981). These include: 1) continental interior types in the central and western parts of the Tibetan Plateau; 2) maritime monsoonal types in the Himalayas and in southeastern Tibet; and 3) continental monsoonal types in eastern and northeastern Tibet (Figures 10.2 and 10.3).

Owen and Dortch (2014) highlight that the maritime and continental monsoonal glaciers of the Himalaya and southeastern and eastern Tibet are warm-based where the ice pressure-temperature is near to its melting point and have summer accumulation and ablation. They are also highly diverse, and include avalanche- and snowfall-fed cirque and valley glaciers, and very steep hanging glaciers, and may have surface velocities up to several hundred meters per year (Owen & Dortch, 2014). In contrast, alpine continental glaciers in the western part of the Tibet and Trans-Himalaya are significantly smaller with high and cold accumulation areas, while the continental ice caps of central and western Tibet have basal ice temperature below 0°C with low surface velocities, usually between 2 and 10 $m.a^{-1}$.

This large variety of Himalayan and Tibetan glacier types results in a complexity of landforms, including impressive latero-frontal moraines, hummocky moraines, lateral moraine valley complexes, glacially eroded bedrock surfaces and deeply entrenched valleys (Figures 10.2 and 10.3). These are described in considerable detail in Benn and Owen (2002). Understanding the nature and formation of these glacial landforms is essential for accurate reconstructions of the former extent of glaciation. Misinterpretations of landforms have in some instances led to erroneous reconstructions of former glacier extents. This is discussed in considerable depth in Derbyshire and Owen (1997), Hewitt (1999), Hewitt *et al.* (2011) and Benn and Owen (2002).

Reconstructing the Timing and Extent of Former Glaciers

Reconstructing the extent of glaciation through the Himalayas and Tibet has been challenging because of the logistical and political inaccessibility of the region.

Despite these challenges, there has been a long history of study of the extent of glaciation in the Himalayas and Tibet. Most of these studies have concentrated on examining the field evidence, mainly geomorphic, for the position of former glaciers (Owen & Dortch, 2014 and references therein). In recent years, remote sensing has greatly assisted these studies (Duncan *et al.*, 1998; Heyman *et al.*, 2008; Moren *et al.*, 2011). Owen and Dortch (2014) summarized the majority of the studies that have been undertaken, highlighting the important compilations of Klute (1930), Frenzel (1960), Li *et al.* (1991) and Shi (1992), which show expanded ice caps and extensive valley glaciers through the region (Figure 10.1a).

Owen and Dortch (2014) also discussed the main approaches used to reconstruct the former extent of glaciers. These include mapping glacial and associated landforms and assigning relative ages to landforms based on their respective positions within valleys and on the weathering and erosional characteristics. Once glacial landforms have been mapped and relative ages defined, most researchers assign a glacial stage name or number to a particular suite of glacial and associated landforms that they recognize to have formed during distinct glacial advances. Applying glacial stage names and/or a numbering scheme allows landforms to be tentatively correlated across study areas and regions. The next step researchers undertake is to date the moraine using numerical dating methods. This allows the researcher to determine the timing of the formation of landforms and hence the timing of a glacial advance.

Numerically dating moraines is also challenging, and the laboratory work is intensive and costly. Radiocarbon dating has long been the most common method used in glacial environments. The method relies on the presence of organic material incorporated into the moraine or associated sediments. This can include trees overridden during the glacial advance that became incorporated into the glacial sediment, providing an age for advance, or organic material buried in sediment beneath the moraine, which provides a maximum age for the timing of glaciation. Organic material may also be present in sediment on top of the moraine, which provides an age that post-dates the timing of moraine formation. Unfortunately, studies are few because organic material is not commonly found in glacial sediments in the Himalayas and Tibet. However, newly developing techniques such as terrestrial cosmogenic nuclide (TCN) surface exposure and optically stimulated luminescence (OSL) dating can now be employed to date moraines devoid of organic material. Furthermore, these dating methods allow moraines older than the limit of radiocarbon dating (~50 ka) to be dated. OSL dating provides a method to date the deposition of sediment. Like radiocarbon dating, OSL dating can provide maximum, concurrent and/or minimum ages for glacial advances, dependent on where the sediment is sampled. TCN dating is normally used to date glacial boulders and rock surfaces, and generally provides minimum ages for a glacial advance. All these techniques have inherent problems, and the uncertainty associated with the dating methods may range from a few per cent to more than 50% of the age of the landform. Owen and Dortch (2014) provided a comprehensive discussion of the various dating techniques and summarized all the radiocarbon, TCN and OSL dating undertaken on Himalayan and Tibetan glacial succession.

Once the former extent of a glacier has been determined, the past ELA can be calculated from the shape of the reconstructed glacier. The numerical dating provides the temporal framework for the ELA reconstructions. Figure 10.1b provides an example based on a study for the valleys on the northern and southern side of Mount Everest for three different glacial advances (Owen *et al.*, 2009).

The Tibetan Ice Sheet Hypothesis

Reconstructions of the extent of glaciation across the Himalaya and Tibet have ranged from localized valley glaciation and ice cap expansion to the view that an ice sheet covered the whole of the Tibetan Plateau and adjacent mountains. Kuhle (e.g. 1985, 1987, 1988, 1991, 1995) was the main proponent of the idea that a Late Quaternary ice sheet covered the Tibetan Plateau, which spread over large areas of the adjacent mountains and had outlet glaciers that advanced into the deep valleys of the Himalayas. This view was contentious and led to much research during recent decades. Numerous researchers have, however, presented evidence to show that an ice sheet could not have existed across Tibet during at least the whole of the past 500 ka (Owen & Dortch, 2014 and references therein).

The geologic evidence shows that glaciers in the mountains of Tibet and the Trans-Himalaya advanced, but they were rather limited in extent, while those in the Great Himalaya were more extensive and were predominantly of valley glacier type (Figure 10.1a). Moreover, the numerical dating undertaken in Tibet shows that moraines for the majority of the most extensive glaciations formed during the penultimate glacial cycle or much earlier (more than 100 ka).

Glacier Fluctuations during the Last Glacial

There is abundant evidence in the mountains of the Himalayas and Tibet for multiple glacial advances throughout the last glacial (~100 ka to 11.6 ka; Owen & Dortch, 2014). Owen and Dortch (2014) showed, for example, that glaciers in semiarid regions of the Himalayas and Tibetan Plateau during the last glacial reached their maximum extent early in the last glacial, and that during the global last glacial maximum (gLGM), glacier advances were significantly less extensive. The gLGM is broadly defined as the time of maximum global ice volume at approximately 18–24 ka (Mix *et al.*, 2001). This was essentially when the Northern Hemisphere ice sheets were at their maximum extent and were most voluminous. Numerous researchers in other mountain regions of the world have also suggested that mountain glaciers reached their maximum extent earlier in the last glacial (e.g. Gillespie & Molnar, 1995; Thackray *et al.*, 2008). In these cases, the maximum extent of glaciation is often referred to as the *local last glacial maximum*. In the Himalayas, however, the pattern is even more complex as along the more monsoon-influenced Greater Himalaya, increasing evidence suggests that glaciation was more extensive later in the last glacial cycle (Owen & Dortch, 2014).

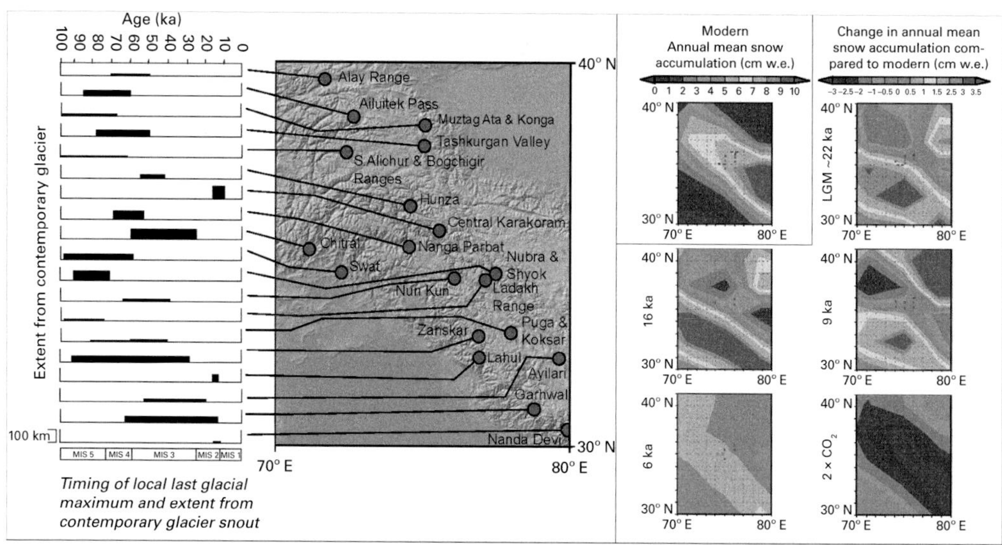

Figure 10.4. Glacial geologic study areas at the western end of the Himalayas and Tibet, showing the ages of local last glacial maxima with the approximate extent of valley glaciation shown by the height of each black bar, after Owen and Dortch (2014) who used data from Owen *et al.* (1992, 2001, 2002a, 2002b, 2006, 2010), Sharma and Owen (1996), Phillips *et al.* (2000), Richards *et al.* (2000), Taylor and Mitchell (2000), Barnard *et al.* (2004a, 2004b), Zech *et al.* (2005), Chevalier *et al.* (2005, 2011), Abramowski *et al.* (2006), Seong *et al.* (2009), Dortch *et al.* (2010), Hedrick *et al.* (2011), Rohringer *et al.* (2012) and Lee *et al.* (2014). The area shown in the right panel is the same area shown in the right frame which presents modelling results from Bishop *et al.* (2010), who used a 70-year simulation with a fully coupled atmosphere-ocean general circulation model for the global LGM, 16 ka, 9 ka, 6 ka and a doubling of CO_2. (A black-and-white version of this figure will appear in some formats. For the colour version, please refer to the plate section.)

The complexity associated with reconstructing the timing and extent of glaciation is illustrated in Figure 10.4a (adapted from Owen & Dortch, 2014). Figure 10.4 highlights the main study areas at the western end of the Himalayas and Tibet, and the timing of the local last glacial maximum is indicated on the left of the figure, with approximate maximum extents of glaciation. Owen and Dortch (2014) showed that the pattern that emerges is very complex, with the local last glacial maximum occurring at very different times across the western end of the Himalayas and Tibet. These data were compared with the climate modelling presented in Bishop *et al.* (2010) that covers the same spatial extent (Figure 10.4b). Owen and Dortch (2014) pointed out that the variation in snowfall across the region and between different time periods is striking, and therefore it should not be surprising that the style and timing of glaciation across the mountains is complex.

Dortch *et al.* (2013) examined the glacial successions across the semiarid regions at the western end of the Himalaya and Tibet to develop a regional framework for glaciation across these dryland regions. They recognized 19 regional glacial advances (regional glacial stages), 13 of which occurred during the last glacial and Holocene, which they named semiarid western Himalayan-Tibetan stages (SWHTS:

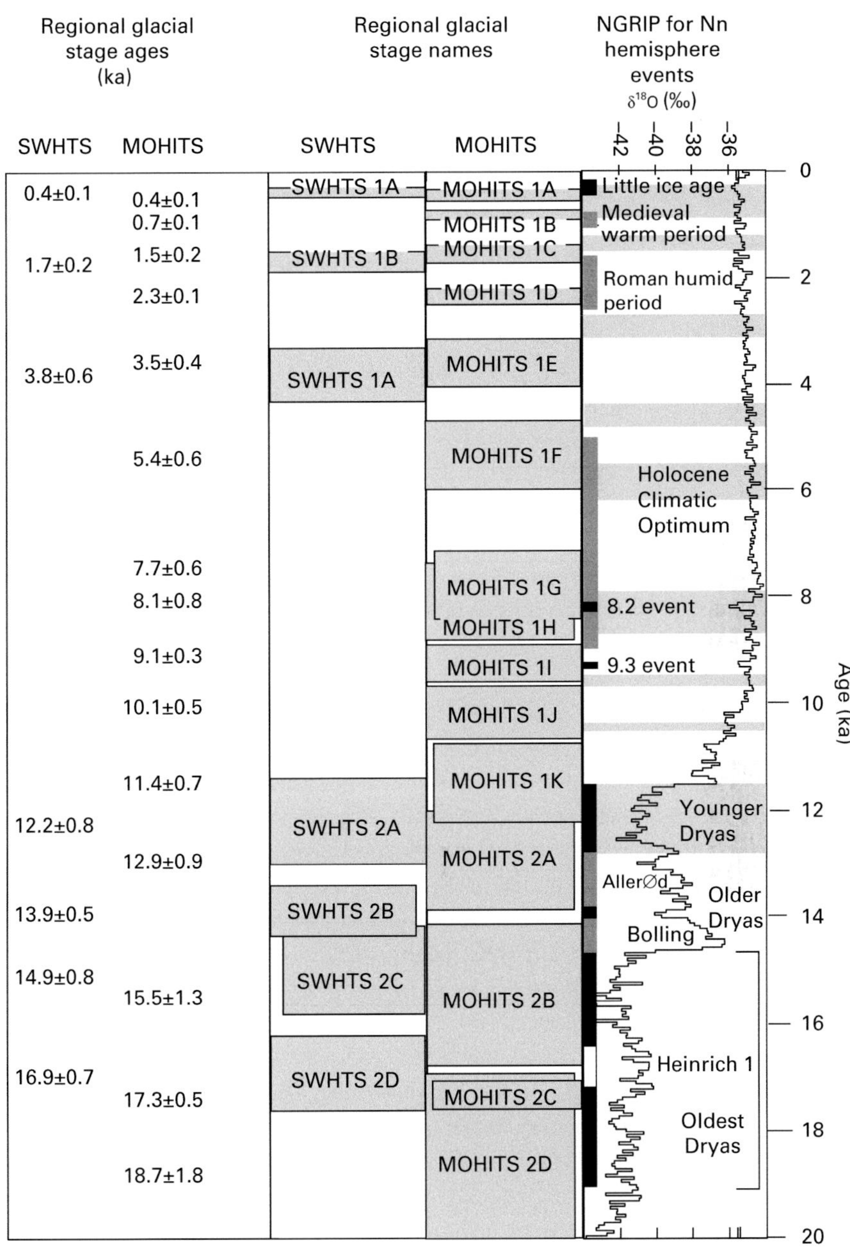

Figure 10.5. Regional glacial stages for the past 20,000 years for the semiarid regions at the western end of the Himalaya and Tibet (semiarid western Himalaya-Tibetan Stage: SWHTS) defined by Dortch *et al.* (2013) and for the monsoon-influenced Himalayan-Tibetan orogen (monsoon Himalayan-Tibetan stages: MOHITS) defined by Murari *et al.* (2014) (adapted from Owen & Dortch, 2014). The $\delta^{18}O$ curve NGRIP (2004) is provided for comparison, and the duration of specific climatic events are marked by black and grey bars in the far right column.

Figure 10.5). Dortch *et al.* (2013) suggested that SWHTS older than 21 ka are broadly correlated with a greater monsoonal influence, and that the 10 SWHTS that are 21 ka or younger broadly correlate with global ice volume and Northern Hemisphere climatic events (Oldest Dryas, Older Dryas, Younger Dryas, Roman Humid Period and Little Ice Age). They also speculated that the arid regions responded to small fluctuations in precipitation controlled by variations in the mid-latitude westerlies (see Bookhagen, Chapter 11).

Similarly, Murari *et al.* (2014) examined the glacial successions in the monsoon-influenced Himalayan-Tibetan regions and identified 29 regional glacial stages, 24 of which occurred during the last glacial and Holocene, which they called monsoon-influenced Himalayan-Tibet stages (MOHITS). Murari *et al.* (2014) suggested that there are strong correlations with both periods of strong monsoons and northern hemisphere events with 16 MOHITS linked to the monsoon and 11 MOHITS linked to the mid-latitude westerlies, and two unassigned, demonstrating a complex pattern of glaciation influenced by two climatic systems in the Late Quaternary. Owen and Dortch (2014) suggested that some of the MOHITS and SWHTS might be correlated, but that many were asynchronous due to the different climatic forcing factors associated with different regions across the Himalaya and Tibet.

Holocene Glacier Fluctuations

Yi *et al.* (2008), Owen (2010), Owen and Dortch (2014) and Solomina *et al.* (2015) provided summaries of the nature of Holocene glacier fluctuations in the Himalaya and Tibet. Of particular note is the early work of Röthlisberger and Geyh (1985a, 1985b), who studied glaciers in Pakistan, India and Nepal and dated moraines using radiocarbon methods to determine the timing of glacial advances to ~8.3 ka, 5.4–5.1 ka, 4.2–3.3 ka and 2.7–2.2 ka with relatively small extensions at 2.6–2.4 ka, 1.7–1.4 ka, 1.3–0.9 ka, 0.8–0.55 ka and 0.5–0.1 ka. Yi *et al.* (2008) compiled 53 radiocarbon ages for Holocene glacial advances and argued that glaciers advanced at 9.4–8.8 ka, 3.5–1.4 ka and 1.0–0.13 ka in Tibet. Owen and Dortch (2014) compiled all published radiocarbon, TCN and OSL ages for the Himalaya and Tibet, and suggested that glaciers were responding to periods of Holocene rapid climate change in the North Atlantic, which were teleconnected via mid-latitude westerlies to Central Asia where they influenced glaciation.

The foregoing may explain the observations that in some regions, there is evidence for glacial advance at particular times in the Holocene, while in other areas there is no evidence for a glacier advance at that time, and even within a region, glaciers may have advanced in some valleys while in other valleys they did not (Owen & Dortch, 2014). Dortch *et al.* (2013), Murari *et al.* (2014) and Owen and Dortch (2014), defining regional glacial advances and stages throughout the Himalayan-Tibetan region, showed that there have been at least 4 SWHTS and 11 MOHITS during the Holocene in the semiarid and monsoonal areas, respectively (Figure 10.5).

Some broad patterns are evident, and the most extensive glacial advances in the region generally occurred during the early Holocene, between ~11.5 and ~8.0 ka (Owen & Dortch, 2014 and references therein). However, as Owen and Benn (2005) and Owen and Dortch (2014) highlight, the extent of glaciation and difference between the present and past equilibrium line, the ΔELA, during the early Holocene varies considerably between regions.

Evidence for mid-Holocene (~8.0–3.0 ka) and late Holocene (Neoglacial, <3.0 ka) glacier advances is also very evident throughout the region (Owen *et al.*, 2009; Owen & Dortch, 2014; Solomina *et al.*, 2015). During these times, glaciers advanced to within a kilometre of the present glacier margins (Owen & Dortch, 2014; Solomina *et al.*, 2015). Dortch *et al.* (2013) recognized three SWHTS during the latter part of the Holocene in the semiarid western areas dated using the build-up of the TCN ^{10}Be on rock surfaces with time, at: 3.7±0.6 ka, 1.6±0.3 ka and 0.4±0.1 ka, and Murari *et al.* (2014) recognized five MOHITS for the monsoon-influenced regions at 3.5±0.4 ka, 2.3±0.1 ka, 1.5±0.2 ka, 0.7±0.1 ka and 0.4±0.1 ka (Figure 10.5). Owen and Dortch (2014) suggested correlations between the MOHITS and SWHTS during this time, and suggested that the changes in Northern Hemisphere climate were helping to drive glaciation.

Little Ice Age (LIA) glacial advances in the Himalayas and Tibet are not well defined, but seem to suggest that glaciers advanced at various times during the past few hundred years (Owen & Dortch, 2014). Yi *et al.* (2008) argued that there were three glacial advances during the LIA in Tibet between 1.0 to 0.13 ka. In most areas of the Himalaya and Tibet, glaciers began to retreat after the LIA at the beginning of the twentieth century and have continued to retreat since then (Owen & Dortch, 2014).

Recent Glacier Fluctuations

Much contention exists over the degree to which glaciers are currently retreating and losing volume (mass). Much of this owes its origin to erroneous estimates of glacier retreat published in the Fourth Assessment Report of the Intergovernmental Panel on Climate Change (Intergovernmental Panel on Climate Change, 2007), which stated that Himalayan glaciers are receding faster than those in other parts of the world and are likely to disappear by at least CE 2035. The accepted consensus, however, is that Himalayan and Tibetan glaciers are generally retreating at rates similar to glaciers elsewhere in the world in much of the Himalaya and Tibet, but west of the Sutlej River the glaciers are experiencing little or no retreat (Bagla, 2009; Bolch *et al.*, 2012; National Academy, 2012; Bookhagen, Chapter 11).

Cogley (2011) suggested that up to about one-fifth of the glaciers present in 1985 may have disappeared by ~2008, and if mass loss were to remain constant at the average rate for 1975–2008, between 3000 and 13,000 more glaciers might disappear by 2035. Moreover, if mass loss were to continue to accelerate as inferred for 1985–2008, only a few thousand to a few hundred glaciers might remain in 2035. Cogley (2011) acknowledged that these projections are uncertain and neglect some possibly important mitigating controls, but they demonstrate the need for more complete analyses to help define the reductions and assess their potential effects. The rate of retreat between 1985 and 2015 was, however, lower in some regions than during the earlier part of the twentieth century, and certain glaciers have begun to stabilize and/or advance in recent years (Copland *et al.*, 2011). Some glaciers have even been remarkably stable such as Baltoro Glacier in northern Pakistan, which has fluctuated no more than 200 m from its present position since the 1850s (Mayer *et al.*, 2006), and in the Karakoram, individual glaciers have advanced and retreated during the past century, sometimes even surging (Copland *et al.*, 2011; Bolch *et al.*, 2012; Bookhagen, Chapter 11).

Owen and Dortch (2014) highlighted that predicting how human-induced global climate change may force glaciers to melt, retreat and eventually disappear is challenging. This is because of the complexity of modelling Himalayan and Tibetan glacial systems, with their large altitudinal range, and the complex topographic setting and microclimatic diversity. Owen and Dortch (2014) stressed that in monsoon-influenced regions, the predicted increased snowfall and increased cloudiness may lead to positive glacier mass balances, which may cause glaciers to thicken and advance. These glaciers may eventually retreat as the increase in mean annual temperature out-competes the increased glacier growth due to enhanced snowfall and cloudiness; this change may be rapid if summer snows become summer rain. This uncertainty makes it even more difficult to predict the hydrological response to changing glacier extent, which in turn has profound implications for environmental risk assessment and management.

Conclusion

The glaciers of the Himalaya and Tibet are very varied, ranging from sub-polar continental to warm maritime types, across a climatically and topographically very diverse region. Defining the extent and timing of Quaternary glaciation throughout the mountains of the Himalaya and Tibet is challenging. Recent studies are, however, providing new insights into the complex nature of glacial change driven by a combination of natural forcing factors related to the South Asian monsoon and mid-latitude westerlies climatic systems, and in recent years, probably due to human-induced climate change. The maximum extent of glaciation has been broadly defined, characterized by extensive valley glaciers and expanded ice caps. The timing of maximum glacier extent is asynchronous throughout the region, with some areas experiencing maximum glaciation prior to the last glacial (> 100 ka), while in other regions it was during the early part of the last glacial (~30–70 ka) and in some regions it was possibly coincident with the gLGM at ~18–24 ka. Glacier advances since the gLGM have been limited to a few kilometres beyond their present position in most regions, with the maximum advance occurring during the early Holocene at ~9–8 ka. There have been numerous minor glacial advances throughout the Holocene with the most recent, the LIA, starting several hundred years ago and ending at approximately the beginning of the twentieth century. Glaciers retreated in most areas between 1915 and 2015, but in some regions such as the

Karakoram there has been little change in extent. This suggests a complex pattern of future glacier fluctuations in response to human-induced climate change. Changes in glacial extent in the past and probably rapid changes in the future have and will have a profound effect on the rivers, lakes and wetlands fed by glacier meltwaters. This in turn will affect the environments and biota of the Himalaya to varying degrees. Most evidence appears to suggest that during the past 500,000 years, corridors for migrating birds have been available, especially in the west and central areas of the Tibetan-Himalayan region.

References

Abramowski, U., Bergau, A., Seebach, D., *et al.* (2006). Pleistocene glaciations of Central Asia: results from ^{10}Be surface exposure ages of erratic boulders from the Pamir (Tajikistan) and the Alay-Turkestan range (Kyrgyzstan). *Quaternary Science Reviews*, **25**, 1080–1096.

Bagla, P. (2009). No sign yet of Himalayan meltdown, Indian report finds. *Science*, **326**, 924–925.

Barnard, P.L., Owen, L.A. & Finkel, R.C. (2004a). Style and timing of glacial and paraglacial sedimentation in a monsoonal influenced high Himalayan environment, the upper Bhagirathi Valley, Garhwal Himalaya. *Sedimentary Geology*, **165**, 199–221.

Barnard, P.L., Owen, L.A., Sharma, M.C. & Finkel, R.C. (2004b). Late Quaternary (Holocene) landscape evolution of a monsoon-influenced high Himalayan valley, Gori Ganga, Nanda Devi, NE Garhwal. *Geomorphology*, **61**, 91–110.

Benn, D.I. & Owen, L.A. (1998). The role of the Indian summer monsoon and the mid-latitude westerlies in Himalayan glaciation: review and speculative discussion. *Journal of the Geological Society*, **155**, 353–363.

Benn, D.I. & Owen, L.A. (2002). Himalayan glacial sedimentary environments: a framework for reconstructing and dating former glacial extents in high mountain regions. *Quaternary International*, **97**/98, 3–26.

Benn, D.I., Owen, L.A., Osmaston, H.A., Seltzer, G.O., Porter, S.C. & Mark, B. (2005). Reconstruction of equilibrium-line altitudes for tropical and sub-tropical glaciers. *Quaternary International*, **138**/139, 8–21.

Bishop, M.P., Bush, A., Copland, L., *et al.* (2010). Climate change and mountain topographic evolution in the Central Karakoram, Pakistan. *Annals of Geography*, **100**, 1–22.

Bolch, T., Kulkarni, A., Kääb, A., *et al.* (2012). The state and fate of Himalayan glaciers. *Science*, **336**, 310–314.

Chevalier, M.-L., Hilley, G., Tapponnier, P., *et al.* (2011). Constraints on the late Quaternary glaciations in Tibet from cosmogenic exposure ages of moraine surfaces. *Quaternary Science Reviews*, **30**, 528–554.

Chevalier, M.-L., Ryerson, F.J., Tapponnier, P., *et al.* (2005). Slip-rate measurements on the Karakoram Fault may imply secular variations in fault motion. *Science*, **307**, 411–414.

Cogley, J.G. (2011). Present and future states of Himalaya and Karakoram glaciers. *Annals of Glaciology*, **52**, 69–73.

Copland, L., Sylvestre, T., Bishop, M.P., *et al.* (2011). Expanded and recently increased glacier surging the Karakoram Arctic. *Alpine and Antarctic Research*, **43**, 503–516.

Derbyshire, E. (1981). Glacier regime and glacial sediment facies: a hypothetical framework for the Qinghai-Xizang Plateau. In: *Proceedings of Symposium on Qinghai-Xizang (Tibet) Plateau, Beijing, China.* Geological and Ecological Studies of Qinghai-Xizang Plateau. Vol. **2**. Beijing: Science Press, 1981, pp. 1649–1656.

Derbyshire, E. & Owen, L.A. (1997). Quaternary glacial history of the Karakoram Mountains and northwest Himalayas: a review. *Quaternary International*, **38**/39, 85–102.

Dortch, J.M., Owen, L.A. & Caffee, M.W. (2010). Quaternary glaciation in the Nubra and Shyok valley confluence, northernmost Ladakh, India. *Quaternary Research*, **74**, 132–144.

Dortch, J.M., Owen, L.A. & Caffee, M.W. (2013). *Timing and Climatic Drivers for Glaciation across Semi-arid Western Himalayan-Tibetan Orogen. Quaternary Science Reviews*, 78, 168–208.

Duncan, C.C., Klein, A.J., Masek, J.G. & Isacks, B.L. (1998). Late Pleistocene and modern glaciations in Central Nepal from digital elevation data and satellite imagery. *Quaternary Research*, **49**, 241–254.

Ehlers, J. & Gibbard, P. (2004). Quaternary glaciations – extent and chronologies. Part III: South America, Asia, Africa, Australia, Antarctica. Developments. *Quaternary Science*, **2**, 380 pp.

Elhers, J., Gibbard, P. & Hughes, P.D. (2011). *Quaternary Glaciations – Extent and Chronology: A Closer Look. Developments in Quaternary Science*, vol. **15**, Elsevier, Amsterdam, 2nd Edition, pp. 929–942.

Fielding, E., Isacks, B., Barazangi, M. & Duncan, C. (1994). How flat is Tibet? *Geology*, **22**, 163–167.

Frenzel, B. (1960). Die Vegetations- und Landschaftszonen Nordeurasiens während der letzten Eiszeit und während der Postglazialen Warmezeit. *Akademie der Wissenschaften und der Literatur in Mainz, Abhandlungen der Mathematisch-Naturwissenschaftlichen Klasse*, **13**, 937–1099.

Gillespie, A. & Molnar, P. (1995). Asynchronous maximum advances of mountain and continental glaciers. *Reviews of Geophysics*, **33**, 311–364.

Hedrick, K.A., Seong, Y.B., Owen, L.A., Caffee, M.C. & Dietsch, C. (2011). Towards defining the transition in style and timing of Quaternary glaciation between the monsoon-influenced Greater Himalaya and the semi-arid Transhimalaya of Northern India. *Quaternary International*, **236**, 21–33.

Hewitt, K. (1999). Quaternary moraines vs catastrophic avalanches in the Karakoram Himalaya, northern Pakistan. *Quaternary Research*, **51**, 220–237.

Hewitt, K., Gosse, J. & Clague, J.J. (2011). Rock avalanches and the pace of late Quaternary development of river valleys in the Karakoram Himalaya. *Geological Society of America Bulletin*, **123**, 1836–1850.

Heyman, J., Hattestrand, C. & Stroeven, V. (2008). Glacial geomorphology of the Bayan Har sector of the NE Tibetan plateau. *Journal of Maps*, **2008**, 42–62.

Intergovernmental Panel on Climate Change (2007). *Climate Change 2007: Impacts, Adaptations and Vulnerability.* Parry, M., Canziani, O., Palutikof, J., Van der Linden,

P. & Hanson, C. eds., Contribution of Working Group II to the Fourth Assessment Report of the Intergovernmental Panel on Climate Change. Cambridge, UK: Cambridge University Press, p. 976.

Kamp, U. & Owen, L.A. (2011). Late Quaternary Glaciation of Northern Pakistan. In *Quaternary Glaciations – Extent and Chronology: A Closer Look*. Developments in Quaternary Science, vol. **15**, Amsterdam: Elsevier, pp. 909–927.

Klute, F. (1930). Verschiebung der Klimagebiete der letzten Eiszeit. *Petermanns Mitteilungen Ergänzungsheft*, **209**, 166–182.

Kuhle, M. (1985). Ein subtropisches Inlandeis als Eiszeitauslöser, Südtibet un Mt. Everest expedition 1984. *Georgia Augusta, Nachrichten aus der Universität Gottingen*, **May**, 1–17.

Kuhle, M. (1987). The Problem of a Pleistocene Inland Glaciation of the Northeastern Qinghai-Xizang Plateau. In Hövermann J & Wang, W., eds., *Reports of the Qinghai-Xizang (Tibet) Plateau*. Beijing: Science Press, pp. 250–315.

Kuhle, M. (1988). Geomorphological findings on the build-up of Pleistocene glaciation in southern Tibet and on the problem of inland ice. *GeoJournal*, **17**, 457–512.

Kuhle, M. (1991). Observations supporting the Pleistocene inland glaciation of High Asia. *GeoJournal*, **25**, 131–231.

Kuhle, M. (1995). Glacial isostatic uplift of Tibet as a consequence of a former ice sheet. *GeoJournal*, **37**, 431–449.

Lee, S.Y., Seong, Y.B., Owen, L.A., *et al.* (2014). Late Quaternary glaciation in the Nun-Kun massif, northwestern India. *Boreas*, **43**, 67–89.

Li, B., Li, J. & Cui, Z. (1991). Quaternary glacial distribution map of Qinghai-Xizang (Tibet) Plateau 1:3,000,000. Shi Y. (Scientific Advisor), Quaternary Glacier, and Environment Research Center, Lanzhou University.

Lowe, J. & Walker, M. (2015). *Reconstructing Quaternary Environments*. London: Routledge, 3rd edition, p. 538.

Mayer, C., Lambrecht, A., Belò, M., Smiraglia, C. & Diolaiuti, G. (2006). Glaciological characteristics of the ablation zone of Baltoro glacier, Karakoram, Pakistan. *Annals of Glaciology*, **43**, 123–131.

Mix, A.C., Bard, E. & Schneider, R. (2001). Environmental processes of the ice age: land, ocean, glaciers (EPILOG). *Quaternary Science Reviews*, **20**, 627–657.

Morén, B., Heyman, J. & Stroeven, A.P. (2011). Glacial geomorphology of the central Tibetan Plateau. *Journal of Maps*, **2011**, 115–125.

Murari, M.K., Owen, L.A., Dortch, J.M., *et al.* (2014). Timing and climatic drivers for glaciation across monsoon-influenced regions of the Himalayan-Tibetan orogen. *Quaternary Science Reviews*, **88C**, 159–182.

National Academy (Committee on Himalayan Glaciers, Hydrology, Climate Change, and Implications for Water Security) (2012). Himalayan Glaciers: Climate Change, Water Resources, and Water Security. Washington, DC: The National Academies Press, p. 156.

NGRIP members (2004). High-resolution record of Northern Hemisphere climate extending into the last interglacial period. *Nature*, **431**, 147–151.

Owen, L.A. (2010) Landscape development of the Himalayan-Tibetan orogen: a review. *Special Publication of the Geological Society of London*, **338**, 389–407.

Owen, L.A. & Benn, D.I. (2005). Equilibrium-line altitudes of the Last Glacial Maximum for the Himalaya and Tibet: an assessment and evaluation of results. *Quaternary International*, **138**/139, 55–78.

Owen, L.A., Caffee, M., Bovard, K., Finkel, R.C. & Sharma, M. (2006). Terrestrial cosmogenic surface exposure dating of the oldest glacial successions in the Himalayan orogen. *Geological Society of America Bulletin*, **118**, 383–392.

Owen, L.A., Caffee, M.W., Finkel, R.C. & Seong, B. S. (2008). Quaternary glaciation of the Himalayan–Tibetan orogen. *Journal of Quaternary Science*, **23**, 513–532.

Owen, L.A. & Dortch, J.M. (2014) Quaternary glaciation of the Himalayan-Tibetan orogen. *Quaternary Science Reviews*, **88**, 14–54.

Owen, L.A., Finkel, R.C., Caffee, M.W. & Gualtieri, L. (2002a). Timing of multiple glaciations during the Late Quaternary in the Hunza Valley, Karakoram Mountains, northern Pakistan: defined by cosmogenic radionuclide dating of moraines. *Geological Society of America Bulletin*, **114**, 593–604.

Owen, L.A., Gualtieri, L., Finkel, R.C., Caffee, M.W., Benn, D.I. & Sharma, M.C. (2001). Cosmogenic radionuclide dating of glacial landforms in the Lahul Himalaya, northern India: defining the timing of Late Quaternary glaciation. *Journal of Quaternary Science*, **16**, 555–563.

Owen, L.A., Kamp, U., Spencer, J.Q. & Haserodt, K. (2002b). Timing and style of Late Quaternary glaciation in the eastern Hindu Kush, Chitral, northern Pakistan: a review and revision of the glacial chronology based on new optically stimulated luminescence dating. *Quaternary International*, **97**–98, 41–56.

Owen, L.A., Robinson, R., Benn, D.I., *et al.* (2009). Quaternary glaciation of Mount Everest. *Quaternary Science Reviews*, **28**, 1412–1433.

Owen, L.A., White, B., Rendell, H. & Derbyshire, E. (1992). Loessic silts in the western Himalayas: their sedimentology, genesis and age. *Catena*, **19**, 493–509.

Owen, L.A., Yi, C., Finkel, R.C. & Davis, N. (2010). Quaternary glaciation of Gurla Mandata (Naimon'anyi). *Quaternary Science Reviews*, **29**, 1817–1830.

Phillips, W.M., Sloan, V.F., Shroder Jr., J.F., Sharma, P., Clarke, M.L. & Rendell, H.M. (2000). Asynchronous glaciation at Nanga Parbat, northwestern Himalaya Mountains, Pakistan. *Geology*, **28**, 431–434.

Porter, S.C. (1970). Quaternary glacial record in the Swat Kohistan, West Pakistan. *Geological Society of America Bulletin*, **81**, 1421–1446.

Richards, B.W.M., Owen, L.A. & Rhodes, E.J. (2000). Timing of Late Quaternary glaciations in the Himalayas of northern Pakistan. *Journal of Quaternary Science*, **15**, 283–297.

Röhringer, I., Zech, R., Abramowski, U., *et al.* (2012). The late Pleistocene glaciation in the Bogchigir Valleys (Pamir, Tajikistan) based on ^{10}Be surface exposure dating. *Quaternary Research*, **78**, 590–597.

Röthlisberger, F. & Geyh, M.A. (1985a). *Gletscherschwankungen der letzten 10.000 Jahre – Ein Verleich zwischen Nord- und Südhemisphäre (Alpen, Himalaya, Alaska, Südamerika, Neuseeland)*. Aarau: Verlag Sauerländer.

Röthlisberger, F. & Geyh, M. (1985b). Glacier variations in Himalayas and Karakoram. *Zeitschrift für Gletscherkunde und Glazialgeologie*, **21**, 237–249.

Rutter, N.W. (1995). Problematic ice sheets. *Quaternary International*, **28**, 19–37.

Schäfer, J.M., Tschudi, S., Zhao, Z., *et al.* (2002). The limited influence of glaciations in Tibet on global climate over the past 170000 yr. *Earth and Planetary Science Letters*, **194**, 287–297.

Seong, Y.B., Owen, L.A., Bishop, M.P., *et al.* (2007). Quaternary glacial history of the central Karakoram. *Quaternary Science Reviews*, **26**, 3384–3405.

Seong, Y.B., Owen, L.A., Bishop, M.P., *et al.* (2008). Reply to comments by Matthias Kuhle on Seong, Y.B., Owen, L.A., Bishop, M.P., Bush, A., *et al.* 2007. Quaternary glacial history of the central Karakoram. *Quaternary Science Reviews*, **27**, 1656–1658.

Seong, Y.B., Owen, L.A., Yi, C. & Finkel, R.C. (2009). Quaternary glaciation of Muztag Ata and Kongur Shan: evidence for glacier response to rapid climate changes throughout the Late Glacial and Holocene in westernmost Tibet. *Geological Society of America, Bulletin*, **121**, 348–365.

Sharma, M.C. & Owen, L.A. (1996). Quaternary glacial history of NW Garhwal Himalayas. *Quaternary Science Reviews*, **15**, 335–365.

Shi, Y. (1992). Glaciers and glacial geomorphology in China. *Zeitschrift für Geomorphologie*, **86**, 19–35.

Shi, Y., Zheng, B. & Li, S. (1992). Last glaciation and maximum glaciation in the Qinghai-Xizang (Tibet) Plateau: a controversy to M.Kuhle's ice sheet hypothesis. *Zeitschrift für Geomorphologie*, **84**, 19–35.

Shiraiwa, T. & Watanabe, T. (1991). Late Quaternary glacial fluctuations in the Langtang Valley, Nepal Himalaya, reconstructed by relative dating methods. *Arctic and Alpine Research*, **23**, 404–416.

Solomina, O., Bradley, R.S., Hodgson, D.A., *et al.* (2015). Holocene glacier fluctuations. Invited review. *Quaternary Science Reviews*, **111**, 9–34.

Spencer, J.Q. & Owen, L.A. (2004). Optically stimulated luminescence dating of Late Quaternary glaciogenic sediments in the upper Hunza valley: validating the timing of glaciation and assessing dating methods. *Quaternary Sciences Reviews*, **23**, 175–191.

Taylor, P.J. & Mitchell, W.A. (2000). Late Quaternary glacial history of the Zanskar Range, north-west Indian Himalaya. *Quaternary International*, **65**/66, 81–100.

Thackray, G.D., Owen, L.A. & Yi, C. (2008). Timing and nature of late Quaternary mountain glaciation. *Journal of Quaternary Science*, **23**, 503–508.

Watanabe, T., Shiraiwa, T. & Ono, Y. (1989). Distribution of periglacial landforms in the Langtang Valley, Nepal Himalaya. *Bulletin of Glacier Research*, **7**, 209–220.

Yi, C., Chen, H., Yang, J., *et al.* (2008). Review of Holocene glacial chronologies based on radiocarbon dating in Tibet and its surrounding mountains. *Journal of Quaternary Science*, **23**, 533–558.

Zech, R., Abramowski, U., Glaser, B., Sosin, P., Kubik, P.W. & Zech, W. (2005). Late Quaternary glacier and climate history of the Pamir Mountains derived from cosmogenic ^{10}Be exposure ages. *Quaternary Research*, **64**, 212–220.

Zech, W., Glaser, B., Sosin, P., Kubik, P.W. & Zech, W. (2003). Evidence for long-lasting landform surface instability on hummocky moraines in the Pamir Mountains (Tajikistan) from ^{10}Be surface exposure dating. *Earth and Planetary Science Letters*, **237**, 453–461.

Zheng, B. & Rutter, N. (1998). On the problem of Quaternary glaciations, and the extent and patterns of Pleistocene ice cover in the Qinghai-Xizang (Tibet) plateau. *Quaternary International*, **45**/46, 109–122.

Zhou, S., Li, J., Zhao, J., Wang, J. & Zheng, J. (2011). Quaternary glaciations: extent and chronology in China. In Elhers, J., Gibbard, P. & Hughes P.D., eds., *Quaternary Glaciations – Extent and Chronology: A Closer Look*. Developments in Quaternary Science, vol. **15**, Amsterdam: Elsevier, 2nd Edition, pp. 981–1002.

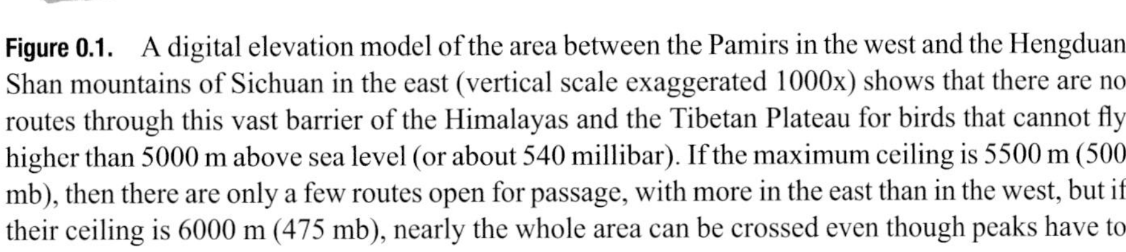

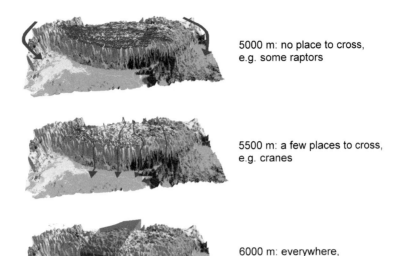

Figure 0.1. A digital elevation model of the area between the Pamirs in the west and the Hengduan Shan mountains of Sichuan in the east (vertical scale exaggerated 1000x) shows that there are no routes through this vast barrier of the Himalayas and the Tibetan Plateau for birds that cannot fly higher than 5000 m above sea level (or about 540 millibar). If the maximum ceiling is 5500 m (500 mb), then there are only a few routes open for passage, with more in the east than in the west, but if their ceiling is 6000 m (475 mb), nearly the whole area can be crossed even though peaks have to be circumvented.

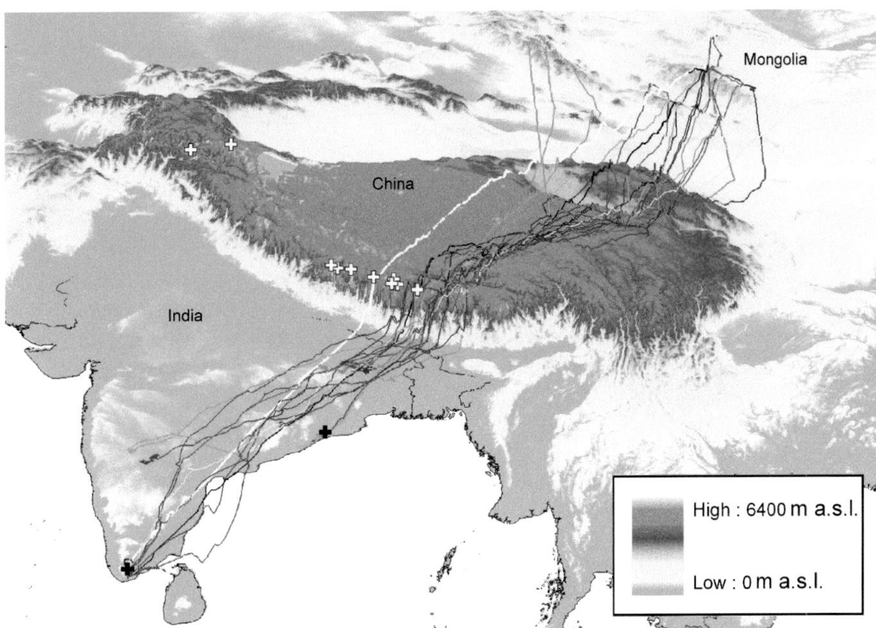

Figure 1.3. Three-dimensional tracks of 16 Bar-headed Geese from CL, KT and TT crossing the Himalayas. Solid white line shows the great circle route, and white plus signs indicate locations of peaks > 8000 m in elevation. Adapted from Hawkes *et al.* (2011).

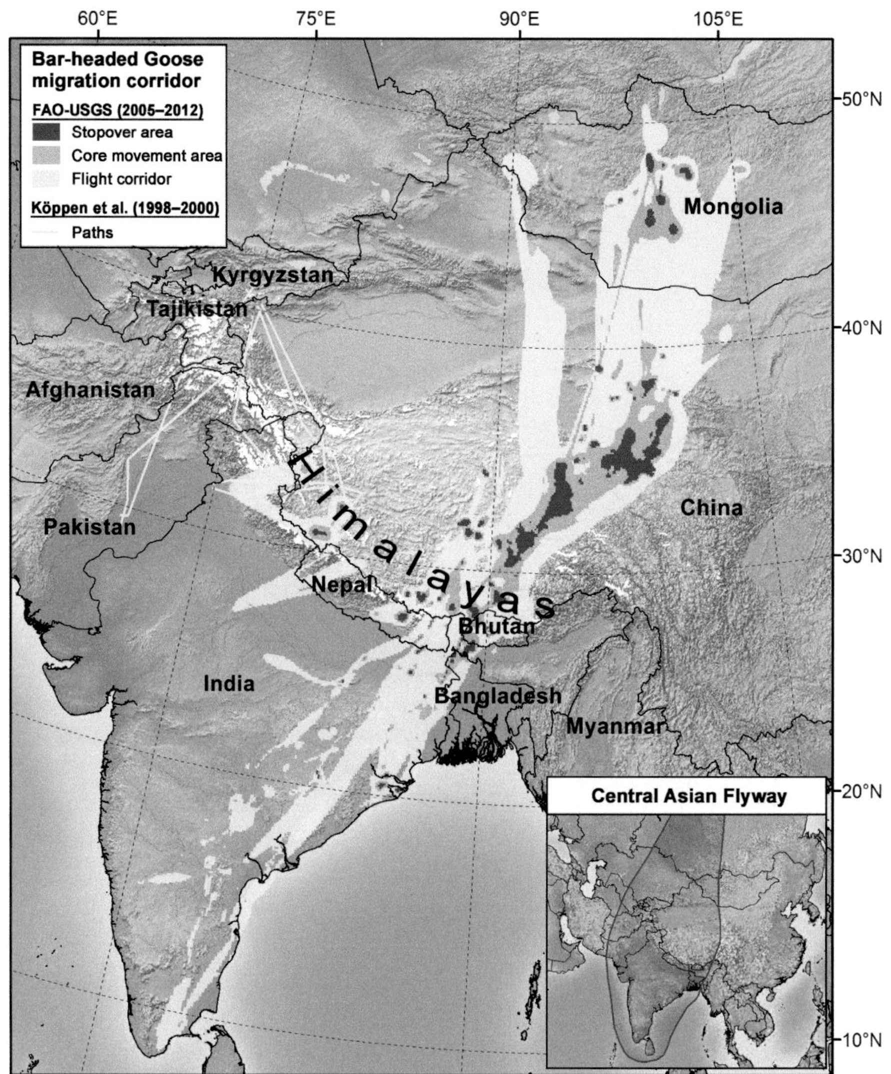

Figure 1.1. Extent of the Central Asian Flyway (inset) and migration routes and relative use of stopover areas for 44 satellite-marked Bar-headed Geese (BHGO). Colours, from darkest to lightest, represent 50%, 75% and 99% cumulative probability contours from a dynamic Brownian Bridge movement model. Adapted from Palm *et al.* (2015) and Köppen (2010).

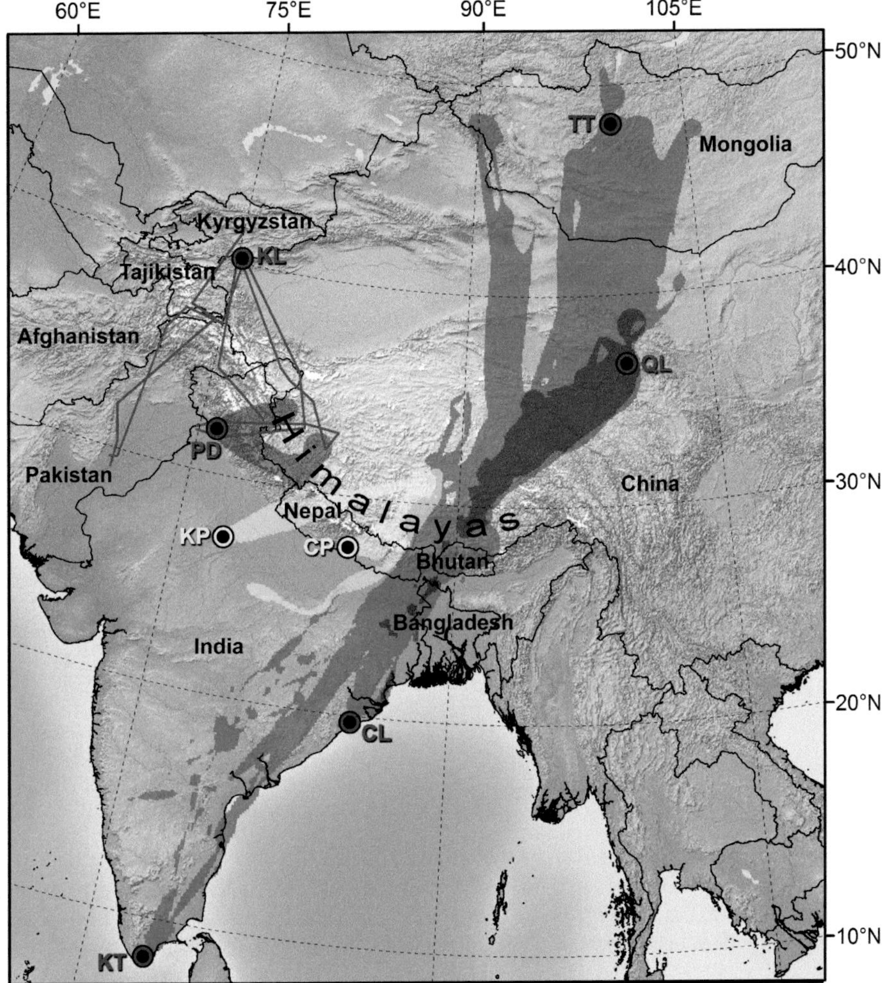

Figure 1.2. Distinct migration patterns of Bar-headed Geese from different marking sites of Terkhiin Tsagaan Lake, Mongolia (TT), Kyrgyzstan Lakes, Kyrgyzstan (KL), Qinghai Lake, China (QL), Chitwan National Park, Nepal (CP), Pong Dam, India (PD), Keoladeo National Park, India (KP), Chilika Lake, India (CL) and Koonthankulam, India (KT). Colours of marking site labels correspond to shaded areas highlighting separate migration routes. Shaded areas represent 99% cumulative probability contours of a dynamic Brownian bridge movement model.

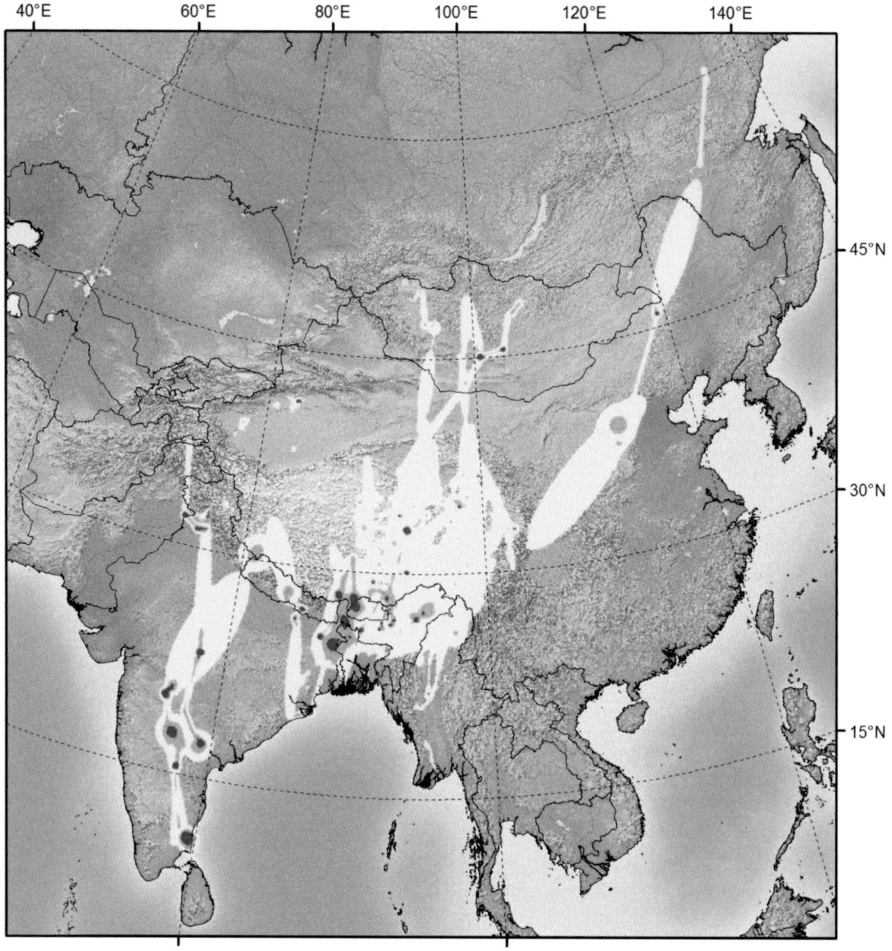

40°E 60°E 80°E 100°E 120°E 140°E
45°N
30°N
15°N

Figure 2.1. Estimated migration routes and utilization distributions of ducks in the Central Asian Flyway. From darkest to lightest, colours represent 50%, 75% and 99% cumulative probability contours (adapted from Palm *et al*., 2015). Note that ringing locations in eastern and central India introduce bias into the mapped distributions, which do not include migration routes from western lowland India. Map data: Google earth.

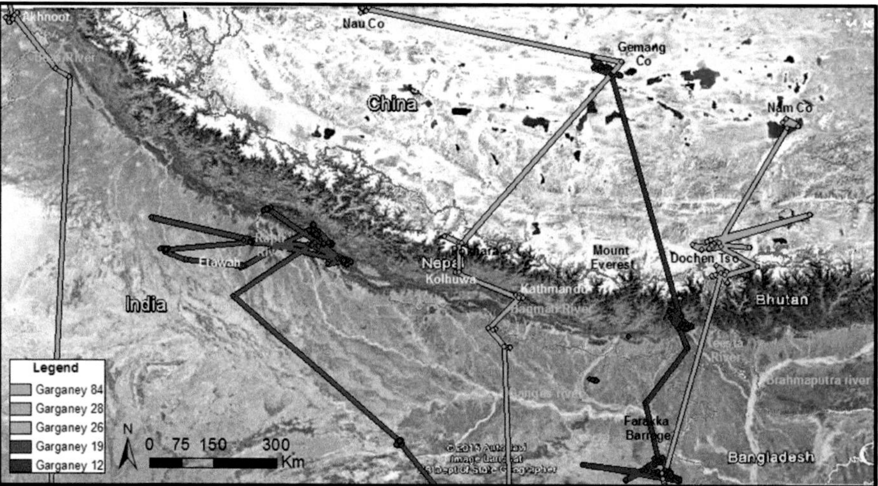

Figure 2.7. Movement paths of Ruddy Garganey migrating between Central Asia and the Indian Peninsula, crossing the Himalayas. Map data: Google earth.

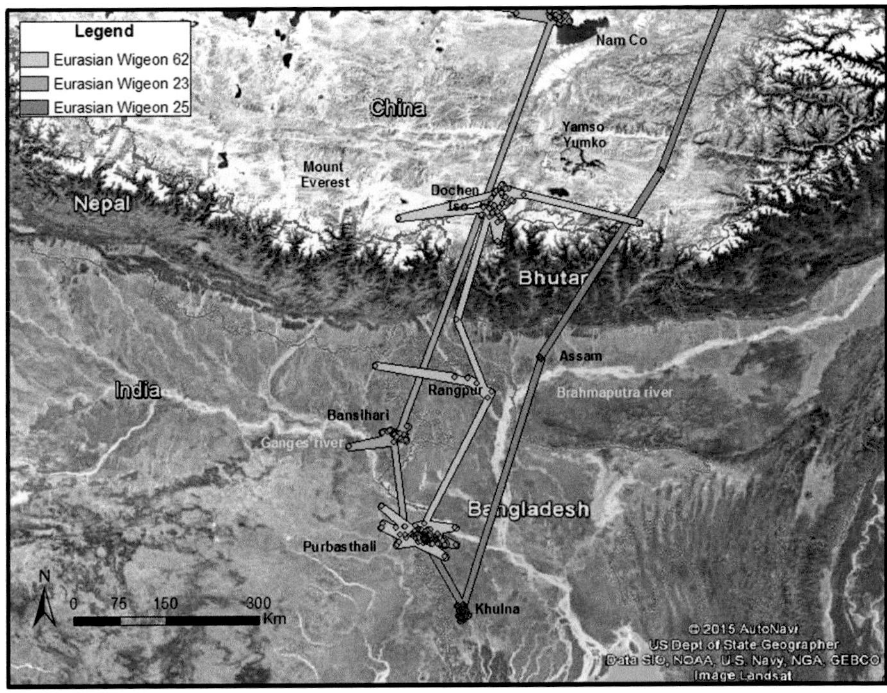

Figure 2.2. Movement paths of Eurasian Wigeon migrating between Central Asia and the Indian Peninsula, crossing the Himalayas. Map data: Google earth.

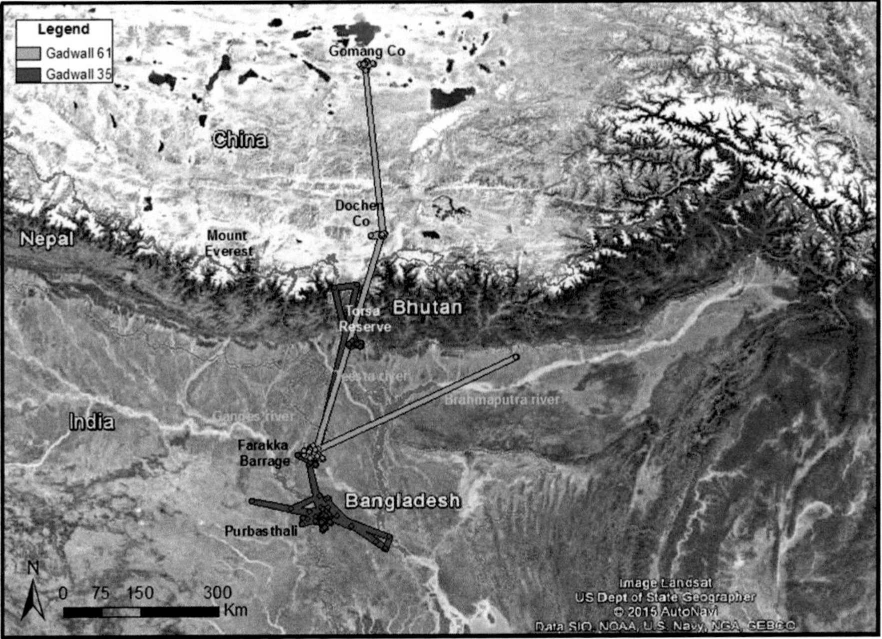

Figure 2.3. Movement paths of Gadwall migrating between Central Asia and the Indian Peninsula, crossing the Himalayas. Map data: Google earth.

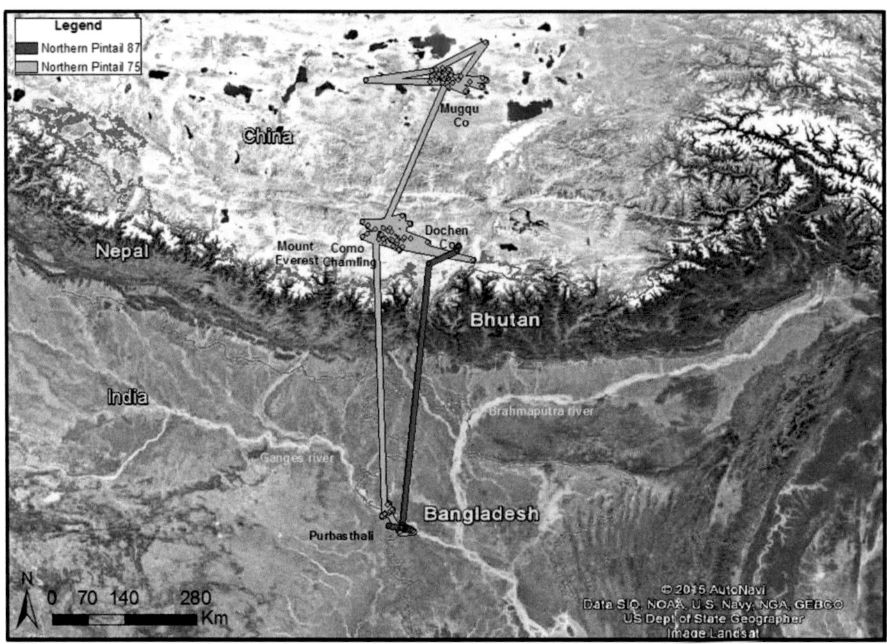

Figure 2.4. Movement paths of Northern Pintail migrating between Central Asia and the Indian Peninsula, crossing the Himalayas. Map data: Google earth.

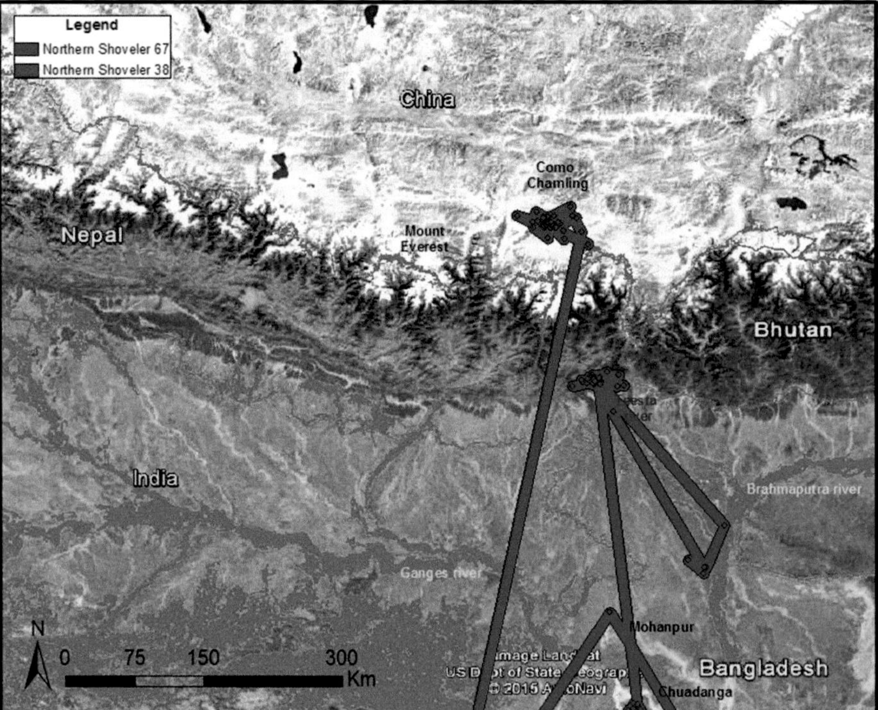

Figure 2.5. Movement paths of Northern Shoveler migrating between Central Asia and the Indian Peninsula, crossing the Himalayas. Map data: Google earth.

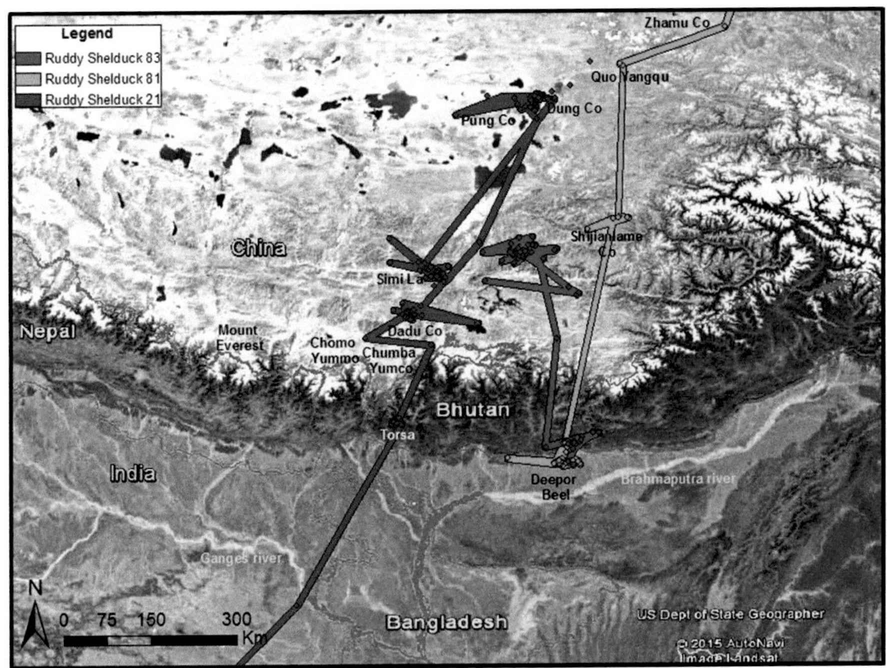

Figure 2.6. Movement paths of Ruddy Shelduck migrating between Central Asia and the Indian Peninsula, crossing the Himalayas. Map data: Google earth.

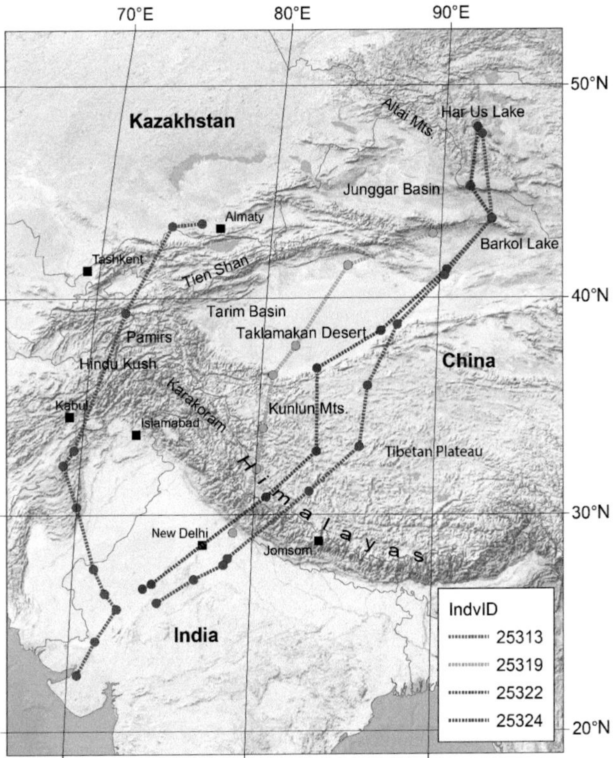

Figure 3.2. Map showing the satellite-tracked migration routes of Demoiselle Cranes from India onto the western Tibetan Plateau. Based on Kanai *et al.*, 2000.

Figure 4.2. Guldenstadt's Redstart, Thikse, Ladakh, October 1980. A sparsely distributed, high-altitude breeder throughout the Himalayas and ranges to the north, large numbers migrate altitudinally to lower-lying river valleys for the winter, where they are associated with thickets of Sea Buckthorn, *Hippophaë rhamnoides*. Photo, by John Norton.

Figure 3.1. A flock of Demoiselle Cranes in flight around Mount Nilgiri (western Himalayas) on 7 October 2013. Photo by Rajendra Suwal.

Figure 4.1. The recovery of a ringed Hume's Lesser Whitethroat suggests a direct migratory route across the Zanskar and Great Himalaya ranges from the Indus Valley.

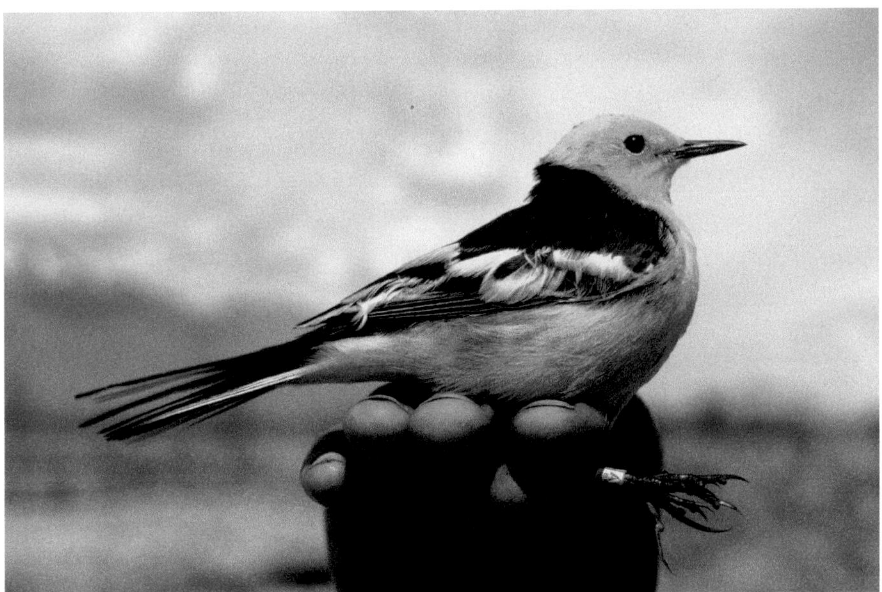

Figure 4.3. Citrine Wagtail, Thikse, Ladakh autumn 1981. A common summer visitor to the Tibetan-Himalayan region, wintering widely in southern Asia. Photo, by Clare Sulston.

Figure 5.3. Green Sandpiper, Thikse, Ladakh, autumn 1981. The most frequently recorded wader in the Tibetan-Himalayan region, this species is very regular on spring and autumn migration and infrequent in other seasons. Photo, by Clare Sulston.

Figure 4.4. Sedge Warbler, Thikse, Ladakh, 11 October 1981. One of only two recorded in the Himalayan region, both by Southampton University expeditions. This was one of five warbler species discovered migrating between Asia and Africa via the northwest Himalayas. Photo, by Clare Sulston.

Figure 5.1. Ibisbill, Thikse, Ladakh, January 1982. A sparsely distributed resident of gravel-bottomed rivers throughout the Tibetan-Himalayan region, birds from the highest altitudes migrate to lower-lying river valleys for the winter. Photo, by Clare Sulston.

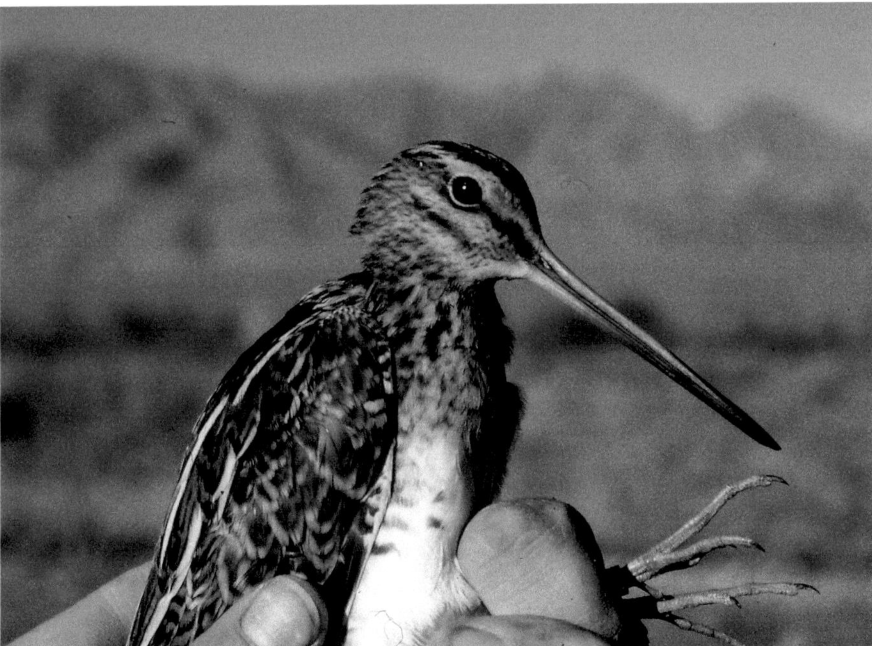

Figure 5.2. Common Snipe, Thikse, Ladakh, autumn 1980. First recorded in the Himalayas by Hodgson (1833), this is a regular long-distance migrant throughout the Himalayas in small numbers. Photo, by John Norton.

Figure 6.1. Adult Steppe Eagle on 11 November 2014 in Thulakharka (Nepal). Photo credit Hawk Mountain Sanctuary Archives.

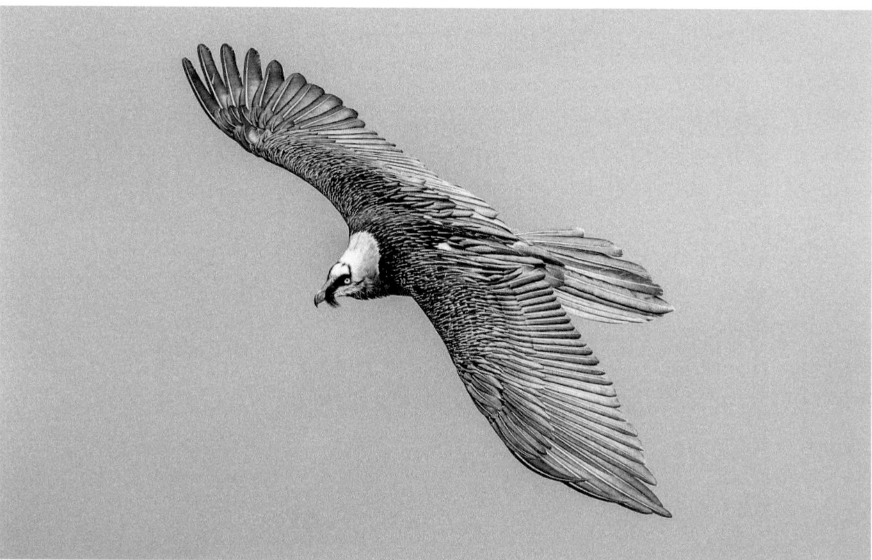

Figure 6.4. Soaring adult Lammergeyer on 23 November 2014 in Thoolakharka (Nepal; on maps referred to as Thuli Kharka: 28°22 N, 82°19'E), which is a well-known raptor migration site. Photo credit Hawk Mountain Sanctuary Archives.

Figure 6.5. Adult Lammergeyer on 26 November 2014 in Thoolakharka (Nepal). For more information on this spot, see http://raptorsofnepal.blogspot.nl/ Photo credit Hawk Mountain Sanctuary Archives.

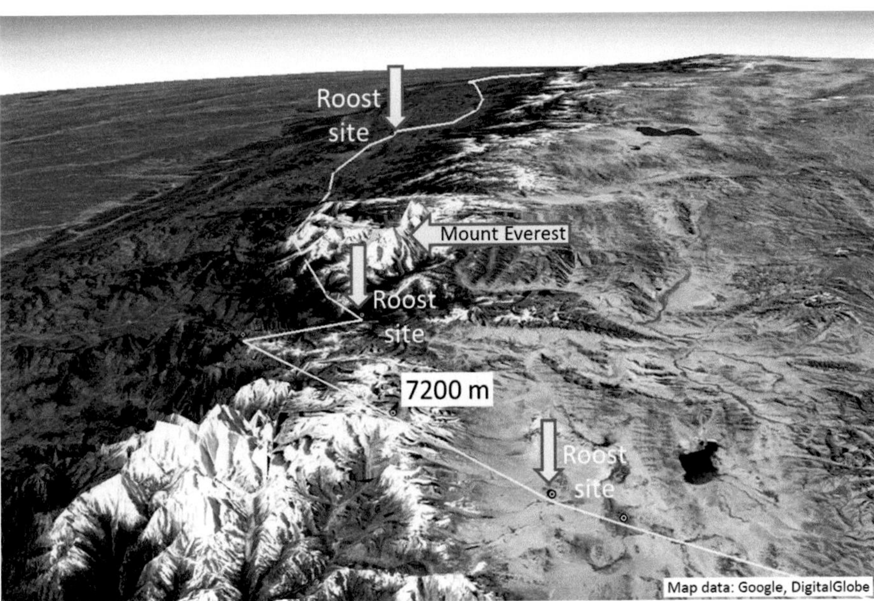

Figure 7.4. Migration path and roost locations of an immature Steppe Eagle before, during and after the crossing of the Himalayan mountain ridge.

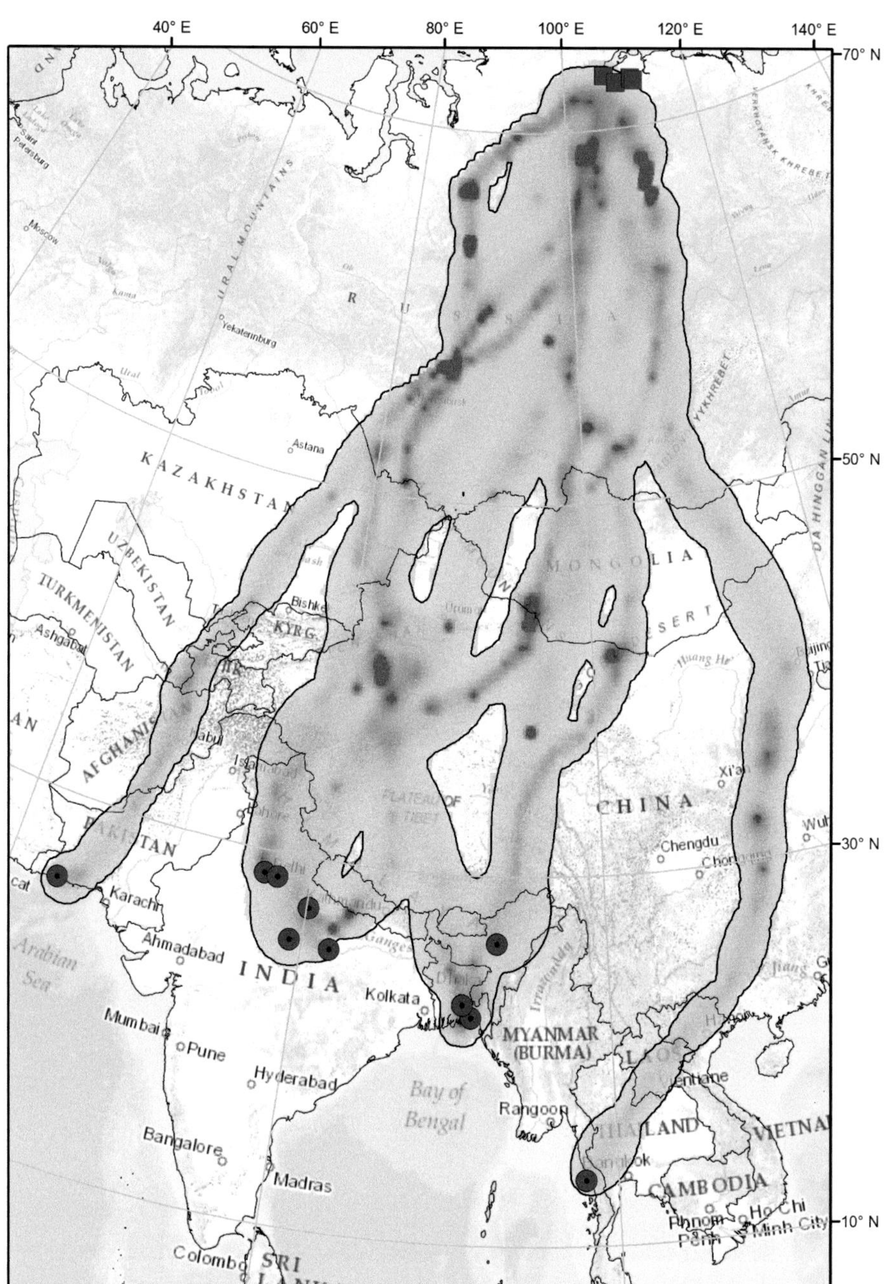

Figure 8.3. Autumn migration pathways of 10 Peregrines in 2011 from the Popigai River, Russia, plotted using dynamic Brownian bridge movement modelling based on best of day satellite locations. We calculated utilization distribution (UD) output grids for all birds at 50 km^2 resolution. We used a window length of approximately 7 days. The mean number of locations during migration period was 29 (range: 13–48). We weighted each individual UD by multiplying all pixel values by the total number of days for its migration event, and then summed the pixel values for all their weighted UDs and then re-scaled their cumulative pixel values to the sum of 1. The resulting UD represented the proportional amount of time each pixel was occupied across an individual's migration route.

Figure 9.1. The Tibetan Plateau (to the right) and the Gangetic Plains of India (to the left) and in between the snow-clad Himalayas as photographed by the NASA Space Shuttle (Photograph courtesy of NASA).

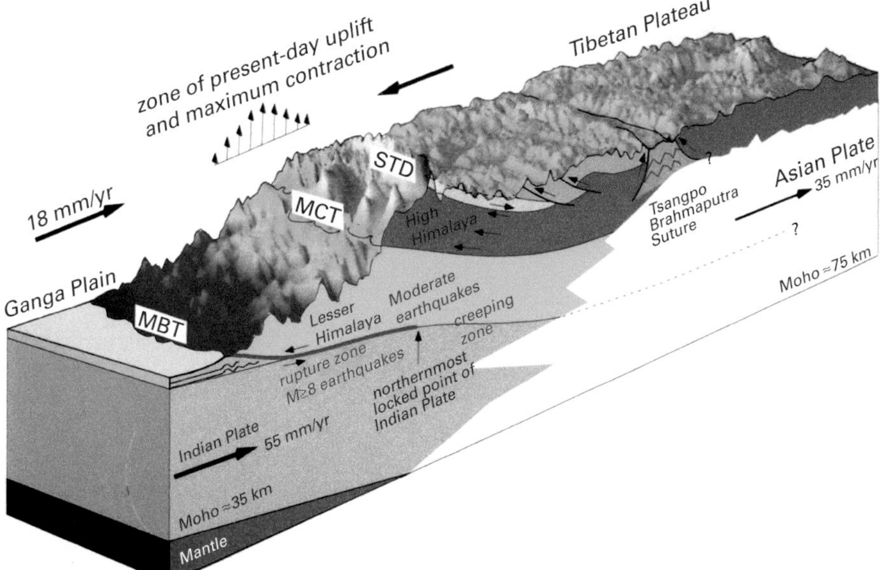

Figure 9.2. Block diagram of the Indian Plate dipping under the Asian Plate. MBT stands for 'Main Boundary Thrust', MCT for 'Main Central Thrust' and STD for 'South Tibetan Detachment'.

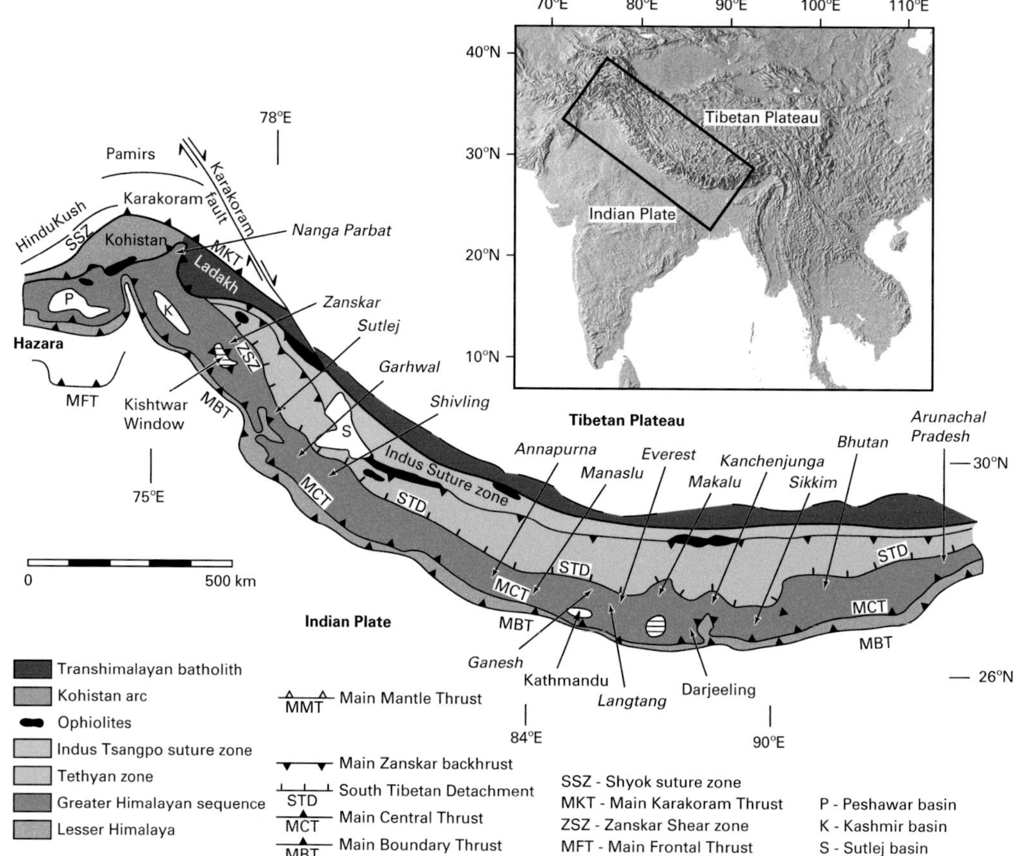

Figure 9.3. Simplified geology of the Himalayas between the Asian Plate (with the Tibetan Plateau) and the Indian Plate (with the Gangetic Plain).

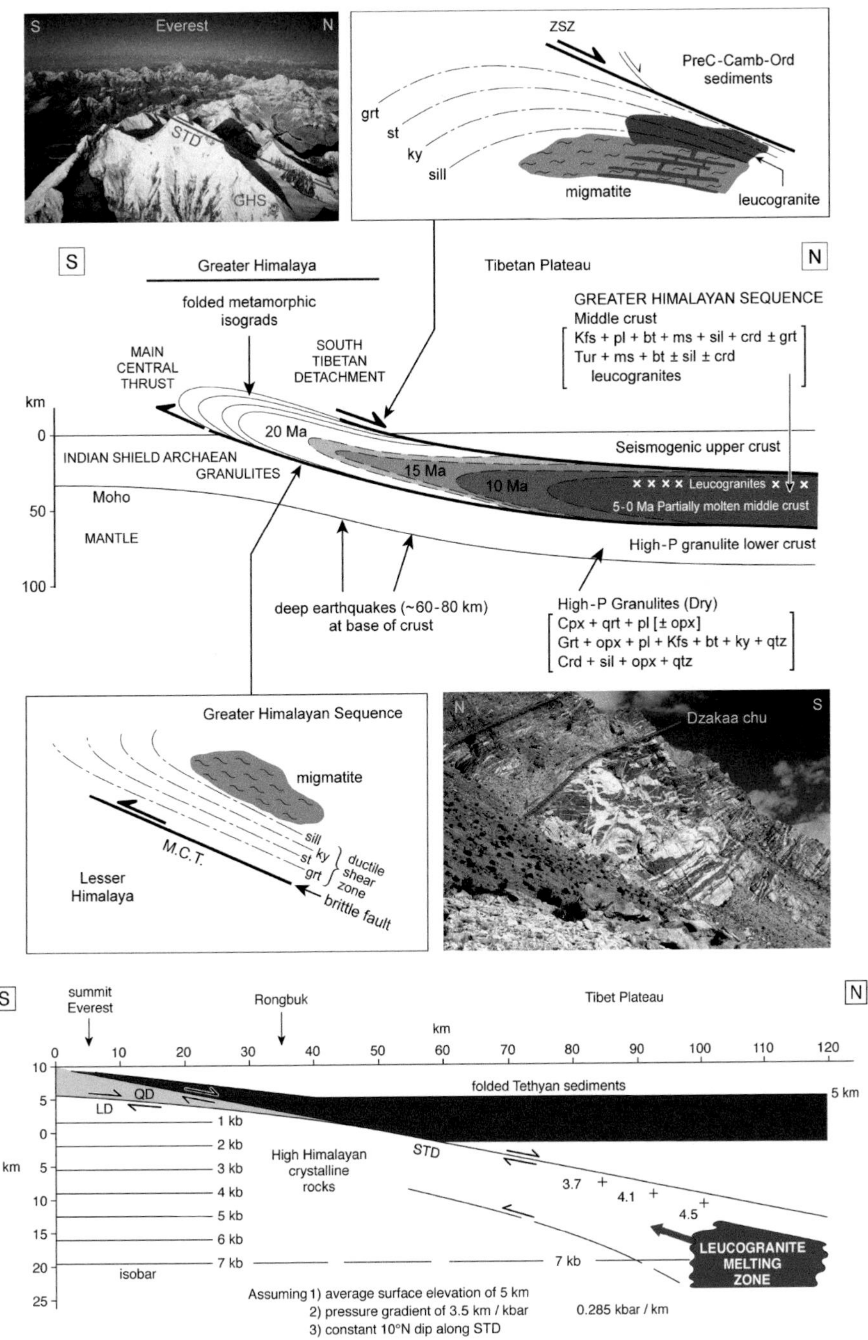

Figure 9.4. The Himalayan channel flow after Searle *et al.* (2006) showing right way up metamorphic isograds along the top of the Greater Himalayan slab and inverted isograds along the base wrapping around a southward extruding partially molten layer of middle crust. The upper fault of the South Tibetan Detachment clips the summit region of Everest and dips north beneath Tibet.

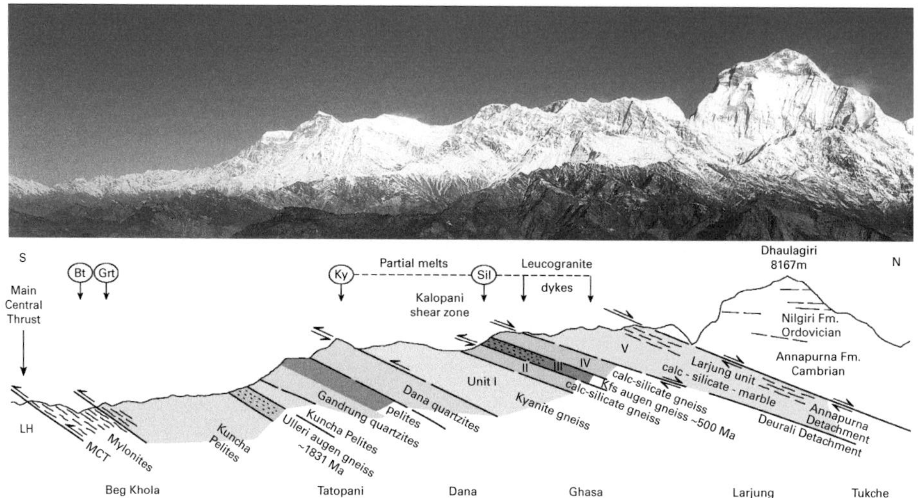

Figure 9.5. Panorama of the Dhaulagiri Himalaya above the Kali Gandaki River, after Searle (2010). (For the full caption, please see page 151.)

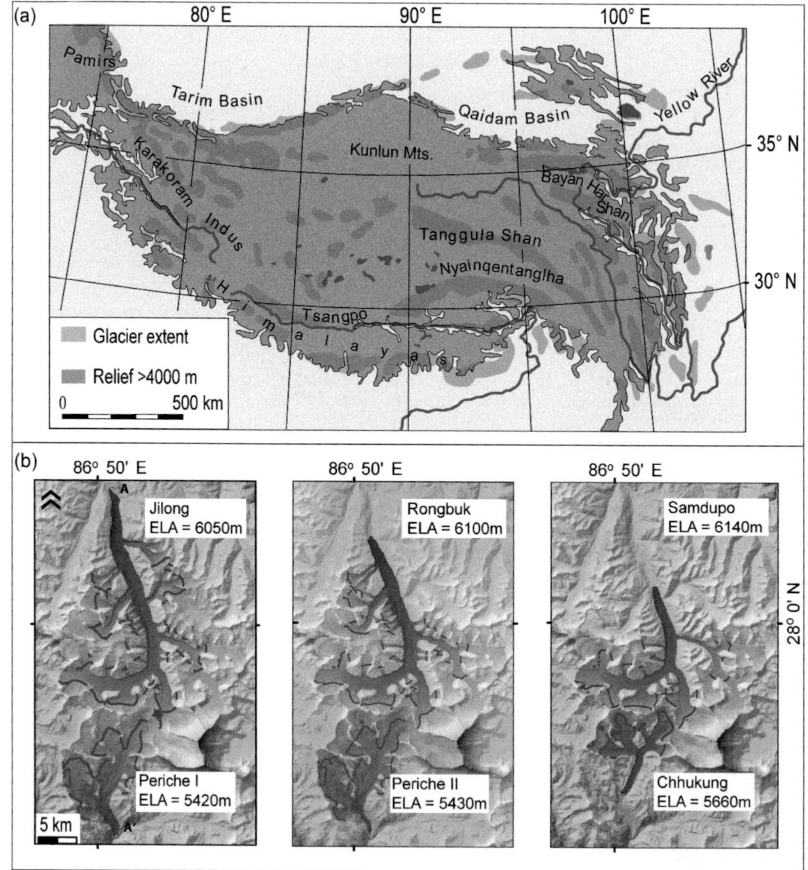

Figure 10.1. a) Reconstruction of the maximum extent of glaciation (from Owen *et al.*, 2008 and Owen, 2010, after Shi, 1992 and Li *et al.*, 1991). b) Example of detailed reconstruction for the Late Quaternary glaciation of a selected area (after Owen *et al.*, 2009). (For the full caption, please see page 157.)

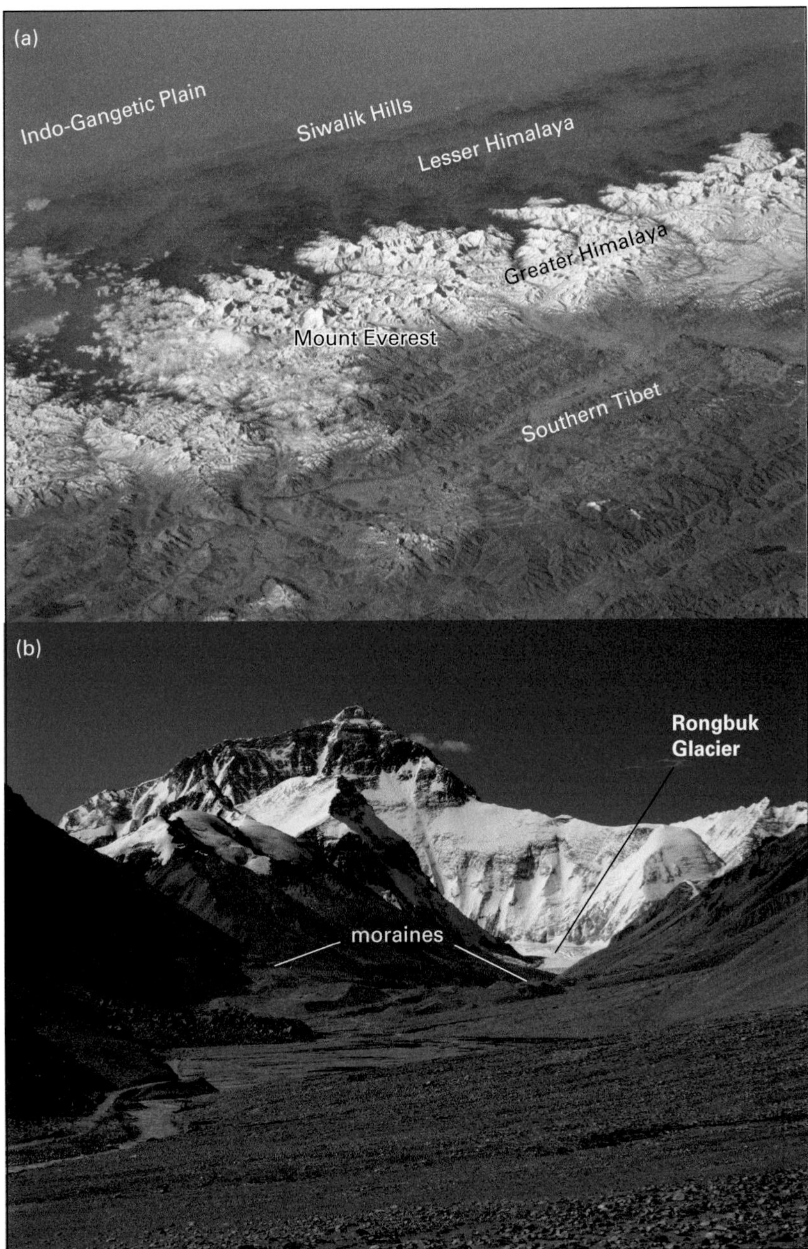

Figure 10.2. a) Low-oblique photograph looking southwest across a stretch of the Himalayas and southern Tibet, with Mount Everest in the centre of the frame, taken from the Space Shuttle in March 1996 (NASA image STS076-727-080). This illustrates typical topography, and glacier and snow cover across a stretch of the mountain belt and plateau. Note the green humid Indo-Gangetic Plain in the top left corner and the brown semiarid interior of Tibet essentially devoid of snow and glacier cover in the lower right corner. b) View of the north face of Mount Everest rising to 8848 m a.s.l. and Rongbuk Glacier and its moraines in the foreground.

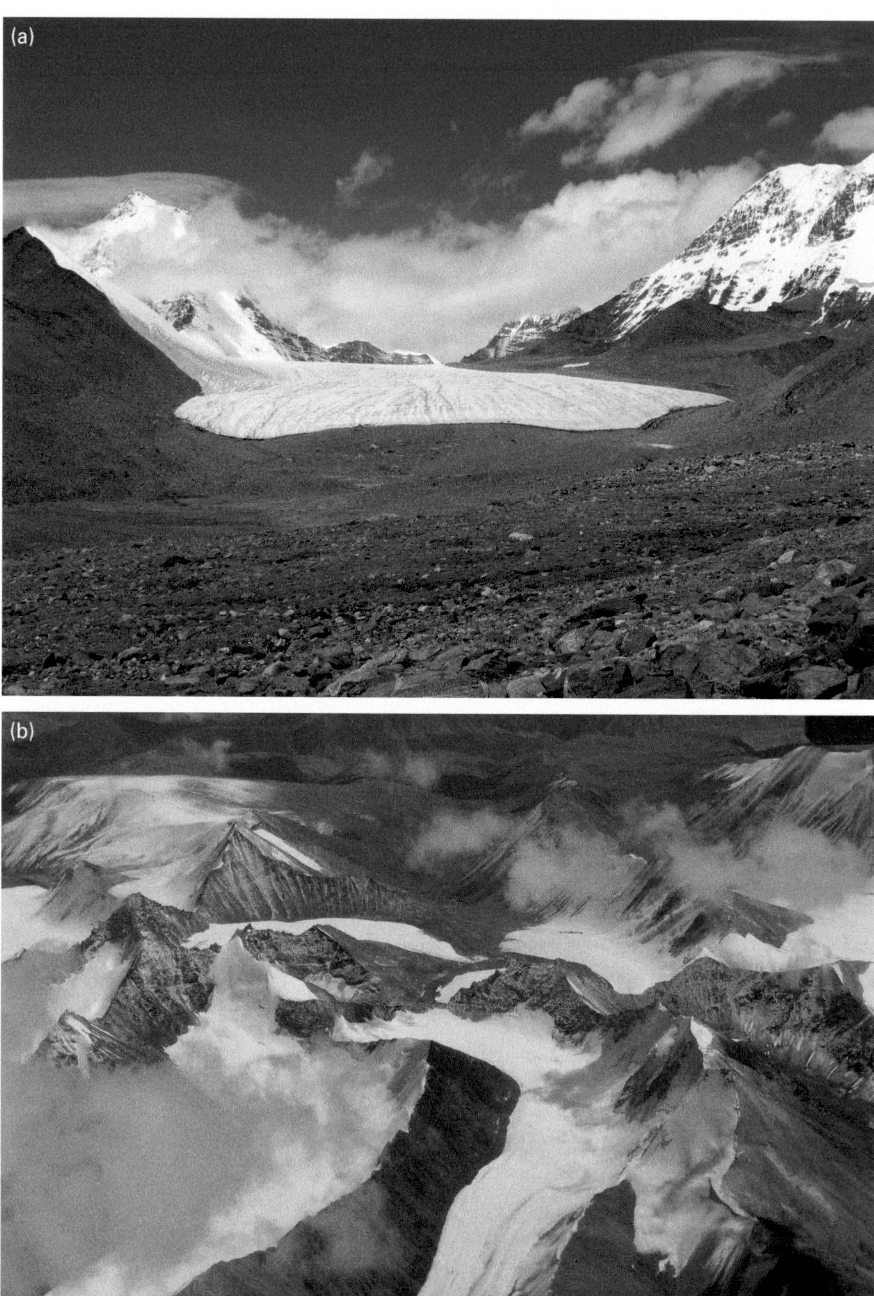

Figure 10.3. Examples of glaciers and associated terrain of the Himalayas and southern Tibet illustrating the wide variety of forms. a) Relatively debris-free continental glacier advancing from the Gurla Mandhata (Naimon'anyi; the peak (left) is at 7694 m a.s.l.) down the Namarodi Valley in southern Tibet. Near its snout, the glacier is expanding to become almost a piedmont-type glacier. b) Small, cold, continental v-shaped valley glaciers in Zanskar, northern India. c) View across the debris-covered Baltoro Glacier at Concordia in northern Pakistan with K2 at 8611 m a.s.l. (left)

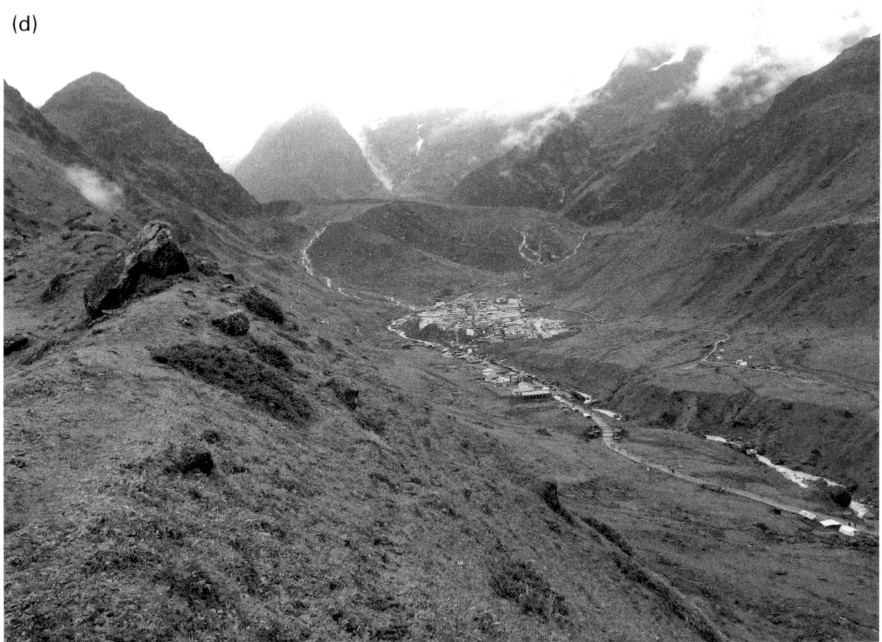

Caption for Figure 10.3. (cont.)

and Broad Peak at 8047 m a.s.l. (right) in the background. At > 60 km long, the Baltoro Glacier is one of the longest mountain glaciers outside of the polar regions and at Concordia the glacier is ~ 3 km wide. d) View north at Kedarnath town (at ~3600 m a.s.l.) with Chorabari Glacier in the background somewhat obscured by the large latero-frontal moraines that define the margins of this maritime debris-mantled glacier (photograph taken in September 2010). The ridges on the left and right sides of the view are moraines that formed during the early Holocene. Kedarnath was tragically destroyed during a flood in June 2013 during a multi-day cloudburst and drainage of a lake associated with the Chorabari Glacier.

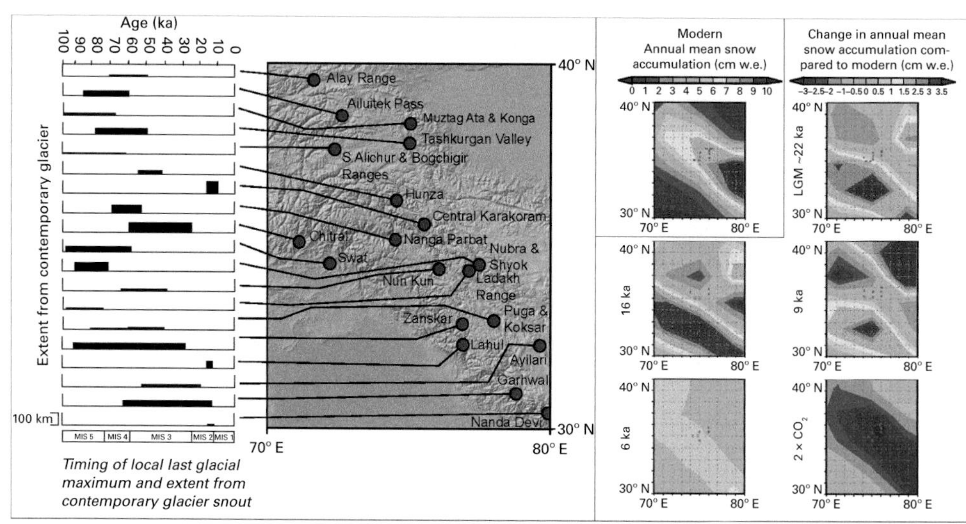

Figure 10.4. Glacial geologic study areas at the western end of the Himalayas and Tibet, showing the ages of local last glacial maxima with the approximate extent of valley glaciation shown by the height of each black bar, after Owen and Dortch (2014) who used data from Owen *et al.* (1992, 2001, 2002a, 2002b, 2006, 2010), Sharma and Owen (1996), Phillips *et al.* (2000), Richards *et al.* (2000), Taylor and Mitchell (2000), Barnard *et al.* (2004a, 2004b), Zech *et al.* (2005), Chevalier *et al.* (2005, 2011), Abramowski *et al.* (2006), Seong *et al.* (2009), Dortch *et al.* (2010), Hedrick *et al.* (2011), Rohringer *et al.* (2012) and Lee *et al.* (2014). The area shown in the right panel is the same area shown in the right frame which presents modelling results from Bishop *et al.* (2010), who used a 70-year simulation with a fully coupled atmosphere-ocean general circulation model for the global LGM, 16 ka, 9 ka, 6 ka and a doubling of CO_2.

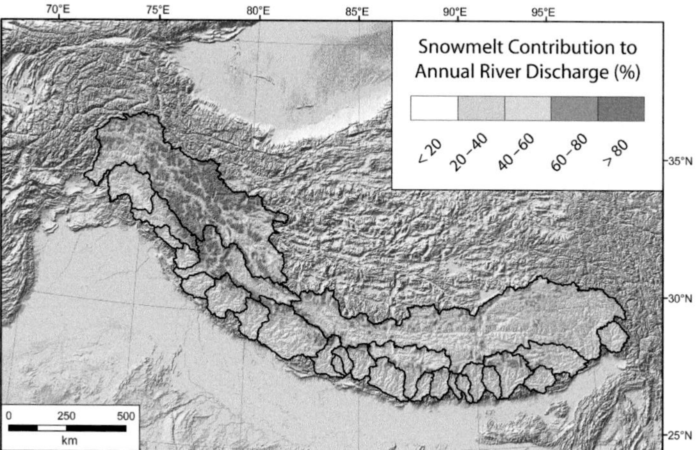

Figure 11.2. The geographic distribution of snowmelt contribution to annual river discharge for 27 main Himalayan catchments draining to the south (catchment outlines are shown by black lines). Model results for each catchment were derived from calibrated and validated satellite products and degree-day runoff modelling at monthly temporal and 1-km spatial resolution (modified after Bookhagen & Burbank (2010)).

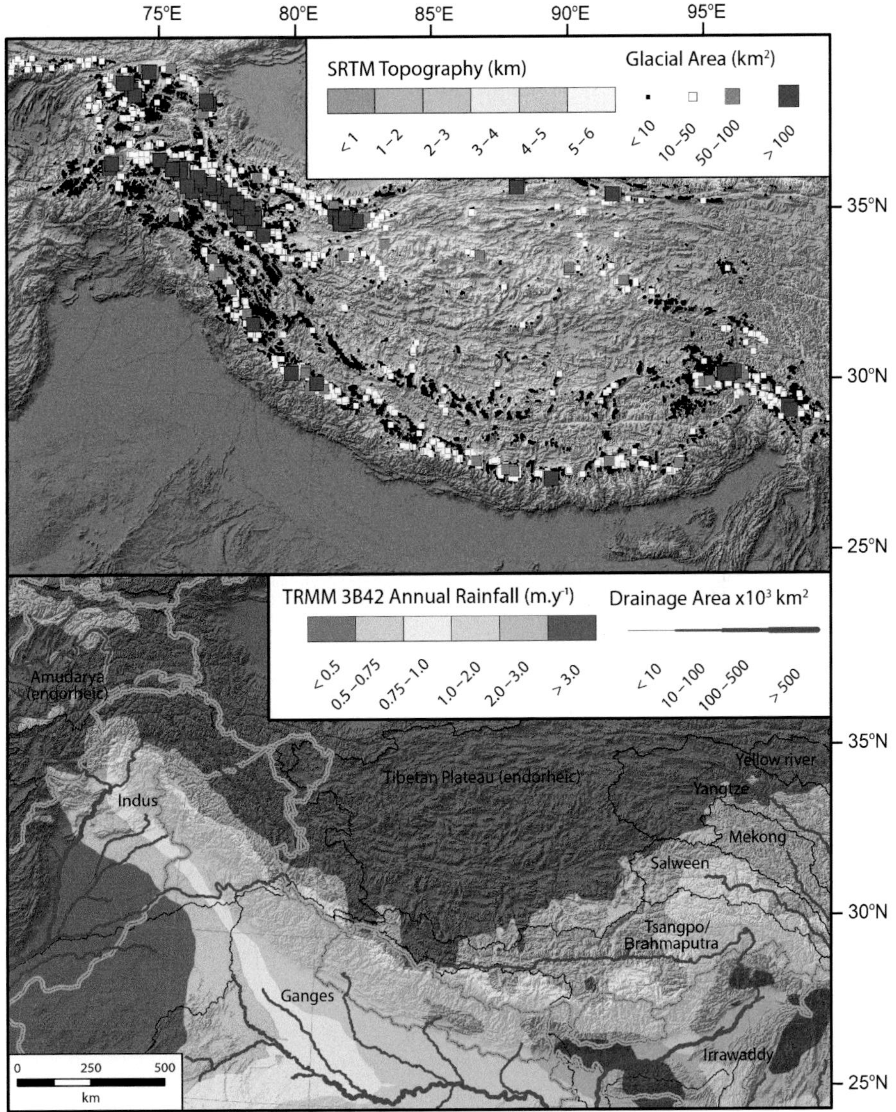

Figure 11.1. Top panel shows topography based on data from the Shuttle Radar Topography Mission (SRTM) with elevations in kilometres and present-day glacial distribution following the Randolph Catalog V3.2 (Arendt *et al.*, 2012). Each mapped glacier is colour-coded by their respective areas. Note the high glacial density and large glacial areas in the western and northwestern Himalaya. Bottom panel shows mean annual rainfall based on TRMM 3B42 satellite data (1998–2014) (Huffman *et al.*, 2007; Bookhagen & Burbank, 2010). Rivers are delineated in blue with line width corresponding to catchment area (based on hydrologically corrected SRTM data). Catchments are outlined in black with bold catchment names and international borders are grey.

Figure 12.1. Different landscapes of the Himalayas. Photo credits for 1a, 1c and 1d: Gopal S. Rawat; for 1b Herbert H.T. Prins.

Figure 12.1. (cont.)

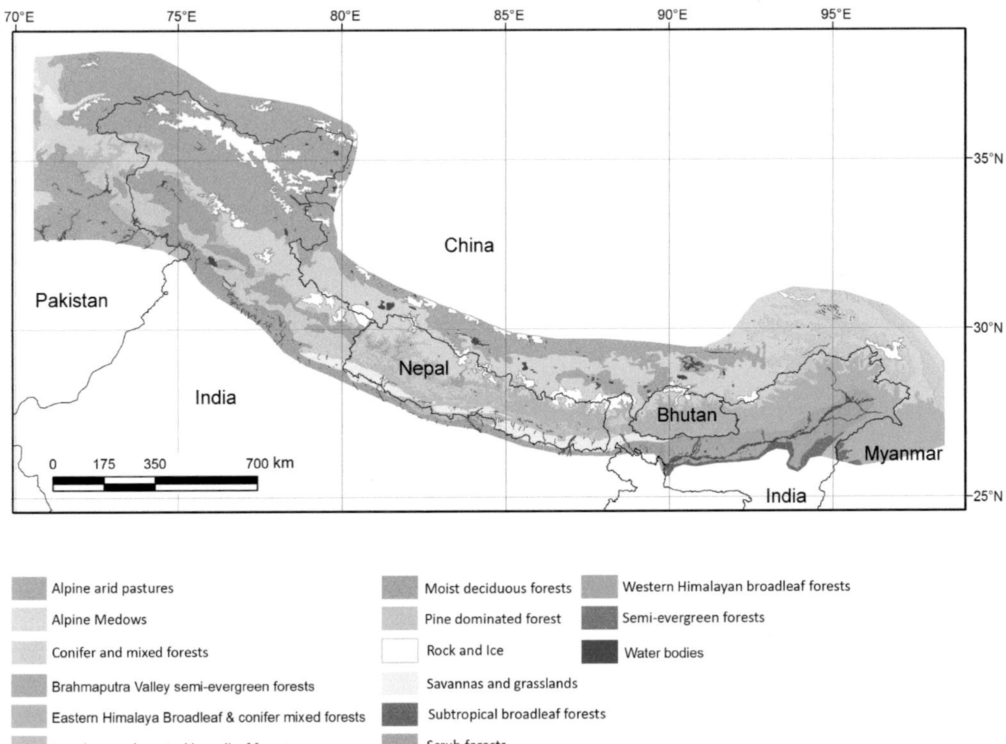

Figure 12.2. A generalized map of Himalayan vegetation, modified from Schweinfurth (1957), Champion and Seth (1968) and Mani (1978).

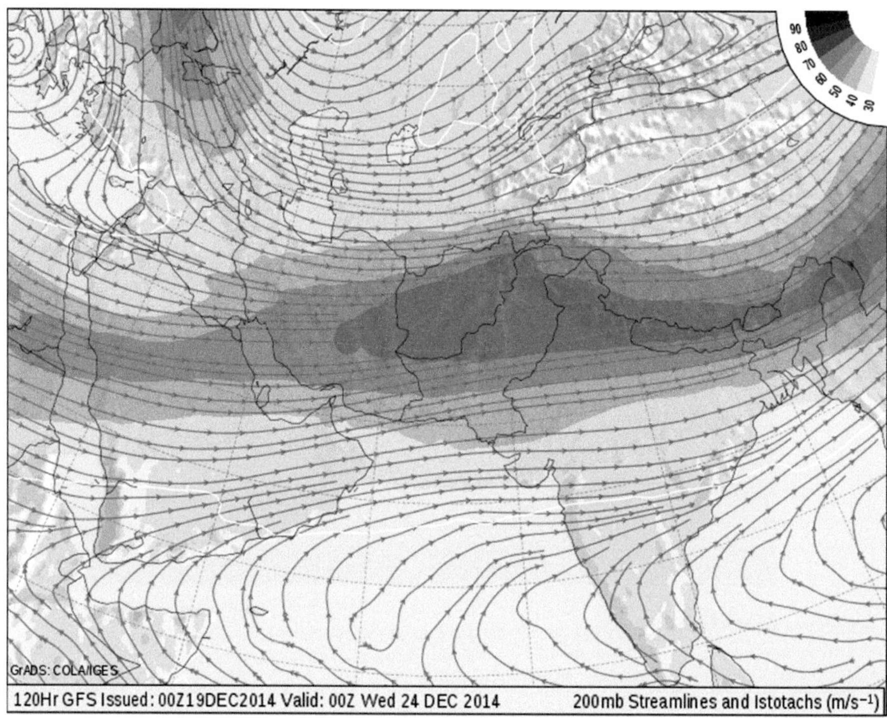

Figure 14.4. Subtropical jet stream above the Himalayas early winter 2014 (24 December 2014; source: www.wxmaps.org from the Centre for Ocean-Land-Atmosphere Studies). The isotachs are lines on the map connecting points of equal wind speed. In the top-right corner the wind speeds in m.s$^{-1.}$

Figure 14.5. Surfing with a glider along the Annapurna range at 8500 m. View to Dhaulagiri and Kali Gandaki Low stratocumulus aspired by a convective low into the Kali Gandaki Gorge. The puff of clouds shows strong rotating vortex clouds in the lee zone of Dhaulagiri. Photo credit: Klaus Ohlmann.

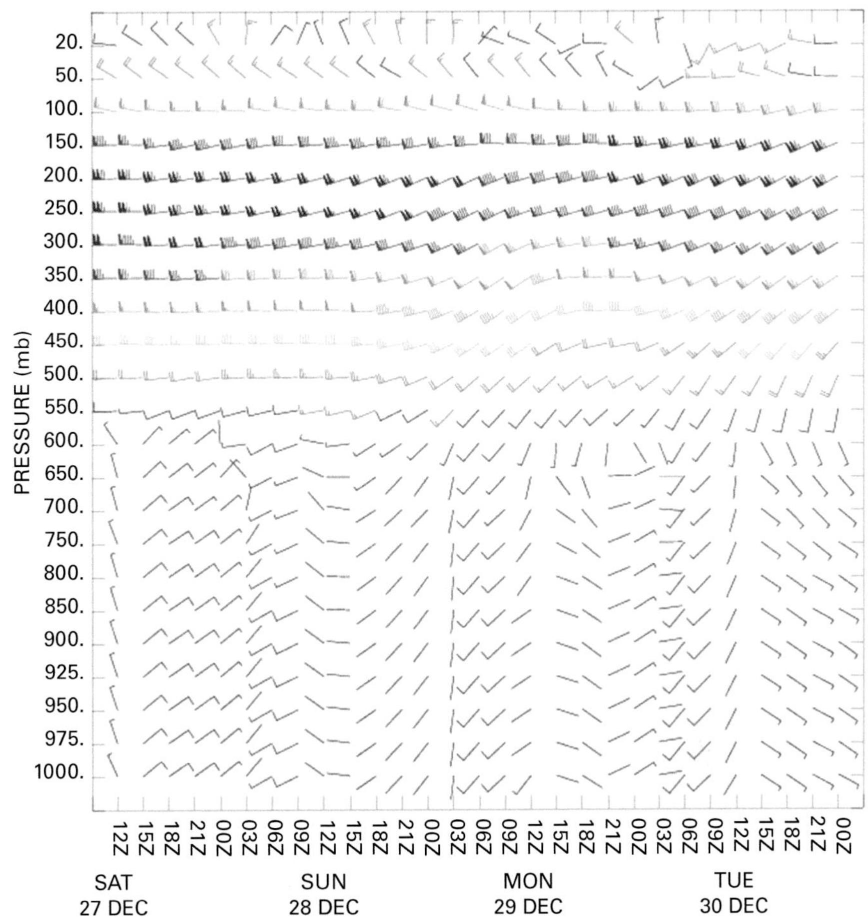

Figure 14.6 a. A 'windgram' for a typical subtropical jet stream above the Himalayas in winter (end of December), with wind speeds differing at different altitudes. The y-axis gives pressure in millibar: 500 mb is about 5600 m, 400 mb about 7200 m, 300 mb about 9200 m, 200 mb to 12,000 m and 100 mb to 15,800 m. A windgram is a forecast for a single spot on the map over the course of the day (in this case for the morning of 8 January 2016 at Dolpa Airfield near the Annapurna.

Figure 15.4. The research aircraft Stemme S10 VTX with underwing pods in the vicinity of the Annapurna Mountain Range (Flight crew Keimer/Heise 23 Jan 2014; Photo credit: DLR/Daniel Hein (CC-BY3.0)) during a measurement schedule of the Mountain Wave Project Nepal in 2014.

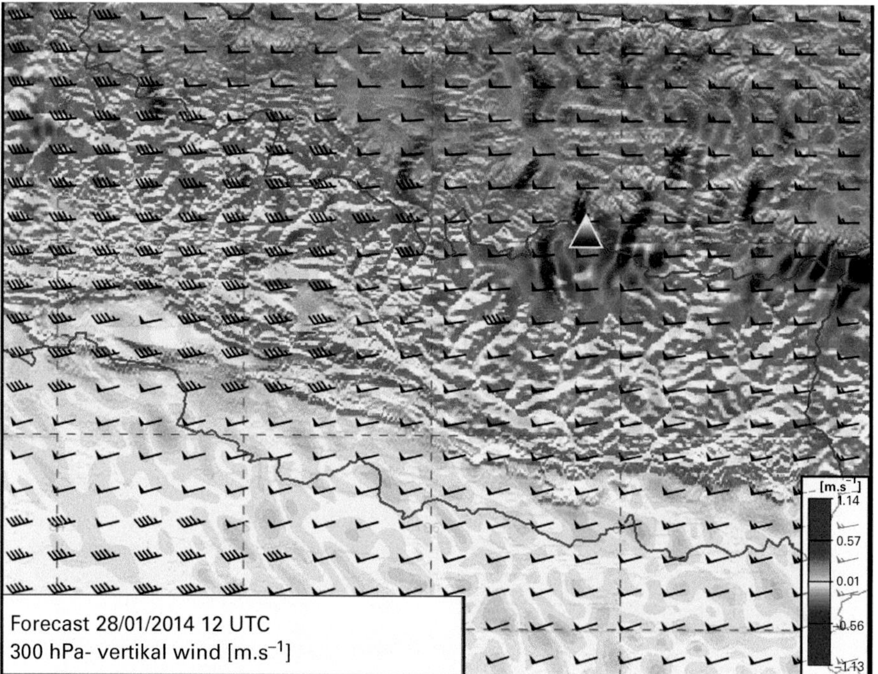

Figure 15.5. Numerical weather forecast of a new high-resolution model (grid spacing 2.8 km) with parametric vertical wind areas (updrafts – red, down – blue), and the horizontal wind symbol at a height of 300 hPa (~9000 m a.s.l.). The weather symbol shows the wind direction (here westerly winds), and the speed of the wind is denoted by the barbs. Each long barb represents 5 m.s^{-1}. At 25 m.s^{-1}, the barbs change into a pennant. During the measurement flight with the motor-glider over the glaciers of Mount Everest (marker Δ) on 28 Jan 2014, pilots used the updraught area of waves for safety flights of up to 9000 m a.s.l. The birds can use these updraught areas too, and the pattern helps improve understanding of migration routes and especially the passage over the Himalayan crest.

Figure 16.1. The Bar-headed Goose is thought to be one of the highest-flying birds in the world, and it has been directly tracked on one occasion as high as 7290 m a.s.l. (Photo: Coke Smith).

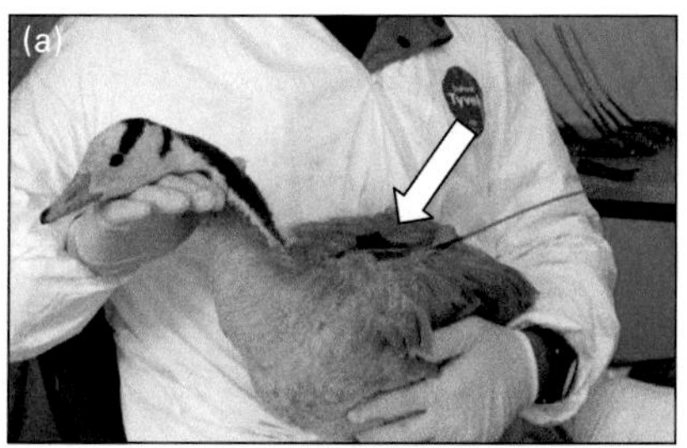

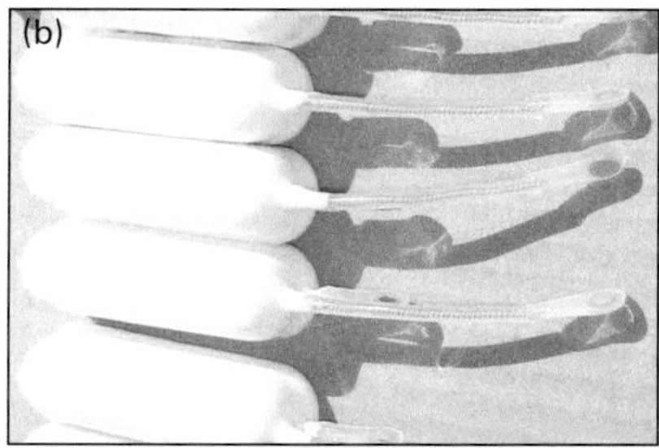

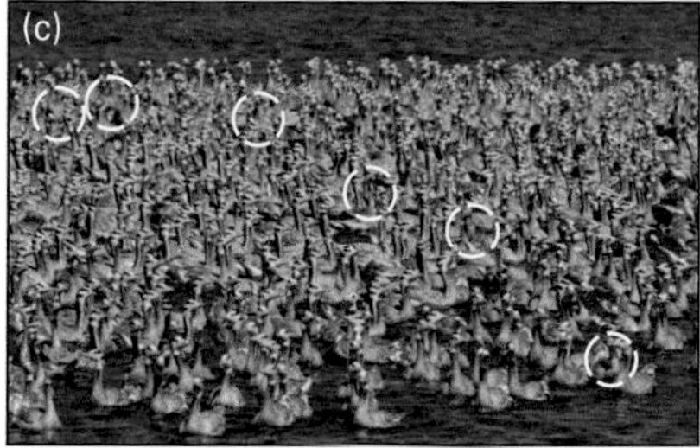

Figure 16.4. Studying the flight physiology of Bar-headed Geese. Photos show (a) a Bar-headed Goose with a satellite transmitter (attached using a Teflon harness, satellite tag indicated by white arrow); (b) archival heart rate recording data loggers (62 x 19 mm, 32 g) that were implanted to collect physiological data; (c) a flock of Bar-headed Geese on a lake in western Mongolia. Geese carrying archival loggers deployed in previous study years can be identified by their green neck collars, indicated here for six geese by white circles. Photo credits: (a, b) the authors, (c) Bruce Moffat.

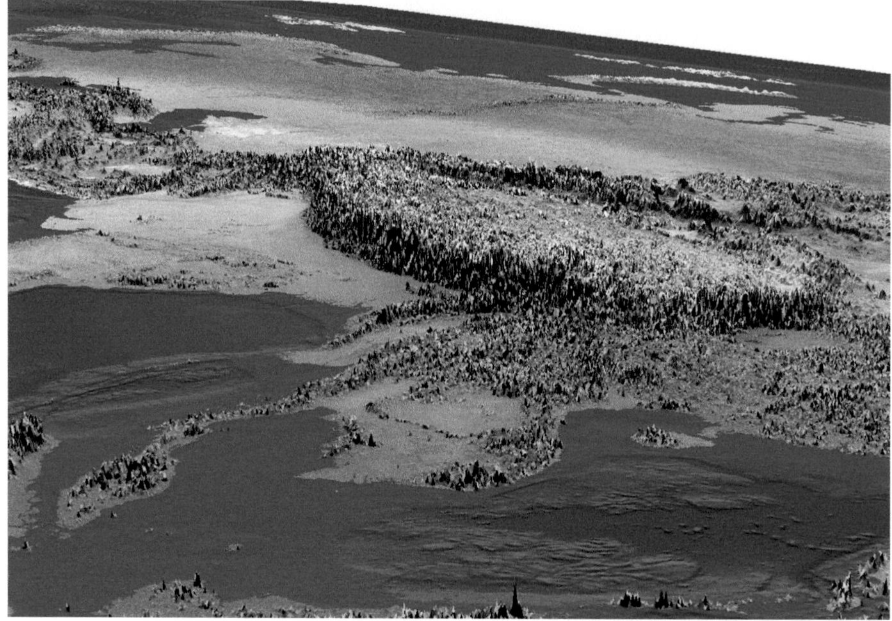

Figure 17.2. View over the Himalayas and its surroundings from the southeast, based on the compiled digital elevation model (DEM). At the top right the Arctic Ocean is visible; at the bottom left is Borneo. The Tibetan Plateau and the Himalayas form a high-altitude barrier for avian migration. Vertical scale ~30x the horizontal scale.

Figure 21.2. View of the Zanskar Gorge upstream of the Zanskar River confluence with the Indus in Ladakh, India (Photo credit: M. Vernier).

Figure 21.4. Rock engraved with four heron-like birds arranged around a circle, a possible pond or lake (left corner). This group is surrounded by several other images of flying birds and *khyungs*, Yaru rock art site (Ladakh) (Photo credit: L. Bruneau, Image enhanced using Adobe Photoshop©).

Figure 23.4. Brown-headed Gull feeding among garbage at Tsomoriri (Photo credit: Blaise Humbert-Droz).

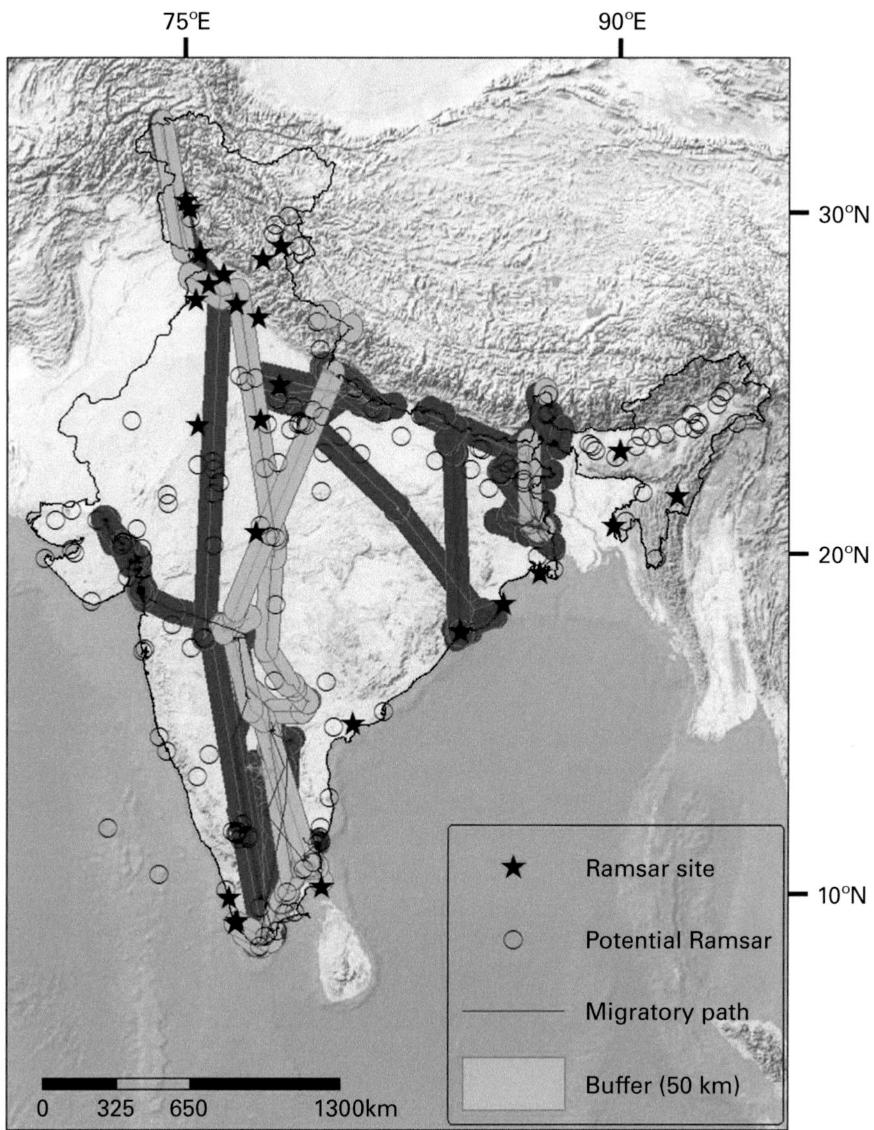

Figure 25.1. Northward migration routes (lines and buffers) of Northern Pintail (light grey) and Garganey (dark grey) within India, plotted in relation to existing Ramsar sites (stars) and potential Ramsar sites (circles).

11 The Influence of Hydrology and Glaciology on Wetlands in the Himalayas

Bodo Bookhagen

Climate and Hydrology in the Himalayas

River runoff or discharge in mountain rivers depends on precipitation and transiently stored water, such as groundwater and permafrost. Liquid precipitation (i.e. rainfall) generates runoff with short lag times, whereas solid precipitation (e.g. snow) can have lag times of up to six months – for example snow accumulated in the winter melts during spring and summer. The significant lag time of winter precipitation is a key process to maintain year-round runoff in mountain rivers and surrounding environments. A change from solid to liquid precipitation will change the timing of runoff, although the total annual runoff may remain the same. The time lag between winter snowfall and snowmelt sustains runoff during the drier summer months.

A warmer climate could change the timing of melt and the volume of the snowpack, and would have significant consequences for both humans and migratory birds, particularly in year-round water provisioning. Among the world's snow-dominated regions, the western Himalayas and Central Asia are particularly susceptible to changes in the timing of snowmelt, as reservoir capacity is currently not sufficient to buffer large seasonal shifts in the hydrograph (e.g. Barnett *et al.*, 2005; Immerzeel *et al.*, 2013; Hijioka *et al.*, 2014). The climatic zones in a mountain range are thus an integral part of the hydrology with cascading effects on downstream areas, including wetlands (e.g. Urrutia & Vuille, 2009; Immerzeel *et al.*, 2010).

During the past century, glaciers around the world have shrunk, with acceleration of glacial losses during the last decade of the twentieth and the first decade of the twenty-first century (e.g. Kaser, 1999; Oerlemans, 2005; Baraer *et al.*, 2012; Bolch *et al.*, 2012; Price & Weingartner, 2012; Vaughan *et al.*, 2013). Glacial runoff and snowmelt are important sources of fresh water and provide a significant part of the annual hydrologic budget in some regions (e.g. Archer & Fowler, 2004; Bookhagen & Burbank, 2010; Immerzeel *et al.*, 2010; Radic & Hock, 2011). However, the contribution of transiently stored moisture to runoff is difficult to determine and varies from year to year. Our understanding of mountain hydrology is limited due to the harsh conditions in high mountain regions, especially in the Himalayan range which has large areas above 5000 m elevation. Remote-sensing studies can help to assess general trends in snowmelt contribution, glacial area, elevation changes and velocities, but *in situ* fieldwork adds crucially important measurements unavailable at the scale of most remotely sensed data sets (e.g. Bookhagen & Burbank, 2006; Quincey *et al.*, 2007; Bookhagen & Burbank, 2010; Scherler *et al.*, 2011a; Boers *et al.*, 2014; Draganits *et al.*, 2014).

There is a clear distinction between runoff generated from snowmelt and that from glaciers: snowmelt runoff generally occurs early in the season with the increase in solar radiation during spring, while glacial-derived runoff occurs during the (late) summer season when most snow has melted and glacial ice is exposed. While glaciers are important for water resources in some tropical regions, especially the central Andes, the winter snow cover in the Himalayas generally provides a larger water volume storage system (e.g. Bookhagen & Burbank, 2010; Kaser *et al.*, 2010; Jeelani *et al.*, 2012; Immerzeel *et al.*, 2013). For example, rivers in the western and northwestern Himalayas, such as the Indus, derive more than 50% of their annual runoff from snowmelt water (e.g. Archer & Fowler, 2004; Bookhagen & Burbank, 2010; Immerzeel *et al.*, 2013), but only a small percentage of the runoff is derived from glacial melt water (Jeelani *et al.*, 2012). The Himalayas have an extensive winter snow cover, especially in the western and northwestern Himalayas, at elevations above 4000 m or 5000 m (Immerzeel *et al.*, 2009).

Measuring the volume of water stored in snow is difficult. Snow cover measurements based on optical or near-infrared satellite imagery indicate only the spatial extent, but not snow volume (depth). For example, a thin, persistent snow layer may give the same signal as thick, seasonal cover, but their water equivalents differ significantly. Snow water equivalent measurements give a better estimate of the amount of water stored in high elevation areas. There are two primary methods used to derive snow water equivalents: (1) direct measurements of remote sensing derived snow water equivalent; and (2) snow water equivalent modelling using a variety of optical remote sensing and field-based input parameters. Direct measurements are useful, but they do not have high spatial resolution, and are often hampered by technical difficulties (e.g. Pulliainen & Hallikainen, 2001; Tedesco *et al.*, 2004; Tedesco *et al.*, 2015; Smith and Bookhagen, 2016), while modelled outputs rely on insufficient *in situ* data, and rarely provide real time estimates. Both approaches have difficulties that limit understanding of snow water equivalents in the Himalayas, but the few robust studies clearly show that snowmelt and glacier melt play a key role in the hydrology and climate of the Himalayas.

Data Sets and Methods

The analysis and synthesis presented in this chapter relies on several field and remotely sensed data sets. Glacial extents were derived from the Randolph Glacier Inventory (RGI), a community-based data set of global glacier outlines (Version 3.2) (Arendt *et al.*, 2012). These data are referred to as RGI V3.2. Rainfall data were based on the Tropical Rainfall Measurement Mission (TRMM) product 3B42 (e.g. Huffman *et al.*, 2007; Bookhagen, 2010; Bookhagen & Strecker, 2010; Boers *et al.*, 2014). This product has a three-hour temporal resolution (data were aggregated to daily time steps) and a spatial resolution of 0.25° x 0.25° (at this latitude about 25 x 25 km) with an observational range from 1998 to 2014. While the temporal resolution of TRMM 3B42 is useful for monitoring the temporal dynamics of rainstorms, the data do not provide sufficient spatial resolution to resolve orographic rainfall and smaller-scale processes. Therefore, in addition to TRMM 3B42, high-spatial resolution TRMM 2B31 data were used, based on satellite-mounted active precipitation radar that allows identification of orographic rainfall. These data are based on the raw orbital observations that have been interpolated to regularly spaced 5000 m grid cells (Bookhagen & Burbank, 2006; Bookhagen & Strecker, 2008; Bookhagen & Burbank, 2010; Bookhagen & Strecker, 2012). A comparison of station data and gridded rainfall data for the Himalayas and South America indicate that TRMM 3B42 and TRMM 2B31 perform reasonably well (e.g. Bookhagen & Strecker, 2008; Bookhagen & Burbank, 2010; Andermann *et al.*, 2011; Carvalho *et al.*, 2012; Boers *et al.*, 2014).

Climatic, Topographic and Hydrologic Gradients in the Himalaya

Despite the uniformly elevated topography of the Himalayas, they are a climatologically and hydrologically diverse system (Figure 11.1). This section describes climatic and hydrologic gradients, that is, the changes in these parameters in space and time, and summarizes hydrologic results. The hydrology and its seasonality have profound impacts on wetland formation and their spatial and temporal extent (Bunn & Arthington, 2002; Tockner & Stanford, 2002).

Three principal climate regimes dominate the Himalayas: the Indian summer monsoon, the East Asian monsoon and the winter westerly disturbances. During the summer months, the Indian monsoon is driven by the temperature difference between ocean and land, resulting in an atmospheric pressure gradient (Clemens *et al.*, 1991; Webster *et al.*, 1998). During the Indian monsoon season (June to September), wind systems transport moisture-laden air from the Bay of Bengal towards the northwest along the Himalayan front. The Himalayan front is the first topographic rise and abuts the Ganges plain: when air masses interact with topography, they are forced to rise and cool; they subsequently lose the ability to store moisture and create the typical heavy monsoonal rainfall (Bookhagen & Burbank, 2006; Bookhagen & Burbank, 2010; Boers *et al.*, 2015). In the western Himalayas, monsoonal precipitation is significantly less than in the eastern and central Himalayas, primarily because of the increasing distance from the Bay of Bengal (e.g. Bookhagen *et al.*, 2005; Wulf *et al.*, 2010). The East Asian monsoon provides rainfall only to the far eastern Himalaya and eastern Tibet.

During winter, the pronounced temperature gradient between land and ocean reverses, with now cold land (Tibetan Plateau) and warm oceans. This allows the atmospheric dynamics to change and the westerly wind systems are now able to propagate into the Himalayas and Tibetan Plateau. The storm systems that bring in large amounts of moisture from the west – sometimes from as far as the

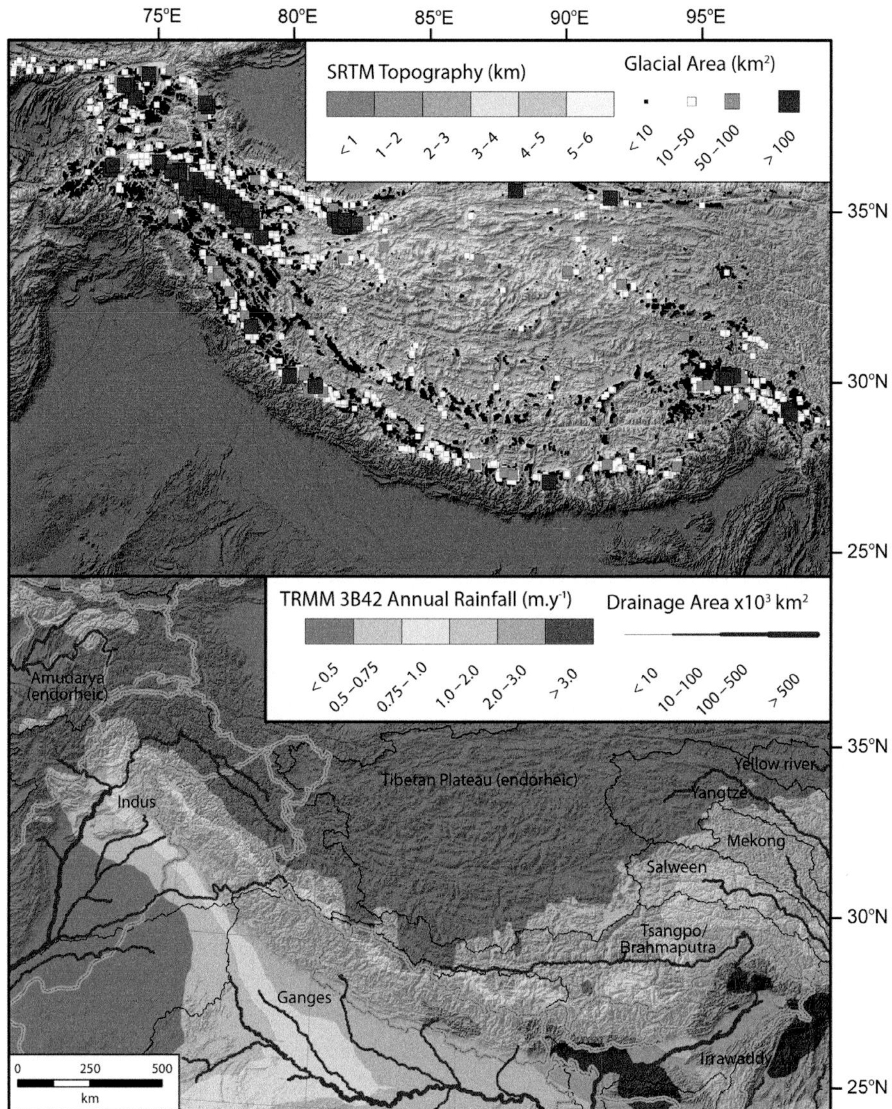

Figure 11.1. Top panel shows topography based on data from the Shuttle Radar Topography Mission (SRTM) with elevations in kilometres and present-day glacial distribution following the Randolph Catalog V3.2 (Arendt *et al.*, 2012). Each mapped glacier is colour-coded by their respective areas. Note the high glacial density and large glacial areas in the western and northwestern Himalaya. Bottom panel shows mean annual rainfall based on TRMM 3B42 satellite data (1998–2014) (Huffman *et al.*, 2007; Bookhagen & Burbank, 2010). Rivers are delineated in blue with line width corresponding to catchment area (based on hydrologically corrected SRTM data). Catchments are outlined in black with bold catchment names and international borders are grey. (A black-and-white version of this figure will appear in some formats. For the colour version, please refer to the plate section.)

Mediterranean or Arabian Seas – are referred to as *winter westerly disturbances*. The winter westerly disturbances are large atmospheric waves in the upper troposphere that travel towards the east at elevations near or above the Himalayan peaks (Lang & Barros, 2004; Winiger *et al.*, 2005; Wulf *et al.*, 2010; compare Ohlmann, Chapter 14 and

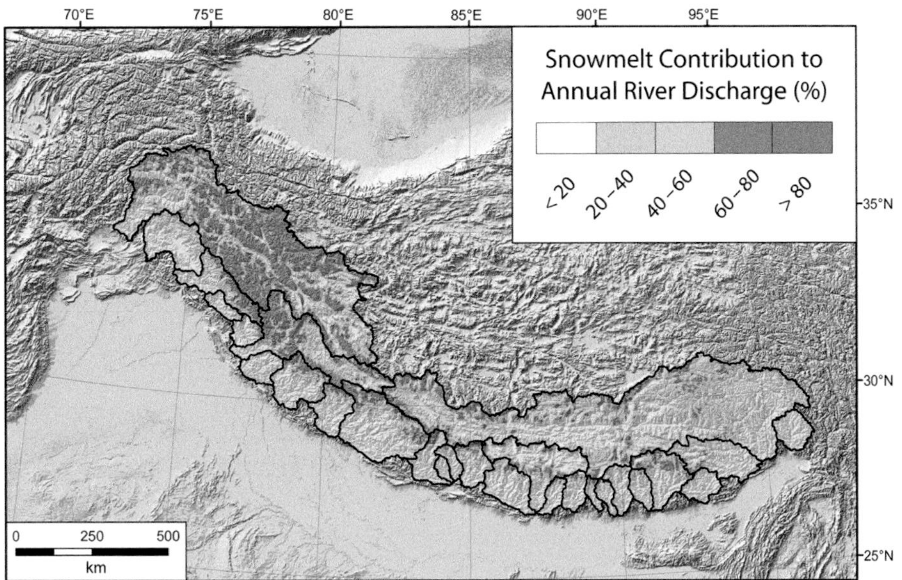

Figure 11.2. The geographic distribution of snowmelt contribution to annual river discharge for 27 main Himalayan catchments draining to the south (catchment outlines are shown by black lines). Model results for each catchment were derived from calibrated and validated satellite products and degree-day runoff modelling at monthly temporal and 1-km spatial resolution (modified after Bookhagen & Burbank (2010)). (A black-and-white version of this figure will appear in some formats. For the colour version, please refer to the plate section.)

Heise, Chapter 15). These eastward-travelling waves follow the main wind direction at this latitude and bring moisture from the west into the interior of the Himalayan-Tibetan mountains without being affected by orographic barriers and are not forced by mountain peaks to lose much of their moisture (e.g. Wulf *et al.*, 2010; Cannon *et al.*, 2015). The winter westerly disturbances are responsible for much of the winter precipitation in the western and northwestern Himalayas. These regions receive higher snowfall than the central or eastern Himalayas, as demonstrated by the significantly greater snow covered area (Immerzeel *et al.*, 2009; Wulf *et al.*, 2010). This leads to a higher density of glaciers which are larger in the western and northwestern Himalayas (Figure 11.1). Consequently, snowmelt contributions to annual river runoff in the western Himalayas are considerably greater than in the eastern and central Himalayas, where monsoonal rainfall is the dominant source of river runoff (e.g. Immerzeel *et al.*, 2009; Bookhagen & Burbank, 2010; Jeelani *et al.*, 2012) (Figure 11.2).

The winter western disturbances are responsible for much of the winter snow accumulation in the northwestern Himalayas. The Indian summer monsoon can still generate high-elevation snow in the central and eastern Himalayas and the Tibetan Plateau (Putkonen, 2004; Winiger *et al.*, 2005; Bookhagen & Burbank, 2010; Wulf *et al.*, 2010). Winter snow cover is, however, more spatially extensive and longer lasting in the western Himalayas than in the central and eastern Himalayas, and snow volume also peaks much later in the western Himalayas (Immerzeel *et al.*, 2009).

Furthermore, snowlines are lower in the western Himalayas (Scherler *et al.*, 2011b). These findings are consistent with the differences in hypsometry (a larger area at higher elevation) between the western and northwestern Himalayas, and with the general storm tracks of the winter western disturbances. Throughout the Himalayas, winter snow cover has decreased over the past 15 years, but has increased to the northwest in the Karakoram (Immerzeel *et al.*, 2009; Tahir *et al.*, 2011). This finding has been debated, but appears to be consistent with the so-called Karakoram glacier anomaly – a region of positive glacier mass balance resulting either from increased wintertime precipitation or decreased summer temperature (e.g. Hewitt, 2005; Scherler *et al.*, 2011b; Bolch *et al.*, 2012; Gardelle *et al.*, 2013; Immerzeel *et al.*, 2013). Positive mass balances indicate that glaciers are gaining mass, that is, water volume, and are often correlated with glacial area.

The combination of changes in topography and atmospheric influence from west to east along the Himalayas results in different hydrologic compartments (Figures 11.1–11.3). The northwestern (Karakoram) and western Himalayas are characterized by large areas above 5000 m elevation, and thus have a large potential area for glacial coverage and snow water storage. These areas are currently glaciated and covered with heavy snow (Figure 11.1), but were even more so in the past (e.g. Scherler *et al.*, 2010; Amidon *et al.*, 2013). This area shows the highest glacial and snowmelt runoff in the Himalaya and Karakoram mountain ranges, with contributions to annual discharge exceeding 50% (Immerzeel *et al.*, 2009; Bookhagen and Burbank, 2010; Jeelani *et al.*, 2012) (cf. Figure 11.2). To demonstrate the large-scale climatic and topographic gradients and their impacts on glaciers and snow cover, a west-to-east profile was constructed that averages values along the Himalayan arc in a north–south direction (Figure 11.3). The profile includes all areas above 500 m elevation (i.e. above the Indus and Ganges plains in the south) and up to the drainage divide along the southern edge of the Tibetan Plateau in the north (the drainage divide is shown in Figure 11.2). This analysis reveals that the maximum elevations along the Himalayan arc remain roughly similar and vary between 6000 m and 8000 m, but the area above 5000 m varies widely and hence modifies conditions for cryospheric processes (Figure 11.3).

A clear west-to-east gradient exists for snow-covered areas and snow water equivalent, with high levels of both in the west. There, about half of the annual precipitation falls as snow during winter western disturbances, resulting in significant snow cover and depth. In addition, the potential area that can be glaciated, for example, delineated by the area above 5000 m elevation, is much larger in the western Himalayas than in the east. Rainfall in the Himalayan foreland shows a clear east-to-west gradient with more rainfall in the eastern regions closer to the moisture source of the Bay of Bengal (Bookhagen & Burbank 2010), but rainfall in the mountainous Himalayas is more evenly distributed and does not show a strong gradient, although rainfall to the west of the Shillong Plateau at 93°E is higher than elsewhere in the Himalayas (Figure 11.3) (Bookhagen *et al.*, 2005; Bookhagen & Burbank, 2010).

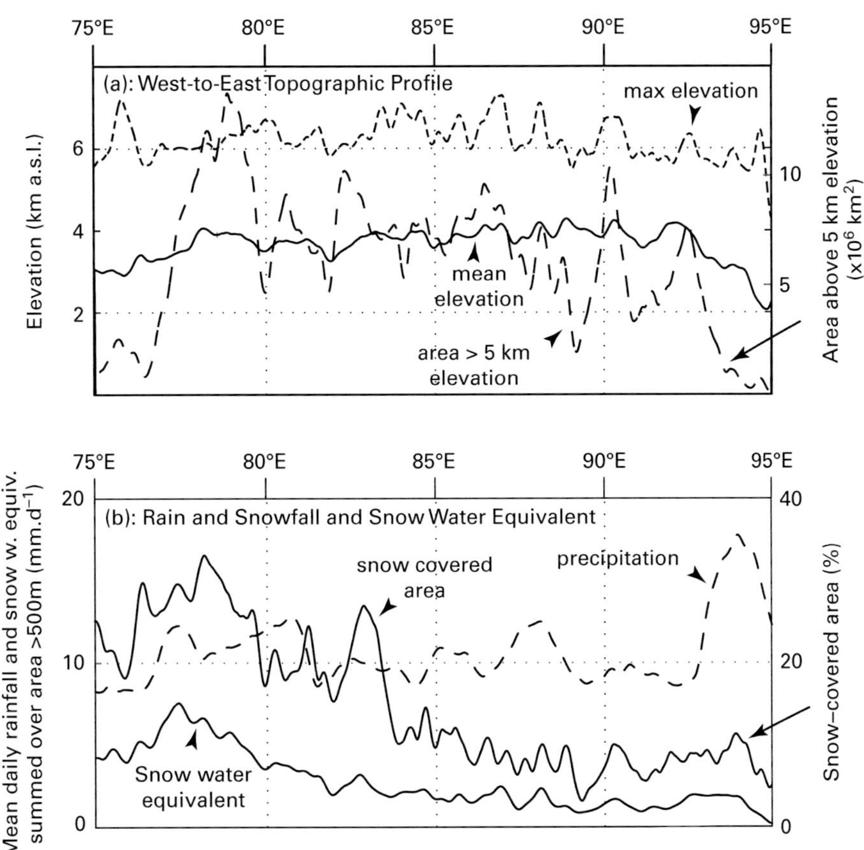

Figure 11.3. West-to-east profile along the Himalayan arc (profile location described in text) showing topographic and climatic data. Top panel shows maximum and mean topographic elevation (peaks have been smoothed and show 5 km running means) in dashed and solid black lines, respectively. Blue line indicates the area above 5 km elevation, which serves as a rough estimator for snow cover and glacial extent. Note the large continuous area of mean elevation above 5 km to the west of 80°E, which corresponds to the Karakoram. Bottom panel shows climatic data for the same region with mean daily rainfall and snow water equivalent in a dashed black line. Snow water equivalent is significantly larger in the Karakoram and western Himalaya than in the central or eastern Himalaya. Similarly, the snow-covered area exhibits a steep east-to-west gradient with higher year-around snow cover in the western areas.

Environmental Changes and Trends in Snow Cover and Glacial Areas

The dynamics of snow and glacial melt waters in the Himalayas is complex, but significant changes have been observed during recent decades (but see Bagchi *et al.*, Chapter 13). In this section, I will first focus on trends in the glacial realm and then attempt to highlight changes in snow cover, although available data and observations are scarce.

The release of glacial melt water peaks in the summer and early autumn, and can be critical for both agricultural practices and natural ecosystems (e.g. Ficke *et al.*, 2007;

Valentin *et al.*, 2008; Sultana *et al.*, 2009; Alford & Armstrong, 2010; Wulf *et al.*, 2010; Bolch *et al.*, 2012; Menon *et al.*, 2013; Kapnick *et al.*, 2014). As a result, changes in the melt water regime due to climate warming will have consequences for environment and ecosystem services, particularly for the western and northwestern Himalayas. Melting glaciers can also increase the risk of ice and snow avalanches and glacial lake outburst floods (Richardson & Reynolds, 2000; Quincey *et al.*, 2007). It is unlikely, however, that significant changes in annual runoff will occur soon, although glacial shrinkage outside the Karakoram is predicted to increase the seasonality of runoff, with impacts on agriculture and hydropower generation (e.g. Kapnick *et al.*, 2014). Glaciers in the western Himalayas are larger than those in the central or eastern Himalayas, and thus will have a slower response time to climatic shifts (Figure 11.1).

Most Himalayan glaciers are losing mass at rates similar to glaciers elsewhere around the globe, except for the Karakoram area (e.g. Hewitt, 2005; Scherler *et al.*, 2011b; Bolch *et al.*, 2012; Kääb *et al.*, 2012; Gardelle *et al.*, 2013); see Owen, chapter 10. Despite recent efforts, the climatic and cryospheric processes in the high-elevation Himalayas are still poorly understood. This is partly due to the difficulty inherent in accessing this region, but also due to the size and topographic complexity of glaciers, and the multi-country setting of this region (Hewitt, 2014). In the western Himalayas, glaciers are, in general, receding, but are not responding uniformly to climate warming (Kargel *et al.*, 2011; Scherler *et al.*, 2011b; Hewitt, 2014; Owen, Chapter 10). Regional patterns have been detected, but even these have inconsistencies as a result of local variations in climate. Current observations suggest that most glaciers have been retreating since about 1850 in the central and eastern Himalayas and the outer Tien Shan, however, slower retreat, standstill, and even advances have been observed in the Karakoram (e.g. Scherler *et al.*, 2011b; Bolch *et al.*, 2012; Gardelle *et al.*, 2012; Kääb *et al.*, 2012; Smith *et al.*, 2014). The causes for the Karakoram anomaly are debated, with speculations including seasonality differences (Kapnick *et al.*, 2014) or increased winter precipitation and/or cooler summers that might be responsible for glacier stability or expansion (Gardelle *et al.*, 2012; Hewitt, 2014).

Snowmelt and snow cover changes in some parts of the Himalayas have been analyzed (e.g. Immerzeel *et al.*, 2009; Gurung *et al.*, 2011; Tahir *et al.*, 2011), and predictions have been attempted using various approaches (e.g. Palazzi *et al.*, 2013; Kapnick *et al.*, 2014; Terzago *et al.*, 2014). In general, the observations indicate a snow cover reduction between 1995 and 2015, although the significance of the trend remains under discussion. Similarly, modelling the results of snow depth predictions for the Hindu Kush-Karakoram and central Himalaya region indicates a strong future decrease in water storage in the form of snow. Much of the snow that falls today may fall as rain in the future due to temperature increases, and the winter snow at high elevations will melt earlier.

Impact of Water Resources on Wetlands

The interplay of snowmelt and glacial melt waters, and runoff generated from rainfall, are the key hydrologic processes in this region and are the principal factors controlling wetland formation. Wetland areas in the western and northwestern Himalayas depend heavily on seasonal water storage in the form of snow and ice. They therefore rely on moisture generated during the pre-monsoon and post-monsoon season, in the form of snowmelt generated in the pre-monsoon season and glacial melting in the post-monsoon season. The wetland areas at high elevations are not heavily influenced by rainfall, and depend on transiently stored waters, so that environmental changes will have significant impacts. Predicted climate changes suggest earlier supply of snowmelt waters, which may leave less runoff during the summer season, when wetland environments are most active. A century ago, higher glacial runoff probably resulted in larger areas being covered by wetlands for longer durations.

In contrast, wetland areas in the central and eastern Himalayas receive their runoff from rainfall during the monsoon season and thus exhibit different seasonality and a high dependence on monsoon variability. Prediction of monsoonal behaviour on decadal to centennial time scales is complex, but recent research indicates that rainfall will become more extreme with more intense rainfall events, and an increasing duration of consecutive dry days in the Indian monsoon domain (IPCC, 2013). In other words, rainfall during the monsoon season is predicted to become more intense and more sporadic with direct impacts on rainfall-sensitive environments. For example, the stochastic water input to wetlands from intense rainfall events in high-elevation areas may lead to increased duration of wetland dryness and increased sediment flux.

Wetland areas in the low-elevation regions to the south of the Himalayas show similar dependence on their water resources, but are generally less sensitive to cryospheric changes, because of their larger drainage areas and higher groundwater influx.

From a spatiotemporal perspective, wetlands in the high-elevation western and northwestern Himalayas receive most of their waters during snowmelt in the spring season. Water levels remain high until the end of the summer, when glacial melting contributes to discharge. In contrast, wetlands in the high-elevation central and eastern Himalayas experience only a small wetland water level rise during the spring, and are generally less extensive in the mountains than in the west. Wetlands at low elevations in the Ganges plain (central and eastern Himalayas) are dominated by monsoonal rainfall with high water levels from June to October. One of the key differences in wetland area and extent from west to east along the Himalayas is the high number and large extent of wetlands in the western and northwestern mountain regions at high elevations. These are sustained by snow and glacial melt waters. Large wetlands in the central and eastern Himalaya are mainly found at low elevations.

An additional important – but often overlooked – factor affecting snow and glacial melt waters is their sediment-free water. Runoff generated from heavy rainfall events is often associated with heavy suspended sediment concentration (Wulf *et al.*, 2010; Wulf *et al.*, 2012). The sediment is often deposited in low-slope environments, such as wetlands, and while they form a key habitat for aquatic invertebrates important for migratory birds in these areas, a rapid increase in sediment flux can lead to sediment infilling of wetlands and to a decrease in wetland biological activity (Johnston, 1991).

The highly seasonal water fluxes in the wetlands along the Himalayas are highly variable, with the western part of the Himalayas experiencing peak water discharges in pre-monsoon (snowmelt) and late monsoon (glacial melt) periods, and with the central and eastern parts experiencing peak water discharge in the monsoon (rainfall) period. The wetlands at the foot of the Himalayas are therefore postulated to function in a variable way for migrating birds. Also, in the western parts of the Himalayas, water levels of wetlands are (or were, before artificial lakes were built) more predictable, because their reliance on glacier meltwaters in late summer is dependent to a lesser degree on variations in annual precipitation. In contrast, water levels in the eastern wetlands of the Himalayas, being heavily dependent on monsoonal rainfall, fluctuate much more widely. It is therefore not unlikely that the natural wetlands at the foot of the Himalayas were suitable for migratory waterbirds before humans modified them. The wetlands would have been more suitable for ducks and other waterbirds, especially in the autumn, when the water level was higher due to glacial melt during summer. Furthermore, the wetlands were more dependable in the western Himalayas than in the eastern Himalayas because of the consistency in glacial melt water supply to the wetlands in the western part. The reliability for migrating waterbirds would have been even higher when the glaciers in the western part of the Himalayas were more extensive one or two centuries ago.

References

Alford, D. & Armstrong, R. (2010). The role of glaciers in stream flow from the Nepal Himalaya. *The Cryosphere Discussion*, **4**, 469–494.

Amidon, W.H., Bookhagen, B., Avouac, J.P., Smith, T. & Rood, D. (2013). Late Pleistocene glacial advances in the western Tibet interior. *Earth and Planetary Science Letters*, **381**, 210–221.

Andermann, C., Bonnet, S. & Gloaguen, R. (2011). Evaluation of precipitation data sets along the Himalayan front. *Geochemistry Geophysics Geosystems*, **12**, 1–6.

Archer, D.R. & Fowler, H.J. (2004). Spatial and temporal variations in precipitation in the upper Indus basin, global teleconnections and hydrological implications. *Hydrology and Earth System Sciences*, **8**, 47–61.

Arendt, A., Bliss, A., Bolch, T. & Cogley, J.G. (2012). *Randolph Glacier Inventory – A Dataset of Global Glacier Outlines: Version 3.2, Edited*. Global Land Ice Measurements from Space, Boulder, CO: Digital Media.

Baraer, M., Mark, B.G., McKenzie, J.M., *et al.* (2012). Glacier recession and water resources in Peru's Cordillera Blanca. *Journal of Glaciology*, **58**, 134–150.

Barnett, T.P., Adam, J.C. & Lettenmaier, D.P. (2005). Potential impacts of a warming climate on water availability in snow-dominated regions. *Nature*, **438**, 303–309.

Boers, N., Bookhagen, B., Marengo, J., Marwan, N., von Storch, J.-S. & Kurths, J. (2015). Extreme rainfall of the South American monsoon system: a dataset comparison using complex networks. *Journal of Climate*, **28**, 1031–1056.

Boers, N., Rheinwalt, A., Bookhagen, B., *et al.* (2014). The South American rainfall dipole: a complex network analysis of extreme events. *Geophysical Research Letters*, **41**, 7397–7405.

Bolch, T., Kulkarni, A., Kaab, A., *et al.* (2012). The state and fate of Himalayan glaciers. *Science*, **336**, 310–314.

Bookhagen, B. (2010). Appearance of extreme monsoonal rainfall events and their impact on erosion in the Himalaya. *Geomatics Natural Hazards & Risk*, **1**, 37–50.

Bookhagen, B. & Burbank, D.W. (2006). Topography, relief, and TRMM-derived rainfall variations along the Himalaya. *Geophysical Research Letters*, **33**, L08405, 1–5.

Bookhagen, B. & Burbank, D.W. (2010). Toward a complete Himalayan hydrological budget: spatiotemporal distribution of snowmelt and rainfall and their impact on river discharge. *Journal of Geophysical Research-Earth Surface*, **115**, F03019, 1–15.

Bookhagen, B. & Strecker, M.R. (2008). Orographic barriers, high-resolution TRMM rainfall, and relief variations along the eastern Andes. *Geophysical Research Letters*, **35**, L06403, 1–6.

Bookhagen, B. & Strecker, M.R. (2010). *Modern Andean Rainfall Variation during ENSO Cycles and Its Impact on the Amazon Basin. Neogene History of Western Amazonia and Its Significance for Modern Diversity.* H.V.C. Hoorn & F. Wesselingh. Oxford: Blackwell Publishing.

Bookhagen, B. & Strecker, M.R. (2012). Spatiotemporal trends in erosion rates across a pronounced rainfall gradient: examples from the southern Central Andes. *Earth and Planetary Science Letters*, **327**, 97–110.

Bookhagen, B., Thiede, R.C. & Strecker, M.R. (2005). Abnormal monsoon years and their control on erosion and sediment flux in the high, and northwest Himalaya. *Earth and Planetary Science Letters*, **231**, 131–146.

Bunn, S.E. & Arthington, A.H. (2002). Basic principles and ecological consequences of altered flow regimes for aquatic biodiversity. *Environmental Management*, **30**, 492–507.

Cannon, F., Carvalho, L.V., Jones, C. & Bookhagen, B. (2015). Multi-annual variations in winter westerly disturbance activity affecting the Himalaya. *Climate Dynamics*, **44**, 441–455.

Carvalho, L.M.V., Jones, C., Posadas, A.N.D., Quiroz, R., Bookhagen, B. & Liebmann, B. (2012). Precipitation characteristics of the South American monsoon system derived from multiple datasets. *Journal of Climate*, **25**, 4600–4620.

Clemens, S., Prell, W., Murray, D., Shimmield, G. & Weedon, G. (1991). Forcing mechanisms of the Indian-Ocean Monsoon. *Nature*, **353**, 720–725.

Draganits, E., Gier, S., Hofmann, C.-C., Janda, C., Bookhagen, B. & Grasemann, B. (2014). Holocene versus modern catchment erosion rates at 300 MW Baspa II hydroelectric power plant (India, NW Himalaya). *Journal of Asian Earth Sciences*, **90**, 157–172.

Ficke, A.D., Myrick, C.A. & Hansen, L.J. (2007). Potential impacts of global climate change on freshwater fisheries. *Reviews in Fish Biology and Fisheries*, **17**, 581–613.

Gardelle, J., Berthier, E. & Arnaud, Y. (2012). Slight mass gain of Karakoram glaciers in the early twenty-first century. *Nature Geoscience*, **5**, 322–325.

Gardelle, J., Berthier, E., Arnaud, Y. & Kaab, A. (2013). Region-wide glacier mass balances over the Pamir-Karakoram-Himalaya during 1999–2011. *The Cryosphere*, **7**, 1263–1286.

Gurung, D.R., Kulkarni, A.V., Giriraj, A., Aung, K.S., Shrestha, B. & Srinivasan, J. (2011). Changes in seasonal snow cover in Hindu Kush-Himalayan region. *The Cryosphere Discussion*, **5**, 755–777.

Hewitt, K. (2005). The Karakoram anomaly? Glacier expansion and the 'elevation effect,' Karakoram Himalaya. *Mountain Research and Development*, **25**, 332–340.

Hewitt, K. (2014). *Glaciers of the Karakoram Himalaya*. Dordrecht: Springer.

Hijioka, Y., Lin, E., Pereira, J.J., *et al.* (2014). Asia. Climate Change 2014: Impacts, Adaptation, and Vulnerability. Part B: Regional Aspects. Contribution of Working Group II to the Fifth Assessment Report of the Intergovernmental Panel on Climate Change. V.R. Barros, C.B. Field, D.J. Dokken, M.D. *et al.* Cambridge, UK and New York: Cambridge University Press, pp. 1327–1370.

Huffman, G.J., Adler, R.F., Bolvin, D.T., *et al.* (2007). The TRMM multisatellite precipitation analysis (TMPA): Quasi-global, multiyear, combined-sensor precipitation estimates at fine scales. *Journal of Hydrometeorology*, **8**, 38–55.

Immerzeel, W.W., Droogers, P., de Jong, S.M. & Bierkens, M.F.P. (2009). Large-scale monitoring of snow cover and runoff simulation in Himalayan river basins using remote sensing. *Remote Sensing of Environment*, **113**, 40–49.

Immerzeel, W.W., Pellicciotti, F. & Bierkens, M.F.P. (2013). Rising river flows throughout the twenty-first century in two Himalayan glacierized watersheds. *Nature Geoscience*, **6**, 742–745.

Immerzeel, W.W., Van Beek, L.P.H. & Bierkens, M.F.P. (2010). Climate change will affect the Asian Water Towers. *Science*, **328**, 1382–1385.

IPCC (2013). *Climate Change 2013: The Physical Science Basis. Contribution of Working Group I to the Fifth Assessment Report of the Intergovernmental Panel on Climate Change*. Cambridge, UK and New York: Cambridge University Press.

Jeelani, G., Feddema, J.J., Van der Veen, C.J. & Stearns, L. (2012). Role of snow and glacier melt in controlling river hydrology in Liddar watershed (western Himalaya) under current and future climate. *Water Resources Research*, **48**, W12508, 1–16.

Johnston, C.A. (1991). Sediment and nutrient retention by freshwater wetlands: effects on surface water quality. *Critical Reviews in Environmental Control*, **21**, 491–565.

Kääb, A., Berthier, E., Nuth, C., Gardelle, J. & Arnaud, Y. (2012). Contrasting patterns of early twenty-first-century glacier mass change in the Himalayas. *Nature*, **488**, 495–498.

Kapnick, S.B., Delworth, T.L., Ashfaq, M., Malyshev, S. & Milly, P.C.D. (2014). Snowfall less sensitive to warming in Karakoram than in Himalayas due to a unique seasonal cycle. *Nature Geoscience*, **7**, 834–840.

Kargel, J.S., Cogley, J.G., Leonard, G.J., Haritashya, U. & Byers, A. (2011). Himalayan glaciers: the big picture is a montage. *Proceedings of the National Academy of Sciences of the United States of America*, **108**, 14709–14710.

Kaser, G. (1999). A review of the modern fluctuations of tropical glaciers. *Global and Planetary Change*, **22**, 93–103.

Kaser, G., Grosshauser, M. & Marzeion, B. (2010). Contribution potential of glaciers to water availability in different climate regimes. *Proceedings of the National Academy of Sciences of the United States of America*, **107**, 20223–20227.

Lang, T.J. & Barros, A.P. (2004). Winter storms in the central Himalayas. *Journal of the Meteorological Society of Japan*, **82**, 829–844.

Oerlemans, J. (2005). Extracting a climate signal from 169 glacier records. *Science*, **308**, 675–677.

Palazzi, E., Von Hardenberg, J. & Provenzale, A. (2013). Precipitation in the Hindu-Kush Karakoram Himalaya: observations and future scenarios. *Journal of Geophysical Research-Atmospheres*, **118**, 85–100.

Price, M.F. & Weingartner, R. (2012). Global change and the world's mountains. *Mountain Research and Development*, **32**(S1), S3–S6.

Pulliainen, J. & Hallikainen, M. (2001). Retrieval of regional snow water equivalent from space-borne passive microwave observations. *Remote Sensing of Environment*, **75**, 76–85.

Putkonen, J.K. (2004). Continuous snow and rain data at 500 to 4400 m altitude near Annapurna, Nepal, 1999–2001. *Arctic Antarctic and Alpine Research*, **36**, 244–248.

Quincey, D.J., Richardson, S.D., Luckman, A., *et al.* (2007). Early recognition of glacial lake hazards in the Himalaya using remote sensing datasets. *Global and Planetary Change*, **56**, 137–152.

Radic, V. & Hock, R. (2011). Regionally differentiated contribution of mountain glaciers and ice caps to future sea-level rise. *Nature Geoscience*, **4**, 91–94.

Richardson, S.D. & Reynolds, J.M. (2000). An overview of glacial hazards in the Himalayas. *Quaternary International*, **65**, 31–47.

Scherler, D., Bookhagen, B. & Strecker, M.R. (2011a). Hillslope-glacier coupling: the interplay of topography and glacial dynamics in High Asia. *Journal of Geophysical Research-Earth Surface*, 116.

Scherler, D., Bookhagen, B. & Strecker, M.R. (2011b). Spatially variable response of Himalayan glaciers to climate change affected by debris cover. *Nature Geoscience*, **4**, 156–159.

Scherler, D., Bookhagen, B., Strecker, M.R., Von Blanckenburg, F. & Rood, D. (2010). Timing and extent of late Quaternary glaciation in the western Himalaya constrained by Be-10 moraine dating in Garhwal, India. *Quaternary Science Reviews*, **29**, 815–831.

Smith, T.T., Bookhagen, B. & Cannon, F. (2014). Improving semi-automated glacial mapping with a multi-method approach: areal changes in Central Asia. *The Cryosphere Discussion*, **8**, 5433–5483.

Smith, T.T. & Bookhagen, B. (2016): Assessing uncertainty and sensor biases in passive microwave data across High Mountain Asia. *Remote Sensing of Environment*, **181**, 174–185.

Sultana, H., Ali, N., Iqbal, M.M. & Khan, A.M. (2009). Vulnerability and adaptability of wheat production in different climatic zones of Pakistan under climate change scenarios. *Climatic Change*, **94**, 123–142.

Tahir, A.A., Chevallier, P., Arnaud, Y. & Ahmad, B. (2011). Snow cover dynamics and hydrological regime of the Hunza River basin, Karakoram Range, northern Pakistan. *Hydrology and Earth System Sciences*, **15**, 2275–2290.

Tedesco, M., Derksen, C., Deems, J.S. & Foster, J.L. (2015). *Remote Sensing of Snow Depth and Snow Water Equivalent. Remote Sensing of the Cryosphere* (ed. M. Tedesco). John Wiley & Sons Ltd, Chichester, UK, pp. 73–98.

Tedesco, M., Pulliainen, J., Takala, M., Hallikainen, M. & Pampaloni, P. (2004). Artificial neural network-based techniques for the retrieval of SWE and snow depth from SSM/I data. *Remote Sensing of Environment*, **90**, 76–85.

Terzago, S., Von Hardenberg, J., Palazzi, E. & Provenzale, A. (2014). Snowpack changes in the Hindu Kush–Karakoram–Himalaya from CMIP5 global climate models. *Journal of Hydrometeorology*, **15**, 2293–2313.

Tockner, K. & Stanford, J.A. (2002). Riverine flood plains: present state and future trends. *Environmental Conservation*, **29**, 308–330.

Urrutia, R. & Vuille, M. (2009). Climate change projections for the tropical Andes using a regional climate model: temperature and precipitation simulations for the end of the 21st century. *Journal of Geophysical Research-Atmospheres*, **114**, D02108, 1–15.

Valentin, C., Agus, F., Alamban, R., *et al.* (2008). Runoff and sediment losses from 27 upland catchments in Southeast Asia: impact of rapid land use changes and conservation practices. *Agriculture Ecosystems & Environment*, **128**, 225–238.

Vaughan, D.G., Comiso, J.C., Allison, I., *et al.* (2013). *Observations: Cryosphere. Climate Change 2013: The Physical Science Basis*. Contribution of Working Group I to the Fifth Assessment Report of the Intergovernmental Panel on Climate Change. Cambridge, UK and New York: Cambridge University Press.

Webster, P.J., Magana, V.O., Palmer, T.N., *et al.* (1998). Monsoons: processes, predictability, and the prospects for prediction. *Journal of Geophysical Research-Oceans*, **103**(C7), 14451–14510.

Winiger, M., Gumpert, M. & Yamout, H. (2005). Karakorum-Hindukush-western Himalaya: assessing high-altitude water resources. *Hydrological Processes*, **19**, 2329–2338.

Wulf, H., Bookhagen, B. & Scherler, D. (2010). Seasonal precipitation gradients and their impact on fluvial sediment flux in the northwest Himalaya. *Geomorphology*, **118**, 13–21.

Wulf, H., Bookhagen, B. & Scherler, D. (2012). Climatic and geologic controls on suspended sediment flux in the Sutlej River Valley, western Himalaya. *Hydrology and Earth System Sciences*, **16**, 2193–2217.

12 The Himalayan Vegetation along Horizontal and Vertical Gradients

Gopal S. Rawat

Background

The Himalayan arc, a complex chain of mountains ranging up to an altitude of 8850 m, and extending 8 degrees (26° to 34° N) in latitude and 28 degrees (69° to 97° E) in longitude, has been of great interest to biogeographers, geohydrologists, naturalists, climatologists, ecologists and social scientists alike. The eastern Himalayas have also been recognized as one of the global biodiversity hotspots (Myers *et al.*, 2000; Mittermeier *et al.*, 2011). The prominent relief features of the Himalayas significantly influence Asia's atmospheric circulation, and experience a regional climate that varies from tropical to alpine and nival types on the vertical, and from semiarid to per-humid types on the horizontal axes (Bookhagen, Chapter 11; Owen, Chapter 10). While the western and northwestern extremities exhibit semiarid and cold arid climates, respectively, the far eastern ranges represent some of the wettest places on earth with > 4000 mm annual precipitation. Although both the western (69° to 84° E) and eastern Himalayas (84° to 97° E) are comparable in terms of altitude, the latter is closer to the equator and has a much narrower range in latitude and precipitation. The effect of the monsoon becomes less pronounced in the western Himalayas, where there is great variation in the annual and diurnal temperatures. Unlike the central and eastern parts, the western Himalayas receive higher precipitation during winter in the form of snow (Bookhagen, Chapter 11; Owen, Chapter 10). Moreover, the western Himalayas are aligned in the north–west to south–east direction representing different eco-climatic zones. Many ranges are separated by deep river valleys, which create climatic conditions with frequent temperature inversions. Additionally, this region differs considerably from the east in terms of history of human habitation and land use practices as it has been under greater influence from Central and West Asia since historical times, and local farmers have adapted various farming practices and methods of exploiting natural resources.

Much research has been carried out to understand the factors influencing the structure and composition of Himalayan vegetation at local and sub-regional scales. Complex topography, diverse climatic conditions and geodynamic processes are manifested in the vegetation types, and a wide range of floral and faunal assemblages has evolved. As early as the 1920s and 1930s, a few studies described the successional patterns in the Himalayan forests along the elevational gradients (Kenoyer, 1921; Champion, 1923; Dudgeon & Kenoyer, 1925). Champion (1936) made an attempt to classify the forest types of the Himalayas, which was later revised (Champion & Seth, 1968). Schweinfurth (1957) published a monumental work on the vegetation types of the Himalaya along vertical and horizontal gradients. This work also presented a general map of major physiognomic types and profiles along altitudinal gradients. Stainton (1972) published an account of climatic and vegetation gradients of Nepal, which was later revised by the Department of Forests, Government of Nepal (Anonymous, 2002). Singh and Singh (1987, 1992) provided a comprehensive review of the studies done on various aspects of Himalayan forest vegetation. Zobel and Singh (1997) discussed the unique features of the Himalayan forests, and concluded that these forests exhibit neither truly tropical nor temperate characteristics.

Ecological studies on the alpine vegetation of the Himalayas are rather limited. Many environmental factors such as precipitation (including snowfall), aspect, altitude and topography, as well as anthropogenic factors, influence the species richness and community structure of the Himalayan flora (Puri *et al.*, 1989; Miehe, 1997; Hartmann, 2009). Rawat (2005a) compared the major physiognomic units of alpine vegetation, viz., alpine scrub, herbaceous meadows, bogs, and fell-fields along altitudinal gradients in the western Himalayas. It has been established that moist meadows of the Greater Himalayas and dry alpine pastures of the Trans-Himalaya exhibit considerable variation in physiognomy, despite apparent similarity in floristic structure (Rawat, 1998, 2005a, 2005b, 2005c; Rawat & Adhikari, 2005). At local scales, the community structure and vegetation types of this alpine region have been described by Kala *et al.* (1998), Samant and Joshi (2005) and Samant *et al.* (2005) (Figure 12.1). At a larger biogeographic scale, Mani (1978) gave a very detailed account of the phytogeography and floral characteristics in the northwest Himalaya.

Despite several studies of the structure and composition of plant communities at local scales in the Himalayan region, and a number of phytogeographic accounts (e.g. Rau 1975; Mani 1978), only limited efforts have been made to synthesize the regional pattern of vegetation in the region(*cf.* Figure 0.4). This chapter gives an overview of the vegetation along vertical and horizontal gradients of the Himalayan region based on the author's own research, and on a review of available literature. Prominent physiognomic classes derived from the prevalent systems of classification (Champion & Seth 1968; Mani 1978) are briefly described, along with the broader floristic structure. The floristic associations along various gradients are based on simple statistical co-occurrences, and not necessarily on the Zurich-Montpellier School (Braun-Blanquet, 1928, 1932). The purpose of such descriptions is to characterize the terrestrial and wetland ecosystems in the region, which serve as important habitat for resident as well as migratory avian communities.

Vegetation along the Horizontal Gradients

The Himalayan region is subjected to increasing amounts of precipitation from west to east, especially on the southern slopes. In the extreme west (74°–76° E), mean annual

Figure 12.1. Different landscapes of the Himalayas. Photo credits for 1a, 1c and 1d: Gopal S. Rawat; for 1b Herbert H.T. Prins. (A black-and-white version of this figure will appear in some formats. For the colour version, please refer to the plate section.)

precipitation is below 500 mm, but at the eastern extremity (91°–96° E), mean annual precipitation ranges between 4000–4500 mm. The west–east gradient can be observed within each of the following parallel zones of vegetation:

(a) Foothill forests along Siwaliks and sub-Himalayan tracts
(b) Himalayan temperate forests
(c) Subalpine forests
(d) Alpine vegetation of the Greater Himalaya
(e) Vegetation of the Trans-Himalaya

Each of these zones supports a number of broad physiognomic classes of vegetation (Figure 12.2). Characteristic features of vegetation along these eco-climatic zones are described later in this chapter:

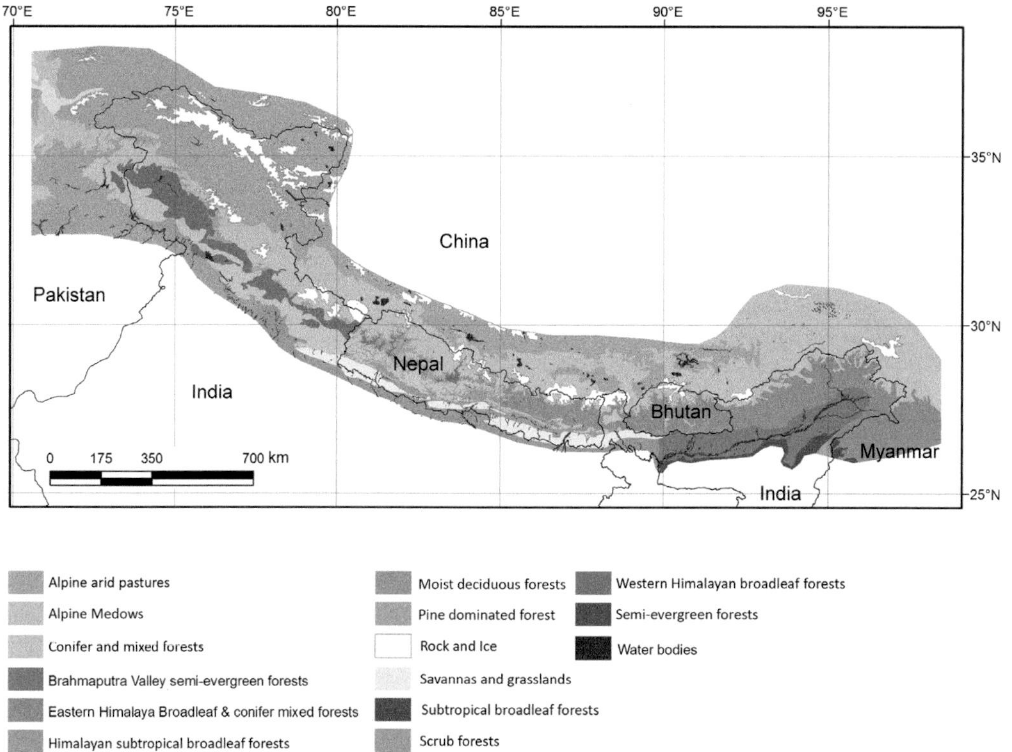

Figure 12.2. A generalized map of Himalayan vegetation, modified from Schweinfurth (1957), Champion and Seth (1968) and Mani (1978). (A black-and-white version of this figure will appear in some formats. For the colour version, please refer to the plate section.)

The Foothill Forests

From west to east, the foothill forests along the sub-Himalayan zones, especially below 1000 m a.s.l., intergrade from tropical dry deciduous to moist deciduous, semi-evergreen and wet-evergreen types. This reflects the increasing gradient of precipitation. In the extreme west, south-facing slopes of the Siwaliks and the outer Himalaya support dry deciduous forests, open scrub or hill savanna depending on the intensity of human use. Several areas are under heavy pressure from transhumant pastoral communities, resulting in open *Dodonea* scrub. Major associations in this area include *Dodonea – Carissa* with increasing preponderance of two alien invasive shrubs, viz., *Lantana camara* and *Prosopis juliflora*. A major association of forest vegetation is *Acacia-Lannea-Anogeissus*. The protected sites represent relatively dense vegetation cover, characteristic tree species being *Acacia catechu, Boswellia serrata, Ougeinia oojenensis, Hymenodictyon excelsum, Lannea coromandelica* and *Anogeissus latifolia*. The steeper slopes above 1000 m are often dotted with scattered trees of *Pinus roxburghii*.

Towards the mid-west (76°–78° E), the Himalayan foothills support tropical, moist, deciduous forests, where mean annual precipitation ranges from 1500 to 2000 mm.

This belt is typically dominated by Sal *Shorea robusta*, a summer deciduous Dipterocarp. The typical floristic association in these forests is *Shorea-Adina-Terminalia*. Sal is a gregarious species prevalent in several wildlife protected areas in this region, viz., Simbalwara, Rajaji and Corbett National Parks in India and Bardiya and Chitwan in Nepal. In addition, this belt supports various other local edaphic and climatic formations such as subtropical swamp forests, subtropical pine forests and subtropical dry evergreen forests (Champion & Seth, 1968).

In the central and eastern part (84°–88° E), the forests tend to become more diverse and lush, marked by an increased profusion of epiphytic orchids, aroids, wild bananas, *Pandanus*, bamboos and ferns (Padmawate *et al.*, 2004; Rawat, 2005c). In the extreme east, around the junction of the Himalayas and the Brahmaputra flood plains, the vegetation is even more diverse, ranging from alluvial grasslands, reed swamps, tropical moist deciduous forests, tropical wet evergreen forest, tropical semi-evergreen forests and subtropical broadleaf forests merging into Himalayan temperate forests in close proximity, as a result of steep eco-climatic gradients. This area is particularly rich in diversity and abundance of bamboos (> 60 species), orchids (> 1000 species), canes and palms (> 20 species) and figs (*Ficus* spp., > 50 species). There are several wild relatives of cultivated plants such as *Camellia caudata* (wild tea), *Coffea benghalensis* (wild coffee), *Musa* spp. (wild banana) and *Mangifera sylvativa* (wild mango). According to Takhtajan (1969), this region can be regarded as the 'cradle of the angiosperm evolution' due to the high abundance of 'primitive' angiosperm families (e.g. Magnoliaceae, Tentracentraceae and Hamamelidaceae).

The foothill forests of the Himalayas have a very rich avifauna which includes many winter as well as summer visitors. A number of Himalayan species descend to the foothills during winter, and some species such as the Indian Pitta (*Pitta brachyura*) from peninsular India also visit these forests during the summer monsoon. These foothill forests in the eastern Himalayas abound in species of hornbills, barbets and green pigeons (Islam & Rahmani, 2004). The open wetlands and riverbanks along Himalayan foothills attract a large number of migratory birds from the Tibetan Plateau and Central Asia, for example, Bar-headed Goose *Anser indicus*, White Wagtail *Motacilla alba* and Ruddy Shelduck *Tadorna ferruginea*. The productivity of these wetland ecosystems and their ability to meet the energy demands of birds migrating long distances across the Himalayas has not been investigated in detail.

The Himalayan Temperate Forests

The second parallel zone, between approximately 1500 to 3000 m a.s.l. in the Himalayas, is quite complex and diverse due to variations in topography, relief features and aspect. Besides the general increase in precipitation from west to east, the local orographic features cause distinct patterns. A large number of physiognomic groups and successional stages occur in this belt, governed strongly by geology, aspect, slope angle, soil depth, altitude, precipitation and, most importantly, anthropogenic activities such as fire, livestock grazing, timber extraction, lopping and collection of non-wood forest products (Singh & Singh, 1987, 1992). Various species of oak (*Quercus* spp.) and conifers form gregarious formations from west to east (Table 12.1). Among oaks, *Q. ilex, Q. leucotrichophora, Q. lanuginosa, Q. grifithii, Q. lamellosa* and *Q. serrata* are prominent. Similarly, along the west-to-east gradient, a few conifers such as Spruce *Picea smithiana*, West Himalayan Fir *Abies webbiana*, East Himalayan Fir *Abies densa* and Larch *Larix grifithii* dominate the cool temperate conifer forests, especially on shady northern aspects. One of the gregarious species of oak, *Q. semecarpifolia*, exhibits disrupted distribution pattern along the west–east gradient. It forms extensive forest in the upper montane zone in the western Himalayas as far east as central Nepal, and reappears in eastern Nepal, Sikkim and Bhutan. Likewise, Eastern Hemlock *Tsuga dumosa* is distributed patchily from 80–90° E. One of the key features of mid-elevation Himalayan temperate broadleaf forests in the eastern Himalaya is the preponderance of Lauraceous taxa (> 80 species in Bhutan and Arunachal Pradesh in India combined, as opposed to fewer than 15 species in the western Himalaya) and the abundance of rhododendrons (> 70 species). These forests are known to support a large number of seasonal migratory birds, especially leaf warblers, and many frugivorous bird species and pheasants (Srinivasan *et al.*, 2010).

Subalpine Forests

The subalpine zone represents a narrow zone of transition terminating at the alpine tree line in the Himalayas. Generally, the zones between 3000–3500 m in the western Himalayas and between 3300–4000 m in the central and eastern Himalayas fall under this category. In the western Himalayas, high altitude fir *Abies spectabilis*, birch *Betula utilis*, occasionally *Q. semecarpifolia, Prunus cornuta, Sorbus vestita* and one species of maple *Acer caesium* form subalpine forests, which terminate at the alpine treeline. Some pockets in the western Himalayas, especially the inner drier ranges, support Blue Pine *Pinus wallichiana*, *Juniperus semiglobosa* and Chilgoza Pine *Pinus gerardiana*, forming the subalpine forests (Aswal & Mehrotra, 1994). Interestingly, one of the junipers, *J. indica*, has a prostrate, shrubby growth form in the western Himalayas while in the central and eastern parts, this species forms extensive woodlands in the subalpine zone as it attains the growth habit of a tree. Among all the species, the Silver Birch *Betula utilis* is most widely distributed from the western to the eastern Himalayas, forming gregarious forests in associations with various species of rhododendrons. These forests support a large number of seasonal altitudinal migrants which breed in these habitats, viz., leaf warblers, flycatchers, tits, pipits, rosefinches, yuhinas and cuckoos.

Table 12.1. Key vegetation types along vertical and horizontal gradients of the Himalaya Vegetation Types are modified after Champion and Seth (1968) and Mani (1978).

SN	Vegetation Type	Characteristic Plant Associations (Single species indicates its dominance in the given area)	Location along Vertical and Horizontal Gradients (Elevation in m a.s.l.)
1.	Alpine Desert Steppe	*Stipa – Oxytropis* *Acantholimon -Thylacospermum*	Trans-Himalaya: > 5000 m; > 34 °N lat
2.	Alpine Dry Scrub	*Artemisia – Kraschenninikovia* *Caragana – Lonicera*	Changthang Plateau Zanskar Range: > 4500 m
3.	Alpine Marsh Meadows	*Kobresa – Carex – Scirpus* *Phragmites – Calamogrostis*	Trans-Himalayan wetlands and peatlands: > 4500 m
4.	Alpine Moist Meadows	*Kobresia – Carex* *Geranium – Potentilla* *Danthonia cachemyriana*	Western to central; Gentle, south facing slopes: 3500–5000 m
5.	Alpine Moist Scrub and *Krummholz*	*Salix – Lonicera* *Rhododendron anthopogon* *Hippophae- Myricaria* *Rhododendron campanulatum* *Rhododendron thomsonii*	Western to eastern; Shady moist slopes upto 4200 m; Riverine scrub: > 3300 m Western: 3300–3600 m Central and eastern: 3000–4000 m
6.	Sub-alpine Forests	*Betula – Rhododendron* *Abies spectabilis – Quercus semecarpifolia* *Abies densa – Juniperus indica*	Western to eastern: 3300–3800 m Western to central: 3000–3600 m Central to eastern: 3800–4200 m
7.	Himalayan (Temperate) Conifer Forests	*Pinus gerardiana* *Cedrus deodara* *Cupressus torulosa* *Pinus wallichiana* *Tsuga dumosa*	North-western, dry: 2500–3000 m Western, moist: 2000–2800 m Central, dry: 2000–2500 m Central to eastern: 2500–3300 m Central to eastern: 2000–3000 m
8.	Himalayan (Temperate) Broadleaf Forests	*Quercus – Rhododendron* *Schima-Castanopsis-Engelhardtia* *Quercus – Lauraceous*	Western to central: 1500–3000 m Central to eastern: 1500–2500 m Central to eastern: 2000–3000 m
9.	Riverine Forests	*Acacia – Dalbergia* *Trewia – Syzygium* *Alnus nepalensis*	Western to central: < 800 m Central to eastern: < 1000 m Central to eastern: 1500–2200 m
10.	Subtropical Forests	*Pinus roxburghii* *Olea – Bauhinia*	Western to central: 1500–2200 m Western, foot-hills: < 1500 m
11.	Tropical Dry Deciduous Forests	*Acacia – Lannea-Anogeissus* *Dodonea – Carissa*	North-west: < 1000 m Western: < 1200 m
12.	Tropical Moist Deciduous Forests	*Shorea – Adina-Terminalia*	Western to central: < 1200 m
13.	Tropical Semi-evergreen	*Syzygium – Altingia- Terminalia*	Central to eastern: 800–1200 m
14.	Tropical Wet Evergreen	*Dipterocarpus-Mesua*	Eastern Himalaya: < 500 m
15.	Hillside Grasslands Tall-wet Grasslands	*Themeda – Arundinella* *Saccharum – Imperata*	Western to central, steep south facing slopes: 1000–3000 m Flood plains, west to east: < 700 m
16.	Reed-beds and Tropical Swamps	*Phragmites – Arundo-Typha*	Wetlands of upper Gangetic and Brahmaputra flood plains: < 300 m

Alpine Vegetation of the Greater Himalaya

The alpine vegetation zone is demarcated by the presence of a distinct treeline at its lower limit that lies around 3300±200 m a.s.l. in the western Himalayas and 3800 ±200 m in the central and eastern parts. The area immediately above the natural tree line is usually covered by *krummholz* (stunted forest or crooked wood) formations. In the western Himalayas, only two species of rhododendron form *krummholz*, viz., *R. campanulatum* and *R. barbatum*, while in the central and eastern Himalayas, this zone is much wider with several constituent species such as *Rhododendron wightii, R. fulgens, R. thomsonii, R. hodgsonii, R. lanatum* and *R. campanulatum*. The *krummholz* zone forms ideal habitat for a number of avian species, for example, Blood pheasant *Ithaginis cruentus*, Satyr tragopan *Tragopan satyra*, Monal pheasant *Lophophorus impejanus* and Hill partridge *Arborophila torqueola*, as well as one of the threatened mammals of the Himalayan region, the Himalayan Musk Deer *Moschus chrysogaster*. Above this zone, the vegetation comprises closely woven and matted strands of shrubs, herbaceous meadows, bogs and fell-fields (Rawat, 2007). Of these, the herbaceous meadows are well known for their species richness and specialized growth forms (dominance of perennial-annual growth habits), many of which bear attractive flowers. Generally, the extent of herbaceous vegetation in the alpine zone of the western Himalayas is much larger compared to the eastern Himalayas due to extensive and wider *krummholz* and alpine scrub in the latter zone where this formation ascends to much higher elevation.

Trans-Himalayan Vegetation along the West-to-East Gradient

The Trans-Himalayan (cold arid) vegetation in parts of northwest Pakistan, India and the Tibetan Plateau includes riverine scrub, high-altitude wetlands, including sedge-dominated marsh meadows, scrub steppe, mixed-alpine herbaceous formations and desert steppe (Miehe, 1997; Dickore & Nusser, 2002; Dickore & Miehe, 2002). Each of these formations vary considerably, depending on the micro-topographic features, moisture availability and altitude (Rawat & Adhikari, 2005; Namgail *et al.*, 2012). In the extreme northwest, below 3000–3200 m, the landscape is more broken and sparsely vegetated (< 5% ground cover) with salt-tolerant taxa such as *Salicornia, Haloxylon ammodendron* and *Zygophyllum xanthoxylon*. Towards the eastern part of the Tibetan Plateau, the steppe formation is gradually replaced by mixed alpine meadow-like conditions and juniper woodland marked by increased species diversity and higher biomass production (Yang *et al.*, 2009). The wetlands on the eastern Tibetan Plateau support a higher diversity of migratory birds as compared to those in the west, also indicated by a higher number of Important Bird Areas in the region (Birdlife International, 2009). Although sparsely vegetated, these areas form important feeding and nesting grounds for several avian species such as the Tibetan Snowcock *Tetraogallus tibetanus*, Himalayan Snowcock *Tetraogallus himalaensis*, Snow Partridge *Lerwa lerwa*, Tibetan Sandgrouse *Syrrhaptes tibetanus*, to name a few. Notable migratory birds inhabiting the wetlands of eastern Ladakh include

Black-necked Crane *Grus nigricollis*, Bar-headed Goose, Black-necked Grebe *Podiceps nigricollis*, Great Crested Grebe *P. cristatus*, Brown-Headed Gull *Larus brunnicephalus*, Common Tern *Sterna hirundo* and several raptor species (Pfister, 2004; Prins *et al.*, Chapter 20).

Vegetation along Vertical Gradients

The vertical zonation of vegetation in the Himalayas is determined by altitude, levels of summer and winter precipitation (the latter in the form of snow), rain-shadow effects and length of growing season. An example of the distribution of resident and migrant birds along such a gradient is provided in Himachal Pradesh by Mahabal (2005). Owing to the varying width of mountain ranges and different relief features, it is difficult to draw parallels between the vegetation of the western, central and eastern Himalayas from south to north, although broad eco-climatic zonation along a single face of the mountains can be done. For example, in the western Himalayas, major mountain ranges, viz., the Siwaliks, Dhauladhar, Pir Panjal, Greater Himalaya, Zanskar, Ladakh and Karakoram ranges influence the distribution of summer and winter precipitation as well as colonization of plant species from Palearctic, Mediterranean and Afro-tropical regions (Mani, 1978). On the other hand, vegetation in the eastern Himalayas reflect the influence of Indo-Malayan, Sino-Japanese and Palearctic regions (see Fig. 0.4). Along the south–north latitudinal and temperature gradients, a few patterns in the general physiognomy and species composition can be seen, as discussed later in this chapter.

Grassland Vegetation at Varying Altitudes

The grassland vegetation along the elevation gradient of the Himalayas includes the most productive hygrophilous grasslands of the Terai, mesophilous grasslands and hill savannas in the Siwaliks, hillside grasslands in the lower and upper montane belts, alpine grasslands of the Greater Himalaya and steppe grasslands in the cold deserts. The grasslands of the Terai belt are among the most productive ecosystems in the world, having annual production potential of up to five tons ha^{-1}, dominated by various species of *Saccharum* (*S. spontaneum, S. benghalense, S. narenga*), *Themeda arundinacea, T. gigantea* and *Imperata cylindrica* (Singh, 1987).

The steeper, south-facing slopes of the western Himalayas, between elevations of 1000 and 3000 m. a.s.l., support temperate grasslands on slopes, which have developed as a result of frequent fires, forest clearance and intensive management by the local communities (Rawat, 1998). A number of tussock-forming perennial grasses dominate steeper slopes, which also harbour several wild ungulates, and a few typical grassland birds such as Cheer Pheasant *Catreus wallichii* and

Black Francolin *Francolinus francolinus*. Dachigam National Park in Kashmir is well known for its extensive temperate grasslands and forest (Naqash & Sharma, 2011). This area supports a variety of long-distance passage migrants, mainly passerines, as well as serving as a breeding area for migrant passerine species that visit during the northern summer (Shah *et al.*, 2013). The area also harbours a globally threatened subspecies of the Red Deer or Hangul *Cervus elaphus hanglu*, of which fewer than 250 individuals survive in the wild. Degradation of these grasslands, especially on gentler slopes due to excessive grazing by heavy livestock, resultant soil erosion and a preponderance of unpalatable species have become major causes of concern (Shah *et al.*, 2011).

Still higher in the alpine zone of the Greater and Trans-Himalaya, several types of grasslands have been reported by Rawat (1998). Of these, *Danthonia* grasslands in the lower alpine regions of the western and central Himalayas (3300–4000 m a.s.l.) and *Stipa – Elymus* and *Festuca – Oryzopsis* type of grasslands deserve special mention as they support not only a considerable biomass of domestic livestock but also a number of wild mammals, and avifauna such as Tibetan Snowcock, Himalayan Snowcock, Snow Partridge and Tibetan Sandgrouse.

The Wetland Vegetation

The Himalayan foothills abound in natural as well as artificial wetlands. Most of these wetlands are of very high conservation significance, especially as winter habitat for the avifauna, both migratory and resident. The Maharana Pratap Sagar or Pong Reservoir on the Beas River in the Siwalik Hills of the western foothills serves as an internationally important site for large numbers of migratory waterbirds, including Bar-headed Geese, Northern Pintail and Ruddy Shelduck, that take advantage of the aquatic vegetation and surrounding crop fields (Pandey, 1993; Jindal et al., 2013, Namgail *et al.*, Chapter 2). The open banks of the Ganges, the Yamuna River and nearby wet grasslands harbour a number of threatened species such as Swamp Francolin *Francolinus gularis*, Bengal Florican *Houbaropsis bengalensis* and Sarus Crane *Grus antigone*. In addition, large flocks of surface-feeding and diving waterbirds such as pochards, grebes, Common Coot *Fulica atra*, Northern Pintail *Anas acuta*, Gadwall *Anas strepera* and Eurasian Wigeon *Anas penelope* occupy these wetlands during various seasons. Loss and degradation of wetlands and alluvial grasslands due to agricultural expansion, ill-planned development activities for habitation, egg predation by feral dogs and heavy use of pesticides are the most significant threats to these species (Sundar & Kittur, 2012).

The wetlands in the foothills of the Eastern Himalayas, especially along the banks of the Brahmaputra and its tributaries, support a large number of migratory birds flying between their breeding grounds in the Palearctic and wintering grounds on the Indian subcontinent. The swamp forests and alluvial grasslands in this area support populations of Globally Threatened White-winged Duck *Cairina scutulata*,

Black-breasted Parrotbill *Paradoxornis flavirostris* and Marsh Babbler *Pellorneum palustre*. One of the important protected areas, well known for its extensive wetland habitats in the foothills of the eastern Himalaya, is Dibru-Saikhowa wildlife sanctuary, with more than 310 species of birds recorded, including the Endangered and elusive White-bellied Heron *Ardea insignis* (Datta, 1996; Choudhury 2002, 2006). This sanctuary faces severe threats due to excessive anthropogenic pressure in the form of fishing, livestock grazing and other forms of habitat encroachment.

The high-altitude wetlands of eastern Ladakh are of particularly high conservation significance as they support considerable breeding populations of the Vulnerable Black-necked Crane, Bar-headed Goose, and Great Crested Grebe (see Prins & Van Wieren, 2004; Prins *et al.*, Chapter 26). These wetlands harbour about 25–30% of the total vascular flora known from the region despite their limited extent (Rawat & Adhikari, 2005). The marsh meadows are dominated by sedges, that is, *Carex* (> 30 species), *Kobresia* (> 7 species), *Blysmus*, rushes (> 12 species of *Juncus*), *Eleocharis* and several species of grasses, for example, *Calamogrostis holciformis, Poa* spp. and *Puccinellia himalaica*. Pools of shallow water support a number of aquatic species such as *Potamogeton pectinatus, Myriophyllum verticillatum, Hippuris vulgaris, Ranunculus natans* and *R. trichophyllus*. Typical herbaceous elements in marsh meadows include species of *Ranunculus, Pedicularis, Gentiana, Gentianella* and *Primula*. Some of the species, typical of saline marshes (halophytes), are *Atriplex tatarica, Pucinellia himalaica, Suaeda olufsenii, Triglochin maritimum* and *Glaux maritima*. Here, Bar-headed Geese and Ruddy Shelducks feed on the sedges and wetland plants (G.S. Rawat, pers. obs.). These wetlands also serve as staging areas for long-distance migrant ducks and other species (see Prins *et al.*, Chapter 26).

The riverine floodplain marshes and lakes in the Kashmir Valley of the lower western Himalayas, associated with the Jhelum River, are recognized as internationally important for large numbers of long-distance passage migrants, and non-breeding populations of Greylag Goose *Anser anser*, Bar-headed Goose, Northern Pintail, Northern Shoveler *Anas clypeata*, Eurasian Wigeon, Common Teal *Anas crecca* and others (Bacha, 1992, 1994; Wetlands International, 2007), and the Wular Lake and its associated marshes have been designated as a Ramsar site. These wetlands support a diversity of submerged and emergent aquatic plant species (Zutshi, 1975). These include *Trapa natans* and *T. bispinosa, Nelumbo nucifera, Nymphoides indica, Phragmites communis* and *Typha augstata* and associations of *Ceratophyllum-Myriophyllum-Potamogeton* in the deeper portions of the lake, and provide an important source of food for waterbirds. However, plantations of willow within the lake and the spread of exotic species such as *Salvinia natans* and *Azolla* sp. in the lakes and marshes have reduced the feeding areas for migratory waterbirds, and the control of these non-native species has been prioritized (Wetlands International, 2007). In addition, urban and rural development and flood control measures in these areas are having impacts on the wetland habitats.

Except for a few wetlands in the Kashmir Valley, most of the wetlands in the Greater Himalayas, especially in the alpine region, are oligotrophic in nature and support relatively low populations of migratory birds (e.g. Prins *et al.*, Chapter 26). Such lakes are largely glacial in origin.

Riverine Forests and Scrub

Most of the Himalayan rivers (except those located in the Trans-Himalayan ranges) flow in a north–south direction. Some of the river valleys create their own climate due to temperature inversion and strong gravity winds, causing higher aridity in the valley bottoms as compared to the higher slopes. Some of the prominent plant associations along riverine areas include *Acacia catechu* – *Dalbergia sissoo, Terminalia myriocarpa* – *Duabanga sonneroides* and *Macaranga pustulata* – *Toona ciliata* in the foothills; *Alnus nepalensis* – *Macaranga pustulata* and *Populus ciliata* – *Hippophae salicifolia* in the middle elevation zone. The riverine forest above 3500 m is replaced by riverine scrub, prominent communities being *Hippophae tibetana* – *Myricaria elegans* and *Salix elegans* – *Myricaria prostrata* up to 4200 m a.s.l. Unlike other vegetation types, the riverine scrub of the Greater and Trans-Himalaya are similar in physiognomy and floristic structure. However, along a few river valleys, especially of the River Indus and the Shayok River, a few characteristic species form the riverine scrub, for example, *Hippophae rhamnoides* ssp. *turkestanica, Populus euphratica*, tamarisk (*Tamarix gallica*) and *Myricaria rosea*.

The riverine forests along the Himalayan foothills, especially those dominated by *Toona ciliata, Terminalia myriocarpa* and *Duabanga*, are rich in vascular epiphytes, especially ferns and orchids (Padmawate *et al.*, 2004). Himalayan alder as well as most of the riverine species form important biological corridors for a large number of avian species, especially altitudinal migrants.

Conclusion

The Himalayan region harbours diverse assemblages of flora and vegetation, strongly influenced by gradients of precipitation and altitude. Along the west–east transition, distinct patterns can be seen in the physiognomy and species assemblages in different eco-climatic zones. For example, the foothill forests range from *Dodonea* scrub to tropical dry deciduous forests, tropical moist deciduous forest, tropical semi-evergreen and tropical wet-evergreen forests from west to east. Similarly, changes in vegetation cover and species assemblages from south to north can be seen in forests, grasslands and wetlands. The tall hygrophilous grasslands of the Himalayan foothills and the wetlands of the Tibetan Plateau are the most important feeding areas for a large number of migratory birds. Other vegetation types in various parts of the Himalayas support altitudinal migrants and sedentary avifaunal assemblages. Riverine vegetation, Terai grasslands and several wetlands are highly threatened due to recent changes in land use and ill-planned developmental activities.

Acknowledgements

I am thankful to Dr Devendra Kumar, Research Associate, Wildlife Institute of India, for helping in the preparation of this chapter and drawing the vegetation map of the Himalayas.

References

Anonymous (2002). *Forest and vegetation types of Nepal.* Department of Forests, Govt. of Nepal. TISC. Document Series No. 105.

Aswal, B.S. & Mehrotra, B.N. (1994). *Flora of Lahaul & Spiti: A Cold Desert in North West Himalaya.* Bishen Singh Mahendra Pal Singh, Dehra Dun.

Bacha, M.S. (1992). *Waterfowl Census in Kashmir Wetlands.* Department of Wildlife Protection Jammu & Kashmir Government.

Bacha, M.S. (1994). *Waterfowl Census in Kashmir Wetlands.* Department of Wildlife Protection Jammu & Kashmir Government.

Birdlife International (2009). *Important Bird Areas in China: Key Sites for Conservation.* Cambridge: BirdLife International.

Braun-Blanquet, J. (1928). Pflanzensoziologie. Grundzüge der Vegetationskunde. Wien: Springer.

Braun-Blanquet, J. (1932). *Plant Sociology: The Study of Plant Communities.* New York: McGraw-Hill.

Champion, H.G. (1923). The interaction between Pinus longifolia (Chir) and its habitat in the Kumaon Hills. *Indian Forester*, **49**, 405–415.

Champion, H.G. (1936). A preliminary survey of forest types of India and Burma. *Indian Forester*, **63**, 613–615.

Champion, H.G. & Seth, S.K. (1968). *A Revised Survey of the Forest Types of India.* Manager of Publications, Delhi.

Choudhury, A. (2002). Globally threatened birds in Dibru-Saikhowa Biosphere Reserve. *Himalayan Biosphere Reserves*, **4**, 49–54.

Choudhury, A. (2006). Birds of Dibru-Saikhowa National Park and Biosphere Reserve, Assam, India. *Indian Birds*, **2**, 95–105.

Datta, S. (1996). Bird watching at Dibru-Saikhowa Wildlife Sanctuary. *Newsletter for Birdwatchers*, **36**, 51–53.

Dickore, W.B. & Miehe, G. (2002). Cold spots in the highest mountains of the world – Diversity patterns and gradients in the flora of the Karakorum. In Ch. Körner & E.M. Spehn, eds., *Mountain Biodiversity: A Global Assessment.* London: Parthenon Publishing, pp. 129–147.

Dickore, W.B. & Nusser, M. (2002). Flora of Nanga Parbat (NW Himalaya, Pakistan): an annotated inventory of vascular plants with remark on vegetation dynamics. *Englera*, **19**, 1–253.

Dudgeon, W. & Kenoyer, L.A. (1925). The ecology of Tehri-Garhwal. A contribution to the ecology of western Himalayas. *Journal of the Indian Botanical Society*, **4**, 233–285.

Hartmann, H. (2009). *A Summarizing Report on the Phytosociological and Floristic Explorations (1976–1997) in Ladakh, India.* Switzerland: Druck.

Islam, M.Z. & Rahmani, A.R. (2004). *Important Bird Areas in India: Priority Sites for Conservation*. Indian Bird Conservation Network. Bombay Natural History Society and Birdlife International U.K., pp. xviii+1133.

Jindal, R., Singh, H. & Sharma, C. (2013). Avian fauna of Pong Dam Wetland – a Ramsar site. *International Journal of Environmental Sciences*, **3**, 2236–2250.

Kala, C.P., Rawat, G.S. & Uniyal, V.K. (1998). *Ecology and Conservation of the Valley of Flowers National Park, Garhwal Himalaya. RR-98/003*. Dehra Dun: Wildlife Institute of India.

Kenoyer, L.A. (1921). Forest formations and successions in the Sat Tal Valley, Kumaon, Himalayas. *Journal of the Indian Botanical Society*, **2**, 236–258.

Mahabal, A. (2005). Aves. In *Fauna of Western Himalaya*. (Ed.: The Director). Kolkata: Zoological Survey of India, pp. 275–339.

Mani, M.S. (1978). *Ecology & Phytogeography of High-Altitude Plants of the Northwest Himalaya*. London: Chapman & Hall.

Miehe, G. (1997). Alpine vegetation types of the Central Himalaya. In F.E. Wielgolaski, ed., *Polar and Alpine Tundra Ecosystems of the World* **3**, pp. 161–184.

Mittermeier, R.A., Turner, W.R., Larsen, F.W., Brooks, T.M. & Gascon, C. (2011). Global biodiversity conservation: the critical role of hotspots. In *Biodiversity Hotspots*. Heidelberg: Springer Berlin, pp. 3–22.

Myers, N., Mittermeier, R.A., Mittermeier, C.G., Da Fonseca, G.A. & Kent, J. (2000). Biodiversity hotspots for conservation priorities. *Nature*, **403**, 853–858.

Namgail, T., Rawat, G.S., Mishra, C.D., Van Wieren, S.E. & Prins, H.H.T. (2012). Biomass and diversity of dry alpine plant communities along altitudinal gradients in the Himalayas. *Journal of Plant Research*, **125**, 93–101.

Naqash, R.Y. & Sharma, L.K. (2011). Management Plan (2011–2016) Dachigam National Park. www.jkwildlife.com/pdf/pub/final_management_plan_DNP_06082011.pdf.

Padmawathe, R., Qureshi, Q. & Rawat, G.S. (2004). Effects of selective logging on vascular epiphyte diversity in a moist lowland forest of Eastern Himalaya, India. *Biological Conservation*, **119**, 81–92.

Pandey, S. (1993). Changes in waterbird diversity due to the construction of Pong Dam Reservoir, Himachal Pradesh, India. *Biological Conservation*, **66**, 125–130.

Pfister, O. (2004). Birding hotspot: Ladakh. *Oriental Bird Club Newsletter*, **2004**, 1–2.

Prins, H.H.T. & Van Wieren, S.E. (2004). Number, population structure and habitat use of Bar-headed Geese Anser indicus in Ladakh (India) during the brood-rearing period. *Acta Zoologica Sinica*, **50**, 738–744.

Puri, G.S., Gupta, R.K., Meher-Homji, V.M. & Puri, S. (1989). *Forest Ecology*. New Delhi: Oxford and IBH Publication Co. Pvt. Ltd., Vol. **II**.

Rau, M.A. (1975). *High Altitude Flowering Plants of West Himalaya*. Howrah: Botanical Survey of India.

Rawat, G. S. (1998). Temperate and alpine grasslands of the Himalaya: Ecology and Conservation. *Parks*, **8**, 27–36.

Rawat, G.S. (2005a). *Alpine Meadows of Uttaranchal: Ecology, Landuse and Status of Medicinal and Aromatic Plants*. Dehra Dun: Bishen Singh Mahendrapal Singh.

Rawat, G.S. (2005b). Vegetation dynamics and management of Rhinoceros habitat in Duars of West Bengal: an ecological review. *National Academy Science Letters*, **28**, 179–186.

Rawat, G.S. (2005c). Terrestrial vegetation and ecosystem coverage within India Protected Areas. *National Academy Science Letters*, **28**, 241–250.

Rawat, G.S. (2007). *Alpine Vegetation of Western Himalaya: Patterns of Species Diversity, Community Structure and Aspects of Conservation*. D.Sc. Thesis. Kumaun University, Nainital.

Rawat, G.S. & Adhikari, B.S. (2005). Floristics and distribution of plant communities across moisture and topographic gradients in Tso Kar Basin, Changthang Plateau, Eastern Ladakh. *Arctic, Antarctic, and Alpine Research*, **37**, 539–544.

Samant, S.S. & Joshi, H.C. (2005). Plant diversity and conservation status of Nanda Devi National Park and comparison with highland national parks of the Indian Himalayan region. *The International Journal of Biodiversity Science and Management*, **1**, 65–73.

Samant, S.S., Dhar, U. & Rawal, R.S. (2005). Diversity, endemism and socio-economic values of the Indian Himalayan Papaveraceae and Fumariaceae. *Journal of the Indian Botanical Society*, **84**, 33–44.

Schweinfurth, U. (1957). Die horizontale und vertikale Verbreitung der Vegetation im Himalaya. *Bonner Geographische Abhandlungen*, **20**, 1–375.

Shah, M., Jan, U., Bhat, B.A. & Ahanger, F.A. (2011). Causes of decline of critically endangered Hangul deer in Dachigam National Park, Kashmir (India): a review. *International Journal of Biodiversity and Conservation*, **3**, 735–738.

Shah, M.G., Jan, U. & Wani, M.R. (2013). Study on distribution of avian fauna of Dachigam National Park, Kashmir, India. *International Journal of Current Research*, **5**, 266–270.

Singh, J.S. & Singh, S.P. (1987). Forest vegetation of the Himalaya. *Botanical Review*, **52**, 80–192.

Singh, J.S. & Singh, S.P. (1992). *Forests of Himalaya: Structure, Functioning and Impact of Man*. Nainital: Gyanodaya Prakashan.

Singh, J.S., Rawat, Y.S. & Chaturvedi, O.P. (1984). Replacement of oak forest with pine in the Himalaya affects the nitrogen cycle. *Nature*, **311**, 54–56.

Singh, P. (1987). *Rangeland Reconstruction and Management for Optimizing Biomass Production*. Jhansi: Indian Grassland and Fodder Research Institute.

Srinivasan, U., Dalvi, S., Naniwadekar, R., Anand, M.O. & Datta, A. (2010). The birds of Namdapha National Park and surrounding areas: recent significant records and a checklist of the species. *Forktail*, **26**, 92–116.

Stainton, J.D.A. (1972). *Forests of Nepal*. London: John Murray.

Sunder, K.G. & Kittur, S.A. (2012). Methodological, temporal and spatial factors affecting modeled occupancy of resident birds in the perennially cultivated landscape of Uttar Pradesh, India. *Landscape Ecology*, **27**, 59–71.

Takhtajan, A. (1969). *Flowering Plants: Origin and Dispersal*. Washington, DC: Smithsonian Institution Press.

Wetlands International. (2007). *Comprehensive Management Action Plan for Wular Lake, Kashmir*. Final Report submitted to the Department of Wildlife Protection, Govt. of Jammu & Kashmir, India.

Yang, Y.H., Fang, J.Y., Pan, Y.D. & Ji., C.J. (2009). Aboveground biomass in Tibetan Rangelands. *Journal of Arid Environments*, **73**, 91–95.

Zobel, D.B. & Singh, S.P. (1997). Himalayan forests and ecological generalization: forests in the Himalayan differ significantly from both tropical and temperate forests. *BioScience*, **47**, 735–744.

Zutshi, D.P. (1975). Associations of macrophytic vegetation in Kashmir lakes. *Vegetatio*, **30**, 61–66.

13 Assessing the Evidence for Changes in Vegetation Phenology in High-Altitude Wetlands of Ladakh (2002–2015)

Sumanta Bagchi, Ekta Gupta, Karthik Murthy and Navinder J. Singh

High-Altitude Wetlands of the Himalayas: Climate and Vegetation Change

Large populations of migratory birds cross the Himalayas during their annual migrations through the Central Asian Flyway, and Himalayan wetlands are critical habitats for many bird species (Namgail *et al.*, 2009; Namgail & Yom-Tov, 2009). In the Ladakh region of the Himalayas, wetlands are important summer breeding grounds for water-birds and other species, and as stopover sites during migrations (Prins *et al.*, Chapter 26). Important wetlands in this region include large lakes such as Tsokar, Tsomoriri and Pangong Tso, and smaller waterbodies such as Riyul Tso and Kyagar Tso (Figure 13.1). Riverine areas of Hanle, Chumur, Choglamsar, Shey, Thikse, Chushul and Loma are also important bird habitats located along important migratory routes in the Himalayas (Pfister, 2004; Delany *et al.*, Chapters 4 & 5).

Over the past few decades, there has been growing understanding and concern about the ecological status of these wetlands, as shown by declining populations of several migratory and resident birds, including the Globally Threatened Black-Necked Crane, *Grus nigricollis* (Hussain, 1985; Hussain & Pandav, 2008). The main socio-ecological changes in the wetlands are (1) agricultural intensification, (2) draining of wetlands for irrigation, (3) intensive use of wetlands for livestock production and (4) proliferation of recreational tourism (Humbert-Droz, Chapter 23), among others. In addition, the region is experiencing altered precipitation and temperature regimes (Bhutiyani *et al.*, 2010; Owen, Chapter 10; Bookhagen, Chapter 11). Collectively, these factors have raised general concerns over environmental degradation, and the future of important biodiversity areas (Bagchi *et al.*, 2013a).

Degradation of these habitats arises from structural changes in vegetation and soil processes, which result in lower levels of ecosystem function, and reductions in ecosystem services (Bagchi *et al.* 2013a). Increasingly, studies around the globe are reporting changes in migration timing, and in propensity and plasticity of timing of

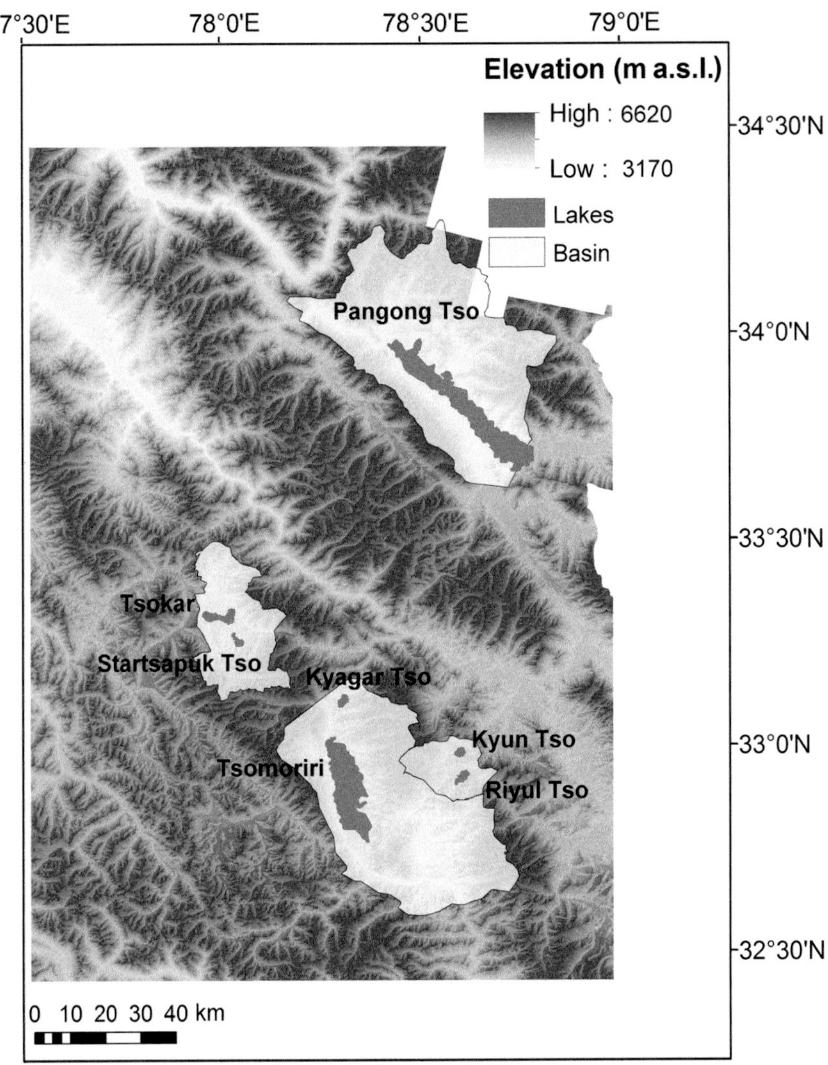

Figure 13.1. Map of the study area, including the boundaries of the study basins.

reproduction in birds. These are often in response to changes in the phenology of vegetation growing seasons linked to climate change, and they have population-level consequences for birds (Visser *et al.*, 1998; Charmantier & Gienapp, 2014). In highly seasonal environments such as the Trans-Himalaya, the optimal window for breeding, migrating and availability of food is short. Performance of migratory populations could be particularly susceptible to changes in vegetation phenology in this region. Here, we attempt to quantify the broad trends of the start, peak and end of the growing season of vegetation in the key watersheds, to provide a baseline in support of monitoring efforts for waterbirds and their habitats. In this chapter, we analyze the changes in vegetation phenology around TsoKar, TsoMoriri and Pangong Tso during the period 2002–2015.

Implications of Change in Vegetation Phenology

Ecological degradation in Ladakh poses complex challenges for conservation. These revolve around conflicting demands of multiple land uses, and mitigating the effects of climatic and anthropogenic factors. Here, we analyze one aspect of regional change in these wetlands, namely trends in vegetation phenology during the period 2002–2015. Vegetation phenology relates to seasonal changes in vegetation status (Cleland *et al.*, 2007; Korner & Basler, 2010). Major phenological parameters are onset of growing season, timing of peak growth, onset of senescence and total duration of growing season (Zhang *et al.*, 2003; Pettorelli *et al.*, 2005; Vrieling *et al.*, 2013). These parameters are influenced by a variety of environmental factors and their associated feedbacks, such as soil fertility, climate and land use. Changes in vegetation phenology may lead to asynchrony between the timing of bird migration and the availability of high-quality habitat (Visser *et al.*, 1998; Stenseth & Mysterud, 2002; Visser & Both, 2005; Visser *et al.*, 2015), with consequences for population status. The results we report here can be used to help understand avian responses to changes in vegetation phenology. Conservation assessment and planning can benefit from empirical information on long-term changes in these parameters, as these can provide insights into the nature, extent and intensity of habitat degradation (Zhang *et al.*, 2003).

We used satellite-based indices, namely the normalized difference vegetation index (NDVI) to assess changes in the vegetation status of three wetlands basins – TsoKar, TsoMoriri and Pangong Tso (Figure 13.1). Spatial and temporal variations in phenological parameters can be estimated through analysis of NDVI. While NDVI in itself is not an index of phenology, temporal patterns in NDVI provide valuable information on phenology (Zhang *et al.*, 2003; Kawamura *et al.*, 2005; Pettorelli *et al.*, 2005). NDVI is a simple index of vegetation phenology, as it relates to reflectance and absorbance of wavelengths broadly associated with photosynthetic activity. It is measured from band-specific sensors as $\text{NDVI} = \frac{\text{NIR} - \text{R}}{\text{NIR} + \text{R}}$ where NIR is reflectance in the near infrared region, and R is reflectance in the red region of the light spectrum (Kriegler *et al.*, 1969). In this way, it indicates the degree to which green plants absorb photosynthetically active radiation. It can be used as an index of vegetation biomass in this ecosystem (i.e. greenness), and temporal changes in NDVI reflect phenological status such as green-up and senescence. The main reason to use NDVI, in preference to more sophisticated indices (e.g. Knox *et al.*, 2011; Ramoelo *et al.*, 2013), is that long-term data are publicly available.

Measuring Changes in Wetland Vegetation

Typical changes in phenology, as captured by corresponding NDVI values, can be conceptualized as a unimodal curve (Figure 13.2). These dynamics in phenological status through time offer an opportunity to observe habitat degradation. Sensitivity of NDVI data to vegetation greenness can be low under high biomass conditions with

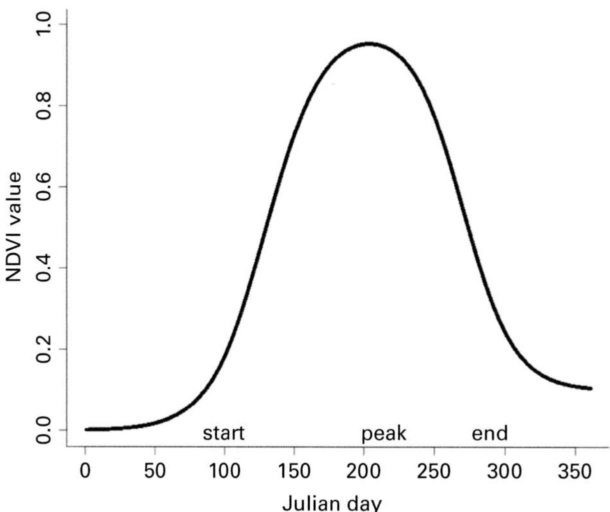

Figure 13.2. Illustrative example of seasonal changes in vegetation that are used for identification of key phenological parameters: 1) start of the growing season, 2) peak growing season, 3) end of the growing season, 4) peak NDVI and 5) end-season NDVI.

a high-leaf area index, such as evergreen forests, or crop fields (Knox *et al.*, 2013). But, in the semiarid Himalayas with a relatively low leaf area index, it allows for easy discrimination between vegetated and barren areas, and also helps in discerning changes such as those in the timing and length of the growing season.

NDVI has been shown to correlate strongly with vegetation productivity and phenology, especially in the semiarid central Asian steppes (Kawamura *et al.*, 2005). The three selected wetlands are known to host a number of migratory and resident bird species, many of which are of global conservation concern. Using long-term time series of NDVI data, we investigated specific questions related to trends in: (1) onset of vegetation growth, (2) peak growth, (3) onset of senescence, (4) length of growing season and (5) peak vegetation biomass.

We used a Geographic Information System (GIS) framework for data handling and nonlinear regressions for data analysis (Figure 13.3). Moderate Resolution Imaging Spectroradiometer (MODIS) sensors aboard the Earth Observing System satellites (EOS) provide NDVI data at 16-day intervals. We used MODIS NDVI data at 250 m resolution (MYD13Q1) from 2002 to 2015 (Figure 13.3). We obtained MODIS data from the Land Processing Distributed Active Archive Center (LP DAAC). Data files were converted into Tagged Image File Format (TIFF) and re-projected from Sinusoidal projection to Universal Transverse Mercator (UTM) projection using the Modis package in R (Mattiuzzi *et al.*, 2014). We used Digital Elevation Model (CartoDEM, from ISRO-Bhuvan) at 30 m resolution to demarcate the basins of the three selected wetlands (Tso Kar, Tso Moriri and Pangong Tso) and extracted relevant pixels using QGIS 2.4 (QGIS Development Team, 2013). We used MODIS product information from the Pixel Reliability Quality Assurance Layer to exclude pixels with

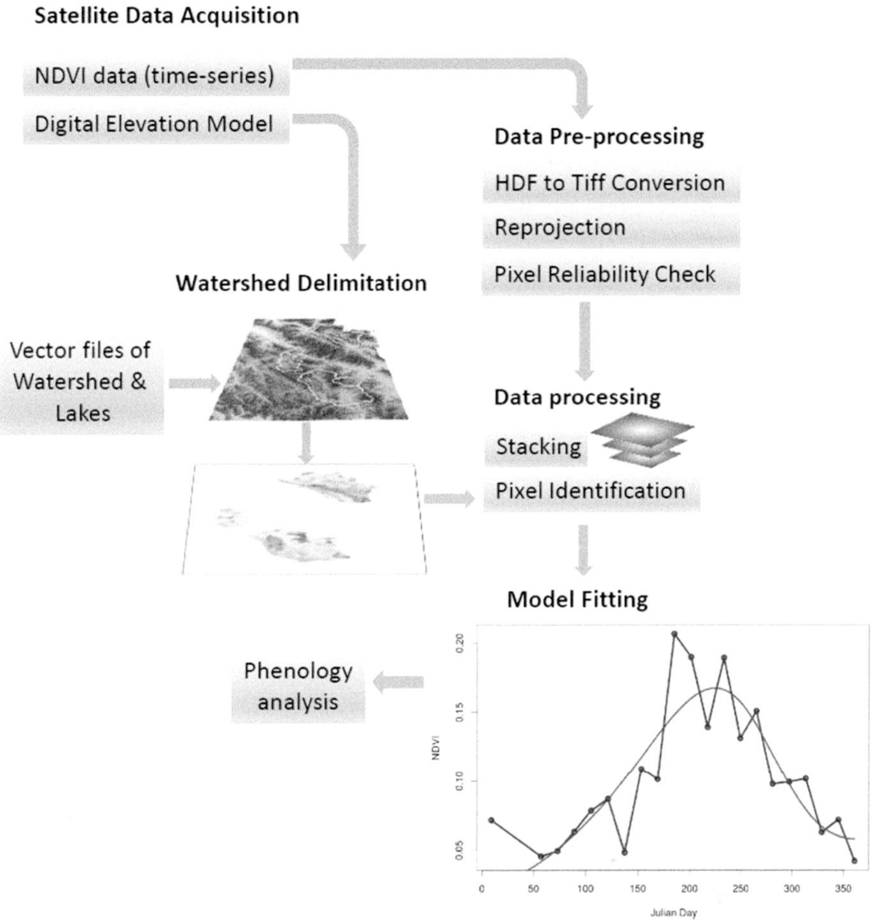

Figure 13.3. Diagrammatic summary of methods used in handling and analysis of remote-sensing data for phenological analysis.

low reliability scores. After the quality check, we only used the reliable pixels from each basin for further analysis. We inspected these pixels, and identified those that were most likely to occur in vegetated patches using the expected unimodal annual pattern in NDVI scores. Pixels that did not show this pattern were deemed to occur in glaciated parts, on barren scree or in other non-vegetation categories, and were not included in the analysis for phenological parameters.

Pixels were geo-referenced, aligned and stacked into raster objects in order to extract NDVI values for the analysis of phenological parameters (Si *et al.*, 2015). The annual unimodal trend in NDVI was estimated with a nonlinear regression in R 3.03 (R Development Core Team, 2014). We used the double logistic model in the phenex package in R (Lange & Doktor, 2014) and 'phenoPhase' function in Phenex to estimate phenology parameters. The next step of analysis was done by fitting an ordinary least squares regression to the yearly phenophase data from 2002 to 2015. For each pixel, on

Table 13.1. Data coverage for three different basins of Ladakh used in NDVI-based phenological analysis with MODIS data (250-m pixels).

Phenology parameter	Pangong Tso	TsoKar	TsoMoriri
Total pixels	6566	2211	8338
Suitable pixels	281	46	101
Peak NDVI (± 95% CI)	0.29 ± 0.02	0.36 ± 0.01	0.35 ± 0.004
End season NDVI (± 95% CI)	0.16 ± 0.002	0.23 ± 0.006	0.21 ± 0.004
Start of season (Julian day) (± 95% CI)	121 ± 1(1 May)	133 ± 3(13 May)	133 ± 2(13 May)
Peak growing season (Julian day) (± 95% CI)	225 ± 1(13 Aug)	240 ± 4(28 Aug)	229 ± 1(17 Aug)
End season (Julian day) (± 95% CI)	279 ± 2(6 Oct)	267 ± 7(24 Sep)	284 ± 1(11 Oct)
Length of growing season (days) (± 95% CI)	158 ± 2	135 ± 8	151 ± 3

any Julian day in the time series, we used change in estimated slope of regression to evaluate trends between years. Additionally, for each Julian day during the growing season (late April to early October) for each pixel, we estimated change in NDVI between years from the slope of a regression.

The basins of Pangong Tso, TsoKar and TsoMoriri in Ladakh cover approximately 410 km^2, 140 km^2 and 520 km^2 in area, respectively. The Pangong Tso basin is larger, but it extends eastward into Tibet (China), and this part was not included. The Tsokar basin included the smaller lake Startsapuk Tso, and the TsoMoriri basin included the smaller Kagyar Tso (Figure 13.1). From the larger basin areas, only a relatively small fraction of pixels suitable for phenological analysis were found, after removing low-quality and non-vegetated pixels (Table 13.1). The final phenological analysis was possible for an area equivalent of 27 km^2 across the three basins (Table 13.1).

Degree of Phenological Changes in the Three Wetlands

Between 2002 and 2015, a majority of the pixels in each of the three basins showed no strong temporal change in green-up date earlier or later in the season. Evidence for advancing or receding date of green up was weak (Figure 13.4). There was no evidence either for a strong temporal change in date of peak NDVI earlier or later in the season. Evidence for advancing or receding date of peak growth was also weak (Figure 13.4). Overall, quite unexpectedly, there was no strong evidence for a shrinking or an expanding growing season (Figure 13.4). These results could also reflect the confounding effects of pixel saturation, but this is unlikely to be a strong factor in our study because leaf area index in these wetlands is expected to be lower than in evergreen forests and crop fields where saturation is of major concern.

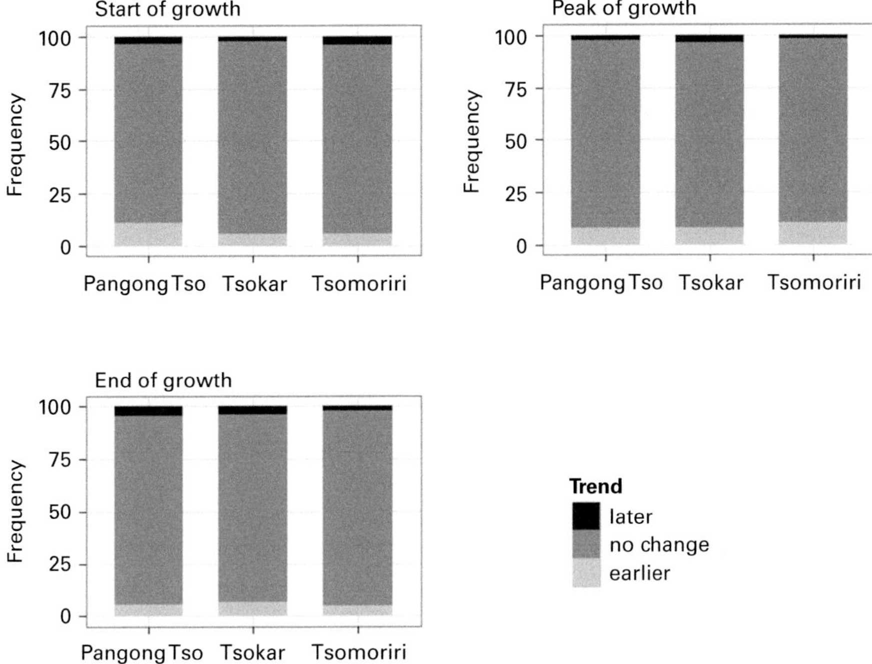

Figure 13.4. Trends in three phenological variables in wetland basins in Ladakh: Pangong Tso (PT), TsoKar (TK) and TsoMoriri (TM). Phenological variables are start of the growing season, date of peak growth and end of the growing season.

Trends in NDVI for different days during the growing season (late April to early October) showed variable patterns across the different wetland basins (Figure 13.5). At Pangong Tso, there were indications for declining NDVI during the early growing season. In all three wetland basins, there were indications for increasing NDVI during the late growing season (Figure 13.5). These trends in slope over time were sorted by statistical significance (at $\alpha = 0.1$). Significantly declining (negative) and increasing (positive) trends were encountered in only a small fraction of pixels (Figure 13.6). While a majority of the pixels showed non-significant trends (Figures 13.5 and 13.6), these trends are partly consistent with observations of reduced winter precipitation and increased summer precipitation for the Trans-Himalayan region (Bhutiyani *et al.*, 2010; Shekhar *et al.*, 2010).

Ecological Resilience of Wetland Vegetation

Although no major changes in indicators of vegetation status were observed in these data, our analyses provide valuable information on phenology and associated parameters that are of high ecological significance. For understanding migratory species, which are known to track changes in vegetation phenology over long

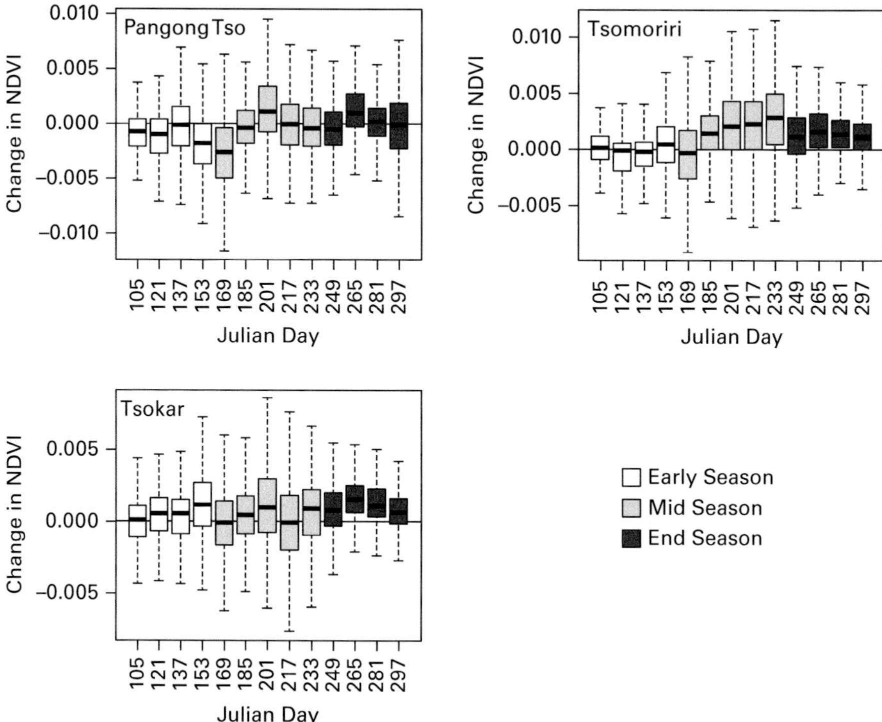

Figure 13.5. Trends in NDVI for different days across years in three wetland basins in Ladakh. Boxplots represent median and inter-quartile range; outliers are suppressed. Change in NDVI is represented as slope calculated across years. Gray-scale colours separate early, peak and end of growing season, covering late April to early October. In Pangong Tso, there are indications of declining NDVI in the early season. Across all wetland basins, there are indications of increasing NDVI in the late season.

distances (Zeng *et al.*, 2010; Si *et al.*, 2015), these estimates may provide vital baseline data for linking migration patterns of waterbirds to food availability and habitat quality (Bauer *et al.*, 2011).

Our results show no remarkable changes in phenology during 2002–2015, and these are consistent with trends reported from adjacent regions of the Tibetan highlands. For example, during the 1980s–1990s, the start of the growing season is estimated to have advanced by two to three days per decade, but it did not change much in the 2000s (Jeong *et al.*, 2011). Such a trend has been found in many other parts of Central Asia (Wang *et al.*, 2015). However, decadal trends for specific days of the year show a pattern of reduced NDVI during the early growing season, and increased NDVI towards the end of the growing season. This may reflect changes in water availability resulting from altered winter and summer precipitation; while winter snowfall (November to March) is decreasing in this region, monsoonal activity (July–August) is strengthening. Therefore, conclusions over habitat degradation issues need to be drawn with caution, and need to accommodate processes and trends that are occurring at long time scales. It is likely that vegetation degradation may have occurred historically in Ladakh's wetlands

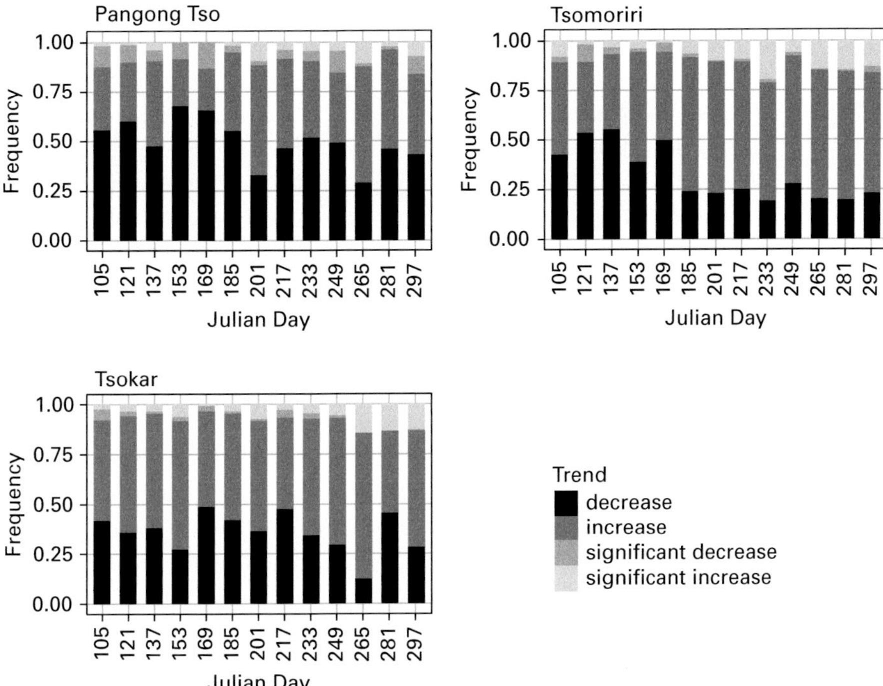

Figure 13.6. Trends in NDVI for different days across years in three wetland basins in Ladakh, covering late April to early October. Change in NDVI is represented as frequency of pixels with positive or negative slope calculated across years. Colours represent whether slope was positive (increasing NDVI over years, i.e. increase), or negative (decreasing NDVI over years, i.e. decrease), and whether changes were statistically significant, or not (P<0.1, i.e. significant decrease and significant increase). In Pangong Tso, there are indications of declining NDVI in the early season. Across all wetland basins, there are indications of increasing NDVI in the late season.

(e.g. in the 1980s and 1990s), but broad vegetation status may not have changed much subsequently during the 2000s and 2010s (Figure 13.4). Older data sets do exist, for example, NOAA AVHRR, but these are at a coarser spatial resolution that precludes specific analysis of individual wetlands. Vegetation status may, however, be responding to changing precipitation patterns in subtle ways (Figures. 13.5 and 13.6). Phenological analyses with moderate resolution remote-sensing data (MODIS 250 m) can still clarify contemporary patterns of vegetation degradation, especially aspects related to land use change in Ladakh (Singh *et al.*, 2013). On one hand, this highlights a limitation of satellite-based interpretation related to data coverage and reliability (Table 13.1). On the other hand, these results underline the importance of primary data, for example, long-term *in situ* vegetation monitoring using permanent plots, that allow interpretation of a wide range of linear and nonlinear ecosystem-level dynamics (e.g. Bagchi *et al.*, 2012; Bagchi *et al.*, 2013b). Our results highlight the importance of initiating long-term vegetation monitoring in ecologically sensitive areas of Ladakh, in addition to remote-sensing approaches.

As outlined earlier, there are several possible reasons why we did not find a shift in phenology in Himalayan wetlands. One reason could be a general lack of response after the 2000s across Central Asia (Jeong *et al.*, 2011; Wang *et al.*, 2015). Another could be the relatively short time frame of our observations, even though there can be no doubt that the climate has changed here in contemporary timescales (Bhutiyani *et al.*, 2010; Shekhar *et al.*, 2010). Furthermore, since wetlands are often not severely water-limited eco-regions compared to the wider landscape, they may show weaker trends and be more resilient. An additional reason why we did not find major phenological shifts may be that day length is more important than temperature in causing the onset of budding and sprouting of plants (Cleland *et al.*, 2007; Korner & Basler, 2010). This interplay between temperature, radiation and day length cannot easily be reduced to simple predictions about the phenology of vegetation development, and a potential mismatch between migration timing and food availability. The importance of these factors has been demonstrated in the movements of some large mammals in high mountains elsewhere in Asia (Wang *et al.*, 2010; Zeng *et al.*, 2010). From this it follows that there is a need to increase the scale and frequency of data collection on a regular basis over a wide range of altitudes and spread over large areas. Satellite data can only be used meaningfully if parameters such as ground radiation, precipitation, snow cover, snow cover duration, snow depth, cloud cover, evaporation and potential evaporation are measured with great precision on the ground over many years. Humanity has ushered the rest of the world into the Anthropocene (Steffen *et al.*, 2011), and we will be unable to manage change if the data are lacking.

Acknowledgements

Our work was supported by the ISRO-IISc STC, DBT-IISc, DST and DST-FIST. We thank the reviewers and editors for improving our early drafts.

References

Bagchi, S., Briske, D.D., Bestelmeyer, B.T. & Wu, X.B. (2013b). Assessing resilience and state-transition models with historical records of cheatgrass *Bromus tectorum* invasion in North American sagebrush-steppe. *Journal of Applied Ecology*, **50**, 1131–1141.

Bagchi, S., Briske, D.D., Wu, X.B., *et al.* (2012). Empirical assessment of state-and-transition models with a long-term vegetation record from the Sonoran Desert. *Ecological Applications*, **22**, 400–411.

Bagchi, R., Crosby, M., Huntley, B., *et al.* (2013a). Evaluating the effectiveness of conservation site networks under climate change: accounting for uncertainty. *Global Change Biology*, **19**, 1236–1248.

Bauer, S., Nolet, B.A., Giske, J., *et al.* (2011). Cues and decision rules in animal migration. In E.J. Milner-Gulland, J.M. Fryxell, and A.R.E. Sinclair, eds., Animal Migration: A Synthesis. Oxford: Oxford University Press, pp. 68–87.

Bhutiyani, M.R., Kale, V.S. & Pawar, N.J. (2010). Climate change and the precipitation variations in the northwestern Himalaya: 1866–2006. *International Journal of Climatology*, **30**, 535–548.

Charmantier, A. & Gienapp, P. (2014). Climate change and timing of avian breeding and migration: evolutionary versus plastic changes. *Evolutionary Applications*, **7**, 15–28.

Cleland, E.E., Chuine, I., Manzel, A., Mooney, H.A. & Schwart, M.D. (2007). Shifting plant phenology in response to global change. *Trends in Ecology & Evolution*, **22**, 357–365.

Hussain, S.A. (1985). Status of the blacknecked crance in Ladakh – problems and prospects. *Journal of the Bombay Natural History Society*, **82**, 449–481.

Hussain, S.A. & Pandav, B. (2008). Status of breeding waterbirds in Changthang Cold Desert Sanctuary, Ladakh. *Indian Forester*, **134**, 469–480.

Jeong, S.-J., Ho, C.-H., Gim, H.-J. & Brown, M.E. (2011). Phenology shifts at start vs. end of growing season in temperate vegetation over the Northern Hemisphere for the period 1982–2008. *Global Change Biology*, **17**, 2385–2399.

Kawamura, K., Akiyama, T., Yokota, H.O., *et al.* 2005. Comparing MODIS vegetation indices with AVHRR NDVI for monitoring the forage quantity and quality in Inner Mongolia grassland, China. *Grassland Science*, **51**, 33–40.

Knox, N.M., Skidmore, A.K., Prins, H.H.T., *et al.* (2011). Dry season mapping of savanna forage quality, using the hyperspectral Carnegie Airborne Observatory sensor. *Remote Sensing of Environment*, **115**, 1478–1488.

Knox, N., Skidmore, A.K., Van der Werff, H.M.K., *et al.* (2013). Differentiation of plant age in grasses using remote sensing. *International Journal of Applied Earth Observation and Geoinformation*, **24**, 54–62.

Korner, C. & Basler, D. (2010). Phenology under global warming. *Science*, **327**, 1461–1462.

Kriegler F.J., Malila, W.A., Nalepka, R.F. & Richardson, W. (1969). Preprocessing transformations and their effects on multispectral recognition. In *Proceedings of the Sixth International Symposium on Remote Sensing of Environment*, pp. 97–131.

Lange, M. & Doktor, D. (2014). Phenex: auxiliary functions for phenological data analysis. http://cran.r-project.org/web/packages/phenex/index.html.

Mattiuzzi, M., Verbesselt, J., Stevens, F., *et al.* (2014). MODIS download and processing package. Processing functionalities for (multi-temporal) MODIS grid data. http://r-forge.r-project.org/projects/modis/.

Namgail, T., Mudappa, D. & Raman, T.R.S. (2009). *Waterbird numbers at high altitude lakes in eastern Ladakh*, India. *Wildfowl*, **59**, 137–144.

Namgail, T. & Yom-Tov, Y. (2009). Elevational range and timing of breeding in the birds of Ladakh: the effects of body mass, status and diet. *Journal of Ornithology*, **150**, 505–510.

Pettorelli, N., Vik, J.O., Mysterud, A., *et al.* (2005). Using the satellite-derived NDVI to assess ecological responses to environmental change. *Trends in Ecology & Evolution*, **20**, 503–510.

Pfister, O. (2004). *Birds and Mammals of Ladakh*. New Delhi: Oxford University Press.

QGIS Development Team (2013). QGIS Geographic Information System. www.qgis.org.

R Development Core Team (2014). *R: a language and environment for statistical computing. R Foundation for Statistical Computing, Vienna, Austria.* www.R-project.org.

Ramoelo, A., Skidmore, A.K., Choa, M.A., *et al.* (2013). Non-linear partial least square regression increases the estimation accuracy of grass nitrogen and phosphorus using in situ hyperspectral and environmental data. *Journal of Photogrammetry and Remote Sensing*, **82**, 27–40.

Shekhar, M.S., Chand, H., Kumar, S., Srinivasan, K. & Ganju, A. (2010). Climate-change studies in the western Himalaya. *Annals of Glaciology*, **51**, 105–112.

Si, Y., Xin, Q., De Boer, W.F., *et al.* (2015). Do Arctic breeding geese track or overtake a green wave during spring migration? *Scientific Reports*, **5**, 8749.

Singh, N.J., Bhatnagar, Y.V., Lecomte, N., Fox, J.L. & Yoccoz, N.G. (2013). No longer tracking greenery in high altitudes: pastoral practices of Rupshu nomads and their implications for biodiversity conservation. *Pastoralism: Research, Policy and Practice*, **3**, 16.

Steffen, W., Grinevald, J., Crutzen, P. & McNeill, J. (2011). The Anthropocene: conceptual and historical perspectives. *Philosophical Transactions of the Royal Society A*, **369**, 842–867.

Stenseth, N.C. & Mysterud, A. (2002). Climate, changing phenology, and other life history traits: nonlinearity and match–mismatch to the environment. *Proceedings of the National Academy of Science, USA*, **99**, 13379–13381.

Visser, M.E. & Both, C. (2005). Shifts in phenology due to global climate change: the need for a yardstick. *Proceedings of the Royal Society B: Biological Sciences*, **272**, 2561–2569.

Visser, M.E., Gienapp, P., Husby, A., *et al.* (2015). Effects of spring temperatures on the strength of selection on timing of reproduction in a long-distance migratory bird. *PLoS Biology*, **13**, e1002120.

Visser, M.E., Van Noordwijk, A.J., Tinbergen, J.M. & Lessels, C.M. (1998). Warmer springs lead to mistimed reproduction in great tits (Parus major). *Proceedings of the Royal Society B: Biological Sciences*, **265**, 1867–1870.

Vrieling, A., de Leeuw, J. & Said, M.Y. (2013). Length of growing period over Africa: variability and trends from 30 years of NDVI time series. *Remote Sensing*, **5**, 982–1000.

Wang, X., Piao, S., Xu, X., *et al.* (2015). Has the advancing onset of spring vegetation green-up slowed down or changed abruptly over the last three decades? *Global Ecology and Biogeography*, **24**, 621–631.

Wang, T., Skidmore, A.K., Zeng, Z., *et al.* (2010). Migration patterns of two endangered sympatric species from a remote sensing perspective. *Photogrammetric Engineering and Remote Sensing*, **76**, 1343–1352.

Zeng, Z., Beck, P.S.A., Wang, T., *et al.* (2010). Plant phenology and solar radiation drive seasonal movements of Golden takin in the Quinling Mountains, China. *Journal of Mammalogy*, **91**, 92–100.

Zhang, X., Friedl, M.A., Schaaf, C.B., *et al.* (2003). Monitoring vegetation phenology using MODIS. *Remote Sensing of Environment*, **84**, 471–475.

Part III

High-Altitude Migration Strategies

14 The Wind System in the Himalayas: From a Bird's-Eye View

Klaus Ohlmann

The Use of Gliders in the Himalayas for Scientific Projects

Having spent more than 30,000 hours of my life flying gliders, I have had a lot of opportunities to study the behaviour of flying birds in mid-air. Sharing the use of the energy offered by the heat of the sun, I learned a lot from their skills and energy management driven by Nature. The flight strategies and the flight efficiency of birds have evolved over millions of years. Increasing understanding of the complexity of daily thermal development, and its interference with the superimposed dynamic weather system, helps understanding of the causal factors that may explain where and how bird migration takes place in mountainous areas. I'm not a scientist, but as a keen observer, I'd like to consider myself a birdwatcher and can hopefully contribute some ideas to the scientific community based on my long experience as a glider pilot.

From October 2013 until February 2014 we carried out an expedition, flying with two Stemme S 10 VT, turbocharged motor-gliders from Berlin to Kathmandu, aiming to carry out several scientific studies in the Himalayas of Nepal. We conducted pollution measurements with the Institute of Technology from Karlsruhe and the International Centre for Integrated Mountain Development (ICIMOD). We carried out 3-D cartography for glacier monitoring and rescue purposes with German Aerospace DLR and University FH Aachen. We also conducted studies of rotor and wave weather patterns in the jet stream–influenced high Himalayas together with the Mountain Wave Project (Heise, Chapter 15). The area covered during our flights went from the west of Dhaulagiri to the east of Mount Everest at altitudes between 1000 m and 9200 m above sea level (a.s.l.). The Kali Gandaki Gorge and the Khumbu Valley from Lukla to Mount Everest were of special interest for the glider, partly due to their orientation perpendicular to the subtropical jet stream. Jet streams are fast-flowing, narrow air currents found in the upper atmosphere or troposphere. The subtropical jet stream typically flows at an altitude of 10 km to 16 km. Because the mountains of the Himalayas are so high, the fast-flowing air mass of a jet stream directly interacts with topography in this part of the world. In jet streams the air circles around the globe at very high speed. This speed is typically faster than 30 $m.sec^{-1}$ (about 110 $km.h^{-1}$ or 60 knots), but wind speeds of more than 107 $m.sec^{-1}$ (385 $km.h^{-1}$ or 210 knots) have been measured (Enfield, 2016).

Relevant Weather Patterns with a Direct Influence on the Flight Strategies of Migrating Birds

Microscale weather patterns: local, isolated thermal events are usually found at the beginning of convection. *Microscale meteorology* refers to phenomena in the atmosphere that last only a short time (minutes or hours), and cover short to reasonably short distances (metres and kilometres). These microscale phenomena are not presented in weather forecasts, and they are not shown on synoptic charts. To us glider pilots, these microscale thermal events are very important, because they provide us with the necessary atmospheric lift. They can include short-duration valley winds (air which moves upward) and mountain breezes (air that moves downward). Of importance in this respect are the stronger anabatic winds and katabatic winds. *Anabatic winds* are warm winds which blow up a steep slope or mountain side, driven by heating of the slope through insolation; they are also known as *upslope flows*. *Katabatic winds* is the generic term for downslope winds flowing from high elevations of mountains, plateaus and hills down their slopes to the valleys or plains below (Figure 14.1).

Mesoscale weather patterns: these are pressure differences due to convective low-pressure systems creating inflow to groups of thermals and strong valley breezes (see Dinges, 2000). I share here my observations at Pokhara Airfield from the beginning of December 2014 to the end of January 2015. The Pokhara Airfield lies at 827 m a.s.l.: very humid air resulting in morning fog close to the ground was observed during our flight operations from the end of November 2013 to the end of January 2014. The fog usually dissipated at around 10.00 to 11.00 local time. After that, thermal activity started, usually with a relatively low cloud base around 1800 m to 2300 m MSL over the hills around Pokhara. In aviation altitude is often measured using mean sea level (MSL; mostly given as above sea level, a.s.l.). In the higher hills, the thermal cloud base can reach between 3000 m to 4500 m MSL. High and dry areas without snow

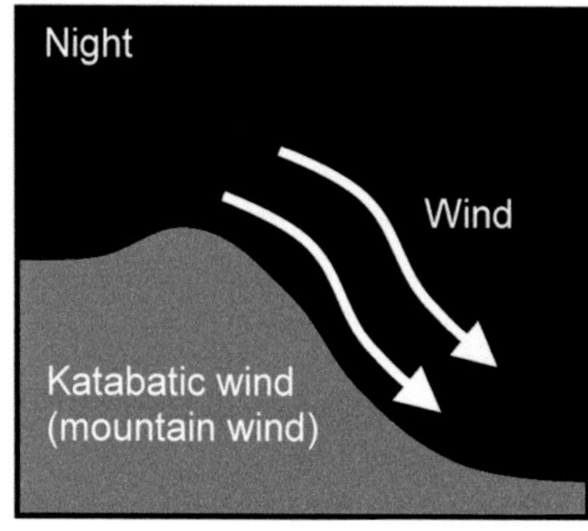

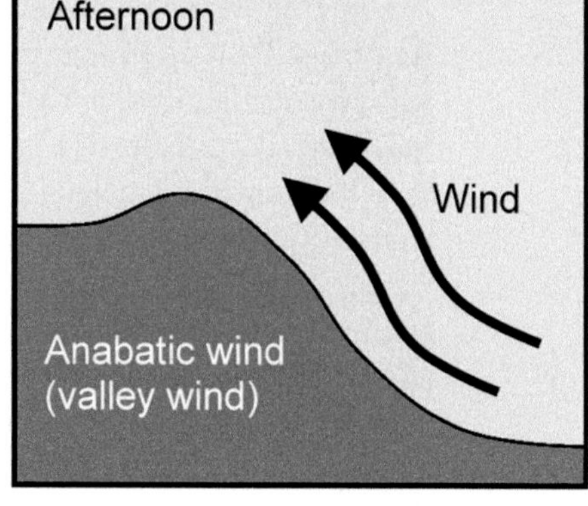

The mountain cools down, the air becomes heavier so it descends

The sun warms the mountain, the air is lighter and ascends

Figure 14.1. Anabatic and katabatic wind flows (redrawn after artinaid.com)

cover, for example, the upper Kali Gandaki Valley near Tibet, often create dry thermals up to 7000 m MSL. These areas create, once convectively active, huge cells of convective low-pressure systems, resulting in a smooth general flow towards these local lows, and strong valley winds needed for pressure balance. The gorge separates the major peaks of Dhaulagiri (8170 m) to the west and Annapurna (8090 m) to the east. Since the gorge floor is only 2520 m a.s.l., the gorge is the world's deepest. During autumn and spring, when, for example, Demoiselle Cranes *Anthropoides virgo* migrate through the Kali Gandaki Gorge (Higuchi & Minton, Chapter 3), the thermal activity should be much more important. Convection probably reaches altitudes of 8000 m a.s.l. and more. The resulting thermal low pressures are greater, and valley winds can reach up to 100 km.h^{-1}. These anabatic valley winds can be seen as an air distribution system. Anabatic winds are upslope winds driven by warmer surface temperatures, reached earlier on sun-exposed mountain slopes than the surrounding air column. They feed the big convection cells in the higher relief with even more energy, collected through contact with overheated surfaces. This is similar to thunderstorm cells in flatlands, where thermal convection feeds the biggest lows, the thunderstorms. The 'distribution channels' in mountain areas are the valleys. These 'channelled winds' do not usually go higher than 850 m above the bottom of the valley. Glider pilots and birds use these thermal-induced diurnal flows to find lift on mountains perpendicular to valley axes. The thermal low-pressure system often sucks low stratus cloud into the entrance of the Kali Gandaki Valley.

On days with higher humidity, this airflow is also responsible for a spreading out of clouds generated by the foothills of the Himalayas. These clouds can sometimes propagate close to 50 km south of the mountains (Figure 14.2).

Figure 14.2. Outspreading stratocumulus on the southern side of the Himalayas (view of Annapurna from out of a glider at 8700 m; view towards the east). Photo credit: Klaus Ohlmann.

At the end of each day, thermal activity 'starves'. Cold air flows from above to the bottom of the valley, creating an inverted valley wind, the katabatic wind system. Katabatic winds arise when the mountain surface is getting colder, cooling down the air in contact with it. It is more strongly affected by gravity, being heavier and more dense than the surrounding air. The result is a down-slope wind, also called a *mountain wind*, reaching the bottom of the valley later (Figure 14.1).

Measurements of the diurnal wind systems of the Kali Gandaki Gorge showed significant differences compared to the European Alps (Egger *et al.*, 2000). Average valley winds during daytime are nearly three times stronger than in the Alps and the depth of flowing air in the valley is twice as great. But there is especially strong asymmetry between the strength of in-blowing valley winds during daylight and the out-blowing mountain winds during the night (Zängl *et al.*, 2001). Figure 14.3a shows the presence of an in-blowing valley wind above Jomsom Airport even as late as midnight (Figure 14.3b). In the morning (Figure 14.3c), even at 06.00 hrs there is no sign of stronger katabatic outblow up to 1500 m above bottom of the valley. The upper part of the Kali Gandaki Valley is famous not only for its depth, but also for its strong diurnal up-valley winds. These winds blow virtually every day and have been noted by travellers as a rather unpleasant feature of the area (Egger *et al.*, 2000).

Very large-scale (i.e. synoptic-scale) weather systems: these are caused by pressure differences, large-scale wind systems due to cyclonic scale (i.e. about 1000 km and more) pressure distribution (Lows and Highs) and the location of the subtropical jet stream. During winter time, this jet stream lies nearly exactly above the Himalayas (Figure 14.4). When the Indian subcontinent heats up in summer, and land surface temperatures are often higher than those at the equator, the subtropical jet stream becomes much weaker than in winter. The Himalayas and the Tibetan Plateau strongly interact with the subtropical jet stream, since its basis normally lies at about 400 millibar (mb) (with a standard height of 7187 m) and it extends in the upper atmosphere to 200 mb (standard height 11,787 m). Autumn bird migration, but especially spring migration is thus frequently affected by high west–east winds over the region.

During our expedition, the day started most of the time with weak winds (15 knots; less than 30 $km.h^{-1}$) on the top of the thermal layer at an altitude of 6000 m, increasing to very strong winds at higher levels, for example, up to 150 knots (280 $km.h^{-1}$) at 12,000 m a.s.l. One knot (kt) is defined as one nautical mile (= 1852 m) per hour and is the measure of speed in aviation (nautical miles are used because 1 mile equals one arcminute on any great circle). The rapidly increasing wind speed at the bottom of the jet stream created severe turbulence due to shearing effects, and had influence at lower levels as well, by creating mountain waves (revealed by apparent stationary clouds downwind of a mountain top) and rotor systems (the ambient wind and the counter-current under it). Friction zone–creating vortexes, separating the turbulent from the laminar flow, usually rotate in the wind direction (see the rotating vortex clouds in the lee zone of Dhaulagiri in Figure 14.5).

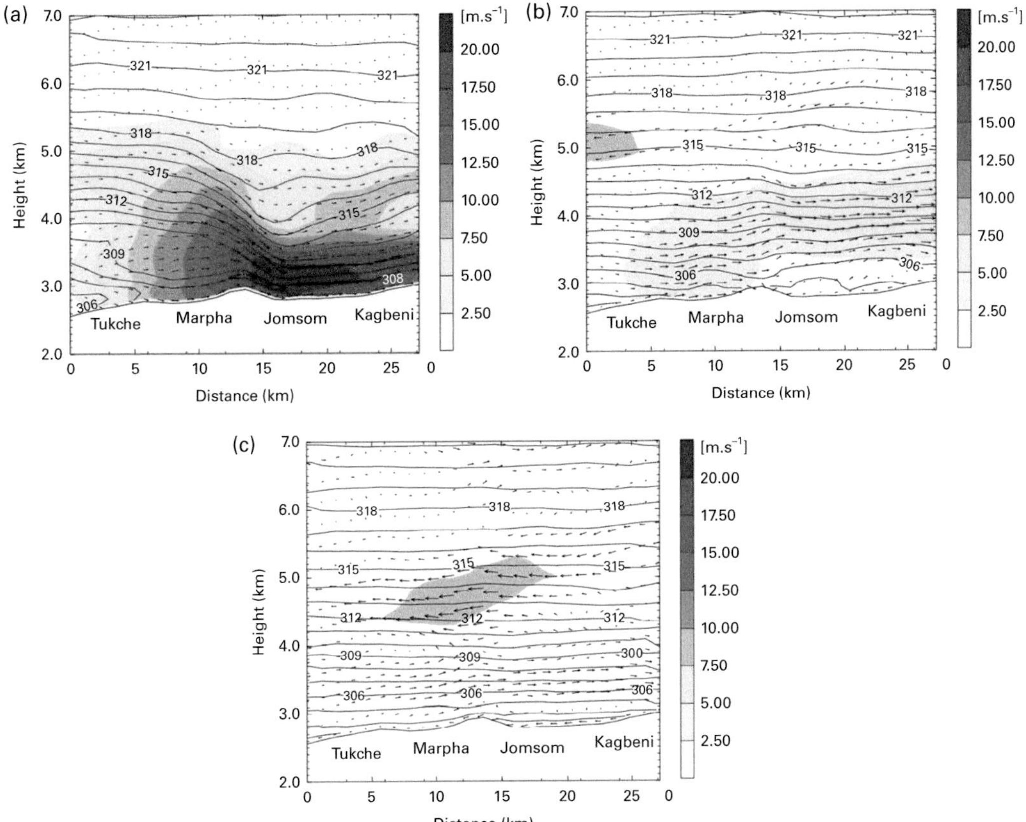

Figure 14.3. a. Vertical cross-section of the atmosphere above Jomsom Airport (Nepal) at 15.00 hrs local time (LT) (source: Zängl *et al.*, 2001). Note the strong winds over the airfield at more than 20 m.s^{-1} (~70 km.h^{-1}) indicative of a strong up-valley wind.
b. Vertical cross-section of Jomsom Airport at midnight (at 24.00 local time, LT) (source: Zängl *et al.*, 2001). The wind is falling off and is now only about 6 m.s^{-1} (~ 20 km.h^{-1}).
c. Vertical cross-section of Jomsom Airport in the early morning at 06.00 local time (LT) (source: Zängl *et al.*, 2001). There is hardly any wind over the airfield, and as compared to the evening before it blows now in the opposite direction one lm above the field, indicative of a very weak down-valley wind.

Figure 14.6 shows the typical jet stream above the Himalayas at the end of December, with different layers of wind speeds. The thermal layer (simplified) from ground to 500 mb has usually weak winds up to 10 kt (~ 20 km.h^{-1}). Above 500 mb (at 5500 m a.s.l.) the wind speed increases rapidly to 100 kt (~ 180 km.h^{-1}) in the core of the jet stream, and then diminishes above. The rapid increase of wind speed from above 5500 m leads to a shearing zone with heavy turbulence. The same situation can be expected at the upper end of the jet stream. The weather forecasts do not take into account the diurnal wind system in the Kali Gandaki Valley.

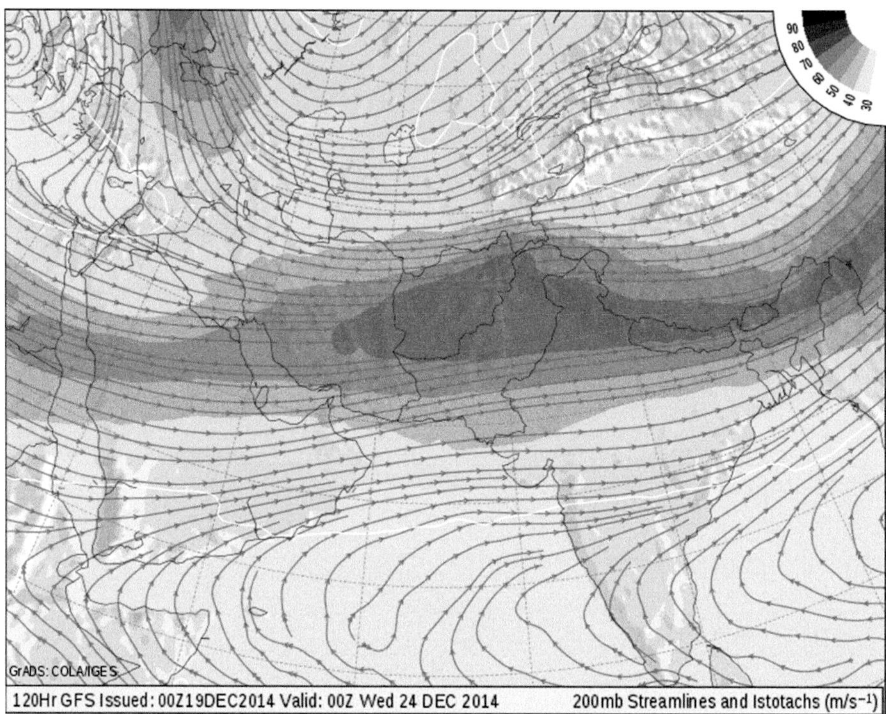

Figure 14.4. Subtropical jet stream above the Himalayas early winter 2014 (24 December 2014; source: www.wxmaps.org from the Centre for Ocean-Land-Atmosphere Studies). The isotachs are lines on the map connecting points of equal wind speed. In the top-right corner the wind speeds in m.s$^{-1.}$ (A black-and-white version of this figure will appear in some formats. For the colour version, please refer to the plate section.)

Figure 14.5. Surfing with a glider along the Annapurna range at 8500 m. View to Dhaulagiri and Kali Gandaki Low stratocumulus aspired by a convective low into the Kali Gandaki Gorge. The puff of clouds shows strong rotating vortex clouds in the lee zone of Dhaulagiri. Photo credit: Klaus Ohlmann. (A black-and-white version of this figure will appear in some formats. For the colour version, please refer to the plate section.)

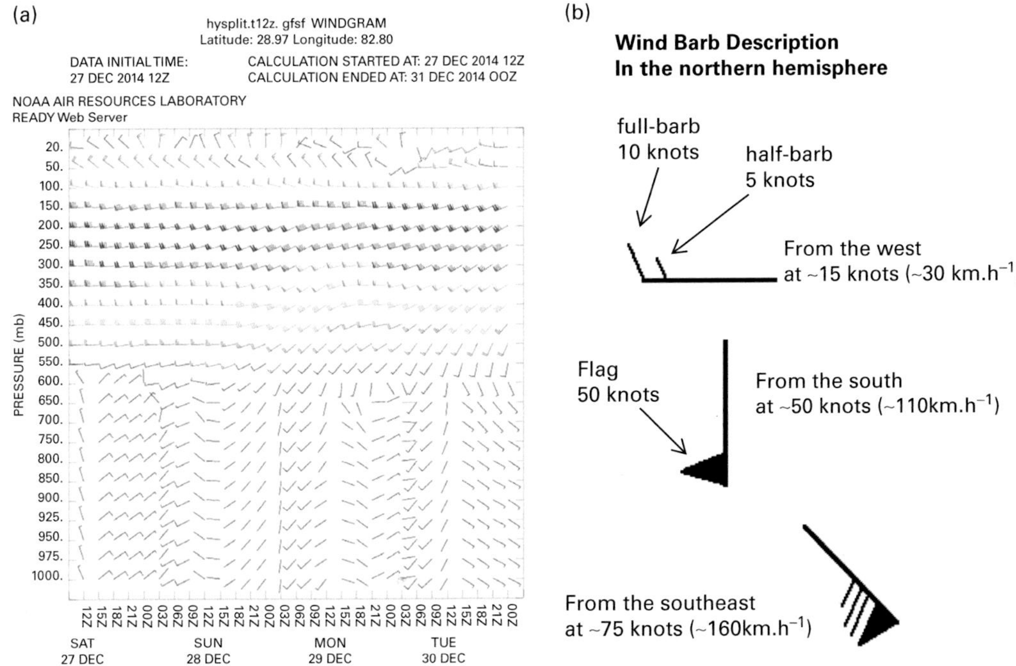

Figure 14.6. a. A 'windgram' for a typical subtropical jet stream above the Himalayas in winter (end of December), with wind speeds differing at different altitudes. The y-axis gives pressure in millibar: 500 mb is about 5600 m, 400 mb about 7200 m, 300 mb about 9200 m, 200 mb to 12,000 m and 100 mb to 15,800 m. A windgram is a forecast for a single spot on the map over the course of the day (in this case for the morning of 8 January 2016 at Dolpa Airfield near the Annapurna.
b. Key to symbols in a 'windgram' (as in Figure 6a) for the northern hemisphere. Wind-barbs show wind speed and direction. Colour shows the lapse rate, or rate of change of temperature with altitude. (A black-and-white version of this figure will appear in some formats. For the colour version, please refer to the plate section.)

Different Bird Species Use Different Flight Strategies

Let me distinguish between different categories of migrating birds according to their different anatomy and, consequently, their different flight styles:

1. Birds that use muscle power (e.g. ducks, waders, passerines)
2. Gliding birds (e.g. vultures, eagles)
3. Birds with a mixed strategy (e.g. cranes, storks)

Birds that Use Muscle Power

Bar-headed Geese *Anser indicus* seem to be the best example of muscle-powered flight. Typically, they do not use lift (currents of ascending air). Their flight needs an optimized flight style similar to that of an aeroplane. The higher one flies, the more energy is needed to climb. Consequently, flying higher than necessary could waste precious energy but may yield precious time (Groen & Prins, Chapter 17). There are a lot of tales about the record-breaking altitudes at which these birds fly. Recent studies showed that Bar-headed Geese fly essentially during the night (Hawkes *et al.*, 2011). From this one may conclude that geese do not use thermal lift during the day, and that they even may avoid it.

My conclusion after numerous flights in the heart of Nepal's Himalayas is simple. During the day you have the usual development of convection and breeze systems. Local anabatic convective systems merge into large cells of low pressure systems, creating strong anabatic (up-valley) wind systems. For example, in the lower Kali Gandaki Valley, between Dhaulagiri and Annapurna, wind speeds up to 55 kt (~ 100 km.h^{-1}) were measured on the ground, creating a lot of turbulence at lower levels (Zängl *et al.*, 2001).

Using convective lift to fly higher does not help to avoid strong turbulence, because at higher altitude one would enter the interference zone between thermals and the mesoscale wind system. Also, because the jet stream is directly above the Himalayas, creating strong wind shear in the rotor systems above the south–north orientated valleys due to mountain (lee) waves, this is a wind system to avoid (Heise, Chapter 15). During our measurement programme in the Kali Gandaki and Khumbu Valleys near Mount Everest, I found some of the worst turbulence in my career as a glider pilot. Pilots hate turbulence because of the ensuing compromises on comfort and control. Why should muscle powered birds like it? In turbulence it costs a lot of extra effort to maintain course and altitude. Airflow over the wing is disturbed, and huge up- or downdrafts would push the birds where they don't want to go.

There is a good chance that vagrant birds, which have been found beyond their normal range at very high altitudes, are probably victims of having been sucked into the strong updrafts of these rotor-systems, and transported to unwanted altitudes. Even if these birds have some extra adaptation, the physiological limits are still there. The higher they fly, the less oxygen is available. No reason to do that.

As the wind above 6000 m a.s.l. is nearly always blowing from the west, often at more than 60 kt (~110 km.h^{-1}), it presents a serious problem for birds heading north or south. Given their relatively low airspeed, the birds would have to angle considerably into the crosswind, making progress along their course very difficult.

During the night, katabatic winds (blowing out of the valley) develop. The bottom of the valley is filled with dense, cold air in a smooth laminar flow, delivering the best conditions for energy-effective flight. Wave induced turbulence is usually less developed, and airflow at higher levels is more laminar, because there is no more interference with convection. From the point of view of an aeroplane pilot, this is the best time to take off because of optimal performance of the motor, short take off distance etc. There is no reason why this should be different for a muscle powered migrating bird.

Gliding Birds

These birds are a glider pilot's best friends. They 'know' most of the time where lift is and they use it extensively. There is no doubt that migrating eagles or vultures have a completely different energy management system compared to the Bar-headed Goose (Groen & Prins, Chapter 17). This difference must be akin to the difference in flight regime between a powered aeroplane and a glider. Both are flying, but in very different ways. You will very rarely see a vulture flying by flapping its wings. They use the energy of the sun, thus from thermals rising along heated ridges. Or they use the wind, which is diverted by hills and mountains, creating large bands of lift, allowing the transformation of kinetic energy into potential energy, and thus altitude. These birds are able to fly hundreds of kilometres without any effort, and a Rüppell's Vulture *Gyps ruepellii* has been recorded in collision with an aircraft at an altitude of more than 11,200 m a.s.l. (Laybourne *et al.*, 1974). I have observed condors in the Argentinian Andes flying as high as 6000 m a.s.l.

Tactics vultures use to optimize flight are very similar to those of well-trained glider pilots. I have often observed that vultures circling in lift simply fly away when the climb is getting weaker. Glider pilots, of course, want to use the strongest lift and thus leave the weak one – and follow the bird. These migrating birds will use any source of lift, thermal or wind generated. It seems likely that their daily migration pattern is completely different from those of birds from the muscle-powered category. Their capacity to find lift, even in dry air, is astonishing. I often wondered if they have specific sensors to see, feel or hear lift at a distance. I believe they fly mostly during the day, using a maximum of daily thermal development.

Birds with a Mixed Strategy

At the very beginning of my glider career I had the opportunity to share a thermal with three white storks *Ciconia ciconia* just north of Braunschweig Airport, northern Germany. Storks are perfect thermal flyers, and it was amazing how they very attentively observed both gliders circling with them. One thing was especially interesting: arriving at cloud base, we had to stop climbing due to flight rules, but they continued circling and disappeared into the cloud. We saw them later, high above us, flying away. Cranes use thermals and lift, generated by wind in mountains, but they can fly long distances without external help as well. I could observe them only from the ground.

Insights about Movement Strategies from an Experienced Glider Pilot

If I were a 'gliding bird' and had to cross the Himalayas from India to Tibet, I would use the daily thermal development by starting with the first thermal lift, later using the strong dynamic southern valley wind as tailwind and finally climbing with the high thermals of the upper Kali Gandaki, and crossing into the Tibetan Plateau and its high thermals. Southbound, I would fly from the high thermals, staying as high as possible, surfing along the wind-exposed side of mountains, avoiding the headwinds of the strong valley breezes. Demoiselle Cranes on their flight through the Kali Gandaki, even though they are classified as 'mixed birds', use this technique.

As a muscle-powered bird, I would use my 'motor' (that is, my muscles), flying as low as possible, in order to avoid any extra effort and also the 'punishment' caused by lack of oxygen. I would prefer a route east of Everest and Makalu, due to the possibility of lower flight levels. If the conditions in these valleys are similar to the Kali Gandaki, the best time of flight would be from midnight to sunrise, in order to use colder air for flight efficiency and to avoid turbulence.

And of course, first of all, I would look for a detailed weather report on the Internet.

References

Dinges, M. (2000). Windsysteme und Thermik im Gebirge. Starnberg: Selbst-verlag.

Egger, J. *et al.* (2000). Diurnal winds in the Himalayan Kali Gandaki Valley, Part I. *Monthly Weather Review*, **128**, 1106–1122.

Enfield, D.B. (2016). Meteorology – climate – jet stream. *Encyclopedia Britannica* www.britannica.com/science/climate-meteorology/Jet-streams#toc53304 (accessed 20 July 2016).

Hawkes, L.A. *et al.* (2011). The trans-Himalayan flights of Bar-Headed Geese (Anser indicus). *Proceedings of the National Academy of Science USA*, **108**, 9516–9519.

Laybourne, R.C. (1974). Collision between a vulture and an aircraft at an altitude of 37,000 feet. *The Wilson Bulletin (Wilson Ornithological Society)*, **86**: 461–462.

Zängl, G., Egger, J. & Wirth, V. (2001). Diurnal winds in the Himalayan Kali Gandaki Valley, Part II. *Monthly Weather Review*, **129**, 1062–1080.

15 Birds, Gliders and Uplift Systems over the Himalayas

René Heise

First Human Flight

Aviation has a long history spanning 2000 years, from attempted kite-flying structures in China around 400 BCE, to the first engineering visions of Leonardo da Vinci in the fifteenth century. But German engineer Otto Lilienthal was the first to design and build a glider in the nineteenth century, which he based on the observations of soaring birds. He studied their aerodynamic behaviour and wing dynamics, and thus developed the first systematic approach to aviation (Lilienthal, 1889). Throughout his career, he built several monoplanes and biplanes, and his designs marked the beginning of modern aviation.

With the first successful human flight by Otto Lilienthal in the year 1891, interest in updraught processes in the atmosphere and their potential utilization for long-distance flight quickly grew. Lilienthal developed charts for wing areas and the resultant lift generated, and proposed that artificial wings should be designed with sustaining surfaces of concave-convex shape. Subsequently, aviation pioneers Wilbur and Orville Wright were inspired by Lilienthal's pioneering and comprehensive work on flight and aerodynamics. At the starting point of modern aviation, all attempts at technical construction to address aerodynamic challenges were based on bird observations.

The study of the flight behaviour of birds is of great importance because of its close association with meteorology – especially the principles underlying the physical laws of the atmosphere. The Organisation Scientifique et Technique du Vol à Voile (OSTIV), the successor to the first international soaring organization, the Internationale Studienkommission für Segelflug (ISTUS), has taken up the challenge 'to encourage and coordinate internationally the science and technology of soaring, and the development and use of the sailplane [i.e. glider] in pure and applied research' (see www.ostiv.org). Indeed, the scientific and meteorological panel of OSTIV is dedicated to investigating atmospheric processes with the aid of gliders as research platforms. The synergies of various scientific domains are thus mobilized and applied to innovations in aviation (e.g. bionics).

In modern soaring and in contemporary air sports, the following principal forces of energy for gliding (OSTIV, 2008) are optimally utilized and implemented: a) thermals, b) thermal waves, c) slope updraughts and d) lee waves and rotors (Figure 15.1).

Bird species use these various ascending air currents in several ways and often in combination. This flight behaviour can be analyzed by glider pilots and used for long-

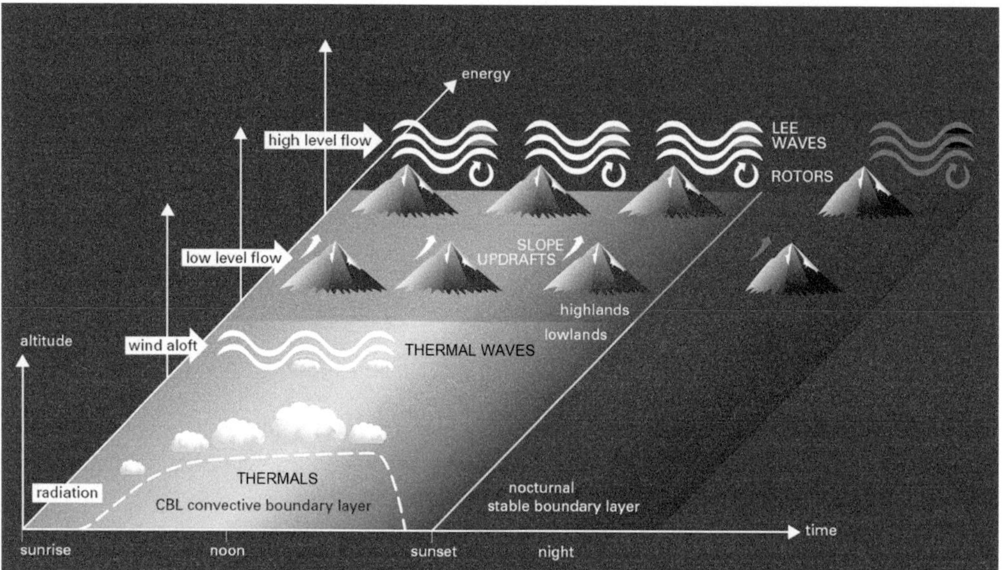

Figure 15.1. Principal sources of energy for gliding. The complex four-dimensional phenomena in nature are reduced here to a sketch where the main regimes are separated in a symbolic manner. The x-axis displays time and illustrates the lifetime of the updrafts in a diurnal cycle. The y-axis displays energy source and the z-axis displays altitude. Solar radiation produces thermals which, in turn, may interact with the upper-level winds to produce thermal waves. Low-level flow can be deflected by orography producing slope updrafts and high-level flow can favour lee-waves (OSTIV, 2009).

distance flight planning and tactical behaviour patterns (e.g. optimized team flight, energy-optimized airspeeds). The instinctive detection of atmospheric patterns can also be adopted and taken further by pilots as guidance for long-distance and competition flights. This is because the actual flight execution can be divided into two processes: a conscious part, which can be achieved by training, and an unconscious ability to adapt situational awareness.

Because of these factors, the national teams that participate in European and world gliding championships use OSTIV experts as advisors, because they are very often experienced competition pilots as well as scientific atmospheric experts (meteorologists) themselves. In this sense, glider pilots can use mesoscale patterns of atmospheric movement, such as convergence lines (updraught areas) of convection, induced ahead of a weather front or by anabatic (ascending) winds due to mountains, sea breeze fronts or convective patterns in the vicinity of thunderstorms. Modern numerical weather forecasting allows for precise prediction of the aforementioned weather phenomena, and allows the pilot to create a tactical mission for long-distance flight and the optimal realization of competition in the same way as a bird. My intention in this chapter is to provide evidence for the possibility of flight in challenging high-altitude regions based on observations of bird flight and actual unmotorized mechanical flight in gliders under extreme atmospheric and environmental conditions. I have therefore used the OSTIV expedition to the Himalayas to analyze the utilization of lift systems by soaring birds. The scarcity of information, and a relative absence of observed data, necessitates measurements with an appropriately high spatiotemporal resolution in order to comprehend the complex meteorological processes over the highest mountain range on earth.

Ups and Downs in the Atmosphere: An Exploration of Gravity Waves – the Mountain Wave Project

In the Mountain Wave Project (MWP), we at OSTIV have undertaken the task of investigating gravity waves or mountain waves from the troposphere up to the stratosphere. A schematic representation of wave updraughts on the lee side of mountains is shown in Figure 15.2. The rotors develop below the waves, and resemble a whirlwind or vortex with the axis tilted horizontally. The story of the discovery of rotor-wave systems

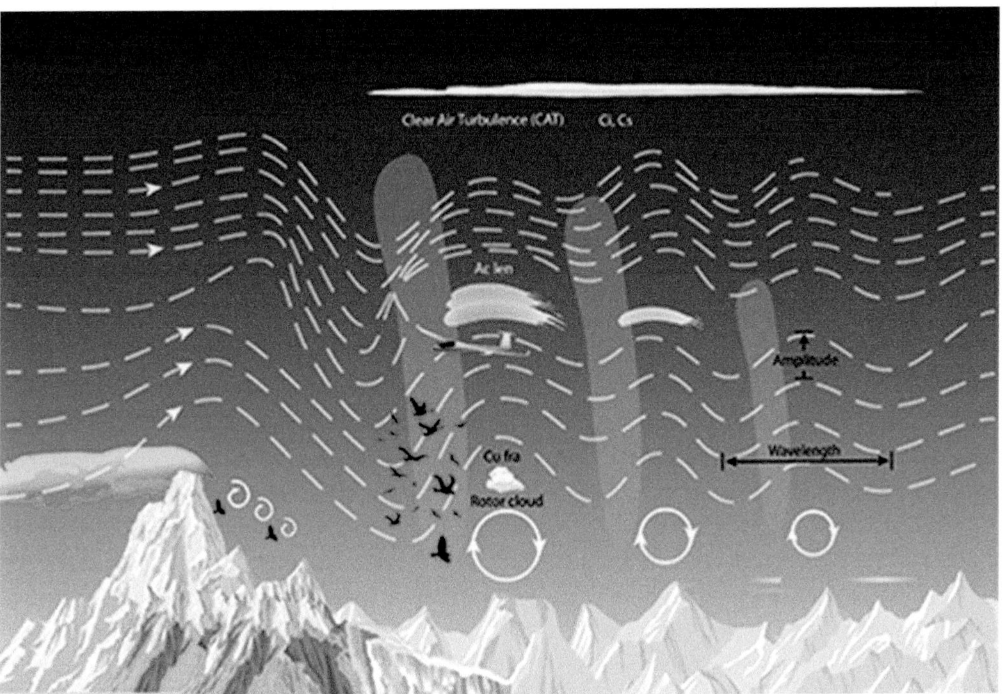

Figure 15.2. Generic two-dimensional wave system with characteristic lift areas (orange), cap, roll (*Cumulus fractus*, Cu fra) and lenticular clouds (*Altocumulus lenticularis*, Ac len) on the lee side of a mountain range. Soaring birds in the Himalayas, especially in the Everest region, use different lift systems: Red-billed Chough and Yellow-billed chough [1] very often the turbulent wake zone; Raven, Steppe Eagle and Lammergeier [2] thermal and rotor lifts. The Bar-headed Goose [3] seems to use the upper lift of wave patterns even though recent research suggests that they stay close to the ground when flying over mountains at high elevations (Hawkes *et al.*, Chapter 16).

by pioneering wave pilot and record-soaring pilot Joachim Kuettner, as well as its implications for aviation, can be found in Doernbrack *et al.* (2006).

Mountain waves develop during a strong airflow over the mountains through the vertical displacement of the molecules in the air. The restoring force (gravity) in a stable, stratified atmosphere on the lee side causes a return of the molecules to the equilibrium level. The presence of quasi-stationary waves will often be revealed by lenticular clouds with a genesis on the front side (windward) of the cloud in an updraught, and dissolving on the leeward side (downdraught) – metaphorically speaking, the wind blowing through the cloud. For modern-day methods of forecasting using numerical weather prediction and the observations from glider pilots of the magnitude of updraughts over mountainous regions, see Heise and Etling (2014).

Kuettner also studied the ridge soaring of Eurasian Jackdaws *Corvus monedula* in strong lee winds off escarpments (Kuettner, 1947). Based on general observations of bird flights, his surveys of wave updraughts, as well as his own record-breaking soaring flights, the questions that occupied him were: 'How far can one fly without a motor in atmospheric updraughts? Is it possible to fly 2000 km or more on the same day?'. These questions were answered for the first time on 23 November 2003, with a world record soaring flight by MWP pilot Klaus Ohlmann from El Calafate to San Juan in the lee of the Argentine Andes – a distance of 2120 km. Prior to that, the Tien Shan (China) (Kaihe, 1999), the Sierra Nevada (Kuettner, 1985) and the Southern Alps were the regions of the earth favoured for the 2000 km long-distance flight discipline involving aviation meteorological and air-sports challenges. The tropical regions near the equator, on the other hand, were hardly considered for these challenges due to their shorter day length. The record altitude for gliders currently stands at 15,460 m (www.fai.org/records) in the stratosphere, and was also achieved through wave soaring. An initial version of wave climatology has been developed using GPS logs and statistical analyses of numerous MWP soaring flights (Ultsch & Heise, 2010).

Research flights, and especially record flights with gliders, enable a space-time scale analysis with atmospheric phenomena to be undertaken in updraught areas. This allows comparison with those migrating birds that primarily use updraughts to manage long-distance flights using gliding (e.g. storks and cranes). It thus becomes possible to analyze and explain behavioural and migratory patterns of these birds in different regions (e.g. usage of lee wind areas in mountain ranges with updraughts, or wind shears with frontal weather situations) allowing for direct comparison and validation (e.g. of gliders with GPS logs with bird migrations). One further point of interest is the way in which birds at high altitudes with thin air might be dependent on atmospheric updraughts.

First Soaring Expedition in the Himalayas

The first expedition with an instrumented motor glider (Taifun 17E) in the Himalayas was conducted in 1985. The goal was to investigate the valley wind circulation in the

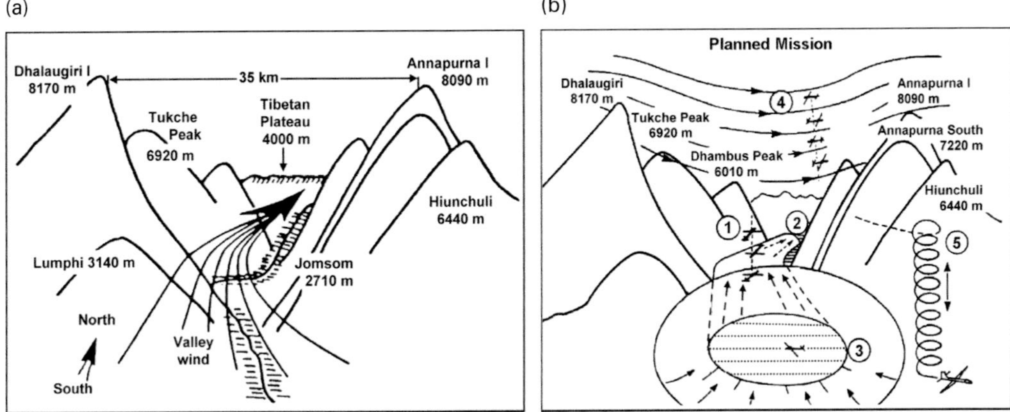

Figure 15.3. Inflow region of the Kali Gandaki Valley with its mountain ranges and planned missions for airborne measurements of the mountain valley circulation (Reinhardt *et al.*, 1985).

Kali Gandaki Gorge (Andha Galchi) – one of the deepest valleys in the world. With its north–south orientation, the gorge literally cuts through the mountain barrier. The formidable mountain slopes of the Dhaulagiri and Annapurna, which are higher than 8000 m a.s.l., as well as the glacier fields, lead to the assumption that there would be considerable mountain and valley wind circulation with high wind velocities throughout the daily cycle. The general explanation for this valley wind circulation – radiation cooling after sunset producing a downhill flow (katabatic wind), and warming up after sunrise producing thermals and upslope flow (anabatic wind) – was detailed in a handbook (OSTIV, 1993, and see Ohlmann, Chapter 14). After sunrise, birds use these often quasi-stationary and pulsating thermals as updraughts. In 1985, the OSTIV scientists first wanted to record the three-dimensional temperature, humidity and wind field along the Kali Gandaki Valley axis (Reinhardt *et al.*, 1985). The missions and measurement schedules are provided in Figure 15.3.

In 1998, another investigation of the extraordinary vertical wind conditions in the Kali Gandaki Gorge, which connects the Tibetan Plateau with the Indian plains, was carried out using pilot-balloon measurements (Egger *et al.*, 2000). The strongest up-valley winds were found over the Mustang Plateau with an anabatic flow (wind speed) of 15 to 20 $m.sec^{-1}$ (about 70 $km.h^{-1}$). During a diurnal cycle, an up-valley wind layer 1000 m to 2000 m in depth formed all the way up to the Tibetan Plateau. In November 2014, at Jomsom Airport, (~2700 m a.s.l.) in Nepal, I observed easterly wind components with 5 to 7 $m.sec^{-1}$ (about 18 to 25 $km.h^{-1}$) before sunrise, then only weak winds from various directions during the morning hours. Subsequently, the wind steadily increased from a westerly direction nearly reaching storm strength by late morning. Thus, air traffic was usually suspended at lunchtime for the rest of the day. According to information from the tower controller at Jomsom Airport, the highest wind velocity (gusts) measured in the past few years was 50 $m.sec^{-1}$ or 185 $km.h^{-1}$ in conjunction with a low-pressure system (cyclone).

Reference is made to the paper 'Soaring Birds of Mt. Everest' (Hindman, 1994), produced within the framework of continuing research at OSTIV. Hindman (1994) remarked that the highest-flying bird appears to be the Yellow-billed Chough *Pyrrhocorax graculus*, just below the South Col at 7925 m. The glide ratio for the Red-billed Chough *P. pyrrhocorax* was also estimated, and it was mentioned that this bird has one of the lowest sink rates (0.66 m.sec^{-1} at 7000 m) of any soaring bird. Common Raven *Corvus corax*, Lammergeier *Gypaetus barbatus* and Steppe Eagle *Aquila nipalensis*, on the other hand, were only found at relatively low altitudes around Mount Everest. The same aerodynamic terms, such as glide ratio and glide polar (polar curve) are used for unmotorized gliders as well as for 'soaring' birds. The glide ratio represents the ratio of distance travelled to altitude descended. Typical values for pigeons are ~9, as well as ~15 for eagles and ~20 for vultures (Nachtigall, 1985). This means that under calm conditions, from an altitude of 1 km, an eagle glides for around 15 km before touching the ground, thanks to its very good aerodynamic wing geometry. Noting an analogue glide feature of modern airliners (Airbus 340–600, glide ratio ~20), aerodynamically optimized gliders have glide ratios of more than 60. The aircraft used in the Himalayan expedition MWP research, the Stemme S10VT, has a glide ratio of 50 and thus allows for safe gliding to the nearest airfield, without danger, even with high wind speeds and strong downdraughts. In addition, the terminus polar curve describes the effect of airspeed on the rate of descent. This allows an assessment of aerodynamic characteristics and performance. Typical measured polar curves of individual species of birds are found in Nachtigall (1985) and Hindman (1994). Due to the high wind speeds and the thin air in the upper air layers above the Himalayas, the good 'bird gliders' (buzzards, eagles and especially vultures) dominate high-altitude, long-distance flights because of their low sink rates at high airspeeds. In contrast to the 'gliders', the Yellow-billed Chough, with the lowest sink rate of soaring birds, uses narrow lift areas in combination with rotors, thermal bubbles and hill lift areas for rising. Through the special manoeuvre of in-flight dynamic soaring, in combination with the use of wind shear, choughs can brave the stormy wind conditions in an extreme environment (Kuettner, 1947). These observations reflect the reported bird sightings in the vicinity of Mount Everest (Hindman, 1994).

Hindman *et al.* (2002) conducted numerous simulations with a convection model on the basis of available atmospheric soundings near Mount Everest, in order to filter out potential days of the year in which a glider would be able to ascend to the summit of Mount Everest in thermal updraughts. During the few days of the year (between the end of April and the beginning of May) with the warmest and driest surface conditions, the model predicted blue thermals (thermal bubbles without cloud convection) with an ascent rate of 1 m.sec^{-1}, to a height of 8600 m MSL. At these times, the convective boundary layer reaches a height of 5000 m above ground level (i.e. not above sea level but above the ground) over the Tibetan Plateau (itself averaging more than 4000 m a.s.l.). This means that birds of prey can ascend in thermals (bubbles of warm air) to an average altitude of around 9000 m a.s.l., and due to the aforementioned good glide ratio, they can glide and cross over the lower ridges where necessary. In the lower atmospheric layers, greater vertical velocities (5 to 7 m.sec^{-1})

can be achieved in the thermals, depending on the stability of the air mass. In a stable air mass, the vertical exchange of molecules in the air is limited or halted, and under unstable air mass conditions, it is reinforced. In the spring, with the increasing influence of the monsoon season and the advection of more humid air masses and therefore lower cloud ceilings – that is to say a lower convective boundary layer, which forms the edge of the thermal lift system for birds – this window of opportunity comes to a close.

Measurement Schedule – MWP Nepal 2014

The investigation of the wave updraughts of the Himalayas began with a visit to the Institute of Tibetan Plateau Research in Lhasa, at the invitation of the Chinese Academy of Sciences in 2010. The goal of the investigation was to examine the thermal and wave updraught conditions, and to identify emergency landing zones on the north side of the Mount Everest region as far as XigazêAir Base.

Regarding wave updraughts, note that the Himalayan ridge with its west–east alignment lies nearly parallel to the flow of the prevailing westerly winds, which inhibits the formation of updraughts. This contrasts with the Andes, which as the longest mountain range in the world acts as a perpendicular obstacle and is therefore ideal for wave formation, and has thereby enabled numerous record wave flights by glider pilots. In order to verify this assumption, however, a flying research expedition in Tibet (China) was necessary.

In preparation, a self-developed forecasting tool for lee waves (OSTIV, 2008), successfully used and validated in the Andes, was deployed in the Himalayas to identify wave locations and the strength of wave updraughts with the seasonal variations in meteorological conditions for this region. The analysis showed regions of strong wave-updraughts in the north-south valleys of Tibet which run perpendicular to the predominantly westerly direction of the upper winds (see also schematic representation of wave updraughts on the lee side of mountains as shown in Figure 15.2), as well as over southeast Tibet (Yunnan Province) with the Hengduan Shan mountain range and the river valleys of the Mekong and the Salween, respectively. At first glance, these updraughts appear to correspond with migratory flights of birds at high altitude described by Hawkes *et al.* (2012), but more closely observed GPS logs of birds are necessary to verify this. A general improvement in the weather conditions necessary for wave flight and wave lift systems in Tibet was forecast in the pre-monsoon period, with the wind rotating progressively to the southwest due to the increasing lee influence of the nearly west–east-aligned Himalayan ridge.

The thermal-lift conditions observed in October 2010 coincided with visually recognizable thermal structures with estimated ceilings as high as 7000 m a.s.l. (here about 3000 m above ground level). Birds of prey use these thermal columns or bubbles, circling within them to ascend to the cloud base (see Juhant & Bildstein,

Chapter 6). The Great Himalaya mountain range separates the region with dry air masses conducive to thermal updraughts and gliding flights in Tibet from the moist and lower convective boundary layer, largely unfavourable for thermal flights on the south side in Nepal. From an energy standpoint, smaller convective and advective processes are enough to form clouds in Nepal, and thus reduce the incident solar radiation necessary for thermal processes and updraughts very quickly. This assessment was a subjective observation on the part of glider pilots participating in the expedition, which would later be confirmed on the southern side of the Himalayas. Here, between November 2013 and January 2014, only weak thermals were observed over the terrain of Pokhara-Mustang in Nepal, which were of little use to gliders during January. Discussion with the paraglider pilots in Pokhara confirmed this assumption of weak thermals, which were detected daily from 12:00 hrs to 16:00 hrs, and often utilized to a height of 4500 m a.s.l. On the basis of optically recognizable thermal structures from photos and film reports, the Ladakh region of northern India, on the other hand, seems to be ideal for thermal soaring. This would explain a mountain crossing and migration of soaring birds over this favourable region based on thermal updraughts, especially in spring time.

Initially, exploratory flights with a glider by the MWP were planned in the spring based on a study (Hindman, 1994) and a simulation (Hindman *et al.*, 2002) of the thermal situation and meteorological rates of climb on Mount Everest. Rates of climb of 1 m.sec^{-1} and thermals were predicted here particularly within the proposed time frame. Where else on Earth are cloud ceilings like these to be found for convective clouds or convective boundary layers? For the purposes of evaluation, the ALPTHERM Weather Prognosis Model (a forecast tool for convective developments and updraughts) employed here was also applied to predict the altitudes of migrating White Storks *Ciconia ciconia* and the convective boundary layer in Israel (Shamoun-Baranes *et al.*, 2003). At the end of 2013, as a result of support by ICIMOD and the Nepali authorities, clearance was granted for research flights with two Stemme Type 10 VT motor gliders in Nepal. Unpowered flights in the thermal updraughts did not take place during the test flights in January 2014. On 23 January 2014, the MWP-team flew in a weak wave over the Annapurna-region at a height of 6500–7000 m a.s.l., which provided the first evidence of the predicted wave updraughts and an explanation for the support of migratory flights of birds at high altitude.

Currently, the scientific MWP co-operation partner, the German Aerospace Centre (Brauchle & Hein, 2014), is compiling a precise, high-resolution 3-D model of the mountain ranges and glacier regions in the Himalayas based on survey flights with a special camera in the MWP research plane (Figure 15.4). These terrain models will be used in the course of safety research (for landslide hazards) and glacier monitoring (assessing climate change). The new, detailed elevation model should also be applied to monitoring bird migration routes, because this model differs substantially in stereoscopic accuracy and precision from elevation models compiled by satellites.

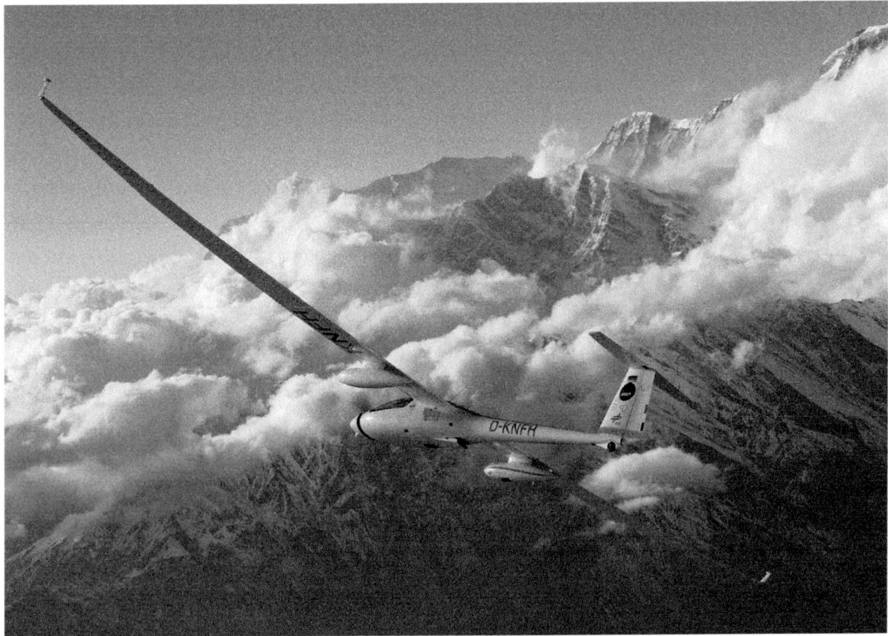

Figure 15.4. The research aircraft Stemme S10 VTX with underwing pods in the vicinity of the Annapurna Mountain Range (Flight crew Keimer/Heise 23 Jan 2014; Photo credit: DLR/Daniel Hein (CC-BY3.0)) during a measurement schedule of the Mountain Wave Project Nepal in 2014. (A black-and-white version of this figure will appear in some formats. For the colour version, please refer to the plate section.)

Potential Exploitation of Updraught Systems by Migratory Birds

During MWP's brief practice sessions with a limited number of flights in the Himalayan region between December 2013 and January 2014, the often recognizable relationship between soaring birds and gliders in exploiting thermals could not be witnessed. Air sports, including cross-country soaring over great distances, have not yet developed in Nepal. Regionally, however, there have been some initiatives, which is in contrast to Tibet on the north side of the Himalayas, where so far there have not been any such activities. Rigorous analyses of the overall strength and structure of thermals are, therefore, not obtainable. I was able to observe paraglider pilots in areas of weak thermals (up to 3000 m) in the afternoons, particularly above Pokhara. Generally, the convective boundary layer had maximum heights of up to 5500 m a.s.l. From 7000 m a.s.l., a steady increase in wind speed occurred, in combination with a turbulent layer. Then, above that, weak waves periodically formed.

Based on extensive GPS-logged flight data from paraglider pilots, preliminary statistical approaches are available within the framework of the Model Output Statistic, to improve the forecasts of thermals in the Pokhara region (Oberson 2015). The MWP was able to observe weak wave lift along the route from Annapurna to the Mount Everest Region and made use of it for soaring. With flow patterns, a band of strong winds in a westerly direction mostly prevails in the upper atmospheric layers (above 7000 m a.s.l.), so that on the south side of the Himalayas in Nepal, favourable wave conditions with updraughts only develop over the mountains during the infrequent northerly and northwesterly winds. On 15 December 2013, the vertical wind velocity in the model forecast reached values of 8 to 10 m.sec^{-1} in northerly winds on the lee side of the mountains – comparable to the speed of a modern elevator – and also comparable to the updraughts of Mistral weather conditions in the Alps. In addition, based on the routine model forecasts, two notable wave regions were identified in the mountain regions of Dhaulagiri and Mount Everest. These were also confirmed by visual observation of the cloud structures (rotor and lenticular clouds) and general pilot accounts. In the Everest region, mountain waves, together with regions of updraughts that reach above 10,000 m, form relatively frequently, depending on the atmospheric stability and wind conditions (a layer of stable air and wind speed that steadily increases with altitude promote wave formation in the atmosphere). In the Himalayas outside of Nepal, a notable area of wave updraughts has been identified only in Yunnan province, China (Hengduan Shan).

Further investigation through GPS tracking of the migration routes across the Himalayas will now have to be made to discover whether the utilization of wave-updraught bands by soaring birds occurs systematically. New, modern weather forecasting models can adequately predict the time and location of these zones of lift in space and time. It is an interesting fact that corresponding with the forecasts and with preliminary flight experience over a wide area, the region around Mount Everest is used by migratory birds for crossing the mountain range (Hawkes *et al.*, 2012; Nyambayar & Lee, Chapter 7). It should be pointed out that the updraughts develop not only on the lee side of the highest peaks, but rather, in combination with rotor waves, also at lower elevations, depending on the activator (that is, the obstacle in the airflow, causing the rotor and waves; see Figure 15.5). Higher rates of climb were predicted at lower levels. Excitation (caused by stimulation of the updraught area in connection waves) from bands of thermals arranged by gravity waves or mountain waves may only be found locally in the Himalayas. In future investigations, it should be noted that in the global weather forecasting models most often employed, the complexity of the vigorous local effects in the Himalayas is only resolved to a limited extent. Therefore, for its research expedition, the MWP used a model with 7 km resolution nested in the global model, and then a high-resolution model with 2.8 km (Figure 15.5). The anabatic and katabatic effects of the high mountain ranges, as well as the local valley breezes, were sufficiently taken into account. Regarding an investigation of the migrations of birds, general prognostic charts of the available global weather model are not sufficient. Local wind effects and updraughts require a special high-resolution weather forecast model with a mesh distance below 3 km in the forecast.

In future, GPS-logged flight paths of gliders (using Global Navigation Satellite System recorders) and GPS position reports of migratory birds, in conjunction with the highest-resolution weather forecasting models, will enable a systematic analysis of the use of updraughts by birds. This will consequently contribute to a better

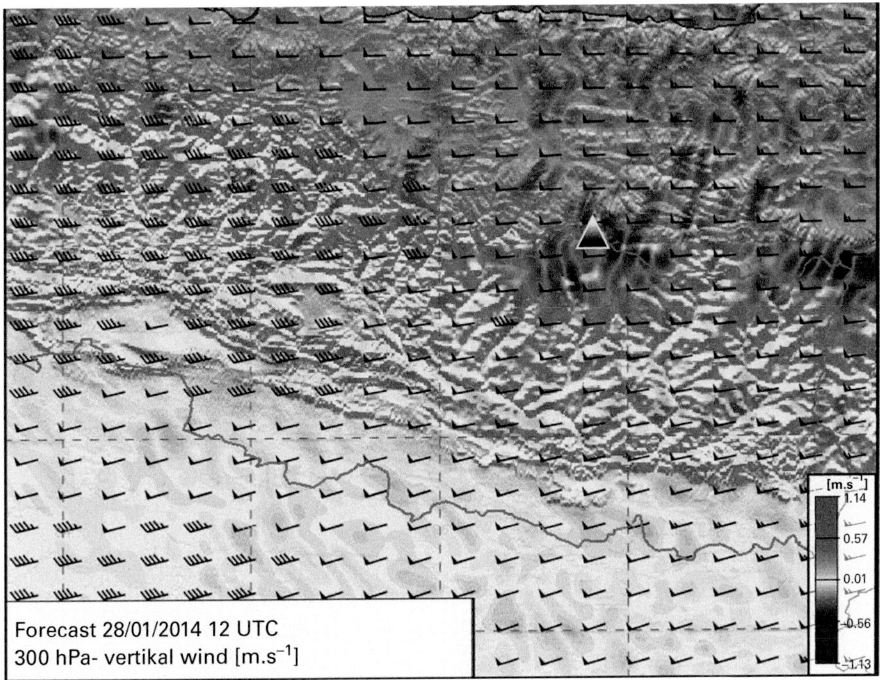

Figure 15.5. Numerical weather forecast of a new high-resolution model (grid spacing 2.8 km) with parametric vertical wind areas (updrafts – red, down – blue), and the horizontal wind symbol at a height of 300 hPa (~9000 m a.s.l.). The weather symbol shows the wind direction (here westerly winds), and the speed of the wind is denoted by the barbs. Each long barb represents 5 m.s^{-1}. At 25 m.s^{-1}, the barbs change into a pennant. During the measurement flight with the motor-glider over the glaciers of Mount Everest (marker Δ) on 28 Jan 2014, pilots used the updraught area of waves for safety flights of up to 9000 m a.s.l. The birds can use these updraught areas too, and the pattern helps improve understanding of migration routes and especially the passage over the Himalayan crest. (A black-and-white version of this figure will appear in some formats. For the colour version, please refer to the plate section.)

understanding of bird migration over the Himalayas. Changes in atmospheric flow patterns over the highest mountains on Earth resulting from global climatic warming are expected to be a part of this endeavour.

The close connection between human gliding activities and the flight of birds explored in this chapter illustrates the importance of research for a better understanding of complex atmospheric processes, for improvements in climate and weather forecasting models and for the study of bird flight during their migrations.

The Himalayan mountain range presents an almost vertical barrier in the lower section of the atmosphere – the troposphere, the atmospheric layer where all weather is generated. For this reason, the complex flow patterns involved in thermal convective processes, and therefore, updraught areas in the Tibetan Plateau, extend exceptionally far to the top of the troposphere. Continuing interdisciplinary cooperation between ornithology, meteorology and aeronautical research is required for better understanding to develop. The synergy effects demonstrated here are only a first step.

References

Brauchle, J. & Hein, D. (2014). On the top of the world. *German Aerospace Magazine*, **142**/143, 41–44.

Doernbrack, A., Heise, R. & Kuettner, J. (2006). Waves and rotors. *Promet*, **32**, 18–24.

Egger, J., Bajracharya, S., Egger, U., *et al.* (2000). Diurnal winds in the Himalayan Kali Gandaki Valley. Part I: Observations. *Monthly Weather Review*, **128**, 1106–1122.

Hawkes, L.A., Balachandran, S., Batbayar, N., *et al.* (2012). The paradox of extreme high-altitude migration in Bar-Headed Geese Anser indicus. *Proceedings of the Royal Society B, Biological Sciences*, **280**, 2012–2114.

Heise, R. & Etling, D. (2014). Gravity waves and rotors. *Promet*, **39**(Suppl. 2), 36–44.

Hindman, E. (1994). Soaring birds of Mt. Everest. *OSTIV-Technical Soaring*, **18**(Suppl. 1), 2.

Hindman, E., Liechti, O. & Lert, P. (2002), Soar Mt. Everest. *OSTIV-Technical Soaring*, **26**(Suppl. 4), 114–116.

Kaihe, Li (1999). A probe of soaring a straight distance of 2000 km. *OSTIV-Technical Soaring*, **23** (Suppl. 1), 7.

Kuettner, J.P. (1947). Über die Flugtechnik einiger Hochgebirgsvögel. *Kosmos*, **43**, 384–389.

Kuettner, J.P. (1985). The 2 000 km wave flight. *Soaring*, **3**, 21–27.

Lilienthal, O. (1889). *Bird Flight as a Basis of Aviation: A Contribution Towards a System of Aviation*, London/New York 1911 (*Der Vogelflug als Grundlage der Fliegekunst: Ein Beitrag zur Systematik der Flugtechnik*, Berlin, R. Gaertners Verlagsbuchhandlung, OLM 9072.

Nachtigall, W. (1985). *Warum die Vögel fliegen?* Hamburg: Rasch und Röhring Verlag.

Oberson, J. (2015). SoarGFS, the output of the famous GFS model adapted to soaring prediction around the world, Poster/Lecture, OSTIV Meteorological Panel Meeting, Winterthur: ZHAW.

OSTIV (1993). *Handbook of Meteorological Forecasting for Soaring Flight*, World Meteorological Organization, WMO-No495, **158**.

OSTIV (2008). *Weather Forecasting for Soaring Flight*, World Meteorological Organization, WMO-No. 1038.

OSTIV (2009). *Weather Forecasting for Soaring Flight*, World Meteorological Organization, WMO-No. 1038, 76 p

Reinhardt, M.E., Neininger, B., Kuettner, J.P., *et.al.* (1985). First results of airborne measurements of the mountain valley circulation in the Kali Gandaki Valley, Nepal, by motorglider, *OSTIV Publication XVIII Rieti, Italy.*

Shamoun-Baranes, J., Liechti, O., Yom-Tov, Y., Leshem, Y. (2003). Using a convection model to predict altitudes of white stork migration over central Israel. *Boundary-Layer Meteorology*, **107**, 673–681.

Ultsch, A. & Heise, R. (2010). Data mining to distinguish wave from thermal climbs in flight data, *Abstract/Lecture 34th Annual Conference of the German Classification Society* (GfKl), Karlsruhe.

(1) www.ostiv.org – website retrieved 2015-01-25

(2) www.fai.org/records – website Class D (Gliding) – retrieved 2015-01-25

16 Goose Migration over the Himalayas: Physiological Adaptations

Lucy A. Hawkes, Nyambayar Batbayar, Charles M. Bishop, Patrick J. Butler, Peter B. Frappell, Jessica U. Meir, William K. Milsom, Tseveenmyadag Natsagdorj and Graham S. Scott

Living in Thin Air

Perhaps as astonishing as the first ascent of Mount Everest (8848 m above sea level (a.s.l.)) by Sir Edmund Hillary and Tenzing Norgay in 1953 was the report of Bar-headed Geese *Anser indicus*; (Figure 16.1) flying over the summit by George Lowe in 1960 (Swan, 1961). The Bar-headed Goose has since been considered a marvel of the natural world, and even holds a place in the Guinness Book of Records. While the validity of this report remains uncertain, recent tracking work (Hawkes *et al.*, 2011, Hawkes *et al.*, 2012; Bishop *et al.*, 2015) found that on one occasion during its migration, a bird reached an altitude of 7290 m a.s.l., although the birds typically remain as close to the ground as possible, often crossing through passes at less than ~6,000 m. Although other species of wildfowl (e.g. Ruddy Shelduck *Tadorna ferruginea*, Northern Pintail *Anas acuta*, Northern Shoveler *Anas clypeata*) are known to fly across the Tibetan Plateau (see Namgail *et al.*, chapter 2), they have not been studied in detail. The Bar-headed Goose, however, has been studied extensively, both in the wild and in captivity, and it clearly has a superior tolerance of high-altitude conditions compared with other species (Scott *et al.*, 2015). It will thus be the focus of this chapter.

One pressing problem for animals at high altitude is supplying sufficient oxygen to the body. At high altitude, there is a reduction in oxygen availability (*hypoxia*), caused by the reduction in ambient barometric pressure (*hypobaria*), resulting in the air being 'thinner' and having a lower oxygen concentration (although the percent of oxygen is the same at ≈ 21%). For many species, reducing oxygen demands by suppressing metabolism is a potent solution to this problem (Boutilier, 2001). This is not always possible, however, for animals that need to maintain high rates of activity and high body temperatures, which require high rates of metabolism, such as migrating birds or mammals that do not hibernate. Indeed, flight may consume some 10 to 12 times more oxygen than resting (Ward *et al.*, 2002), so meeting the oxygen demand for metabolism during flight at high altitude with the reduced supply of oxygen poses a unique problem for flying migrants.

Mammalian species that are exposed to low oxygen at high altitudes for most of their lives are considered *chronically exposed* to hypoxia (e.g. North American deer mice *Peromyscus maniculatus*, Andean Llama *Lama glama* and plateau pika *Ochotona*

Figure 16.1. The Bar-headed Goose is thought to be one of the highest-flying birds in the world, and it has been directly tracked on one occasion as high as 7290 m a.s.l. (Photo: Coke Smith). (A black-and-white version of this figure will appear in some formats. For the colour version, please refer to the plate section.)

curzoniae) and have evolved larger lungs and hearts, for example, to cope with hypoxia (Ramirez *et al.*, 2007; Storz, 2007). Birds such as the Bar-headed Goose, however, generally travel through high-altitude areas over short periods (days to weeks) twice a year, and so are not exposed to high altitude/low oxygen for long (i.e. they are *acutely exposed*). For example, the Bar-headed Goose travels across the whole of the Tibetan Plateau in just 47 days (median value; Hawkes *et al.*, 2012). In addition, because flight permits birds to cover large distances in relatively short time periods, they can rapidly experience dramatic changes in barometric pressure and oxygen availability. In this regard, the Bar-headed Goose is thought to be the world's fastest climber, climbing up to 2.2 km of vertical elevation per hour during its northward migration from India onto the Tibetan Plateau (median value 1.1 km. h^{-1}; Hawkes *et al.*, 2011). These observations suggest that adaptations for performance at high altitude, including absence of any need for acclimatization, can evolve without chronic exposure to high altitude. On the other hand, some Bar-headed Goose populations breed and raise their young at altitudes close to 5000 m a.s.l. (Prins & Van Wieren, 2004) where acclimatization could occur.

High altitude presents the additional challenges of very cold and dry conditions, which we will not deal with explicitly in the present chapter because little research has been directed to these issues in birds. Such factors may pose little challenge to migrating birds, because they produce large amounts of excess metabolic heat during flight with which they can easily maintain their body temperature, and because long-distance avian migrants probably derive much of their water from lipid metabolism during flight (reviewed in Jenni & Jenni-Eiermann, 1998). Bar-headed Geese also make relatively regular stops during which they can probably rehydrate between flights.

In this chapter, we explain the stages of the oxygen transport cascade – from the air at high altitude via the blood to the muscles and mitochondria producing power for flight (Figure 16.2a). Thus, we will sequentially describe adaptations of the Bar-headed Goose in ventilation, pulmonary gas exchange, circulatory oxygen delivery in the blood and oxygen extraction by the active muscles (Figure 16.2b) to support what is probably one of the highest migrations in the world.

Getting Air In (Ventilation)

Mammalian lungs comprise an elaborately branching system in which the trachea, through which air enters the lung system, first branches into two primary bronchi that each enters one of the two lungs. The bronchi themselves then branch into secondary bronchi and then into bronchioles. This part of the lung system is responsible for conducting air into the lung, but is not thought to engage significantly in gas exchange and is therefore known as anatomical dead space. The bronchioles eventually end blindly in alveolar ducts and alveoli, small balloon-like sacs, each surrounded by a net of pulmonary capillaries and where gas exchange primarily takes place. Oxygen travels from the air into the alveoli across the alveolar surface and then into the blood. Carbon dioxide, a metabolic by-product, travels from the blood into the air in the alveoli and is breathed out on exhalation. The diaphragm and muscles of the chest wall and abdomen cause the mammalian lung to expand on inhalation and compress on exhalation as most of the air is forced out. The mammalian lung therefore does two jobs: it serves as a bellows that moves air in and out (**ventilation**), and it provides the interface for **gas exchange** between air and blood (travelling in the pulmonary capillaries). Birds, however, have mechanically separated ventilation and gas exchange. The avian lung is a relatively rigid gas exchange structure, and air is drawn in and out and directed through the lungs by separate structures, air sacs, which are distributed throughout the body. There are usually eight or more air sacs, varying by species. The shape of the air-sac-lung system in birds results in air moving through their lungs unidirectionally (as opposed to tidally in mammals; Scheid *et al.*, 1972). That is, fresh air moves from the air sacs into the avian lung and then through a network of gas-exchanging parabronchi in the same direction during both inhalation and exhalation. This system reduces dead space contribution and has been shown to support higher extraction of oxygen per unit of time than the system found in mammals (Piiper & Scheid, 1972; Powell & Scheid, 1989). In addition, the total amount of air that can be held in the avian respiratory system (lungs and air sacs) is on average double that of mammals, even though the lung itself is proportionally smaller (Powell, 2000).

Ventilation is a particularly important part of the oxygen cascade. A review by Scott and Milsom (2006) predicted that total ventilation would be one of the key factors likely to have the greatest influence on the ability of Bar-headed Geese to extract and consume oxygen from the thin air at high altitude. Total ventilation comprises two factors: how often an animal breathes (the ventilation frequency) and how deeply it breathes (the tidal volume), thus in hypoxia, animals can increase

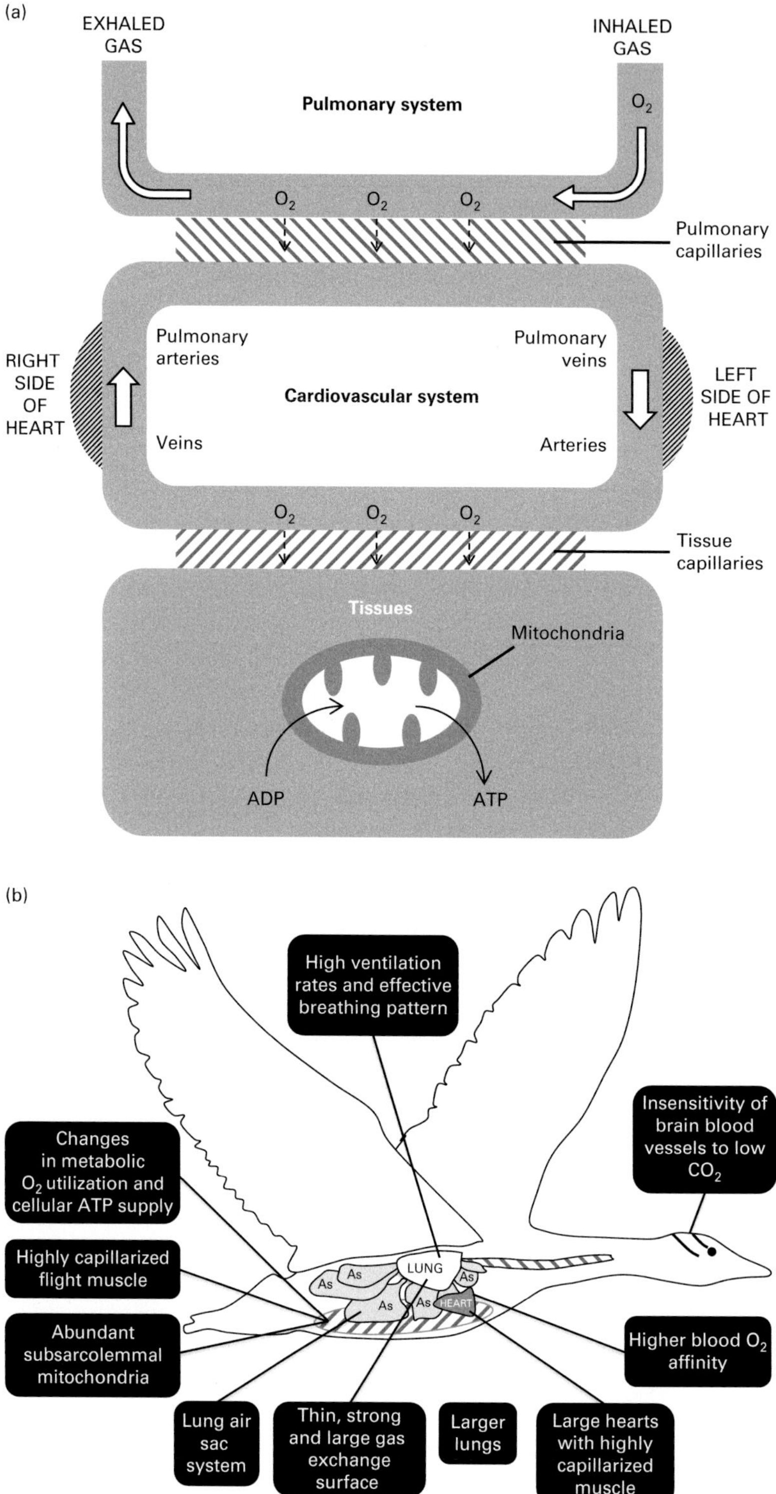

Figure 16.2. Schematic diagram of (a) the oxygen transport cascade, showing oxygen moving from the lung (the pulmonary system) via the circulatory system (veins, arteries and capillaries) to tissues and ultimately, to the mitochondria for aerobic respiration; and (b) the adaptations by Bar-headed Geese to maximize oxygen delivery to the tissues during flight at high altitude.

oxygen intake by breathing more frequently and/or more deeply. This hypoxic ventilatory response (HVR) occurs immediately with the onset of hypoxia, and accounts for the increase in ventilation seen in humans after arrival at high altitude (Martin *et al.*, 2010). Birds in general appear to increase total ventilation more than twice as much as mammals for the same level of hypoxia (Frappell *et al.*, 1992; Scott & Milsom, 2007), and Bar-headed Geese increase theirs even more.

By hyperventilating, however, birds also increase the rate at which they lose carbon dioxide (CO_2). Since CO_2 is an acidic compound in solution, losing it at an increased rate leads to an overall increase in the pH of the blood (it becomes more alkaline, known as *respiratory alkalosis*). This can cause intra-cellular damage (Neubauer *et al.*, 1990). In mammals, such a change in blood pH reduces blood flow to the lungs and brain (*pulmonary and cerebral vasoconstriction*) (Neubauer *et al.*, 1990). In all birds, however, including the Bar-headed Goose, no change in the circulation of blood to the brain seems to occur (Grubb *et al.*, 1978; Faraci & Fedde, 1986). This could partially explain why birds may not suffer the cerebral dysfunction at high altitudes seen in human mountaineers (West, 2006; Imray *et al.*, 2010; Martin *et al.*, 2010). Furthermore, Bar-headed Geese may cope even better with respiratory alkalosis than some other birds. Respiratory alkalosis tends to inhibit breathing, thus, when both oxygen and CO_2 levels fall (*hypocapnic hypoxia*), ventilation does not increase as much as when oxygen falls alone (*isocapnic hypoxia*). This offsetting effect of the respiratory alkalosis is greatly reduced in Bar-headed Geese compared to other birds that have been studied (Scott & Milsom, 2007).

Finally, while most species of birds increase ventilation primarily by breathing more frequently (Powell, 2000), Bar-headed Geese, at least at rest, increase total ventilation through large increases in air system volume (Scott & Milsom, 2007; but cardiac output and rate of diffusion of oxygen over membranes are important too). This appears to be a more effective way of breathing, delivering more fresh air to the parabronchial gas exchange surface for the same level of total ventilation (Scott & Milsom, 2007). Note that the foregoing discussion arises from studies of the avian respiratory system at rest. Gas exchange during free flight is extremely difficult to measure accurately, even with the use of a wind tunnel (Ward *et al.*, 2002; Engel *et al.*, 2010), but see Butler *et al.* (1977). The true capacity of the avian ventilatory system may only be apparent during flight at high altitude, when gas exchange demands are at their greatest, and this remains an area for future study.

Getting Oxygen into the Blood (Pulmonary Oxygen Diffusion)

Birds are inherently more effective at extracting oxygen than other air-breathing vertebrate groups. First, highly specialized lungs evolved in this group that, as described earlier, dramatically increase their capacity for gas exchange compared with mammals. In the avian lung, pulmonary blood flows cross-current to lung air flow, permitting higher oxygen extraction than is seen with mammalian tidal air flow (Piiper & Scheid, 1972). Second, the blood-gas barrier over which oxygen and CO_2 diffuse is very thin (0.05 to 0.56 μm in birds versus 0.3 μm in bats and 0.4 to 0.8 μm in non-flying mammals), dramatically reducing resistance to diffusion and thus increasing diffusion rates, all other factors being equal (Maina & West, 2005; West, 2009). Surprisingly, despite being very thin, the avian blood-gas barrier is also particularly strong. It is thought to be supported by epithelial plates, which may act collectively like spokes in a bicycle wheel to enhance the total strength of the system, and/or by the surface tension of the air flowing through the parabronchi, which creates a reinforcing effect like an inflated beach ball (West, 2009). The avian lung is also relatively rigid; its volume changes by only 1.4% on inspiration, thus ultra-structural stresses caused by lung inflation do not occur (Jones *et al.*, 1985; Maina & West, 2005). Finally, the relative surface areas of the lungs of birds are greater than those of mammals (two to three times greater in birds than mammals: 40 to 100 $cm^2.g^{-1}$ versus 15 to 40 $cm^2.g^{-1}$). These factors (cross-current exchange across a large, thin surface) when combined with the relatively large lungs of Bar-headed Geese compared to other birds (Scott & Milsom, 2006) provide an even greater capacity to extract oxygen from thin air than other birds. In the Bar-headed Goose, extraction of oxygen from the lung into the blood is so effective that even in severely hypoxic conditions of only 4% oxygen (five times less than at sea level) there was no measurable difference between inspired and arterial oxygen (i.e. it seems that most of the oxygen available in the atmosphere was transferred into the blood (Scott & Milsom, 2007).

Delivering Oxygen to the Body (Circulatory Oxygen Delivery)

In nearly all vertebrates, most oxygen is carried in the blood in combination with the iron-based protein haemoglobin (Hb). The propensity of haemoglobin to carry oxygen varies by taxon and life stage (Weber, 2007), and is described by the 'P_{50}' value, the partial pressure of oxygen in the blood at which the Hb is 50% saturated. The relative uptake of oxygen by Hb is sensitive to pH and temperature, such that the amount of oxygen carried in the blood can be higher (lower P_{50}) for a given amount of oxygen in the environment when blood is cooler (such as might occur when the blood flows through lungs ventilated with cold air) and pH is high (alkalotic), although there is no empirical evidence at present that this can occur. The Bar-headed Goose has an amino acid substitution that increases overall Hb affinity for oxygen (Petschow *et al.*, 1977), and a further study has suggested that the Bar-headed Goose's Hb has a greater thermal sensitivity as well (Meir & Milsom, 2013). Thus, relative to other birds, Bar-headed Geese can take up oxygen and saturate their Hb more readily when there is a lower amount of oxygen in the environment. Hawkes *et al.* (2014) showed that even in hypoxic conditions of only 7% oxygen (three times less than at sea level, the equivalent of the summit of Mount Everest), arterial oxygen content remained unchanged during running

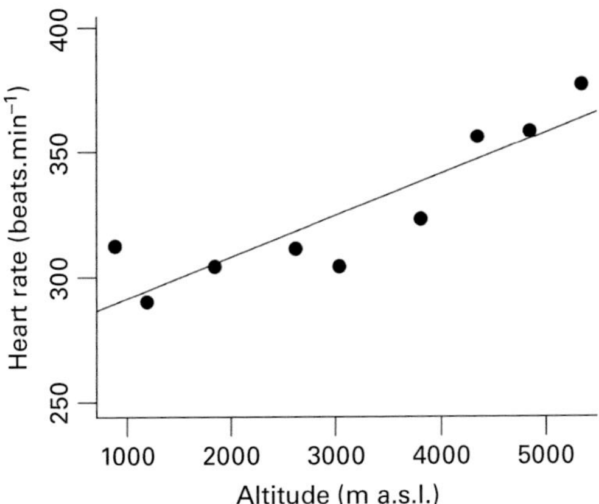

Figure 16.3. Heart rate increases with altitude for seven Bar-headed Geese outfitted with implantable heart rate loggers flying across the Tibetan Plateau (plot from Bishop *et al.*, 2015).

exercise compared with sea level values. In contrast, in humans near the summit of Mount Everest arterial oxygen content roughly halved (Grocott *et al.*, 2009).

Birds are also particularly effective at delivering oxygenated blood to all the tissues of the body compared with non-flying mammals. First, birds have proportionally larger hearts and expel a greater quantity of oxygenated blood per heartbeat (the *stroke volume*) than do most mammals (Bishop, 1997). They also have higher maximum heart rates during exercise than many mammals, largely supported by a greater supply of capillaries to the heart muscle itself (Faraci, 1991). Delivery of oxygen to the tissues may be enhanced in Bar-headed Geese compared to lowland geese, as they can maintain high heart rates during exercise even in severely hypoxic conditions (7% oxygen; Hawkes *et al.*, 2014). This appears to be possible because Bar-headed Geese have more than 30% greater capillary density in the left ventricle of the heart than other lowland wildfowl, which supplies the cardiac muscle with more oxygen and thus enables the heart to distribute oxygenated blood to the systemic circulation, including the working muscles, even at high altitudes (Scott *et al.*, 2011). Indeed, the median heart rates of wild Bar-headed Geese in flight have been shown to increase with altitude, from 300 beats per minute below 2300 m a.s.l. to 364 beats per minute above 4800 m a.s.l. (Bishop *et al.*, 2015; Figure 16.2). Importantly, this is much lower than heart rates recorded in captive Bar-headed Geese running at maximum sustainable speed on a treadmill in severe hypoxia (heart rates average 453 beats per minute; Hawkes *et al.*, 2014), and suggests that wild Bar-headed Geese may not reach their heart rate maxima, and that cardiac performance of wild Bar-headed Geese may not limit their ability to fly at high altitude.

When blood reaches the tissues, the release of oxygen from Hb is also sensitive to pH and temperature, and the amount of oxygen released from the blood is greater when blood is warmer (such as will occur when the blood flows through exercising muscle that is generating heat) and pH is low (acidotic due to the production of CO_2).

Using Oxygen Efficiently (Muscle Oxygen Utilization)

Another key to successful performance at high altitude above the Himalayas is the efficient use of oxygen to produce energy for muscles to use to support flight. Birds have a proportionally greater length of capillaries per unit volume of muscle than do mammals (6000 to 14,000 mm^{-2} in birds versus 1900 to 5700 mm^{-2} in non-flying mammals). Furthermore, muscle fibres are proportionally smaller in birds than they are in non-flying mammals (14–20 μm diameter in birds versus 29–45 μm in mammals; Mathieu-Costello *et al.*, 1992), meaning that there is a shorter distance for oxygen to diffuse between the blood inside the capillaries and the mitochondria inside muscle fibres. Bar-headed Geese are superior to many avian species in this regard as well. First, they have a higher density of capillaries and higher capillary to muscle fibre ratio in their flight muscles than lowland wildfowl (Scott *et al.*, 2009). In addition, the mitochondria of the working muscle are located close to the cell membrane within each flight muscle fibre, reducing the diffusion distance between the oxygenated blood and the mitochondria (although this may have some drawbacks; see Scott *et al.*, 2009). Second, a greater proportion of their muscle fibres are aerobic-type fibres, which are more resistant to fatigue than glycolytic type fibres. This allows for endurance exercise with little build-up of anaerobic by-products such as lactate (Scott *et al.*, 2009) and may be present in a wide variety of birds. Finally, within the mitochondria, the function of a key enzyme in the electron transport chain of the mitochondria, cytochrome-C-oxidase, is altered in a way that could reduce damage from oxidative stress during prolonged migration (Scott *et al.*, 2009).

Flying Smart (Behavioural Adaptations)

Recent work using satellite transmitters and custom heart rate loggers (Hawkes *et al.*, 2011, 2012; Bishop *et al.*, 2015) (Figure 16.4) has suggested that, even though Bar-headed Geese have an excellent tolerance of high altitude, they actually minimize flight altitude wherever possible, remaining close to the ground to exploit the densest air conditions available to them. This is estimated to save up to 12% of the costs of flight, because more lift and oxygen are available in denser air. In addition, air becomes denser in cooler conditions, and thus perhaps not surprisingly, it appears that Bar-headed Geese make the majority of their flights at night and during the early morning (Hawkes *et al.*, 2011, 2012).

The question of how high these birds can fly under their own power remains enigmatic. A modelling study suggested that the maximum altitude to which Bar-headed Geese might be capable of level flight without any wind assistance could be as low as 7500 m a.s.l. (Figure 16.5; Hawkes *et al.*, 2014), well below the summit of Mount Everest. This model relied on data based on the maximum heart rate of captive geese during sustainable running exercise and, on several assumptions, since not all of the physiological data are yet available to make accurate calculations, and other unique

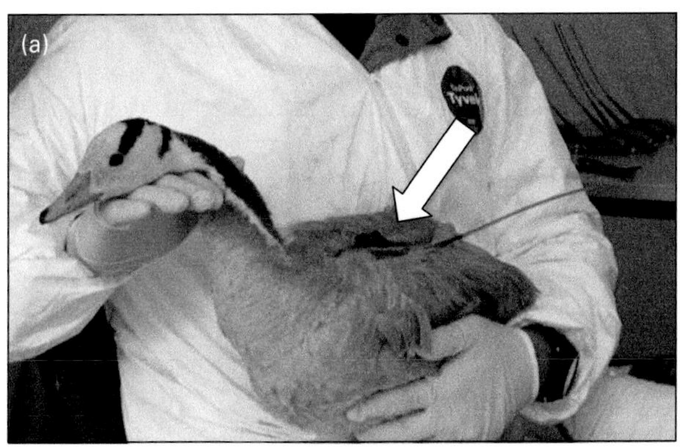

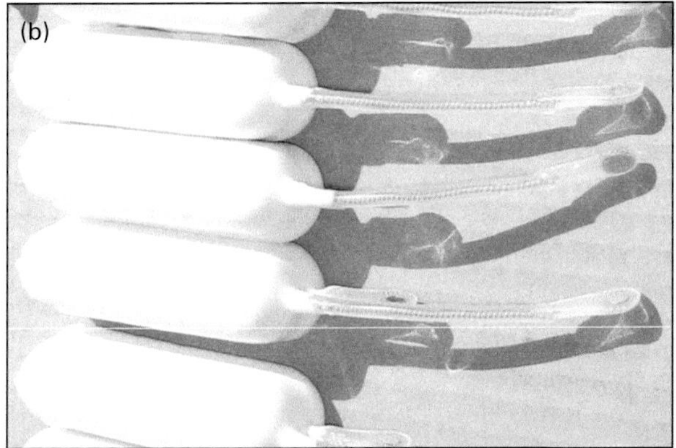

Figure 16.4. Studying the flight physiology of Bar-headed Geese. Photos show (a) a Bar-headed Goose with a satellite transmitter (attached using a Teflon harness, satellite tag indicated by white arrow); (b) archival heart rate recording data loggers (62 x 19 mm, 32 g) that were implanted to collect physiological data; (c) a flock of Bar-headed Geese on a lake in western Mongolia. Geese carrying archival loggers deployed in previous study years can be identified by their green neck collars, indicated here for six geese by white circles. Photo credits: (a, b) the authors, (c) Bruce Moffat. (A black-and-white version of this figure will appear in some formats. For the colour version, please refer to the plate section.)

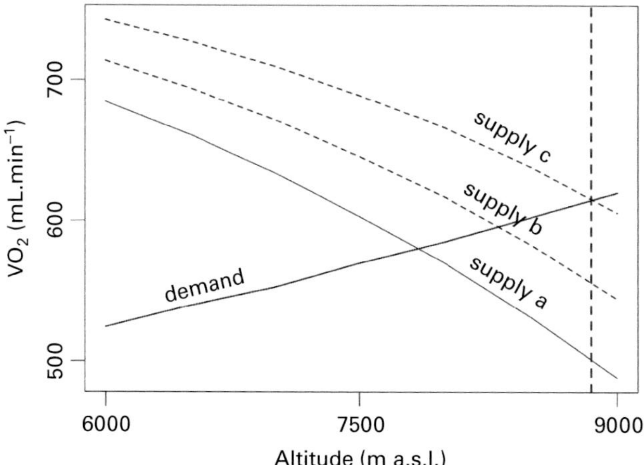

Figure 16.5. As geese fly to ever greater altitudes, the *demand* for oxygen increases as air becomes less dense ('thinner'), resulting in birds needing to flap harder and more frequently to stay aloft. However, the *supply* of oxygen to the working muscles decreases because the ambient pressure decreases, meaning there are fewer molecules of oxygen per unit volume of air. Here, supply and demand have been modelled based on currently available data and several assumptions (see Hawkes *et al.*, 2014, for details). Supply curves a and b are derived from different reports of in vivo oxy-haemoglobin binding (Hawkes *et al.*, 2014). Supply curve c is derived from data from an in vitro study modelling the possible effects of the pH and temperature shifts that may occur at altitude (Meir & Milsom, 2013). These latter data have not been validated in vivo. The model suggests that, depending on which curve is most accurate, level, unassisted flight might only be possible up to roughly 7500 m a.s.l. (curve a) or might be possible up to 8848 m a.s.l., the summit of Mount Everest (shown as a dashed grey vertical line).

aspects of their physiology may not yet have been discovered. When values for oxygen-haemoglobin binding from other studies were incorporated into the model, maximum altitude increased (Figure 16.5), although extrapolation of heart rate data from wild migrant Bar-headed Geese suggested that unaided flights above 8000 m might be very challenging (Bishop *et al.*, 2015). It remains possible that the geese may make use of wind assistance (e.g. thermals, orographic lift and lee waves; see Ohlmann, Chapter 14, and Heise, Chapter 15), for which there is some evidence (Bishop *et al.*, 2015) and which may have permitted Bar-headed Geese to fly to the 7290 m a.s.l. that has been directly measured using satellite telemetry (Hawkes *et al.*, 2012), or to perhaps greater altitudes.

Future Directions

In summary, birds in general have increased capacity, at many steps of the oxygen transport cascade, to improve their ability to cope with conditions in the high Himalayas relative to non-flying mammals (Figure 16.2). Bar-headed Geese in particular have additional specializations that make them superior in this regard to most other species of birds. However, even though we know more about the high-altitude physiology of

Bar-headed Geese than any other species of bird, additional work is required to understand the extent to which variables measured in birds raised in captivity reflect the physiology of birds in the wild. Given their superior tolerance of hypoxia relative to other birds and mammals, both at rest and during exercise (Scott *et al.*, 2015), Bar-headed Geese have become a model study species for the effects of high altitude on vertebrate eco-physiology. Recent breakthrough technologies are being employed to gain sub-second insights into the strategies these incredible birds use to make their remarkable migration across the Himalayas (Bishop *et al.*, 2015) along with studies to determine how widespread the traits exhibited by Bar-headed Geese might be in dozens of other bird species native to high altitude in the Himalayas and other mountain ranges.

References

Bishop, C.M. (1997). Heart mass and the maximum cardiac output of birds and mammals: implications for estimating the maximum aerobic power input of flying animals. *Philosophical Transactions of the Royal Society of London. Series B: Biological Sciences*, **352**, 447–456.

Bishop, C.M., Spivey, R.J., Hawkes, L.A., *et al.* (2015). The roller coaster flight strategy of Bar-headed Geese conserves energy during Himalayan migrations. *Science*, **347**, 250–254.

Boutilier, R.G. (2001). Mechanisms of cell survival in hypoxia and hypothermia. *Journal of Experimental Biology*, **204**, 3171–3181.

Butler P.J., West, N.H. & Jones, D.R. (1977). Respiratory and cardiovascular responses of the pigeon to sustained, level flight in a wind tunnel. *Journal of Experimental Biology*, **71**, 7–26.

Engel, S., Bowlin, M.S. & Hedenström, A. (2010). The role of wind-tunnel studies in integrative research on migration biology. *Integrative and Comparative Biology*, **50**, 323–335.

Faraci, F.M. (1991). Adaptations to hypoxia in birds: how to fly high. *Annual Review of Physiology*, **53**, 59–70.

Faraci, F.M. & Fedde, M.R. (1986). Regional circulatory responses to hypocapnia and hypercapnia in Bar-headed Geese. *American Journal of Physiology – Regulatory, Integrative and Comparative Physiology*, **250**, R499–R504.

Frappell, P.B., Dotta, A. & Mortola, J.P. (1992). Metabolism during normoxia, hyperoxia, and recovery in newborn rats. *Canadian Journal of Physiology and Pharmacology*, **70**, 408–411.

Grocott, M.P.W., Martin, D.S., Levett, D.Z.H., McMorrow, R., Windsor, J. & Montgomery, H.E. (2009). Arterial blood gases and oxygen content in climbers on Mount Everest. *New England Journal of Medicine*, **360**, 140–149.

Grubb, B., Colacino, J.M. & Schmidt-Nielsen, K. (1978). Cerebral blood flow in birds: effect of hypoxia. *American Journal of Physiology – Heart and Circulatory Physiology*, **234**, H230–H234.

Hawkes, L.A., Balachandran, S., Batbayar, N., *et al.* (2011). The Trans-Himalayan flights of Bar-Headed Geese (Anser indicus). *Proceedings of the National Academy of Sciences*, **108**, 9516–9519.

Hawkes, L.A., Balachandran, S., Batbayar, N., *et al.* (2012) The paradox of extreme high altitude migration in Bar-headed Geese Anser indicus. *Proceedings of the Royal Society B.*

Hawkes, L.A., Butler, P.J., Frappell, P.B., *et al.* (2014). Maximum running speed of captive Bar-headed Geese is unaffected by severe hypoxia. *PLoS ONE*, **9**, e94015.

Imray, C., Wright, A., Subudhi, A. & Roach, R. (2010) Acute mountain sickness: pathophysiology, prevention, and treatment. *Progress in Cardiovascular Diseases*, **52**, 467–484.

Jenni, L. & Jenni-Eiermann, S. (1998). Fuel supply and metabolic constraints in migrating birds. *Journal of Avian Biology*, **29**, 521–528.

Jones, J.H., Effman, E.L. & Schmidt-Nielsen, K. (1985). Lung volume changes during respiration in ducks. *Respiration Physiology*, **9**, 15–25.

Maina, J.N. & West, J.B. (2005) Thin and strong! The bioengineering dilemma in the structural and functional design of the blood-gas barrier. *Physiological Reviews*, **85**, 811–844.

Martin, D.S., Levett, D.Z.H., Grocott, M.P.W. & Montgomery, H.E. (2010) Variation in human performance in the hypoxic mountain environment. *Experimental Physiology*, **95**, 463–470.

Mathieu-Costello, O., Suarez, R.K. & Hochachka, P.W. (1992). Capillary-to-fiber geometry and mitochondrial density in hummingbird flight muscle. *Respiration Physiology*, **89**, 113–132.

Meir, J.U. & Milsom, W.K. (2013). High thermal sensitivity of blood enhances oxygen delivery in the high-flying Bar-headed Goose. *Journal of Experimental Biology*, **216**, 2172–2175.

Neubauer, J.A., Melton, J.E. & Edelman, N.H. (1990). Modulation of respiration during brain hypoxia. *Journal of Applied Physiology*, **68**, 441–451.

Petschow, D., Wurdinger, I., Baumann, R., Duhm, J., Braunitzer, G. & Bauer, C. (1977). Causes of high blood O_2 affinity of animals living at high altitude. *Journal of Applied Physiology*, **42**, 139–143.

Piiper, J. & Scheid, P. (1972). Maximum gas transfer efficacy of models for fish gills, avian lungs and mammalian lungs. *Respiration Physiology*, **14**, 115–124.

Powell, F.L. (2000). Respiration. *In* G.C. Whittow (ed.) *Sturkie's Avian Physiology.* London: Academic Press. Pages 301–336.

Powell, F.L & Scheid, P. (1989). Physiology of gas exchange in the avian respiratory system. In A.S. King & J. McLelland (eds.) *Form and Function in Birds Vol 4.* London: Academic Press. Pages 393–437.

Prins, H.H.T. & Van Wieren, S.E. (2004). Number, population structure and habitat use of Bar-headed Geese Anser indicus in Ladakh (India) during the brood-rearing period. *Acta Zoologica Sinica*, **50**, 738–744.

Ramirez, J.M., Folkow, L.P. & Blix, A.S. (2007). Hypoxia tolerance in mammals and birds: from the wilderness to the clinic. *Annual Review of Physiology*, **69**, 113–143.

Scheid, P., Slama, H. & Piiper, J. (1972). Mechanisms of unidirectional flow in parabronchi of avian lungs: measurements in duck lung preparations. *Respiration Physiology*, **14**, 83–95.

Scott, G.R., Egginton, S., Richards, J.G. & Milsom, W.K. (2009). Evolution of muscle phenotype for extreme high altitude flight in the Bar-headed Goose. *Proceedings of the Royal Society B: Biological Sciences*, **276**, 3645–3653.

Scott, G.R., Hawkes, L.A., Frappell, P.B., Butler, P.J., Bishop, P.J. & Milsom, W.K. (2015). Review: How Bar-Headed Geese fly over the Himalayas. *Physiology*, **30**, 107–115.

Scott, G.R. & Milsom, W.K. (2006). Flying high: a theoretical analysis of the factors limiting exercise performance in birds at altitude. *Respiratory Physiology & Neurobiology*, **154**, 284–301.

Scott, G.R. & Milsom, W.K. (2007). Control of breathing and adaptation to high altitude in the Bar-headed Goose. *American Journal of Physiology – Regulatory, Integrative and Comparative Physiology*, **293**, R379–R391.

Scott, G.R., Schulte, P.M., Egginton, S., Scott, A.L.M., Richards, J.G. & Milsom, W.K. (2011). Molecular evolution of cytochrome c oxidase underlies high-altitude adaptation in the Bar-Headed Goose. *Molecular Biology and Evolution*, **28**, 351–363.

Storz, J.F. (2007). Hemoglobin function and physiological adaptation to hypoxia in high-altitude mammals. *Journal of Mammalogy*, **88**, 24–31.

Swan, L.W. (1961). Ecology of the high Himalayas. *Scientific American*, **205**, 68–78.

Ward, S., Bishop, C.M., Woakes, A.J. & Butler, P.J. (2002). Heart rate and the rate of oxygen consumption of flying and walking barnacle geese (*Branta leucopsis*) and Bar-headed Geese (Anser indicus). *Journal of Experimental Biology*, **205**, 3347–3356.

Weber, R.E. (2007). High-altitude adaptations in vertebrate hemoglobins. *Respiratory Physiology & Neurobiology*, **158**, 132–142.

West, J.B. (2006). Human responses to extreme altitudes. *Integrative and Comparative Biology*, **46**, 25–34.

West, J.B. (2009). Comparative physiology of the pulmonary blood-gas barrier: the unique avian solution. *American Journal of Physiology – Regulatory, Integrative and Comparative Physiology*, **297**, R1625–R1634.

17 Distance-Altitude Trade-Off May Explain Why Some Migratory Birds Fly over and not around the Himalayas

Thomas A. Groen and Herbert H.T. Prins

Migration Strategies and Trade-offs between Time and Energy Expenditure

Some bird species fly over, rather than around, the Himalayas during their annual migration. The most iconic species that migrates over these highest mountains on Earth is the Bar-headed Goose *Anser indicus* (Hawkes *et al.*, 2012; Chapter 16). The Bar-headed Goose is known for flying at extreme altitudes, but some soaring birds, such as the condors Ohlmann (Chapter 14) observed, may fly even higher. The altitude record is perhaps held by Rüppell's Vulture *Gyps ruepellii*, which has been recorded soaring as high as 11,300 m above sea level (a.s.l.) in West Africa (Laybourne, 1974). During migration, Bar-headed Geese regularly traverse altitudes approaching 6000 m a.s.l. at some locations (Hawkes *et al.,* 2012; Chapter 16).

Flying at such altitudes comes at a cost. For example, Brent Geese *Branta bernicla* seem unable to fly non-stop over the Greenland ice cap (highest altitude 3200 m, average altitude 2000 m a.s.l.), and they are thought to need regular resting stops due partly to anaerobic flying (Gudmundsson *et al.*, 1995). Hawkes *et al.* (2012) showed that the flight pattern of Bar-headed Geese is also dictated chiefly by altitude, and that the birds generally fly through valleys, avoiding the highest summits (cf. Ohlmann, Chapter 14, Heise, Chapter 15). The birds in their study were fitted with satellite tracking devices, and they showed that these birds flew on average 112 km further than the shortest possible route (i.e. a straight line between departure and arrival point). This suggests that there is a trade-off between altitude and distance covered by the species. On the basis of some rough estimates, Hawkes *et al.* (2012) argued that at 8000 m a.s.l., the minimum mechanical power required for flight is 50% more than that at sea level, whereas the partial pressure of oxygen is 40% lower (West *et al.*, 1983). So, at high altitudes, more energy per unit time is required, but less oxygen is available. One thus wonders why migratory birds do not avoid the Himalayas altogether, since the lowest altitude in that mountain chain is about 3000 m a.s.l (Fig. 0.1). Would flying around these mountain ranges rather than over them not be more efficient, that is, cost less total energy even if the total route length would be longer? A possible answer lies in the consideration that in addition to energy constraints, the migratory birds are constrained by time. Birds have to make sure they arrive in time at the breeding grounds (e.g. Both & Visser, 2001; Stervander *et al.*, 2005; Si *et al.*, 2015), and therefore should not lose too much time travelling.

Energy-Time Trade-off

Here we try to shed light on the trade-off between time and energy by migratory birds flying across the Himalayas (cf. Alerstam, 2001). We investigate whether birds flying at extremely high altitudes minimize their time or energy (cf. Ydenberg, Chapter 19). For that we have modelled the routes across the Himalayas that minimize (a) time or (b) energy expenditure for two bird species (Mallard *Anas platyrhynchos* and Bar-headed Goose), taking into consideration their flight characteristics (Table 17.1) and the topography of the Himalayas. As input for this study, we needed altitude–cost relationships. These relationships were used to calculate least-cost paths through the terrain of the Himalayas using a satellite-based digital elevation model (DEM). GIS algorithms are available that calculate optimal routes given the cost of travel in a two-dimensional environment (e.g. Collischonn & Pillar, 2000; Atkinson *et al.*, 2005). Using a DEM and established altitude-cost models, the cost of traversing at a specific altitude can be calculated for any location in the Himalayas and the least-cost path can be established. Least-cost path analysis generally follows three steps:

- First, a cost surface is calculated that determines the cost of travelling through a grid-cell on the surface.
- Then, a spreading function is applied to derive the accumulated cost of reaching any location on the cost surface, given a fixed point of departure.
- Finally, the route towards a target location is traced over the accumulated cost surface.

The final result is considered the optimal solution (Lee & Stucky, 1998). The main input that is needed for such an analysis is a reliable cost surface for flight that is a function of altitude, that is, altitude-cost functions. From the established least-cost paths, the total duration of non-stop flight as well as the total amount of energy required to arrive at the destination can then be derived: these two are the different 'costs' to be considered in a trade-off. We calculated altitude-cost functions for the two 'model' species, Bar-headed Goose and Mallard. The Bar-headed Goose is known to fly over some of the highest areas in the Himalayas, and Mallards have also been observed at altitudes exceeding 6000 m a.s.l. (Faraci, 1986; Scott, 2011). These species stand as examples for other species that cross the Himalayas during their migration to and from their breeding grounds.

Altitude-Cost Curves

Altitude–cost relationships depend on the 'costs' considered. For species that are optimizing energy, the parameter to consider for optimizing energy use over the total path is the energy consumption per unit of distance travelled. When a species optimizes its time, the speed of travel is the key parameter. Birds can control their own flying speeds, but their optimal flying speeds are influenced to a large extent by the density of the air. In thinner air, more energy and higher flying speeds are required (Pennycuick, 2008).

Cost of Horizontal Flight

There is a well-established theorem describing bird flight and the factors that influence the amount of energy needed and the speed of flight required (Pennycuick, 1969; Alerstam 1991). This theorem can be used to calculate the cost of flying horizontally at any specific altitude. Energy expenditure at a given altitude by a given bird species depends on the air density at that altitude, the morphological traits of the bird's body (wingspan, wing area and body mass are the main input variables in the model of Pennycuick, 1969) and the speed of flight (Alerstam, 1991). When all relevant parameter values are known, power curves can be calculated (Figure 17.1). These curves display the power (P_x in Watt x indicates horizontal displacement, as opposed to z indicating vertical displacement, see 'Cost of ascending' later in this chapter) required to fly as a function of speed (V_x in m s^{-1}) and are typically u-shaped (Pennycuick, 1989). The lowest point in this curve indicates the speed at which minimal power is needed ($V_{x,mp}$, *mp* for minimal power), and thus which allows the bird to fly for the longest period on any given fuel reserve. The speed at which the energy use per covered distance is most efficient can be found by drawing a tangent from the origin to the power curve. The associated speed can

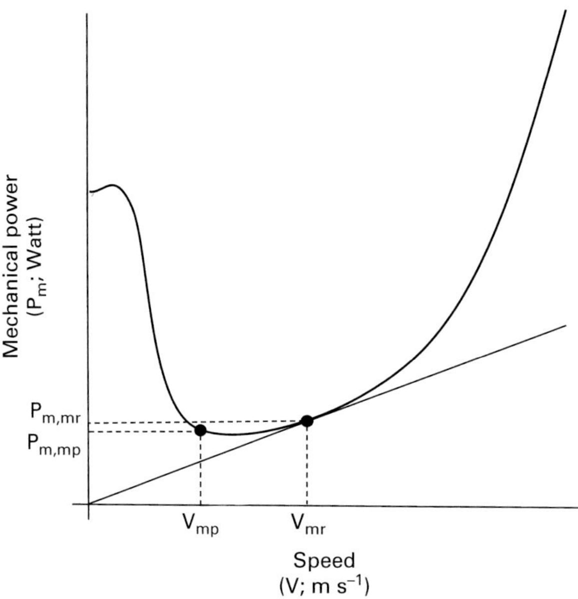

Figure 17.1. Example of a typical power curve (thick line) for flapping flight by birds, and the derivation of V_{mr} and V_{mp}. V_{mr} is found at the point where the tangent line through the origin (thin line) touches the power curve. Notice that Watt is J s^{-1}; V_{mr} gives the speed at which the maximum range (*mr*) can be covered on a given amount of fuel (i.e. at the lowest amount of energy expenditure per covered distance J m^{-1}). V_{mp} gives the speed at which the energy expenditure per second is lowest (minimum power; *mp*).

be considered as the speed at which the maximum distance can be covered on a given fuel reserve ($V_{x,mr}$, mr for maximum range; Alerstam, 1991). These two speeds and associated powers can be used to calculate the total time needed to cover a given distance efficiently, and the mechanical energy required to cover that distance.

Altitudinal Limit

The power curve can be thought of as moving diagonally (to the right and up on Figure 17.1) as air becomes less dense. In other words, V and P will both increase as a bird flies higher. Depending on the capability of the respiratory system of the bird, it may or may not be able to provide sufficient oxygen to the flight muscles to provide the required power. This capability varies greatly between species, as has been shown in several comparative studies, where species were exercised on treadmills under normoxia and hypoxia (Chappel & Dlugosz, 2009; Hawkes *et al.*, 2014; cf. Weber, 2007). So there must be an upper altitudinal limit where the oxygen requirements equal the maximum oxygen supply capacity of bird species. This limit will vary between species and depends on a number of species traits. Modelling studies exist that analyze factors affecting exercise performance of birds at increasing altitudes. For example, total ventilation capacity or haemoglobin O_2 affinity at various partial O_2 pressure has been studied for a number of species (Scott & Milsom, 2006; Scott 2011; Hawkes *et al.*, Chapter 16). But there are, to our knowledge, no databases that can be used to parameterize maximum flying altitudes of different bird species as a function of the required power and the oxygen supply rate of that species.

Cost of Ascending

Apart from the power needed to sustain flight at a certain altitude, additional power is needed to gain altitude. This power can be derived from simple classical mechanics. The power (P_z; z for vertical displacement in m) needed to generate a specific vertical speed (V_z) is:

$$P_z = V_z mg \qquad 17.1$$

with m being the body mass of the bird (kg) and g being the gravitational acceleration (m s^{-2}). It has been shown that the altitude at which Bar-headed Geese cross the Himalayas is mainly dictated by the altitude of the terrain (Hawkes *et al.*, 2012). So it seems reasonable to assume that birds will follow the terrain. Additionally, at least during the night, the air is coldest and densest closest to the surface (Ohlmann, Chapter 14) and also most stable (Heise, Chapter 15). The slope of the terrain (α) can be calculated from a DEM. Given the horizontal speed of the bird ($V_{x,mr}$ or $V_{x,mp}$) a corresponding vertical speed ($V_{z,mr}$ or $V_{z,mp}$) can be calculated with:

$$V_z = V_x tan(\alpha) \qquad 17.2$$

We made the simplifying assumption that when ascending, P_z has to be added to the total energy used, because the bird's potential energy increases with increasing altitude. But this potential energy will be released when descending because it is then converted again to kinetic energy.

Since geese more or less follow the terrain when traversing the Himalayas, as is expected for birds that use active muscle power for flight (see Ohlmann, Chapter 14), we took flying altitude to be equal to the altitude described by the DEM.

Total Cost Calculations

For the two species, an altitude–cost relationship was established in terms of energy use per horizontal metre travelled (E in J m^{-1}) and time needed to travel a metre (T in s.m^{-1}). These relationships were based on the model developed by Pennycuick (Pennycuick, 2008; Version 1.24 downloaded from: www.bio.bristol.ac.uk/people/pennycuick.htm). Power curves can be calculated with this model, and estimates of V_{mp}, V_{mr} and the associated P_{mp} and P_{mr} can also be derived. Inputs into this model are the body mass, wing span, wing area and flying altitude. The resulting V and P values were converted into E by:

$$E = \frac{P_x + P_z}{V_x} \qquad 17.3$$

P_z can be substituted by Equations 17.1 and 17.2, which yields:

$$E = \frac{P_x}{V_x} + \frac{V_x tan(\alpha)mg}{V_x} = \frac{P_x}{V_x} + tan(\alpha)mg \qquad 17.4$$

The last term [$tan(\alpha)mg$] can either be negative (when descending) or positive (when ascending). T is derived from V_x by:

$$T = \frac{1}{V} \qquad 17.5$$

assuming only the horizontal speed. The value of V, P and α were dictated by the altitude of the landscape. The total cost (either in time or in energy) was then obtained by multiplying E and T by the distance across a pixel in the DEM, taking into account whether the grid-cell was crossed straight (i.e. 991 m, see 'Landscape data') or diagonally (i.e. 1401 m).

The estimation of V and P (and derived from that, of T and E) are purely based on the mechanics of flight. There is no consideration of possible energy conversion efficiencies that could be variable between species.

Landscape Data

For the entire Himalayas and its surroundings, the shuttle radar topography mission (SRTM) digital elevation model version 2.1 was downloaded from the USGS website

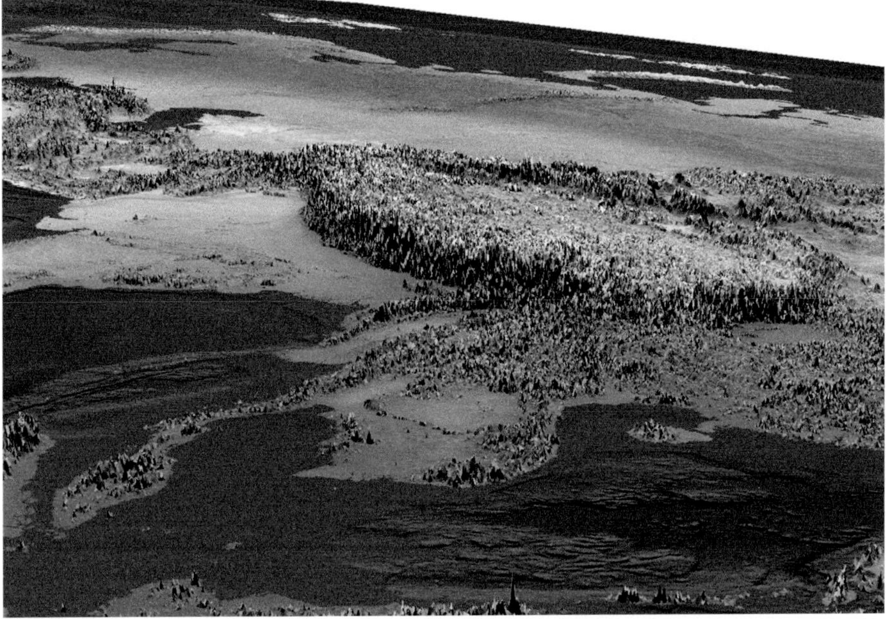

Figure 17.2. View over the Himalayas and its surroundings from the southeast, based on the compiled digital elevation model (DEM). At the top right the Arctic Ocean is visible; at the bottom left is Borneo. The Tibetan Plateau and the Himalayas form a high-altitude barrier for avian migration. Vertical scale ~30x the horizontal scale. (A black-and-white version of this figure will appear in some formats. For the colour version, please refer to the plate section.)

(http://dds.cr.usgs.gov/srtm/version2_1/SRTM30/) with a 30 arc seconds spatial resolution (which corresponds to ~1 km). The surroundings had to be widely defined to include sufficient options for flying around the Himalayas in case that was more cost-efficient. In other words, the extent of the selected area should not limit the possible flyways (Figure 17.2).

The DEM comes in a global geographic coordinate reference system (CRS) with latitude and longitude expressed in degrees. This was converted into an equidistant projected CRS (conical projection) with coordinates expressed in metres to allow for correct distance calculations. The final size of a single pixel after the conversion was 991 m by 991 m.

Least-Cost Path Analysis

Least-cost paths were calculated using ArcGIS 10.3. Costs for crossing a pixel of the DEM were calculated using the equations given earlier, both for flying at minimum power (using V_{mp} and P_{mp} as input) and for flying at maximum range (V_{mr} and P_{mr}). From this, a minimal 'cost distance grid' could be calculated that indicated the total minimal cost needed to reach each cell given a point of departure. From this cost distance grid, the least-cost path to reach a given destination given the point of departure was

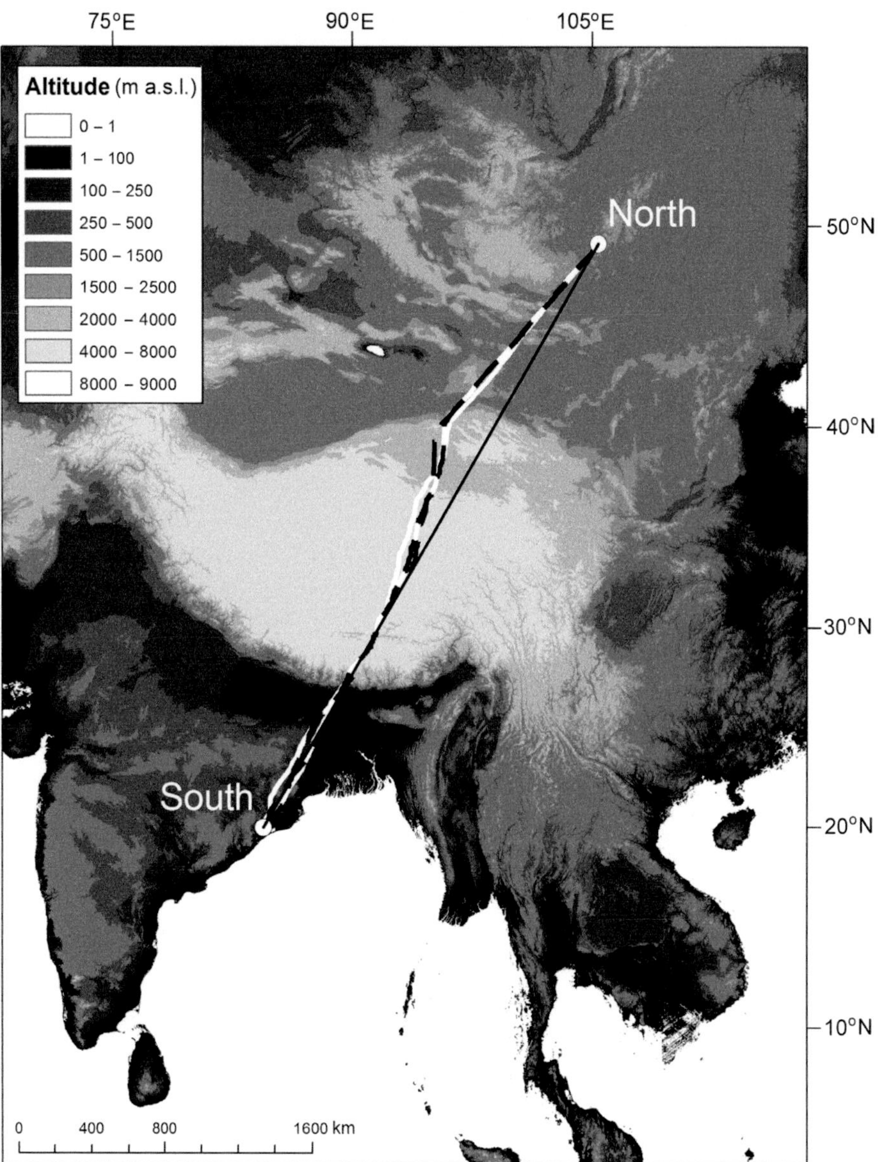

Figure 17.3. Locations used in the analysis, and the shortest route (thin black line) and the routes used by Bar-headed Goose (dashed black) and Mallard (white) as calculated according to the least-cost path analysis.

calculated. We calculated the total time needed to travel the path and the total energy spent while travelling along that path.

For a least-cost path analysis, a point of departure and a point of arrival are needed. We used the locations where Hawkes *et al.* (2012) captured and tagged Bar-headed Geese with satellite transmitters before their flights across the Himalayas (Figure 17.3).

Table 17.1. Bird biometrics used in the flight model (V1.24); the aspect ratio is a dimensionless figure.

	Bar-headed Goose *Anser indicus*	Mallard *Anas platyrhynchos*
Weight (kg)	2.5	1.2
Wing span (m)	1.5	0.9
Wing area (m^2)	0.291	0.099
Aspect ratio (-)	7.73	8.15
Maximum altitude (m a.s.l.)	8800	6400
Source	http://animaldiversity.org/	www.allaboutbirds.org/

As mentioned in the paragraph on altitudinal limits, there will be an altitude at which the partial oxygen pressure will no longer be sufficient to supply the energy demanded by the flight muscle system of the bird. However, there are clear observations of both Bar-headed Goose and Mallard at high altitudes (see earlier in this chapter). We used these highest observations (8800 m for Bar-headed Goose and 6400 m for Mallard; Faraci, 1986; Scott, 2011) to cap the DEM to the maximum altitude that can be negotiated by these birds. This essentially means that Bar-headed Geese can fly over any altitude (because Mount Everest, the highest point on Earth, is 8848 m, and at a 1 km cell size in the DEM this results in a maximum altitude of 8685 m a.s.l.), but the Mallard was thus capped in our calculations. Table 17.1 provides the details used for the parameterization of the two species. Any other values needed in the calculations were based on the default parameter settings in the flight model (v1.24) by Pennycuick (2008).

Alternative Strategies

To compare the results of the least cost path calculations, two alternative flying strategies were analyzed. Firstly, birds can fly in a straight line, having the flying altitude dictated by the terrain. Secondly, birds can ascend to the minimum altitude required to pass all obstacles and fly at a constant altitude in a direct line. For this latter case, the assumption is that there is no additional time needed to reach this cruising speed, however, the additional energy needed to reach this altitude is of course added to the energy calculations.

Model Outputs

The relationship between altitude and flying speeds (Figure 17.4a and b) shows that there is hardly any difference in flight speeds (either at minimum power or at maximum range, V_{mp} and V_{mr}, respectively) between the modelled Bar-headed Goose and Mallard. There is, however, a large difference in the power required to sustain their flights. This is much higher for Bar-headed Geese than for Mallards. This is related to differences in both their weight and their aspect ratio (relative wingspan per wing area, Table 17.1).

When costs are expressed per metre (Figure 17.4c and d), rather than per second, it is remarkable that energy costs are effectively constant over altitudes. This

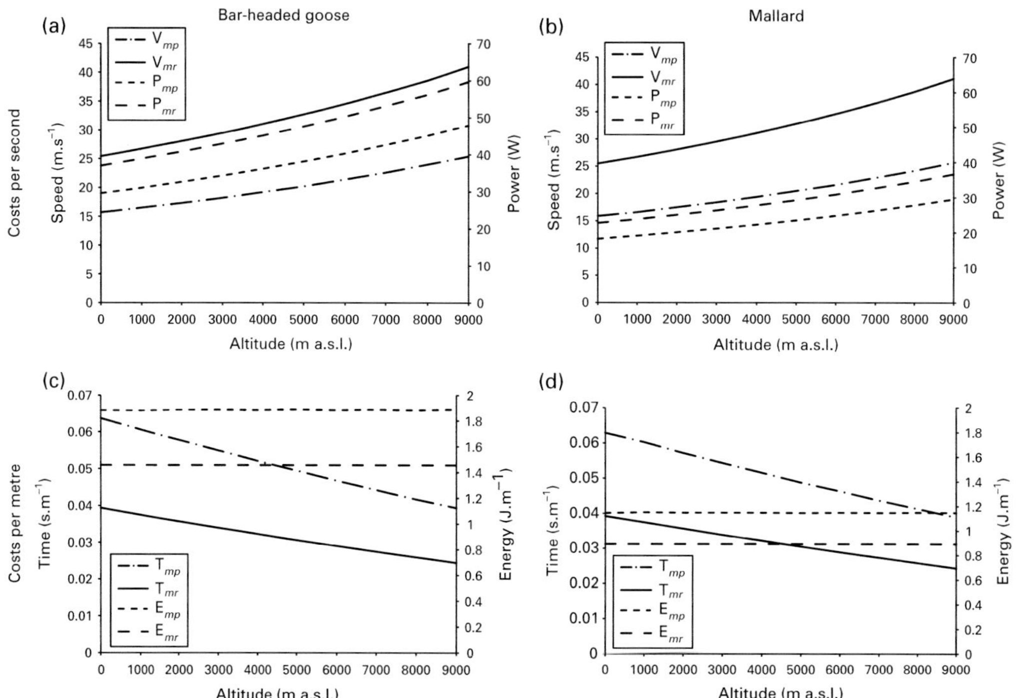

Figure 17.4. The relationships between optimum speeds for Bar-headed Goose (left) and Mallard (right) in relation to altitude (m above sea level). V_{mp} is the speed (m.s^{-1}) at which the energy expenditure per second is lowest. V_{mr} is the speed where the maximum range can be reached given an amount of fuel. P_{mp} is the energy expenditure (J.sec^{-1}) at the speed where power requirement is minimal. P_{mr} is the energy expenditure at the speed that would cover the maximum range. T_{mr} and T_{mp} give the time needed to cover a metre (s.m^{-1}) given V_{mr} and V_{mp} respectively. E_{mr} and E_{mp} give the energy needed to cover a metre given the speed (V_{mr} and V_{mp}) and energy expenditure (P_{mr} and P_{mp}) associated with maximum range and minimum power optimization.

implies that any additional power required to sustain flight at higher altitudes is compensated by the higher speed needed at these altitudes, resulting in a constant energy cost per metre travelled. This means that the only limiting cost for flying at high altitude is the cost of ascending, and there is no additional cost for horizontal flight over and above what is in any case needed to cover that distance. This also means that a least-cost path analysis based on E (either E_{mp} or E_{mr}, excluding costs of ascending) would be the same as crossing a flat plane, as altitude does not cause differences in costs. We can also observe that the cost per metre travelled horizontally is higher for the strategy that minimizes power (E_{mp}). So flying at minimal power expenditure keeps a bird that uses muscle power in the air for the longest amount of time, but brings that bird less far at a higher cost per metre travelled. In other words, optimizing a migratory path following a minimum power strategy would be a very inefficient approach. We therefore excluded least-cost path analyses based on minimum power input (E_{mp} and T_{mp}). For the maximum range optimization, the time needed to cross a given distance is differentiated over

Table 17.2. Energy and time 'costs' for the various optimized routes

			Bar-headed Goose	Mallard
North to south	Energy optimizing	Total energy	5.69 MJ	3.44 MJ
		Total time	1.58 days	1.57 days
		Total distance	3808 km	3803 km
	Straight line	Total energy	7.30 MJ	5.61 MJ
		Total time	1.46 days	1.45 days
		Total distance	3543 km	3543 km
South to north	Energy optimizing	Total energy	5.74 MJ	3.42 MJ
		Total time	1.57 days	1.57 days
		Total distance	3801 km	3798 km
	Straight line	Total energy	7.33 MJ	5.64 MJ
		Total time	1.46 days	1.45 days
		Total distance	3543 km	3543 km
Either direction	Flying at one altitude	Total energy	5.31 MJ	3.22 MJ
		Total time	1.23 days	1.23 days
		Total distance	3543 km	3543 km

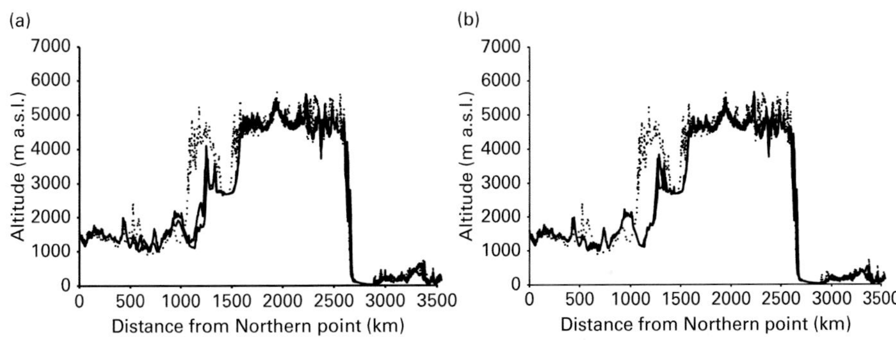

Figure 17.5. Altitude profiles of the optimized paths according to maximum range for the Bar-headed Goose (A) and Mallard (B) and the profile of the shortest distance (dashed line).

altitude, and follows a negative trend. This is an obvious result, because at higher altitudes, faster flying speeds are required, which result in less time spent travelling a given distance. This would suggest that flying at higher altitudes would always be beneficial if the cost of ascent could be ignored.

This simple analysis shows that the main impact of topography on travelling paths of birds might very well not be the altitude, but rather the steepness of the slopes and the cost of ascending. Steepness could form a barrier when a bird cannot generate sufficient power to both sustain horizontal flight and ascend fast enough. Therefore, the performed least-cost path analysis was based on E_{mr} defining the cost for horizontal flight, but with inclusion of the costs of ascent. The resulting path, and the energy and time required to travel this path, are presented in Table 17.2. The paths are depicted in Figure 17.3, and the altitude profiles in Figure 17.5. The energy-optimizing strategy substantially reduces the total energetic costs for travelling across the Himalayas (~22% for the Bar-headed Goose and ~39% for the Mallard in either direction) with only a fraction of additional extra time for the flight (~8% for both the Bar-headed Goose and the Mallard) when compared to the direct flight. The additional distance travelled was fairly constant at around 260 km. This is more than the observed 112 km additional distance, relative to direct flight that was observed by Hawkes *et al.* (2012) for Bar-headed Goose when traversing the Himalayas. This difference can partially be attributed to the resolution at which the altitude information was available. Although a resolution of ~1 km is fairly large for the total extent at which the least-cost path analysis was performed, it could be that more detailed altitude information would result in different paths. From this it may also be inferred that Bar-headed Geese must have a very detailed knowledge of the terrain.

Comparing the calculated energy expenditure when following the most energy efficient path while following the terrain, with flying at one single altitude for the entire path (Table 17.2) shows that this latter strategy is actually more efficient. This latter strategy is also what Bishop *et al.* (2015) call the 'prevailing expectation'. The fact that birds do not follow this strategy could mean that flying at higher altitudes is more complicated than assumed for models based on first principles. Bishop *et al.* (2015) noted that 'Flight cost increase[s] more rapidly than anticipated as air density declines.' This might mean that the established models on flying energy expenditure require updating. This may have to do with the anaerobic costs that increase when the oxygen pressure drops. Anaerobiosis can be tolerated for a short time, and it is known that Bar-headed Geese have adaptations to tolerate severe hypoxia (Hawkes *et al.*, 2014). There must, nevertheless, be an upper ceiling above which anaerobiosis will become limiting, even for Bar-headed Geese. Besides anaerobiosis, very high flight might also be very cold, very dry or more affected by wind (see Ohlmann chapter 14, Heise chapter 15). None of these factors was included in this analysis.

The analysis also showed that maximum flying altitudes do not need to exceed 6000 m a.s.l. (Figure 17.5). This suggests that Mallards should be able to cross the Himalayas. Also for Bar-headed Goose, the least-cost path route suggests that travelling at altitudes above 6000 m is unnecessary because alternative routes are available, with only minimal increases in distance. This contradicts the observations of flights over 7000 m (Hawkes *et al.*, 2012). This discrepancy could be attributed to the fact that apart from cost and time optimization by birds based on costs of horizontal flight and costs of ascending, another important factor, namely wind, was left out of the current analysis. As Bishop *et al.* (2015) showed, perhaps birds follow a roller-coaster strategy, where at times they fly higher than dictated by the terrain, to make use of orographic lift to save energy. However, Bishop *et al.* (2015) recorded very few observations above 6000 m a.s.l.

Insights from the Least-Cost Path Analysis

Our analysis shows that the migration of birds over high mountains is likely to be most constrained by the cost of ascending rather than the cost of sustaining flight at a very high altitude. Indeed, over and above the horizontal distance that birds have to cover anyhow when migrating from north of the Tibetan Plateau to the Indian subcontinent in the south and back, there appears to be hardly any additional energy expenditure apart from the costs of attaining altitude. This might explain the enormous benefit that soaring birds have in the mountains: their costs of flying in a thermal are only in the order of 1.5 times their basal metabolic rate (Duriez *et al.*, 2014), which is equivalent to the costs of watching TV or working on a manuscript (see Newton *et al.*, 2013). A major problem faced by soaring birds such as storks and eagles is the number of hours of good soaring time per day, which they cannot control. Weather conditions across the Himalayas are in fact often unfavourable for thermal soaring (see Ohlmann, Chapter 14; Heise, Chapter 15; Juhant & Bildstein, Chapter 6). When conditions are favourable, however, the daily distance covered may be as large or even larger than that covered by birds that use active propulsion (Duerr *et al.*, 2012).

Birds that use flapping flight have an energy expenditure which is several times their basal metabolic rate (Duerr *et al.*, 2012), and as high as people exerting themselves in endurance sports (Rosenkilde *et al.*, 2015). Because of oxygen demands, there is a ceiling beyond which a species will not be able to sustain flight due to physiological constraints, which is highly variable across species. As shown in our analysis, the costs-altitude curve is mainly a result of the body weight of the species. Heavier species require more energy to traverse, and therefore gain relatively less by optimizing their route than lighter species. In other words, the trade-off between the loss in time and that in energy favours the lighter species. Optimizing can still make sense, because saving substantially on total energy expenditure for the flight comes at the cost of arriving only a little later at the next refuelling place due to travelling a bit more slowly. This may not be relevant for very small birds such as passerines, because it appears that they do not cross the Himalayas much: it appears as if in spring (when flying from India towards Siberia) a considerable majority of Palaearctic passerine nocturnal migrants avoid crossing the highlands of western Central Asia altogether: they appear to fly from northern Afghanistan and northern Pakistan and thence cross the much narrower Pamirs and Hindu Kush towards the northeast. In autumn, crossing the Tibetan Plateau and Himalayas may be more frequent, but appears still to be concentrated in the area of the Pamirs and the Hindu Kush (Bolshakov, 2003; cf. Delany *et al.*, Chapter 4). The causes that have been cited are lack of food to refuel and adverse winds (Bolshakov, 2003).

Several factors lead to the counterintuitive conclusion that large birds cover less distance in a day than smaller ones (Alerstam, 2003), basically because they spend more time re-fattening during migration. Whether soaring birds need refuelling is not very clear: if migrating in thermals is energetically as cheap as suggested from the work of Duerr *et al.* (2012) on Golden Eagles *Aquila chrysaetos*, then the observations of Berthold *et al.* (2001) on White Storks *Ciconia ciconia* make sense: they migrated non-stop, did not overtly engage in refuelling activities and covered large distances per day. Food for refuelling during migration is, however, very important for many other bird species. Alerstam (2003) showed how 'migrating birds alternate between two main phases of (1) flight when distance is covered and energy (fuel) is consumed, and (2) fuel deposition when energy is accumulated by intensive foraging. Fuel deposition takes place during stopover periods between flights'. From studies on Mallard elsewhere in Asia, it is known that they spend much time on 'refuelling' between rather short migratory 'hops' (Yamaguchi *et al.*, 2008). For a number of duck species in the weight range of the Mallard, Namgail *et al.* (Chapter 25) emphasize the numerous lakes and other wetlands that are used by migrating ducks. Bar-headed Geese, being larger than Mallards, would thus be expected to engage even more in refuelling on stopover sites. But do they? Takekawa *et al.* (Chapter 1) emphasize the short distances that Bar-headed Geese fly and mention the relatively high number of stopover sites that the geese use on autumn and spring migration. We thus conclude that Anatidae such as ducks and some geese can cross the Himalayas without too much effort as long as refuelling sites where they can fatten up again for the next hop on their migration are close enough to keep the migration going, and offer sufficiently undisturbed sites with high food quality. Our prediction is that travelling across the Himalayas using flapping flight takes the modelled species roughly one and a half days of non-stop flying at a cost of roughly 3.4 MJ (Mallard) up to 5.7 MJ (Bar-headed Goose) of energy. With an energy density of 40 $kJ.g^{-1}$ for fat (Berg *et al.*, 2002; Weber, 2011) this is the equivalent of 85 g to 140 g of fat, respectively (or 7.1% and 5.6% of their body weights). This suggests that when flying across high mountain ranges such as the Himalayas, birds are energy optimizers, rather than time optimizers.

References

Alerstam, T. (1991). Bird flight and optimal migration. *Trends in Ecology and Evolution*, **6**, 210–215.

Alerstam, T. (2001). Detours in bird migration. *Journal of Theoretical Biology*, **209**, 319–331.

Alerstam, T. (2003). Bird migration speed. In Berthold, P., Gwinner, E. & Sonnenschein, E., eds., *Avian Migration*. Springer Berlin Heidelberg, pp. 253–267.

Atkinson, D.M., Deadman, P., Dudycha, D. & Traynor, S. (2005). Multi-criteria evaluation and least-cost path analysis for an arctic all-weather road. *Applied Geography*, **25**, 287–307.

Berg, J.M., Tymoczko, J.L. & Stryer, L. (2002). *Biochemistry*, 5th edn. New York: W. H. Freeman.

Berthold, P., Bossche, W v d, Fiedler, W., *et al.* (2001). Der Zug des Weisstorchs (Ciconia ciconia): eine besondere Zugform auf Grund neuer Ergebnisse. *J Ornithologie*, **42**, 73–92.

Bishop, C.M., Spivey, R.J., Hawkes, L.A., *et al.* (2015). The roller coaster flight strategy of Bar-headed Geese conserves energy during Himalayan migrations. *Science*, **347**, 250–254.

Bolshakov, C.V. (2003). Nocturnal migration of passerines in the desert-highland zone of western Central Asia: selected aspects. In Berthold, P., Gwinner, E. & Sonnenschein, E., eds., *Avian Migration*. Springer Berlin Heidelberg, pp. 225–236.

Both, C. & Visser, M.E. (2001). Adjustment to climate change is constrained by arrival date in a long-distance migrant bird. *Nature*, **411**, 296–298.

Chappel, M.A. & Dlugoz, E.M. (2009) Aerobic capacity and running performance across a 1.6 km altitude difference in two sciurid rodents. *The Journal of Experimental Biology*, **212**, 610–619.

Collischonn, W. & Pilar, J.V. (2000). A direction dependent least-cost-path algorithm for roads and canals. *International Journal of Geographical Information Science*, **14**, 397–406.

Duerr, A.E., Miller, T.A., Lanzone, M., *et al.* (2012). Testing an emerging paradigm in migration ecology shows surprising differences in efficiency between flight modes. *PLoS ONE*, **7** (4), e35548.

Duriez, O., Kato, A., Tromp, C., *et al.* (2014). How cheap is soaring flight in raptors? A preliminary investigation in freely-flying vultures. *PLoS One*, **9**, e84887.

Faraci, F.M. (1986). Circulation during hypoxia in birds. *Comparative Biochemistry and Physiology*, **85A**, 613–620.

Gudmundsson, G.A., Benvenuti, S., Alerstam, T., Papi, F., Lilliendahl, K. & Akesson, S. (1995). Examining the Limits of Flight and Orientation Performance: Satellite Tracking of Brent Geese Migrating across the Greenland Ice-Cap. *Proceedings of the Royal Society B*, **261**, 73–79.

Hawkes, L.A., Balachandran, S., Batbayar, N., *et al.* (2012). The paradox of extreme high-altitude migration in Bar-headed Geese Anser indicus. *Proceedings of the Royal Society B*, **280**: 20122114.

Hawkes, L.A., Butler, P.J., Frappell, *et al.* (2014). Maximum running speed of captive Bar-headed Geese is unaffected by severe hypoxia. *PLoS ONE*, **9**, e94015.

Laybourne, R.C. (1974). Collision between a vulture and an aircraft at an altitude of 37,000 feet. *The Wilson Bulletin*, **86**, 461–462.

Lee, J. & Stucky, D. (1998). On applying viewshed analysis for determining least-cost paths on digital elevation models. *International Journal of Geographical Information Systems*, **12**, 891–905.

Newton Jr, R.L., Han, H., Zderic, T. & Hamilton, M. (2013). The energy expenditure of sedentary behavior: a whole room calorimeter study. *PLoS ONE*, **8** (5), e63171.

Pennycuick, C.J. (1969). The mechanics of bird migration. *Ibis*, **111**, 525–556.

Pennycuick, C.J. (1989). *Bird Flight Performance: A Practical Calculation Manual.* New York: Oxford University Press.

Pennycuick C.J. (2008) *Modelling the Flying Bird.* London: Elsevier (Academic Press).

Pennycuick, C.J. & Rezende, M.A. (1984). The specific power output of aerobic muscle, related to the power density of mitochondria. *Journal of Experimental Biology*, **108**, 377–392.

Rosenkilde, M., Morville, T., Andersen, *et al.* (2015). Inability to match energy intake with energy expenditure at sustained near-maximal rates of energy expenditure in older men during a 14-d cycling expedition. *The American Journal of Clinical Nutrition*, **102**(6), 1398–1405.

Scott, G.R. (2011). Elevated performance: the unique physiology of birds that fly at high altitudes. *Journal of Experimental Biology*, **214**, 2455–2462.

Scott, G.R. & Milsom, W.K. (2006) Flying high: a theoretical analysis of the factors limiting exercise performance in birds at altitude. *Respiratory Physiology & Neurobiology*, **154**, 284–301.

Si, Y., Xin, Q., Boer, W.F. de, Gong, P., Ydenberg, R.C. & Prins, H.H.T. (2015). Do Arctic breeding geese track or overtake a green wave during spring migration? *Scientific Reports*, **5**, 8749. 10.1038/srep08749.

Stervander, M., Lindström, Å., Jonzén, N. & Andersson, A. (2005). Timing of spring migration in birds: long-term trends, North Atlantic Oscillation and the significance of different migration routes. *Journal of Avian Biology*, **36**, 210–221.

Weber, J.M. (2011). Metabolic fuels: regulating fluxes to select mix. *Journal of Experimental Biology*, **214**, 286–294.

Weber, R.E. (2007). High-altitude adaptations in vertebrate hemoglobins. *Respiratory Physiology & Neurobiology*, **158**, 132–142.

West, J.B., Lahiri, S., Maret, K.H., Peters, R.M. & Pizzo, C.J. (1983). Barometric pressures at extreme altitudes on Mt. Everest: physiological significance. *Journal of Applied Physiology*, **54**, 1188–1194.

Yamaguchi, N., Hiraoka, E., Fujita, M., *et al.* (2008). Spring migration routes of mallards (Anas platyrhynchos) that winter in Japan, determined from satellite telemetry. *Zoological Science*, **25**, 875–881.

18 Refuelling Stations for Waterbirds: Macroinvertebrate Biomass in Relation to Altitude in the Trans-Himalayas

Herbert H.T. Prins, Rob J. Jansen and Víctor Martín Vélez

Macroinvertebrates as Food for Migrating Birds

To cover the long distances between wintering areas and breeding areas, birds typically catabolize body fat reserves and even sacrifice body tissues such as intestines (Piersma *et al.*, 1996; Wojciechowski *et al.*, 2014). Their capacity for 'burning' fat has drawn attention for many years and has been well studied. The strategies that migratory birds use form a range, varying from very long non-stop migrations (e.g. Bar-tailed Godwit *Limosa lapponica* crossing the Pacific; Gill *et al.*, 2005) to a mixture of long flights interspersed with shorter flights and stop-overs (e.g. Brent Geese *Branta bernicla* have their longest uninterrupted flights ranging between 700 km and 1300 km, but the average is only 300 km per 'hop' between the Dutch Waddensea and Taymyr in Arctic Russia; Green *et al.*, 2002). Other birds make shorter 'hops' on their migratory route, such as Barnacle Geese *Branta leucopsis* that more or less precisely follow a green wave of spring food emerging after winter while migrating from the Netherlands to the Arctic (Yali *et al.*, 2015), or Mallard *Anas platyrhynchos* (Yamaguchi *et al.*, 2008). Short stops appear typical for Bar-headed Geese *Anser indicus* (Takekawa *et al.*, Chapter 1) and ducks (Namgail *et al.*, Chapter 2) crossing the Himalayas and the Tibetan Plateau.

The Himalayas and associated mountain ranges present a formidable barrier for birds migrating between the South (the Indian subcontinent and further afield) and the North (the tundra, taiga, bogs and grasslands of Siberia and Mongolia). The mountain ranges and vast deserts (Taklamakan, Gobi) offer few feeding places for migratory birds, while ice caps, bare rocks and scree slopes of the high-altitude rocky deserts do not offer food either. For birds that can cross this 3000 km wide stretch of terrain without stopping to replenish their energy reserves, the barren lands of the Gobi and the high Himalayas are perhaps even tougher than the Sahara. Indeed, birds cross that vast barrier between Europe and sub-Saharan Africa in their millions (Moreau, 1972), and knowledge of migration strategies is still improving (e.g. Sergio *et al.*, 2014). For birds that typically do not cover long distances between successive stop-over sites, there must be places in the landscape where there is sufficient food for them to reliably replenish their fat reserves (cf. Groen & Prins, Chapter 17).

A high proportion of the migratory birds that pass the Himalayas on their way to the North or to the South are dependent on animal food: raptors eat small mammals, reptiles or birds, some species eat fish, but many depend on aquatic invertebrates (Delany *et al.*, Chapter 5). In this chapter, we report on the density of aquatic invertebrate biomass in relation to altitude.

One of the recurrent themes in the study of high altitudes is, of course, the partial oxygen pressure: the higher one gets into the mountains, the lower the amount of oxygen that is available. In summer, this relationship for partial oxygen pressure (PO_2) has been found to be $P_{O2} = e^{(3.07–0.131altitude)}$; in winter: $P_{O2} = e^{(3.07–0.133altitude)}$ (with altitude in km) (Dillon *et al.*, 2006). For humans, the permanent habitation limit is around 4200 m a.s.l. in the latitude of the Himalayas, where the village of Kibber, India, is located (pers. obs.) and for Red Fox *Vulpes vulpes* in the Dhorpatan Hunting Reserve, Nepal, a limit of 4500 m a.s.l. has been observed (Aryal *et al.*, 2010). People and animals are actually severely affected not only by low oxygen availability, but also by low atmospheric pressure; plants are not affected by low partial oxygen pressure, but are mainly limited by aridity and frost at high altitudes (Paul & Ferl, 2006). Likewise, the partial oxygen pressure in surface water decreases with altitude, more or less in direct proportion to that in the atmosphere. The most recent insight is that partial oxygen pressure is not the key determinant for aquatic macro-invertebrates, which is better explained by an oxygen supply index that incorporates both partial pressure and oxygen solubility (Verberk *et al.*, 2011). They found, contrary to normal assumptions, that more oxygen is actually available to these organisms when the temperature is high than when it is low (Verberk *et al.*, 2011). This means that at increasing altitudes (where the partial oxygen pressure is lower and where the temperature on average is lower too – the lapse rate is 0.62°C.100 m^{-1}), aquatic invertebrates do suffer from less and less available oxygen in the water. This generally results in lower species richness of aquatic invertebrates with increasing altitude (Jacobsen, 2008), the idea being that fewer and fewer species possess the specialized adaptations needed to garner dissolved oxygen from the water when concentrations decrease (Jacobsen, 2008). Yet birds are observed at very high altitudes. For example, remains of Common Pintail *Anas acuta* and Black-tailed Godwit *Limosa limosa* were found on the Khumbu Glacier of Mount Everest at 5000 m above sea level (a.s.l.) (Geroudet, 1954), and Black Redstart *Phoenicurus phoenicurus* and Grey Wagtail *Motacilla cinerea* have been observed at 6200 m a.s.l. (Prins *et al.*, Chapters 20 and 26): these animals forage in the wetland habitats of these very high mountains, very often on aquatic macroinvertebrates. Aquatic macroinvertebrates were defined in our surveys as animals which live at least one stage of their life cycle in the water and can be observed without the aid of equipment (Weber, 1973; Rosenberg *et al.*, 1997). 'Macro' implies here that the animal has a body length or width larger than 0.5 mm.

Collecting Samples of Aquatic Macroinvertebrates

We decided to collect aquatic macroinvertebrates in the Trans-Himalaya of northern India, in the province of Ladakh, a part of the northern state of Jammu and Kashmir. Ladakh covers an area of approximately 80,000 km^2 and is bordered in the north by the eastern part of the Karakoram Mountains and in the south by the western extreme of the Great Himalaya Range; the west is bounded by the Hindu Kush Mountains and the east by the Tibetan Plateau. The altitude ranges from approximately 3000 m (lower River Indus and Nubra Valleys) to 7600 m (Great Himalaya and Karakoram Ranges). The mountain ranges block the monsoon clouds so that Ladakh lies in the rain shadow of these mountains, making the climate arid (Namgail *et al.*, 2012). Annual precipitation is especially low in the eastern part (about 100 mm) and mainly comes from snow precipitation during the winter months. The climate is also marked by large annual and diurnal variations in temperature. Winter temperatures may fall to −30°C; in summer, the weather is generally milder, with temperatures ranging from 0°C to 30°C, although sometimes temperatures may rise as high as 35°C. Due to the lack of water, the vegetation does not form forest but is adapted to arid conditions, with low shrubs and herbs. On steep slopes there is no vegetation, and the highest vegetation cover is at about 4500 m a.s.l.: at lower levels, the effect of the rain shadow is too severe, and at higher levels, the vegetation growing season becomes too short to maintain much plant production (Namgail *et al.*, 2012, also see Rawat chapter 12).

The region is characterized by high, steep mountains separated by deep valleys into which the Indus and its tributaries flow (Namgail *et al.*, 2012). In the east there are a number of high-altitude mountain lakes fringed by marshes, principally Tsokar, Startsapuk Tso, Thatsangkaru Tso (also known as Kiagar Tso) and Tsomoriri. Tsokar is saline, as is Kiagar Tso. Startsapuk Tso is a freshwater lake. Tsomoriri is fresh-brackish and is the largest of the four lakes (Guija *et al.*, 2003; Philip & Mathew, 2005). Most of the fringing marshlands have a high vegetation cover during the summer months (Rawat & Adhikari, 2005). Lakes such as Tskar and Tsomoriri have permanent human settlements on their shores while Startsapuk Tso and Kiagar Tso only have temporary camps of pastoralists. Marshland areas around the lakes are also grazed by domestic livestock and this grazing pressure increases during summer (Chandan *et al.*, 2008). At a much lower altitude, the freshwater wetlands of Shey, Sakti and Gayaik occur at 3250 m, 3500 m and 3750 m a.s.l, respectively. Keshar, Mahe-Puga and Puga are riverbank wetlands and are also fresh.

Abiotic factors (such as temperature, pH or turbidity of the water) can be important factors that determine aquatic invertebrate presence or abundance (Havas & Hutchinson, 1982; Füreder *et al.*, 2006). Salinity can also affect the invertebrate distribution in the area (Kefford *et al.*, 2003). Traits unrelated to the water itself may not only contribute to the distribution of aquatic invertebrates, but also to the physical characteristics of the entire water bodies such as the depth, substrate or water velocity (Collier, 1995). The principal interest of our study was to find out whether waterbirds favour particular water bodies in order to maximize their food intake, and to find out at which altitudes they could still refuel for their migrations. We thus measured a total combination of 12 water and physical variables: altitude, vegetation, substrate (or sediment) type presence (mud, sand, granules and pebbles and/or cobbles), depth (related to turbidity), temperature (° Celsius), conductivity (as a measure of salinity ($\mu S.cm^{-1}$), pH, velocity of the flowing water and quantity of organic matter. Using a handheld GPS device we first registered altitude and coordinates of the different sampling points. We measured water temperature, pH and

Table 18.1. Sediments were classified following the Udden-Wentworth grain-size scale (Wentworth, 1922). We only included stones smaller than cobbles in the analyses.

Sediment type	Size	Recognized by
Mud	< 0.06 mm	Clayish/muddy feeling when rubbed between fingers
Sand	0.06–2 mm	Gritty feeling when rubbed between fingers
Granules	2–4 mm	Classified by eye
Pebbles	> 4 mm	Classified by eye
Cobbles	> 60 mm	Classified by eye

conductivity in shallow water (less than 25 cm deep) just above the bottom. Using a conductivimeter with associated thermometer (WTW Cond 315i.) and pH-strips (pH-Fix 0-14, Macherey-Nagel) had the advantage that measurements could be done *in situ*. We measured turbidity with a Secchi disk (25 cm diameter). Percentage of vegetation within the sample plot and habitat type were visually estimated whereas the organic matter was categorized from 0 (nothing) to 4 (high) based on the quantity that remained in the sampling net. Water velocity was determined through the use of a spherical float, measuring tape and chronometer. Measurement was repeated three times to maximize precision. Finally, the sediment type was classified visually (Table 18.1).

Colwell and Taft (2000 and references therein) found that most waterbirds (except for divers) mainly took advantage of food when depth of the water was less than 20 cm to 25 cm (cf. Davis & Smith, 1998; Bancroft *et al.*, 2002). We therefore took each sample at a depth of less than 25 cm. The sample plots were 0.7 x 0.7 m square, which we sampled with a dipnet (0.5 mm mesh size). When the dipnet gathered aquatic plants, we removed and washed these to obtain the fauna hidden in the plants. The content of the dipnet was placed on a white tray, and by using a magnifying glass and with the help of field guides, we identified all insects to Family level and non-insect invertebrates to Class and Order level (cf. Bournard *et al.*, 1996; Jacobsen *et al.*, 1997). Because classifying the animals to species level would go beyond the biological relevance for this study, the animals were classified into the main groups, such as Acari, Amphipoda, Annelida, Anisoptera larvae, Bivalvia, winged aquatic insects (coleoptera/hemiptera), Diptera and similar larvae (Diptera and Coleoptera larvae), Ephemeroptera larvae, Gastropoda, mosquito larvae, Trichoptera larvae and wormlike macroinvertebrates. We determined the number of individuals by visually counting, whereas we measured wet weight by weighing each taxon after drying them with a tissue in order to remove the adherent water. The weighing was done in grams (g) using a mechanical balance in early spring (precision 0.5 g), which we replaced with an electric balance (precision 0.1 g) for the rest of the year.

Because we took multiple samples at the same location (always minimally 10 from which we calculated the means per sample location), we tested every month for spatial dependency. This was done with the program GS+. Based on these results, with the highest r^2 being only 0.137, we assumed that the observations were independent of each other, despite being taken from the same rivulet or wetland.

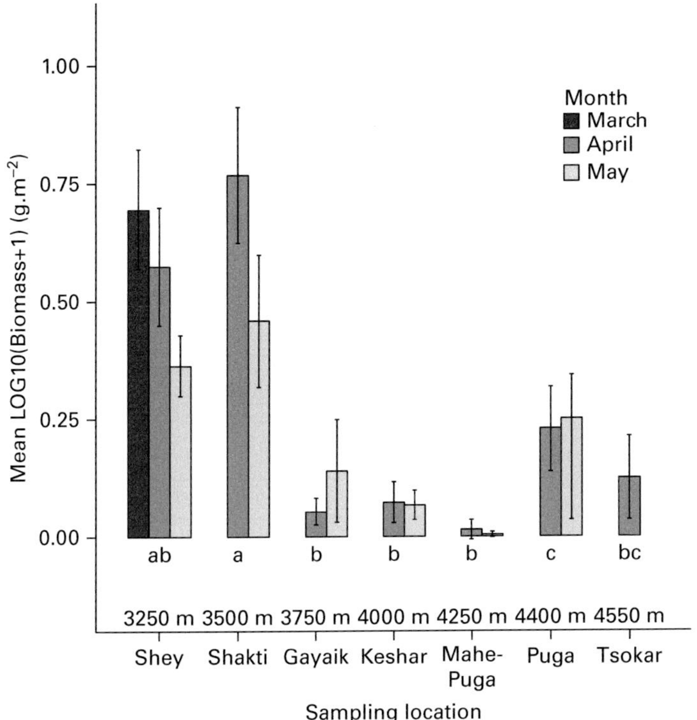

Figure 18.1. Mean ^{10}Log(biomass+1) against altitude for different months (hatched = March, grey = April, white = May). April R^2 Linear = 0.482; May R^2 Linear = 0.428. Error bars represent ±95% confidence interval. Biomass at an altitude of 3250 m to 3500 m above sea level is about 5 g.m^{-2}, between 3750 m and 4250 m about 0.4 g.m^{-2}, and at 4500 m it is about 1.2 g.m^{-2}. The letters a, b and c denote significant differences at $P < 0.05$ (Sidak Pairwise Comparison test). The increase at 4400 m is probably caused by water of a higher temperature than expected for this altitude due to hot springs. The weighing was done using an electric balance (minimum 0.1 g) in grams (g); 9 samples per location per period.

Where Is the Food for Waders and Ducks?

At the end of winter (March), we could only sample at Shey (3250 m a.s.l.): all higher lying waterbodies were still frozen for most of the day or permanently, except around hot springs (where Black-Necked Cranes *Grus nigricollis*, were already foraging). However, in spring (April and May), we could sample a number of areas (see Figure 18.2 for location names), except for some of the high-altitude lakes that were still frozen over (namely, Tsomoriri, Startsapuk Tso and Kiagar Tso). The biomass of macroinvertebrates was highest at low altitudes and reached on average about 5 g.m^{-2} in the wetlands of Shey and Sakti between 3250 m and 3500 m a.s.l., but at higher altitudes it quickly dropped to about 0.2 g.m^{-2} only (Figure 18.1). The amount of biomass was weakly correlated with the temperature of the bottom water in which the macroinvertebrates were sampled (April R^2 linear = 0.082; May R^2 linear = 0.119), but this explained a significant amount of variation (Table 18.2). There were undoubtedly macroinvertebrates

Table 18.2. Linear regression with aquatic macroinvertebrate biomass as dependent and month and temperature as main effects, and their interaction for the months of March, April and May. At the bottom of shallow water, the temperature is important in explaining biomass abundance.

Dependent Variable: LOG (biomass+1)					
Source	Type III Sum of Squares	df	Mean Square	F-value	Significance
Corrected Model	1.333[a]	3	0.444	6.572	0.001
Intercept	0.004	1	0.004	0.058	0.811
Month	0.067	1	0.067	0.990	0.321
TemperatureBottomCelsius	1.191	1	1.191	17.616	0.001
Month * TemperatureBottomCelsius	0.007	1	0.007	0.100	0.752
Error	12.035	178	0.068		
Total	23.772	182			
Corrected Total	13.368	181			

under the ice of the frozen over lakes at high altitude, but these would have been out of reach for migrating waders or ducks. From our results it appears that if long-distance migrants flying from south to north on prenuptial migration need refuelling stop-over sites containing high densities of aquatic macroinvertebrates, they would fail to find these above 3500 m. In other words, it is likely that they depend on lower-lying wetlands and ponds that are heavily impacted by humans. If that is the case, the identification and protection of low-altitude wetlands in the transition zone between the Ganges Plain and the Lesser Himalaya mountain range (see Searle, Chapter 9) would be of utmost importance.

In summer and autumn (July, August and September), the situation was very different. At low altitudes, the biomass of aquatic macroinvertebrates stayed about the same (from about 5 g.m^{-2} in spring to about 3 to 7 g.m^{-2} in summer (Figure 18.2), but the high-altitude lakes, especially the brackish and saline ones, bloomed into production with high biomass densities of about 15 g.m^{-2} (Figure 18.2). The high-altitude lake of Tsomoriri, which is nearly fresh, reached a lower biomass of approximately 6 g.m^{-2}.

To explore the main effects of habitat variables on macroinvertebrate assemblage structure, we conducted a Principal Component Analysis (PCA) (Figure 18.3) and a Redundancy Analysis (RDA) (Figure 18.4). Before undertaking these analyses, we carried out a Spearman Correlation to detect co-variation among variables. The variables 'granules presence', 'pH' and 'organic matter' were excluded due to high correlation with other variables. We chose a Principal Component Analysis over a Correspondence Analysis because macroinvertebrate variation was better explained by linear models than unimodal models. In the PCA, the first and second axes represent the most important environmental gradients along which macroinvertebrate families are distributed. The direction of each environmental vector represents the maximum rate of change for that particular environmental variable and its length indicates its relative importance to the ordination (Miserendino, 2001). Based on the results of the PCA, we carried out

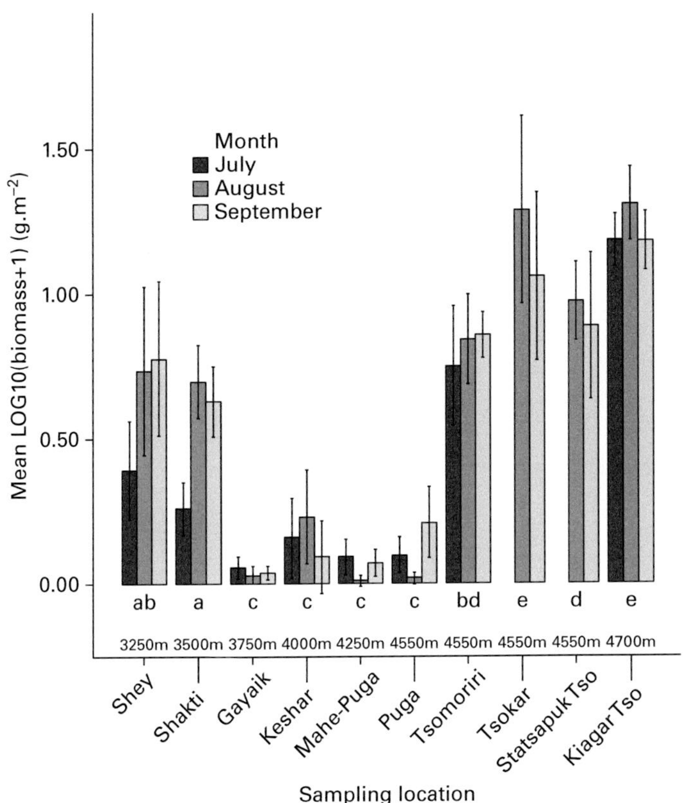

Figure 18.2. Bar graph showing the biomass of aquatic macroinvertebrates in relation to location (hatched = July, grey = August, white = September). The highest biomass occurs at locations in the brackish to saline lakes that are found at the highest altitudes (Tsokar and Kiagar Tso: about 15 g.m^{-2}), and the lowest at fresh water locations (Gayaik, Keshar, Mahe-Puga and Puga: about 0.2 g.m^{-2}), which are sample points of the River Indus or its tributaries. The biomass density at Shey and Sakti is about 1.7 g.m^{-2} in summer. Tsomoriri with about 5 g.m^{-2} is intermediate. Different letters (a to e) denote significantly different biomass (Sidak Pairwise Comparison test). The locations are ordered from low altitudes (Shey: 3250 m above sea level) to high ones (Kiagar Tso: 4700 m). Error bars represent 95% confidence intervals.

a redundancy analysis with forward selection using the most representative variables. We conducted data analyses using SPSS Statistics 20 and CANOCO 5 statistical packages.

The most important environmental variables to explain macroinvertebrate assemblages in the Trans-Himalayan region of Ladakh were found to be altitude and the amount of aquatic vegetation (Figure 18.4). High-altitude freshwater lakes contain little aquatic vegetation (in Tsomoriri, at 4550 m, there are some *Potamogeton* species) and the brackish or saline lakes do not contain aquatic vegetation at all, even though there is fringing marsh vegetation. The much lower-lying wetlands of Shey and Sakti have considerably more aquatic vegetation. The pattern of decreasing macroinvertebrate biomass with altitude, and increasing biomass with aquatic vegetation has also been found elsewhere (e.g. in the

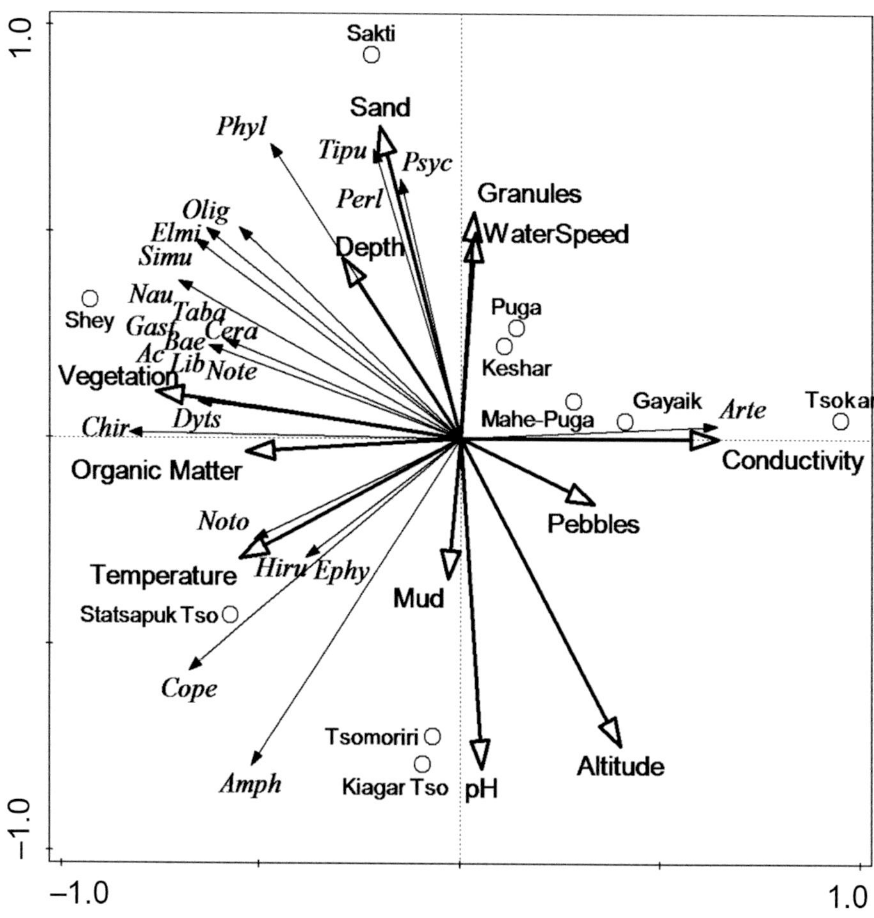

Figure 18.3. PCA ordination diagram for 23 macroinvertebrate families, 10 locations and 12 unconstrained variables in lakes and rivers of the Trans-Himalaya of Ladakh in summer and autumn. Locations (open dots), macroinvertebrate families (thin-lined vectors) and unconstrained environmental variables (thick-lined vectors) are shown in the diagram. Family codes: Acarina (Ac), Amphipoda (Amph), Artemiidae (Arte), Baetidae (Bae), Ceratopogonidae (Cera), Chironomidae (Chir), Copepoda (Cope), Dytiscidae (Dyti), Elmidae (Elmi), Ephydridae (Ephy), Gastropoda (Gast), Hirudinea (Hiru), Libelludidae (Lib), Naucoridae (Nau), Noteridae (Note), Notonectidae (Noto), Oligochaeta (Olig), Perlidae (Perl), Phylopotimidae (Phyl), Psychodidae (Psyc), Simulidae (Simu), Tabanidae (Taba) and Tipulidae (Tipu).

Argentine Andes, Scheibler *et al.*, 2014). Other important environmental variables that were positively associated with aquatic macroinvertebrate richness were mud and organic material (Figure 18.3). The rivers and rivulets that form the water system of the Indus generally had little mud and organic matter in summer and autumn, but many pebbles, cobbles and coarse sand (Gayaik, Keshar on the Indus, and Sakti, Mahe-Puga and Puga on tributaries of the Indus), and they were characterized by low levels of biomass of aquatic macroinvertebrates (Figure 18.3). These waterbodies in generally rocky environments were all relatively low in bird numbers during summer (pers. obs.).

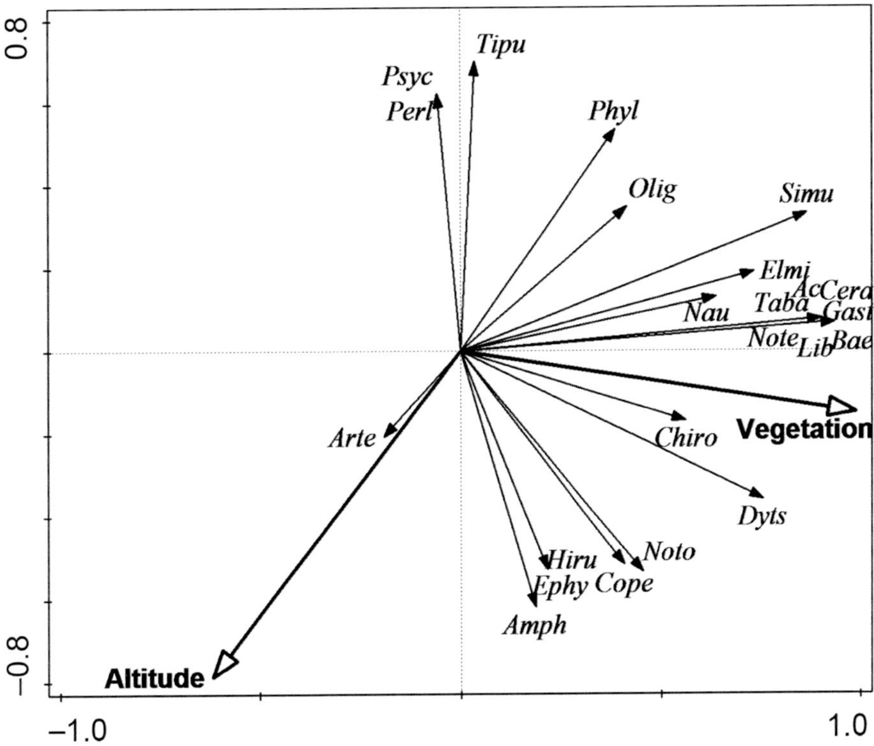

Figure 18.4. RDA ordination biplot for 23 macroinvertebrate families (thin-lined vectors) and two variables (thick-lined vectors): Altitude and Aquatic vegetation, in summer and autumn. Aquatic vegetation together with altitude determines macroinvertebrate family distribution in the Trans-Himalaya of Ladakh. For the meaning of the Family codes, see the legend of Figure 18.3.

The most important finding of our surveys for discovering the best refuelling sites for migratory waders and ducks (and also other species such as wagtails) seems clear. The effects of macroinvertebrate abundance and biomass on determining waterbird distribution are strongly supported by studies elsewhere (Mitchell & Grubaugh, 2005; Schummer *et al.*, 2008; Patra *et al.*, 2010). In early spring (March), the waters are still too cold, or are even still frozen, at altitudes higher than about 3500 m above sea level and hold little macroinvertebrate life. Later in spring, in April and May, waters are still cold and the level of macroinvertebrate biomass is still low, but at lower altitudes water temperatures are higher and more macroinvertebrate biomass was found. Biomass and abundance decreased with altitude, and at low altitude, with the progression of spring (Figure 18.1). At lower altitudes, larvae develop earlier due to higher water temperatures, but the nutrient flush from the previous autumn and winter probably also becomes depleted. This peak for aquatic macroinvertebrate abundance and biomass in spring and late winter is in line with other studies, and has been attributed to both temperature and nutrient availability (Duffy & LaBar, 1994; Romaniszyn *et al.*, 2007; Scheibler *et al.*, 2014). The macroinvertebrate populations at lower altitudes would therefore be at an earlier phase of development compared to higher altitudes. This is supported, for example, by observations of fewer large Trichoptera larvae caught in the dipnet and many empty Trichoptera pupae on rocks in the wetland at 3500 m (own observations). Higher macroinvertebrate population biomasses with increasing temperature have also been found in marine (Tumbiolo & Downing, 1994), as well as in freshwater ecosystems (Plante & Downing, 1989). Dissolved oxygen may set a limit on productivity of these macroinvertebrates as there is a very significant negative relationship between altitude and macroinvertebrate biomass (Figure 18.2). However, the high-altitude wetlands in the Himalayas (and probably also in Tibet) do not present many options for refuelling during a spring stop-over simply, we think, because ice melt is too late (ice shoals can still be observed at the end of July in Lake Tsomoriri; pers. obs.). In late summer and autumn, however, the high-altitude brackish and salt lakes form a very rich source of food for dabbling ducks and waders (Figure 18.2). The sources of food are not very diverse, and mainly comprise *Artemiidae, Amphipoda* and some *Chironomidae* (cf. Loayza-Muro *et al.*, 2013).

It is very striking, however, that birds do breed at these extreme altitudes of 4500 m a.s.l. and higher. We have observed the breeding of Bar-headed Geese here (Prins & Van Wieren, 2004), of which the young eat animal food in the littoral zone of the high-altitude lakes (pers. obs.). But we also saw Redshank *Tringa totanus*, Ruddy Shelduck *Tadorna ferruginea* and Black-necked Crane *Grus nigricollis* successfully raising their young at this altitude. The combination of high macroinvertebrate abundance and biomass with the presence of suitably nutritious taxa makes these lakes appropriate places to stop over. Reproduction demands high amounts of energy, which makes food availability a critical factor. Furthermore, food resources must not only cover reproduction demands, but also chick-rearing requirements (Namgail & Yom-Tov, 2009). Additionally, many of these breeding birds are migratory and must accumulate enough energy to cope with post-nuptial migration. Although energy is needed for such processes, high amounts of protein are also essential for survival, reproduction and migration of waterfowl, waders and passerines (Anderson & Smith, 1998). Kaminski and Prince (1981) showed that dabbling ducks do not consume water boatmen *Notonectidae*, despite their high energy and protein levels, because they are difficult to catch, and thus reduce the efficiency of foraging by dabbling ducks. Nevertheless, they foraged on *Chironomidae* and crustaceans better as they required less energy to catch and offered a high energy yield (Schummer *et al.*, 2008). The biomass densities in the saline high-altitude lakes are in the same order as those found, for example, in bivalves in the foraging areas of Red Knots *Calidris canutus* in the Wadden Sea in the Western European Flyway (Piersma *et al.*, 1993; Piersma *et al.*, 1994: about 2 to 6 g ash-free dry weight per m^2). We found about 15 $g.m^{-2}$ fresh weight in the high-altitude saline lakes, which converts to about 2 to 3 $g.m^{-2}$ ash-free dry weight (Ricciardi & Bourget, 1998). We thus conclude that the high food abundance, possibly combined with the right mix of macroinvertebrate taxa in the high-altitude lakes compared with the low altitudinal locations in summer, would accommodate energy and nutrition demands for wetland birds both for breeding and for refuelling during autumn migration. This may explain why the highest avian species richness is found around these brackish and saline high-altitude lakes in summer and autumn (cf. Prins *et al.*, Chapter 20).

The Functioning of Saline Water Bodies at High Altitudes

A perhaps unexpected outcome of our surveys was that the very high brackish and saline lakes did not conform to the expected relationship of decline in the density and biomass of aquatic macroinvertebrates with increasing altitude (see Figures 18.3 and 18.4). Altitude produces physiological stress because of the energy and nutrient limitations caused by the ever shorter growing season with increasing altitude (Füreder *et al.*, 2006). Moreover, higher altitudes result in ever lower partial oxygen pressures, thus further increasing the stress on aquatic organisms (Jacobsen, 2008). The strongest combination is high salinity and low temperature, because that provides the lowest oxygen supply index (Verberk *et al.*, 2011). The relationships between macroinvertebrate community structure and environmental variables have been the subject of numerous investigations (e.g. Walker & Mathewes, 1989; Hoback & Stanley, 2001; Miserendino, 2001). The effects of abiotic variables become greater than biotic interactions when altitude increases (Füreder *et al.*, 2006), and the general outcome appears to be that with increasing altitude, taxon richness decreases substantially, just as we found at the end of winter and spring in Ladakh. The very high saline and brackish lakes at around 4500 m a.s.l. buck the trend, however: the fact that these lakes are rich in salts apparently offsets in some way the lack of oxygen, even though oxygen solubility is further depressed by salinity. This was also found for bacterial communities in hyper-saline lakes in Tibet (Wang *et al.*, 2011). Phytoplankton biomass, however, decreases with increasing altitude and with salinity on the Tibetan Plateau (Wen *et al.*, 2005).

High in the Himalayas and on the Tibetan Plateau a number of basins without outflow to the sea were formed accidentally through geological processes. These random outcomes at the landscape level resulted in the creation of habitats vital for the development and maintenance of the Central Asian Flyway. Without the high concentrations of salts in these lakes, it seems unlikely that cranes, waders and ducks and associated raptors could make their migrations in such numbers over the highest mountains on earth, and over the deserts and glaciers of Central Asia.

References

Anderson, J.T. & Smith, L.M. (1998). Protein and energy production in playas: implications for migratory bird management. *Wetlands*, **18**, 437–446.

Aryal, A., Sathyakumar, S. & Kreigenhofer, B. (2010). Opportunistic animal's diet depends on prey availability: spring dietary composition of the red fox (Vulpes vulpes) in the Dhorpatan hunting reserve, Nepal. *Journal of Ecology and the Natural Environment*, **2**, 59–63.

Bancroft, G.T., Gawlik, D.E. & Rutchey, K. (2002). Distribution of wading birds relative to vegetation and water depths in the northern Everglades of Florida, USA. *Waterbirds*, **25**, 265–277.

Bournaud, M., Cellot, B., Richoux, P. & Berrahou, A. (1996). Macroinvertebrate community structure and environmental characteristics along a large river: congruity of patterns for identification to species or family. *Journal of the North American Benthological Society*, **15**, 232–253.

Chandan P., Chatterjee, A. & Gautam, P. (2008). Management of Himalayan high altitude wetlands. A case study of Tso Moriri and Tso Kar Wetlands in Ladakh, India. The 12th World Lake Conference: 1446–1452.

Collier, K.J. (1995). Environmental factors affecting the taxonomic composition of aquatic macroinvertebrate communities in lowland waterways of Northland, New Zealand. *New Zealand Journal of Marine and Freshwater Research*, **29**, 453–465.

Colwell, M.A. & Taft, O.W. (2000). Waterbird communities in managed wetlands of varying water depth. *Waterbirds*, **23**, 45–55.

Davis, C.A. & Smith, L.M. (1998). Ecology and management of migrant shorebirds in the Playa Lakes Region of Texas. *Wildlife Monographs*, **140**, 3–45.

Dillon, M.E., Frazier, M.R. & Dudley, R. (2006). Into thin air: physiology and evolution of alpine insects. *Integrative and Comparative Biology*, **46**, 49–61.

Duffy, W.G. & LaBar, D.J. (1994). Aquatic invertebrate production in southeastern USA wetlands during winter and spring. *Wetlands*, **14**, 88–97.

Füreder, L., Ettinger, R., Boggero, A., Thaler, B. & Thies, H. (2006). Macroinvertebrate diversity in Alpine lakes: effects of altitude and catchment properties. *Hydrobiologia*, **562**, 123–144.

Geroudet, P. (1954). Des oiseaux migrateurs trouvés sur la glacier de Kumbu dans l'Himalaya. *Nos Oiseaux*, **22**, 254.

Gill, R.E., Piersma, T., Hufford, G., Servranckx, R. & Riegen, A. (2005). Crossing the ultimate ecological barrier: evidence for an 11000-km-long non-stop flight from Alaska to New Zealand and eastern Australia by Bar-tailed Godwits. *The Condor*, **107**, 1–20.

Green, M., Alerstam, T., Clausen, P., Drent, R. & Ebbinge, B.S. (2002). Dark-bellied Brent Geese Branta bernicla bernicla as recorded by satellite telemetry, do not minimize flight distance during spring migration. *Ibis*, **144**, 106–121.

Guija, B., Chatterjee, A., Gautam, P. & Chandan, P. (2003). Wetlands and lakes at the top of the world. *Mountain Research and Development*, **23**, 219–221.

Havas, M. & Hutchinson, T.C. (1982). Aquatic invertebrates from the Smoking Hills, N.W.T.: effect of pH and metals on mortality. *Canadian Journal of Fisheries and Aquatic Science*, **39**, 890–903.

Hoback, W.W. & Stanley, D.W. (2001). Insects in hypoxia. *Journal of Insect Physiology*, **47**, 533–542.

Jacobsen, D. (2008). Low oxygen pressure as a driving factor for the altitudinal decline in taxon richness of stream macro invertebrates. *Oecologia*, **154**, 795–807.

Jacobsen, D., Schultz, R. & Encalada, A. (1997). Structure and diversity of stream invertebrate assemblages: the influence of temperature with altitude and latitude. *Freshwater Biology*, **38**, 247–261.

Kaminski, R.M. & Prince, H.H. (1981). Dabbling duck activity and foraging responses to aquatic macroinvertebrates. *The Auk*, **98**, 115–126.

Kefford, B.J., Papas, P.J. & Nugegoda, D. (2003). Relative salinity tolerance of macroinvertebrates from the Barwon River, Victoria, Australia. *Marine and Freshwater Research*, **54**, 755–765.

Loayza-Muro, R.A., Marticorena-Ruiz, J.K., Palomino, E.J., *et al.* (2013). Persistence of Chironomids in metal polluted Andean high altitude streams: does melanin play a role? *Environmental Science & Technology*, **47**, 601–607.

Miserendino, M.L. (2001). Macroinvertebrate assemblages in Andean Patagonian rivers and streams: environmental relationships. *Hydrobiologia*, **444**, 147–158.

Mitchell, D.W. & Grubaugh, J.W. (2005). Impacts of shorebirds on macroinvertebrates in the Lower Mississippi Alluvial Valley. *The American Midland Naturalist*, **15**, 188–200.

Moreau, R. (1972). *The Palearctic-African Bird Migration Systems*. London: Academic Press.

Namgail, T., Rawat, G.S., Mishra, C., Van Wieren, S.E. & Prins, H.H.T. (2012). Biomass and diversity of dry alpine plant communities along altitudinal gradients in the Himalayas. *Journal of Plant Research*, **125**, 93–101.

Namgail, T. & Yom-Tov, Y. (2009). Elevational range and timing of breeding in the birds of Ladakh: the effects of body mass, status and diet. *Journal of Ornithology*, **150**, 505–510.

Patra, A., Santra, K.B. & Manna, C.K. (2010). Relationship among the abundance of waterbird species diversity, macrophytes, macroinvertebrates and physico-chemical characteristics in Santragachi Jheel, Howrah, W.B., India. *Acta Zoologica Bulgarica*, **62**, 277–300.

Paul, A.L. & Ferl, R.J. (2006). The biology of low atmospheric pressure – implications for exploration mission design and advanced life support. *Gravitational and Space Biology*, **19**, 3–17.

Piersma, T., Bruinzeel, L., Drent, R., *et al.* (1996). Variability in basal metabolic rate of a long-distance migrant shorebird (Red knot, *Calidris canutus*) reflects shifts in organ sizes. *Physiological Zoology*, **69**, 191–217.

Piersma, T., Hoekstra, R., Dekkinga, A., *et al.* (1993). Scale and intensity of intertidal habitat use by knots Calidris canutus in the western Waddensea in relation to food, friends and foes. *Netherlands Journal of Sea Research*, **31**, 331–357.

Piersma, T., Verkuil, Y. & Tulp, I. (1994). Resources for long-distance migration of Knots Calidris canutus islandica and *C. c. canutus*: how broad is the temporal exploitation window of benthic prey in the western and eastern Wadden Sea? *Oikos*, **71**, 393–407.

Philip, G. & Mathew, J. (2005). Climato-tectonic impression on trans Himalayan lakes: a case study of Kyun Tso basin of the Indus Suture Zone in NW Himalaya using remote sensing techniques. *Current Science*, **89**, 1941–1947.

Plante, C. & Downing, J.A. (1989). Production of freshwater invertebrate populations in lakes. *Canadian Journal of Fisheries and Aquatic Sciences*, **46**, 1489–1498.

Prins, H.H.T. & Van Wieren, S.E. (2004). Number, population structure and habitat use of Bar-headed Geese Anser indicus in Ladakh (India) during the brood-rearing period. *Acta Zoologica Sinica*, **50**, 738–744.

Rawat, G.S. & Adhikari, B.S. (2005). Floristics and distribution of plant communities across moisture and topographic gradients in Tso Kar basin, Changthang plateau, eastern Ladakh. *Arctic, Antarctic, and Alpine research*, **37**, 539–544.

Ricciardi, A. & Bourget, E. (1998). Weight-to-weight conversion factors for marine-benthic macroinvertebrates. *Marine Ecology Progress Series*, **163**, 245–251.

Romaniszyn, E.D., Hutchens, J.J., & Bruce Wallace, J. (2007). Aquatic and terrestrial invertebrate drift in southern Appalachian Mountain streams: implications for trout food resources. *Freshwater Biology*, **52**, 1–11.

Rosenberg, D.M., Davies, I.J., Cobb, D.G., & Wiens, A.P. (1997). *Protocols for Measuring Biodiversity: Benthic Macro-invertebrates in Fresh Waters*. University Crescent, Manitoba.

Scheibler, E.E., Claps, M.C. & Roig-Juñent, S.A. (2014). Temporal and altitudinal variations in benthic macroinvertebrate assemblages in an Andean river basin of Argentina. *Journal of Limnology*, **73**, 92–108.

Schummer M.L., Petrie, S.A. & Bailey, R.C. (2008). Interaction between macroinvertebrate abundance and habitat use by diving ducks during winter on Northeastern Lake Ontario. *Journal of Great Lakes Research*, **34**, 54–71.

Sergio, F., Tanferna, A., De Stephanis R., *et al.* (2014). Individual improvements and selective mortality shape lifelong migratory performance. *Nature*, **515**, 410–413.

Tumbiolo, M.L., & Downing, J.A. (1994). An empirical model for the prediction of secondary production in marine benthic invertebrate populations. *Marine Ecology-Progress Series*, **114**, 165–174.

Verberk, W.C.E.P., Bilton, D.T., Calosi, P. & Spicer, J.I. (2011). Oxygen supply in aquatic ectotherms: partial pressure and solubility together explain biodiversity and size patterns. *Ecology*, **92**, 1565–1572.

Walker, I.R. & Mathewes, R.W. (1989).Chironomidae (Diptera) remains in surficial lake sediments from the Canadian Cordillera: analysis of the fauna across an altitudinal gradient. *Journal of Paleolimnology*, **2**, 61–80.

Wang, J., Yang, D., Zhang, Y., *et al.* (2011). Do patterns of bacterial diversity along salinity gradients differ from those observed for macroorganisms? *PLoS ONE* **6**(11): e27597.

Weber, C. (1973). *Biological field and laboratory methods for measuring the quality of surface waters and effluents*. EPA-670/4-73-001.

Wen, Z., Mian-Ping, Z., Xian-Zhong, X., Xi-Fang, L., Gan-Lin, G. & Zhi-Hui, H. (2005). Biological and ecological features of saline lakes in northern Tibet, China. *Hydrobiologia*, **541**, 189–203.

Wentworth, C.K. (1922). A scale of grade and class terms for clastic sediments. *The Journal of Geology*, **30**, 377–392.

Wojciechowski, M.S., Yosef, R. & Pinshow, B. (2014). Body composition of north and southbound migratory blackcaps is influenced by the lay-of-the-land ahead. *Journal of Avian Biology*, **45**, 264–272.

Yali S., Xin, Q., de Boer, W.F., Gong, P. Ydenberg, R.C. & Prins, H.H.T. (2015). Decision rules based on plant phenology in timing the spring migration of Arctic breeding geese. Scientific Reports: SREP-14-06997B.

Yamaguchi, N., Hiraoka, E., Fujita, M., *et al.* (2008). Spring migration routes of mallards (Anas platyrhynchos) that winter in Japan, determined from satellite telemetry. *Zoological Science*, **25**, 875–881.

19 The Himalayas as an Ecological Barrier for Avian Migrants: High and Dry, but also Dangerous?

Ron C. Ydenberg

Rationale

On 14 May 2011, the editors of this book, Herbert Prins and Tsewang Namgail, along with their colleague Sip van Wieren of Wageningen University, were trekking from Tsokar to Nuruchen at an altitude of 4650 m, when they observed two Golden Eagles *Aquila chrysaetos* launch an attack on some migrating Bar-headed Geese *Anser indicus*. It was an exciting spectacle – all the more so because it was the only such attack they witnessed in about 2500 km of trekking on six Himalayan expeditions. Naturally, they wondered if such events are rare. Are predators scarce in these high mountains, so that safety is a benefit weighing against the otherwise formidable obstacles faced by Trans-Himalayan migrants? Or, is the danger higher than they would face on alternative, more roundabout routes between breeding and non-breeding areas? These questions led to this chapter.

I warn readers at the outset that there are as yet few answers, even for such basic questions. While studies make clear that danger from predators is an important ecological consideration for some migrant species, it is unclear as to how general this is, and Himalayan migration systems are more poorly known than others. Most research on avian migrants focusses on questions such as scheduling, energetics, navigation and physiology; and the potential role of predation danger has, until recently, received relatively little attention. Hence, my purpose here is to outline for ecologists interested in the migratory birds in the Himalayan region how questions about predation risk might be framed and approached.

Is Predation Risk Important in Migration?

Alerstam and Lindström (1990) identified safety as one of the prime selective forces shaping the evolution of avian migration strategies and behaviour. There certainly is no doubt that raptors prey on avian migrants, and several studies estimate significant mortality rates for migrants due to predators during their southward passage (Kerlinger, 1989; Lindström, 1989; Moore *et al.*, 1990). But safety has received much less interest from ecologists compared to that received by time and energy, and so we do not yet know its importance relative to the other selective factors.

The potential importance of predation mortality is evident in the analysis Sillet and Holmes (2002) carried out for a neotropical migrant, the Black-throated Blue Warbler *Dendroica caerulescens*. They studied these warblers on their breeding grounds (May–August) in New Hampshire in the United States, and in their winter quarters (October–March) in Jamaica. Using annual and seasonal survival estimates from both locales, they calculated warbler survival for the in-between migratory periods from the difference. More than 85% of the apparent annual mortality occurs during the two migratory periods (a total of about two months), making the mortality rate during migration at least 15 times higher than that during the rest of the year. But the agents of all this mortality are unknown. Lindström (1990) estimated that as many as 10% of migrant bramblings *Fringilla montifringilla* might be killed by Eurasian Sparrowhawks *Accipiter nisus*, and so implicates predators as an important factor. Other studies point to factors such as weather (Drake *et al.*, 2014; Xu *et al.*, 2015). Newton (2008) identified many possible sources of mortality, including predation, in his review of what is known of the population limitation of migrants in the 976-page book *The Migration Ecology of Birds*.

The ways in which danger and 'anti-danger' (that is, the anti-predator behaviour of animals) interact is key to the potential ecological significance of predators for migration. The way most ecologists think about this is illustrated by a paper titled *Predation pressure by avian predators suggests summer limitation of small-mammal populations in the Canadian Arctic*, by Therrien *et al.* (2014), in which it was found that the cumulative predation mortality exceeded the lemming population growth rate. However, the authors did not consider that the danger posed by the many predators probably caused lemmings to become cautious and adjust their reproductive investment downwards – lowering the population growth rate well below that of which they are capable. Hence, it remains unclear to what extent summer limitation is due to predation mortality (termed a 'lethal' or 'consumptive' effect), and to what extent the behavioural response of lemmings to predation danger (a 'non-lethal' or 'non-consumptive' effect) is responsible. A large number of studies make clear that the latter may have as large an impact on life history and population parameters as the former (Preisser *et al.*, 2005). I focus this chapter on the concept of behavioural responses to predation danger, and how this might influence migration.

A Primer of Predation Danger

We must first define and understand 'predation risk'. The review titled *Behavioural decisions made under the risk of predation: a review and prospectus* by Lima and Dill (1990) has been cited thousands of times, suggesting that ecologists have considered this carefully. Unfortunately, the term has been used in such different ways (Lank & Ydenberg, 2003) that confusion remains. To illustrate, imagine a situation in which ground-feeding birds feed close to cover, into which they are able to escape when a predator attacks. An ecologist studying this situation might never observe a successful predatory attack, even after many days in the field, and conclude that the risk of predation is low. But a second ecologist could argue that predators would easily catch the ground-feeding birds if they were not vigilant, or if they fed further away from cover. She concludes that the risk of predation is high. What is the difference?

An analogy with pedestrians (as prey) and cars (as predators), made by both Lima and Dill (1990) and by Lank and Ydenberg (2003), will help to clarify. Just as exemplified by the ground-feeding bird example, one could observe a busy city street for a long time and never observe a 'predation event' (i.e. pedestrian struck by car). The first ecologist would conclude that the situation is safe because the level of mortality is low. But every city dweller knows that traffic is dangerous. The basis for our second ecologist's claim is that the mortality rate would be much higher if pedestrians were careless and jaywalked without looking. It is therefore useful to distinguish between the actual mortality (our first ecologist's concern, referred to here as 'risk') and the mortality that would occur were pedestrians heedless (our second ecologist's concern, referred to here as 'danger', or when more convenient, as 'safety'). The difference between 'risk' and 'danger' depends on the level of caution that pedestrians exercise.

An important corollary is that situations with low mortality are not necessarily safe. We all know from personal experience that city traffic is dangerous and that pedestrians must be cautious, but this is usually far less clear for animals in natural settings (e.g. Prins & Iason, 1989). For example, the mortality attributable to Eleonora's Falcon *Falco eleonorae* among the billions of southbound avian migrants annually crossing the Mediterranean is < 0.05%, or less than 1 in 2000. Perhaps the crossing is really not dangerous, or perhaps the danger these falcons pose is significant, and migrants have evolved traits to reduce the risk (Ydenberg *et al.*, 2007). Analogously, the reason that the editors of this volume observed only a single eagle attack in six Himalayan expeditions could be that there is little danger; or it could be the Himalayas are dangerous, and that migrating geese – like pedestrians in cities – exercise enough caution to keep such encounters rare.

Measuring Danger

A powerful experiment to distinguish between these alternative hypotheses would be to somehow 'turn off' the anti-predator behaviour in a sample of animals and measure whether the subsequent mortality increases. The classic examples in ecology are experiments lowering the degree of crypsis of moths resting on tree trunks during daytime. Kang *et al.* (2012) provide a recent example. Such experiments clearly show that camouflage provides a survival advantage. Individual moths that are less well camouflaged are more likely to be found and eaten by visually searching birds, proving that the 'danger' (i.e. the mortality that would be observed if moths were not camouflaged) is much higher than the 'risk' (i.e. the realized mortality). Analogous experiments with behaviour are more difficult, but the principle is exactly the same. Would the realized mortality change accordingly if we were able to alter the degree of alleged anti-predator behaviour?

Such experiments are not possible on a large scale, and certainly not on the scale of an entire migratory route. (And for avian migrants at least, manipulations that render individuals more susceptible to predators may also be unethical.) Fortunately, animals reveal by their behaviour whether they perceive that a situation is safe or dangerous. The analogy with pedestrians and traffic is again helpful. On a side street in a quiet neighbourhood where there is little traffic, pedestrians routinely cross in the middle of the block, rather than at the corner. But on a busy downtown street, almost everyone crosses only at a corner, where crosswalks and stoplights help to make the crossing safe. A systematic observer would soon note that the jaywalking tendency diminishes with factors such as the volume and speed of traffic, darkness, rainfall and so forth – things that make jaywalking more dangerous. In fact, even if she were blind to the traffic, by measuring street-crossing behaviour our observer would be able to evaluate how dangerous is the situation.

Ecologists studying predation risk in nature are usually in the situation of an observer unable to see the traffic, because predators and their behaviour are generally difficult to observe and measure. Even when they can be seen, the danger they pose to prey varies with many factors. Habitat features such as ambush cover, the presence of refuges or moonlight may make places more or less dangerous, even if the number of predators is unaltered. Fortunately, a large volume of literature demonstrates that prey animals are very attentive to danger, and adjust the level of caution appropriately. Their assessment of the danger is revealed when the observer measures behaviour analogous to the jaywalking tendencies of pedestrians.

A well-developed theory of anti-predator behaviour exists (Brown & Kotler, 2007; Kotler & Brown, 2007). The basic underlying idea is that of a trade-off, in which the time and effort for behaviour to gain extra safety from predators comes at the expense of reducing something else, such as access to food or mates. For example, an animal may increase the time devoted to vigilance or spend more time hiding – but this reduces the time available for other activities. Ydenberg and Dill (1986) put forward one of the first and most basic models of anti-predator behaviour. In it, a predator approaches a foraging prey animal. The danger climbs the closer the predator comes. The prey can escape by fleeing, but the sooner it flees, the more food it gives up. At what point should it flee?

It was long thought that the prey should flee as soon as it notices ('perceives') the threat, but many studies show that this is not what happens. Instead, the prey behaves as though it is weighing the benefits of continuing to feed against the rising danger. This framework lends itself to a convenient experimental paradigm in which observers gradually raise the risk by walking towards animals and measuring the 'flight initiation distance' – the distance at which a prey initiates escape from an approaching predator. Anders Pape Møller and his colleagues (Møller & Ibáñez-Álamo, 2012; Díaz *et al.*, 2013; Møller, Grim & Ibáñez-Álamo, 2013) have measured flight initiation distances of lizards and birds to compare the relative danger of urban and rural settings, and the danger across latitudes on a continental scale. Cooper and Blumstein (2015) review a large body of work that uses this paradigm. Frederick and Cooper (2007) discuss mathematical details of this cost-benefit comparison, but this technical issue need not concern us here.

Brown and Kotler (2007) address questions of how apprehensive foragers should be, which they term the 'ecology of fear'. Their basic model also lends itself readily to an experimental technique, called 'giving-up density'. To measure a giving-up density, the experimenter sets out feeders (e.g. trays or buckets) in which food items (e.g. seeds) are mixed into an inedible medium (e.g. sand or sawdust) through which the forager must search. Initially, the forager finds food items quickly, but the rate slows as the feeder is depleted. At a certain point, the cost of continued feeding exceeds the benefit and the feeder is abandoned. The experimenter measures the food density remaining: this is the giving-up density. This technique has been used in many settings. For example, small desert rodents are more fearful (higher giving-up density) in the open than under bushes when the main predators are owls, but more fearful under bushes when the predators are snakes (Brown & Kotler, 2007). The theory for both the flight initiation distance and giving-up density frameworks states that the feeding site is abandoned ('flight initiation' or 'giving up') when the benefits (reducing or escaping exposure to danger) exceed the costs (the opportunity lost when feeding ceases). Either benefit or cost could vary between species, sex or age classes, individuals or situations, and hence the degree of anti-predator behaviour would be affected.

When and How Are Avian Migrants Exposed to Predation Danger?

The energetic requirements of long migratory flights have been extensively investigated. Alerstam *et al.* (2003) give an overview, and Pennycuick (1989) gives a practical guide to calculating flight energetics. Of importance here is that a typical avian migrant spends most of its migration time at stopover sites, foraging to gain the fuel it needs to power migratory flight. Hedenström and Alerstam (1997) estimate the ratio of time at stopover to time in flight for a typical small bird at about 7:1. The ratio is even higher for large birds. In contrast, a migrant spends most of its energy on flights between stopover sites.

The body mass of a bird rises and falls in the course of a migration. Drent *et al.* (2003; see their figure 1) give a graphic representation of the body mass trajectory for a typical migrant, showing periods of fuel deposition at stopover sites alternating with (much shorter) periods of rapid fuel expenditure on flights between stopovers. Fat is used as the primary fuel because it has a higher energy density than other metabolic fuels, and weight economy is important for long-distance flyers. The flight musculature is enlarged prior to departure, and organs such as the gut that are not used during the flight may atrophy (Piersma, 1998; Piersma & Gill, 1998). The extent of these changes depends on the length of the flights, but can be very large. Many species routinely launch crossings of large ecological barriers (oceans, deserts) at double their non-migratory body mass, or even more (e.g. Maillet & Weber, 2006).

These basic facts about avian migration provide the context for understanding how long-distance migration creates exposure to predation danger. First, migrants must forage to acquire the fuel they need. Building fuel reserves quickly requires intense foraging (i.e. low vigilance), in places with high food densities, which (it turns out) are generally more dangerous (see later in this chapter). Beauchamp (2014) describes many aspects of the vigilance and foraging behaviour of Semipalmated Sandpipers *Calidris pusilla* in the Bay of Fundy as they fuel up for their Atlantic crossing to South America. Building fuel reserves less quickly is not as dangerous per unit time, but of course requires more time.

The fat reserve that powers migration may impair escape ability. The idea that a larger fat load increases predation danger has been invoked in theoretical treatments of body mass regulation during migration (Alerstam & Lindström, 1990; Houston, 1998; cf. Lima, 1986). Some tests of this hypothesis have found no effect of body mass on take-off speed (Kullberg, 1998; Veasey *et al.*, 1998; Van der Veen & Lindström, 2000), and Van der Veen (1999) points out that tests finding an effect did so with artificially weighted birds (Witter *et al.*, 1994) or with 'non-alarmed' take-offs (Metcalfe & Ure, 1995; cf. Veasey *et al.*, 1998). She (Van der Veen, 1999) suggested that the effect may be more important when heavy fat reserves are carried (as in migrants) than when relatively small daily changes in mass are undergone by birds. In support of this claim, Kullberg *et al.* (1996) and Lind *et al.* (1999) were able to show an effect of fat reserves on escape speed in two migratory passerine species. Western Sandpipers *Calidris mauri* on stopover in British Columbia, Canada, carry 3–10 g of fat, or even more. Every gram of fat measurably slows take-off performance (Burns & Ydenberg, 2002).

Migrants may concentrate at certain places (e.g. wetlands), where their large numbers and relative vulnerability attract predators. In some areas, there might be just a few wetlands that offer possible stopover sites. This factor is likely to be of great importance in the Himalayas, where there are only a few scattered lakes and wetland complexes at which migrants can gather.

Many raptors are themselves migratory, and their journeys often coincide with those of their prey. Falcons, notably Peregrines *Falco peregrinus* and Merlins *Falco columbarius*, and in the Himalayas, the Amur Falcon *Falco amurensis*, Western Osprey *Pandion haliaetus* and Steppe Eagle *Aquila nipalensis* are especially migratory. Unlike larger raptors which migrate largely using energy-efficient soaring and gliding, falcons mainly use powered flight, and therefore must capture prey regularly while on migration (e.g. Hunt *et al.*, 1975; Wiedner *et al.*, 1992). This creates danger for their prey. Raptor migrations are generally broad-front phenomena (Bednarz & Kerlinger, 1989; Juhant & Bildstein, Chapter 6; Dixon *et al.*, Chapter 8), with concentrations occurring along some mountain ranges and coastlines (see Heintzelman, 1975). In mountains, migrants often use passes, which may be of extra importance in the very high Himalayas, where the thin air creates an advantage to keeping flight as low as possible, even if this lengthens the journey (Bishop *et al.*, 2015; Groen & Prins, Chapter 17).

A fifth factor is also significant. It was noted earlier that on a busy street, pedestrians take the time to walk to the corner where the street can be crossed safely. But when pressed for time – late for a meeting with his boss, or lunch with his fiancée, for example – our pedestrian might jaywalk to save a few minutes. Timely arrival at destinations is also important for migrants, and an individual that is behind schedule has to take some chances if it is to catch up. In fact, it may continually take small risks during the migration to offset the chance of falling into a situation in which a large risk has to be taken. The dynamic model of Clark and Butler (1999), for example, demonstrates that northbound Western Sandpipers maintain a higher fuel load than needed to power flights. This enables them to take advantage of conditions favourable for further migration sooner after arrival at a stopover than would otherwise be the case. Because favourable conditions occur irregularly, missing a good migration window due to low fuel reserves might result in a long wait for the next window.

Avian migrants can be depredated while in migratory flight (e.g. Walter, 1979; Ibanez *et al.*, 2001), but the considerations listed previously – the great amount of migratory time spent at stopovers, the need for intensive foraging, the heightened vulnerability resulting from large fuel reserves, the co-occurrence of raptors in migratory time and space and the requirement for risk-taking to keep to time – combine to make stopovers especially dangerous for migrants.

Predatory Behaviour

Another important aspect of predation danger lies in when and how predators actually catch migrants. Many raptor species occur in the Himalayas, including accipiters and falcons that catch small birds, buzzards and harriers that are rodent specialists but can also catch birds, and many eagle species that pose a threat to larger birds, such as geese. A useful concept is the 'predator landscape', a dynamic map representing the danger created by the presence of non-migratory predators, in combination with the large spatial and temporal variations generated by the movements of migratory predators. Ydenberg *et al.* (2007) describe how the migratory movements of Peregrines and other predators of small birds generate continental-scale predator landscapes with features unique to each of the main flyways in the Americas and in Europe. These features help to explain broad patterns in the migrations and life histories of their prey that differ between these flyways.

Beyond their presence, the manner whereby raptors catch prey is also important, for this can influence, on a fine scale, which places are more dangerous and which defensive tactics are effective. Raptors catch birds using several techniques, including ambush or short pursuit from cover (falcons, accipiters), on the wing in aerial hunts (some falcons), by low quartering and pouncing (harriers, owls), by soaring and hovering, descending on prey from a great height (eagles, buzzards, falcons), or by 'perch-hunting', in which they sit, often in an exposed position, waiting and scanning for opportunities. Co-operative hunting has been described for several species (Ellis *et al.*, 1993), and may involve prolonged pursuit.

Surprise is often an important element for success (Cresswell, 1996; Dekker & Ydenberg, 2004; Beauchamp, 2014), and raptors can be very stealthy. They range widely, appearing unpredictably at any particular location. They conceal their approaches to prey behind cover, behind visual obstructions on the horizon or in the glare of the sun. As emphasized throughout this chapter, prey have behavioural counter-tactics that include wide-ranging themselves, avoiding cover or lingering near refuges, vigilance, flocking and aggregation, and, upon attack, escape into cover, wildly erratic flight or rapid climbing flight (Hedenström & Rosén, 2001). Each of these has its own costs and benefits, and are more or less effective against different types of predators.

The presence of a variety of predators affects the defensive measures prey can take against each, a phenomenon called 'risk allocation' (Lima & Bednekoff, 1999). When there are several predatory species, prey have a broader range of predatory tactics to guard against, and so have to allocate time and effort. This may lower the effectiveness of each type of defence, or, if one type of predator or period is especially dangerous, prey may be forced to lower its defences against less dangerous predators.

Safety-Selected Features of Avian Migration

Natural selection for safety has acted on many aspects of migration behaviour, altering basic attributes such as its timing and routing, as well as giving individuals the flexibility to adjust behaviourally. In this section, I outline how some important aspects of avian migration are influenced by safety considerations.

Timing: In keeping with its focus on energetics, most literature has emphasized the relation between scheduling and the availability of food at stopover sites along the migration route. For example, the migration of geese from wintering to breeding areas is hypothesized to follow the northward progression of spring growth (the 'green wave' Si *et al.*, 2015; see also Lok *et al.*, 2012). Food is of course important for migratory birds, but safety considerations are also important, and migrants avoid times with high danger. For example, Lank *et al.* (2003; see their figure 1) assert that the strong presence of migratory Peregrines along the migration route in August and September has selected for shifts in the migratory timing of Western Sandpipers and the Pacific Dunlin *Calidris alpina pacifica*. These ecologically similar species breed in identical habitats in Alaska, and migrate along the Pacific coast to their non-breeding ranges. Their timing is very different, however: Western Sandpipers depart Alaska in late June or early July, while the days are still long and food is abundant enough that Pacific Dunlins are able to continue breeding for up to two more weeks. The early departure allows this long-distance migrant to travel the 6000–11,000 km to its tropical wintering areas and moult prior to the arrival there of migratory Peregrines. As with much else in migration ecology, there is a trade-off, and Western Sandpipers give up some potential breeding time in order to make an early departure (Jamieson *et al.*, 2014). Pacific Dunlins, in contrast, reside in Alaska until long after breeding has finished. They move to coastal areas to moult, and do not migrate south until October. An earlier departure would place them on their temperate wintering areas ahead of Peregrine passage. Lehikoinen (2011) records analogous shifts in European short- and long-distance migrants in response to climate-change-driven changes in the migratory timing of the Eurasian Sparrowhawk .

The timing changes made by prey may be due either to natural selection or to behavioural flexibility (Sutherland, 1998), but in the case of recent changes in

Barnacle Goose *Branta leucopsis* migration, flexibility is clearly responsible. In this species, the growth in the population of White-tailed Eagles *Haliaeetus abicilla* in the Baltic Sea has led to a major alteration of the northward schedule, made specifically to avoid this dangerous stopover location (Eichhorn *et al.*, 2009; Jonker *et al.*, 2010).

Speed of migration: Long migrations consist of successive flights between stopover sites at which fuel reserves are replenished. As described earlier in this chapter, most of the journey is spent at the stopover sites, and so the speed of migration (in kmd^{-1}, measured over the entire migration) depends strongly on the stopover duration. As migrants load fuel, the potential flight range increases, but at a decelerating rate due to the drag created by the fuel load itself. Migrants with very large fuel loads are able to make very long flights, but the overall speed of migration is low because the necessary fuelling time is disproportionately long. Migrants with small fuel loads can make only short flights, and the speed of migration is slow due to the amount of 'upfront' time required to settle at each site and begin loading fuel. The fastest migration speeds are attained at intermediate fuel loads (Alerstam & Lindström, 1990; Alerstam & Hedenström, 1998, Houston, 1998).

Many studies, including some experimental work, show that migrants are sensitive to predation danger. In response, they choose safer habitats, slow the rate of fuel deposition or reduce fuel loads (Lindström, 1990; Cresswell, 1994; Ydenberg *et al.*, 2002, 2004; Schmaljohann & Dierschke, 2005; Cimprich *et al.*, 2005; Pomeroy, 2006; Pomeroy *et al.*, 2006). Predation danger can thus slow the speed of migration. But there are complexities: if the situation is safe but about to become more dangerous, migrants may increase the rate of fuel deposition in order to ramp up the speed of migration and so keep ahead of the migration of the predators themselves (Hope *et al.*, 2011, 2014). And in very safe situations, migrants may put on very large fuel loads to enable a long flight and thus jump over ecological barriers, or over portions of the route where higher danger lurks.

Routing: An ecological barrier, by definition, offers few stopover possibilities. In an important contribution, Alerstam (2001) noted that avian migration routes often detour around ecological barriers, even though birds could fly directly across the barrier. His analysis of the evolutionary reasons considers the energetics of these longer flights, especially the need to acquire and carry fuel over great distances.

The Himalayas form such a barrier for birds migrating between Tibet, Mongolia and Siberia and the Indian subcontinent. The mountain ranges themselves extend approximately 10 degrees of latitude (1100 km), which is not especially long by the standards of avian migration, but they are of course very high. (The physiological demands of high-altitude flight are described by Hawkes *et al.*, Chapter 16.) They are also barren and arid. Available water (and hence the organic production that migrants need to refuel) is confined to just a few large wetland complexes, and to the extensive network of very small, dispersed wetlands between 4500 m and 5200 m (Prins *et al.*, Chapter 26). Dry grasslands and scree plains cover much of the area, with water unavailable in the hot or frozen deserts at lower and higher altitudes. Further, they are bordered to the north by the extensive arid lands of the Tibetan Plateau, and the Taklamakan and Gobi Deserts. A detour would add substantial distance, as the Himalayas stretch over about 35 degrees of longitude

(~2400 km). Some birds such as the Houbara Bustard *Chlamydotis macqueenii* are known to skirt around the Himalayan Range (Tourenq *et al.*, 2004), but evidently the penalty the detour imposes is not worthwhile for many birds, which make the crossing in spite of the altitude and the distance required (cf. Groen & Prins chapter 17).

As with migration timing, the danger attendant on the alternative routes is probably important. For example, Lank *et al.* (2003) calculated the energetic costs to Western Sandpipers of migrating between southwest British Columbia and Alaska via the coastal route (which they use northbound) and via the shorter trans-Gulf of Alaska route (used southbound). They conclude that the trans-Gulf route is energetically cheaper in both directions, but requires a much larger fuel load, as it is made in a single long 'jump' rather than in three shorter 'skips'. Exposure to predators at the launch site for the trans-Gulf flight is consequently crucial because it is here that the fuel reserve and hence the vulnerability it creates is largest. The data show that this exposure is high when northbound, when sandpipers use the longer coastal route. The exposure is much lower when southbound, enabling use of the shorter trans-Gulf route.

Studying Migratory Danger in the Himalayas

Here then, at last, is the question with which this chapter began. Is the predation risk accompanying a direct crossing of the Himalayas greater or smaller than would be incurred on a necessarily much longer detour? As indicated at the outset, this question has no answer as yet, but I hope it has become clear that it has many facets: a measure of the number of deaths attributable to predators cannot provide the answer. The key is the degree to which migrants express anti-predator behaviour, and the extent to which this reduces the risk.

Advances in tracking technology are beginning to reveal details of how, where and when migrants move through the Himalaya (e.g. Takekawa *et al.*, 2009; Bishop *et al.*, 2015, Chapters 1, 2, 3, 7 and 8), and data on the locations and durations of stopovers will no doubt continue to accumulate. But just as important are natural history data on the presence of species, and the timing and routing of their migratory movements. Here even basic observations can yield important insights. For example, Worcester and Ydenberg (2008) derived cross-continental patterns in North American Peregrine migration based on the data collected at sites where volunteers simply counted the numbers of migratory raptors passing over (see Inzunza, 2005). The data on Peregrine danger presented in Lank *et al.* (2003) were collected by one of the co-authors (John Ireland) who recorded over 16 years the numbers seen on a standardized lunchtime walk. Data like these, simply and systematically collected, can be assembled to describe the predator landscape of the Central Asian Flyway (see Juhant & Bildstein, Chapter 6). As was the case for the flyways in the Americas and in Europe described by Ydenberg *et al.* (2007), there will probably be features that will help to explain the significance of timing, speed of migration and routing of various species.

Of special interest is the question of whether Himalayan stopovers are dangerous. Many birds congregate at the few wetland complexes, and hence raptors may be drawn to these places. It seems likely that the food availability in these high mountains is low, and hence that the attainable refuelling rate is also low. There are, however, no data with which to evaluate this. Measuring the rate at which migrants deposit fuel can be done by assessing blood metabolite profiles (Williams *et al.*, 1999). These can be simply analyzed from small amounts of blood drawn from captured individuals (see Ydenberg *et al.*, 2002 for an example). Even basic information on whether the plasma contains triglycerides or glycerol and 13-hydroxybutyrate would indicate whether migrants are gaining or expending fat.

Behavioural observations of foraging intensity would help evaluate whether migrants are working to gain fuel at stopover locations, or whether they stop only to rest and drink. A simple and informative project would be to collect data on the vigilance of geese and other birds at the large wetlands. These could be compared with geese at other sites (e.g. Kurvers *et al.*, 2014), and would help to assess whether birds regard the situation as safe or dangerous. The 'flight initiation' and 'giving-up density' measures described earlier could be implemented in the field to assess whether migrants at stopover sites perceive local differences in the danger level at different stopover sites.

The weight of the fuel required for a long-distance flight renders migrants more vulnerable to predators, because it reduces flight acceleration and agility. Hence, large fuel loads are acquired only at times and in places where the predation danger is low. Much could be learned with basic information on how much fuel Himalayan migrants carry. An interesting feature in high mountains is that low air density could exacerbate this effect. Bishop *et al.* (2015) found that Bar-headed Geese flew as low as the terrain allowed, and that to keep flight costs down they readily gave up the altitude they had had to gain to cross ranges.

Conclusion

The question addressed in this chapter is whether predation danger is a cost exacerbating the already formidable obstacles avian migrants face on the Himalayan route, or whether this route is safer than more roundabout routes between breeding and non-breeding areas. The answer is unknown, but it is becoming clearer that the mortality attributable to predators will not by itself yield the answer. The reason lies in the distinction between 'danger' and 'risk'. The latter refers to the actual mortality that occurs in any ecological situation, while the former refers to the mortality that would occur were migrants heedless of the danger. The difference depends on the level of caution that migrants exercise: actual depredation could be low even in a very dangerous situation if the anti-predator tactics of migrants were effective, and inexpensive enough to allow migrants to invest the requisite time and effort.

I discuss what is known of how predation danger can influence the timing, routing and speed of migration. A key concept is the 'predator landscape', which describes on a flyway scale when and where it is especially dangerous. Migration schedules across predator landscapes must balance access to the food that powers the high energy demand of long-distance migration with exposure to predation danger. Stopovers are important, because of the great amount of migratory time spent on stopovers, the need for intensive foraging and the heightened vulnerability resulting from large fuel reserves.

References

Alerstam, T. (2001). Detours in bird migration. *Journal of Theoretical Biology*, **209**, 319–331.

Alerstam, T. & Hedenström, A. (1998). The development of bird migration theory. *Journal of Avian Biology*, **29**, 343–369.

Alerstam, T., Hedenström, A. & Akesson, S. (2003). Long-distance migration: evolution and determinants. *Oikos*, **103**, 247–260.

Alerstam, T. & Lindström, Å. (1990). Optimal bird migration: the relative importance of time, energy and safety. In E. Gwinner, ed., *Bird Migration: Physiology and Ecophysiology*, Berlin: Springer-Verlag, pp. 331–351.

Beauchamp, G. (2014). *Social Predation: How Group Living Benefits Predators and Prey*. New York: Academic Press.

Bednarz, J.C. & Kerlinger, P. (1989). Monitoring hawk populations by counting migrants. *National Wildlife Federation Scientific and Technical Series Supplement No*. **13**, 328–342.

Bishop, C.M., Spivey, R.J., Hawkes, L.A., *et al.* (2015). The roller coaster flight strategy of Bar-headed Geese conserves energy during Himalayan migrations. *Science*, **347**, 250–254.

Brown, J.S. & Kotler, B.P. (2007). Foraging and the ecology of fear. In D.W. Stephens, J.S. Brown & R.C. Ydenberg, eds., *Foraging: Behaviour and Ecology*. Chicago: University of Chicago Press, pp. 437–482.

Burns, J.G. & Ydenberg, R.C. (2002). The effects of wing loading and gender on the escape flights of Least Sandpipers (Calidris minutilla) and Western Sandpipers (Calidris mauri). *Behavioural Ecology Sociobiology*, **52**, 128–136.

Cimprich, D.A., Woodrey, M.S. & Moore, F.R. (2005). Passerine migrants respond to variation in predation risk during stopover. *Animal Behaviour*, **69**, 1173–1179.

Clark, C.W. & Butler, R.W. (1999). Fitness components of avian migration: a dynamic model of Western Sandpiper migration. *Evolutionary Ecology Research*, **1**, 443–457.

Cooper, W.E. & Blumstein, D.T., eds. (2015). *Escaping from Predators: An Integrative View of Escape Decisions*. Cambridge: Cambridge University Press.

Cresswell, W. (1994). Age-dependent choice of redshank (Tringa tetanus) feeding location – profitability or risk? *Journal of Animal Ecology*, **63**, 589–600.

Cresswell, W. (1996). Surprise as a winter hunting strategy in Sparrowhawks Accipiter nisus, peregrines Falco peregrinus and Merlins F. columbarius. *Ibis*, **138**, 684–692.

Dekker, D. & Ydenberg, R.C. (2004). Raptor predation on wintering Dunlins in relation to the tidal cycle. *Condor*, **106**, 415–419.

Díaz, M., Møller, A.P., Flensted-Jensen, E., *et al.* (2013). The geography of fear: a latitudinal gradient in anti-predator escape distances of birds across Europe. *PLoS One*, **8**(5), e64634.

Drake, A.E.G., Rock, C.A., Quinlan, S.P., Martin, M. & Green, D.J. (2014). Wind speed during migration influences the survival, timing of breeding, and productivity of a Neotropical migrant, Setophaga petechia. *PLoS One*, **9**, e97152.

Drent, R., Both, C., Green, M., Madsen, J. & Piersma, T. (2003). Pay-offs and penalties of competing migratory schedules. *Oikos*, **103**, 274–292.

Eichhorn, G., Drent, R.H., Stahl, J., Leito, A. & Alerstam, T. (2009). Skipping the Baltic: the emergence of a dichotomy of alternative spring migration strategies in Russian Barnacle Geese. *Journal of Animal Ecology*, **78**, 63–72.

Ellis, D.H., Bednarz, J.C., Smith, D.G. & Flemming, S.P. (1993). Social foraging classes in raptorial birds. *BioScience*, **43**, 14–20.

Frederick, W.G. & Cooper, W.E. (2007). Optimal flight initiation distance. *Journal of Theoretical Biology*, **224**, 59–67.

Hedenström, A. & Alerstam, T. (1997). Optimum fuel loads in migratory birds: distinguishing between time and energy minimization. *Journal of Theoretical Biology*, **189**, 227–234.

Hedenström, A. & Rosén, M. (2001). Predator versus prey: on aerial hunting and escape strategies in birds. *Behavioural Ecology*, **12**, 150–156.

Heintzelman, D.S. (1975). *Autumn Hawk Flights: The Migrations in Eastern North America*. New Brunswick, NJ: Rutgers University Press.

Hope, D.D., Lank, D.B., Smith, B.D. & Ydenberg, R.C. (2011). Migration of two calidrid sandpiper species on the predator landscape: how stopover time and hence migration speed vary with proximity to danger. *Journal of Avian Biology*, **42**, 523–530.

Hope, D.D., Lank, D.B. & Ydenberg, R.C. (2014). Mortality-minimizing sandpipers vary stopover behaviour dependent on age and geographic proximity to migrating predators. *Behavioural Ecology Sociobiology*, **68**, 827–838.

Houston, A.I. (1998). Models of optimal avian migration: state, time and predation. *Journal of Avian Biology*, **29**, 395–404.

Hunt, W.G., Rogers, R.R. & Slowe, D.J. (1975). Migratory and foraging behaviour of Peregrine Falcons on the Texas coast. *Canadian Field-Naturalist*, **89**, 111–123.

Ibanez, C., Juste, J., Garcia-Mudarra, J.L. & Agirre-Mendi, P.T. (2001). Bat predation on nocturnally migrating birds. *Proceedings of the National Academy of Science (US)*, **98**, 9700–9702.

Inzunza, E.R. (2005). The raptor population index (RPI) project in its second year. *Hawk Migration Studies*, **32**, 46.

Jamieson, S.E., Ydenberg, R.C. & Lank, D.B. (2014). Does predation danger on southward migration curtail parental investment by female Western Sandpipers? *Animal Migration*, **2**, 34–43.

Jonker, R.M., Eichhorn, G., Van Langevelde, F. & Bauer, S. (2010). Predation danger can explain changes in timing of migration: the case of the Barnacle Goose. *PLoS One*, e11369.

Kang, C.-K., Moon, J.-Y., Lee, S.-I. & Jablonski, P.G. (2012). Camouflage through an active choice of a resting spot and body orientation in moths. *Journal of Evolutionary Biology* **25**, 1695–1702.

Kerlinger, P. (1989). *Flight Strategies of Migrating Hawks*. Chicago: University of Chicago Press.

Kotler, B.P. & Brown, J.S. (2007). Community ecology. In D.W. Stephens, J.S. Brown & R.C. Ydenberg, eds., *Foraging: Behaviour and Ecology*. Chicago: University of Chicago Press, pp. 397–436.

Kullberg, C. (1998). Does diurnal variation in body mass affect take-off ability in wintering willow tits? *Animal Behaviour*, **56**, 227–233.

Kullberg, C., Fransson, T. & Jacobsson, S. (1996). Impaired predator evasion in fat blackcaps (*Sylvia atricapilla*). *Proceedings of the Royal Society, London, Series B*, **265**, 1659–1664.

Kurvers, R.H.J.M., Straates, K., Ydenberg, R.C., Van Wieren, S.E., Swierstra, P. & Prins, H.H.T. (2014). Social information use by Barnacle Geese *Branta leucopsis*, an experiment revisited. *Ardea*, **102**, 173–180.

Lank, D.B., Butler, R.W., Ireland, J. & Ydenberg, R.C. (2003). Effects of predation danger on migratory strategies of sandpipers. *Oikos*, **103**, 303–319.

Lank, D.B. & Ydenberg, R.C. (2003). Death and danger at migratory stopovers: problems with 'predation risk'. *Journal of Avian Biology*, **34**, 225–228.

Lehikoinen, A. (2011). Advanced autumn migration of Sparrowhawk has increased the predation risk of long-distance migrants in Finland. *PLoS ONE*, **6**(5), e20001.

Lima, S.L. (1986). Predation risk and unpredictable feeding conditions: determinants of body mass in birds. *Ecology*, **67**, 377–385.

Lima, S.L. (1998). Stress and decision making under the risk of predation: recent developments from behavioural, reproductive, and ecological perspectives. *Advances in the Study of Behaviour*, **27**, 215–290.

Lima, S.L. & Bednekoff, P.A. (1999). Temporal variation in danger drives antipredator behaviour: the predation risk allocation hypothesis. *American Naturalist*, **153**, 649–659.

Lima, S.L. & Dill, L.M. (1990). Behavioural decisions made under the risk of predation: a review and prospectus. *Canadian Journal of Zoology*, **68**, 619–640.

Lind, J., Fransson, T., Jacobsson, S. & Kullberg, C. (1999). Reduced take-off ability in robins due to migratory fuel load. *Behavioural Ecology and Sociobiology*, **46**, 65–70.

Lindström, A. (1989). Finch flock size and risk of hawk predation at a migratory stopover site. *Auk*, **106**, 225–232.

Lindström, A. (1990). The role of predation risk in stopover habitat selection in migrating bramblings, Fringilla montifringilla. *Behavioural Ecology*, **1**, 102–106.

Lok, E.K., Esler, D., Takekawa, J.Y., *et al.* (2012). Spatiotemporal associations between Pacific herring spawn and surf scoter spring migration: evaluating a 'silver wave' hypothesis. *Marine Ecology Progress Series*, **457**, 139–150.

Maillet, D. & Weber, J.M. (2006). Performance-enhancing role of dietary fatty acids in a long-distance migrant shorebird: the semipalmated sandpiper. *Journal of Experimental Biology*, **209**, 2686–2695.

Metcalfe, N.B. & Ure, S.E. (1995). Diurnal variation in flight performance and hence potential predation risk in small birds. *Proceedings of the Royal Society, London, Series B*, **261**, 395–400.

Møller, A.P., Grim, T., Ibáñez-Álamo, J.D., Markó, G. & Tryjanowski, P. (2013). Change in flight initiation distance between urban and rural habitats following a cold winter. *Behavioural Ecology*, **24**, 1211–1217.

Møller, A.P. & Ibáñez-Álamo, J.D. (2012). Escape behaviour of birds provides evidence of predation being involved in urbanization. *Animal Behaviour*, **84**, 341–348.

Moore, F.R., Kerlinger, P. & Simons, T.R. (1990). Stopover on a gulf-coast barrier island by spring trans-gulf migrants. *Wilson Bulletin*, **102**, 487–500.

Newton, I. (2008). *The Ecology of Bird Migration*. London: Academic Press.

Pennycuick, C.J. (1989). *Bird Flight Performance: A Practical Calculation Manual.* Oxford: Oxford University Press.

Piersma, T. (1998). Phenotypic flexibility during migration: optimization of organ size contingent on the risks and rewards of fueling and flight? *Journal of Avian Biology*, **29**, 511–520.

Piersma, T. & Gill Jr., R.E. (1998). Guts don't fly: small digestive organs in obese Bar-Tailed Godwits. *Auk*, **115**, 196–203.

Pomeroy, A.C. (2006). Trade-offs between food abundance and predation danger in spatial usage of a stopover site by Western Sandpipers, Calidris mauri. *Oikos*, **112**, 629–637.

Pomeroy, A.C., Butler, R.W. & Ydenberg, R.C. (2006). Experimental evidence that migrants adjust usage at a stopover site to trade off food and danger. *Behavioural Ecology*, **17**, 1041–1045.

Preisser, E.L., Bolnick, D.I. & Benard, M.F. (2005). Scared to death? The effects of intimidation and consumption in predator-prey interactions. *Ecology*, **86**, 501–509.

Prins, H.H.T. & Iason, G. (1989). Dangerous lions and nonchalant buffalo. *Behaviour*, **108**, 262–296.

Schmaljohann, H. & Dierschke, V. (2005). Optimal bird migration and predation risk: a field experiment with Northern Wheatears Oenanthe oenanthe. *Journal of Animal Ecology*, **74**, 131–138.

Si, Y., Xin, Q., de Boer, W.F., Gong, P., Ydenberg, R.C. & Prins, H.H.T. (2015). Do arctic breeding geese track or overtake a green wave during spring migration? *Scientific Reports*, **5**, 8749.

Sillet, T.S. & Holmes, R.T. (2002). Variation in survivorship of a migratory songbird throughout its annual cycle. *Journal of Animal Ecology*, **71**, 296–308.

Sutherland, W.J. (1998). Evidence for flexibility and constraint in migration systems. *Journal of Avian Biology*, **29**, 441–446.

Takekawa, J.Y., Heath, S.R., Douglas, D.C., *et al.* (2009). Geographic variation in Bar-headed Geese Anser indicus: connectivity of wintering area and breeding grounds across a broad front. *Wildfowl*, **59**, 100–123.

Therrien, J.-F., Gauthier, G., Korpimaki, E. & Béty, J. (2014). Predation pressure by avian predators suggests summer limitation of small-mammal populations in the Canadian arctic. *Ecology*, **95**, 56–67.

Tourenq, C., Combreau, O., Pole, S.B., *et al.* (2004). Monitoring of Asian Houbara Bustard Chlamydotis macqueenii populations in Kazakhstan reveals dramatic decline. *Oryx*, **38**, 62–67.

Van der Veen, I.T. (1999). Trade-off between Starvation and Predation: Weight-Watching in Yellowhammers. PhD thesis, Uppsala, Sweden: Uppsala University.

Van der Veen, I.T. & Lindström, K.M. (2000). Escape flights of yellowhammers and greenfinches: more than just physics. *Animal Behaviour*, **59**, 593–601.

Veasey, J.S., Metcalfe, N.B. & Houston, D.C. (1998). A reassessment of the effect of body mass upon flight speed and predation risk in birds. *Animal Behaviour*, **56**, 883–889.

Walter, H. (1979). *Eleonora's Falcon: Adaptations to Prey and Habitat in a Social Raptor*. Chicago: University of Chicago Press.

Wiedner, D.S., Kerlinger, P., Sibley, S.A., *et al.* (1992). Visible morning flight of neotropical landbird migrants at Cape May, New Jersey. *Auk*, **109**, 500–510.

Williams, T.D., Guglielmo, C.G., Egler, O. & Martyniuk, C.J. (1999). Plasma lipid metabolites provide information on mass change over several days in captive Western Sandpipers. *Auk*, **116**, 994–1000.

Witter, M.S., Cuthill, I.C. & Bonser, R.H.C. (1994). Experimental investigations of mass-dependent predation risk in the European starling, Sturnus vulgaris. *Animal Behaviour*, **48**, 201–222.

Worcester, R. & Ydenberg, R. (2008). Cross-continental patterns in the timing of southward Peregrine Falcon migration in North America. *Journal of Raptor Research*, **42**, 13–19.

Xu, C., Barrett, J., Lank, D.B. & Ydenberg, R.C. (2015). Large and irregular population fluctuations in migratory Pacific (Calidris alpina pacifica) and Atlantic (C. a. hudsonica) Dunlins are driven by density-dependence and climatic factors. *Population Ecology*, **57**, 551–567.

Ydenberg, R.C., Butler, R.W. & Lank, D.B. (2007). Effects of predator landscapes on the evolutionary ecology of routing, timing and molt by long-distance migrants. *Journal of Avian Biology*, **38**, 523–529.

Ydenberg, R.C., Butler, R.W., Lank, D.B., Guglielmo, C.G., Lemon, M. & Wolf, N. (2002). Trade-offs, condition dependence, and stopover site selection by migrating sandpipers. *Journal of Avian Biology*, **33**, 47–55.

Ydenberg, R.C., Butler, R.W., Lank, D.B., Smith, B.D. & Ireland, J. (2004). Western Sandpipers have altered migration tactics as peregrine falcon populations have recovered. *Proceedings of the Royal Society of London B*, **271**, 1263–1269.

Ydenberg, R.C. & Dill, L.M. (1986). The economics of fleeing from predators. *Advances in the Study of Behaviour*, **16**, 229–249.

Zimmerman, J.L. (1990). *Cheyenne Bottoms: Wetland in Jeopardy.* Lawrence: University Press of Kansas.

20 Bird Species Diversity on an Elevational Gradient between the Greater Himalaya and the Tibetan Plateau

Herbert H.T. Prins, Sipke E. van Wieren and Tsewang Namgail

Hypotheses Concerning Biological Diversity and Elevational Gradients

The Himalayas form a massive and globally unique altitudinal gradient from the Gangetic Plain nearly at sea level to the world's highest mountain peaks. The high areas of Ladakh in northern India, straddling the lands between the Himalayas and the Tibetan Plateau, constitute a good terrain to test some general ecological principles. The relation between altitude and the distribution of organisms has been perhaps one of the oldest challenges in ecology, and it had already been studied before Ernst Haeckel coined the term 'ecology' in 1866 by, for example, MacGillivray (1832). The distribution of birds and bird diversity in relation to altitude has been pivotal in the development of general theory concerning niche size, niche overlap and community saturation (MacArthur *et al.*, 1966; Terborgh, 1977). Understanding bird diversity patterns along altitudinal gradients helps in understanding the underlying processes of diversity because gradients are relatively short, thus making it plausible that every species can potentially occur everywhere: historical explanations can safely be ignored (cf. Hawkins & Porter, 2003). It is a well-known but less well-understood fact that species diversity decreases with increasing latitude (e.g. Tramer, 1974; Hawkins *et al.*, 2007); perhaps the best explanation is the effect of random-walk extinctions due to variable weather at higher latitudes (Tilman *et al.*, 1993). Even though the climate at high elevations is very reminiscent of that in the Arctic, a major difference between the elevational gradient in Ladakh and that of a latitudinal gradient between the tropics and the arctic is that local extinctions can hardly be the cause of a decrease in avian diversity with elevation, simply because Ladakh lies squarely on the long-distance migratory routes of many bird species (cf. McLure, 1974; Williams & Delany, 1985, 1986; Takekawa *et al.*, 2009; Bishop *et al.*, 2015; Chapters 1–8) so that local extinctions are hardly an issue because recolonization is likely and endemics hardly occur (Namgail & Yom-Tov, 2009). This fact also makes it possible to reject out of hand Lomolino's (2001) hypothesis concerning an elevational pattern for the

Himalayas *sensu lato* (namely, the Himalaya, Karakoram, Pamir, Zanskar, Hindu Kush and Kailash Ranges), since it assumes that migration is higher at the base of the mountains than at the top. Indeed, with very few exceptions (the Booted Eagle *Hieraaetus pennatus* and other eagles are a case in point because many of them migrate east to west: Den Besten, 2004, Juhant & Bildstein, Chapter 6), migration is across the mountains of the Himalaya and not parallel to them: migration occurs everywhere although often funnelled through mountain passes at altitudes between 5000 m and 6500 m above sea level (a.s.l.) (Namgail et al., Chapter 2). Of course, short-distance altitudinal migrants, such as Guldenstadt's Redstart *Phoenicurus erythrogastrus*, Brown Accentor *Prunella fulvescens* or Eastern Great Rosefinch *Carpodacus rubicilloides*, are exceptions to this reasoning.

The Trans-Himalayan ranges themselves are as dry as the deserts further north. For example, the annual precipitation in Leh, the capital of Ladakh (at an altitude of 3620 m a.s.l.; in the State of Jammu Kashmir, India), is about 80 mm. The recorded precipitation in the village of Kibber (4200 m a.s.l. in Spiti, part of the State of Himachal Pradesh, India), just to the south of the Greater Himalayan range, is about 50 mm (Charudutt Mishra, pers. comm.). Summer temperature may reach 35°C while snowfall in winter is low (in the order of 50 cm, which is equivalent to 50 mm of rain). At higher altitudes, temperature and hence evaporation are much lower, while both summer rainfall and winter snowfall are higher. Climate (Bookhagen, Chapter 11) and geology (Searle, Chapter 9) resulted in the formation of wetlands at altitudes between 4500 m and 5200 m a.s.l.; the saline lakes occurring at those altitudes are especially rich in invertebrate life, which forms a good source of food for many waterbirds (Prins *et al.*, Chapter 18). At even higher altitudes, the summer season is so short that, even though water may be available, there is no appreciable plant cover and wetlands in the normal sense of the word do not exist (see Mani, 1978; Mishra, 2001; cf. Fig. 0.3).

Bird species richness over an elevational gradient has been shown to conform to any of four general patterns (McCain 2009), namely (a) decreasing diversity, (b) low-elevation plateaus, (c) low-elevation plateaus with mid-peaks and (d) unimodal mid-elevation peaks. The main drivers of these patterns are climate, including temperature and moisture level. Since the Trans-Himalayan mountains are very dry, McCain (2009) predicted that pattern (d) should prevail. Note that although her meta-analysis also included the western Himalayas, through analyzing the data from the Chamba district in Himachal Pradesh, India (see her supporting material), we could independently test this as a hypothesis because the study she referred to was on the monsoon side of the Himalayas with an average annual rainfall of 785 mm, while our study area, in the rain shadow of the Himalaya, only receives an average of about 80 mm per year.

McCain (2009) formulated hypotheses that predict the emerging patterns of avian diversity along elevational gradients. Fundamentally ecological are (1) temperature, (2) presence of water, (3) species-area effect and (4) the mid-domain effect; these are (partly) competing hypotheses that can explain the patterns that have been found at a global scale. Note that McCain (2009) did not include primary production in her conclusions concerning the emerging patterns of avian diversity along elevational

Table 20.1. Short description of the hypotheses used in this chapter. They concern the predicted pattern of avian diversity over an elevation gradient in the western Trans-Himalayan region of Ladakh.

	Prediction concerning avian species diversity	Underlying causation
Hypothesis 1	Decreases with altitude	Temperature decreases with altitude and energy expenditure goes up so fewer species can cope.
Hypothesis 2	Peaks around 4700 m	Avian diversity peaks where water availability is highest, which happens to be around 4700 m.
Hypothesis 3	Peaks around 4900–5000 m	Avian diversity peaks where most habitat is available, so where an elevational band has its widest distribution, which happens to be at 4900–5000 m
Hypothesis 4	Dips around 4500 m	Fewest species occur on steepest slopes, and most steep slopes are found around 4500 m; on steep slopes little water infiltration and little plant growth.
Hypothesis 5	Decreases with altitude	Vegetation structure (and thus resource availability for birds) decreases with altitude.
Hypothesis 6	Decreases with increasing avian predator diversity	Increased diversity of avian predators leads to an increased number of predation strategies and thus fewer prey species that are free of predation.

gradients; we will return to that later. Temperature is closely associated with elevation since the lapse rate is $-0.56°C.100\ m^{-1}$ gain; the elevational gradient of 3 km that we studied thus represents on average a temperature gradient of 16.8°C (as big a gradient as between, for example, Belem in Brazil and Boston in the United States, or Jakarta in Indonesia and London in the United Kingdom). Associated with the steady decline in temperature along an elevational gradient is the decline in atmospheric pressure (from 70 kilopascal (kPa) at 3200 m, the base of our elevational gradient, to 48 kPa at the top of the gradient at 6200 m), and hence partial oxygen pressure declines (from about 21.5 kPa to about 14.5 kPa, which is about 50% of the partial pressure at sea level) (www.altitude.org/air_pressure.php). Temperature and oxygen levels are closely associated with energy expenditure of birds, and hence the prediction is that fewer bird species can cope with these adverse conditions, and thus that avian diversity steadily declines with altitude (Hypothesis 1 in this chapter. See Table 20.1 for a summary of all hypotheses considered.) Bird species such as Bar-headed Goose, with adaptations such as larger lungs and higher capillary densities in the heart to withstand low oxygen levels, fly over mountain passes higher than 5000 m (Scott & Milsom, 2007; Scott *et al.*, 2011; Bishop *et al.*, 2015; Hawkes *et al.*, Chapter 16), and species that apparently cannot cope with these environmental pressures have been recorded taking a detour, for example, Houbara Bustard *Chlamydotis undulata* (Crombreau *et al.*, 1999).

The Trans-Himalayan area lies in the rain-shadow of the Himalayan Range because the monsoon comes from the south. At relatively low altitudes (3200–3800 m), this effect is profound and the ecosystem is very arid. At the high end of the gradient, permanent snow can be found (due principally to non-monsoonal precipitation brought by the westerlies; see Bookhagen, Chapter 11): on north-facing slopes, snow rarely disappears above altitudes of about 5500 m. Permanent snow and ice can be found on summits at 6500 m, but glacier tongues stretch down to about 5000 m (Owen, Chapter 10). This implies that free-flowing water for plant growth dwindles with increasing altitude. Hence, the highest availability of moisture for plant growth occurs at around 4700 m (Namgail *et al.*, 2012), thus predicting that there should be a mid-elevational peak in water availability for plant growth at 4700 m (Hypothesis 2 in this chapter). Since moisture determines primary production, one would expect the highest production there too, but note that this may be restrained by short growing seasons.

The species-area hypothesis predicts that avian diversity should be highest where a particular altitude band is most spatially extensive. Because mountain ranges are defined by peaks and valleys, the geometry of the landscape leads to a small total aggregated surface area of summits, but also of valley bottoms. The most spatially extensive altitude belt is available somewhere halfway between, and species diversity would be expected to peak there (our Hypothesis 3). Note that our Hypotheses 2 and 3 are based on different reasoning, and that the predicted peaks depend on measurements that can lead to different exact predicted biodiversity peaks.

The mid-domain effect makes the prediction that the peak of avian species richness lies where the elevational ranges of most species overlap (McCain, 2009, following Colwell & Lees, 2000). This assumption is not safe in Ladakh, because especially at mid-elevations, slopes occur which are too steep to support vegetation (Namgail *et al.*, 2012); slopes without vegetation are typically nearly devoid of bird life (pers. obs.). This means that if we did not record a bird species in a particular elevation band, then there is a fair chance that it was truly absent from that elevation band. This is also related to detectability, which is high in the sparsely vegetated, open and well-lit habitats that prevail at high altitudes. From this it follows that we predict that avian diversity was lower on steep slopes than on gentle ones. So instead of testing McCain's hypothesis concerning the mid-domain effect, we can test one associated with it, namely the hypothesis that on steep slopes, avian diversity is lower than that on gentle slopes, and that steep slopes more frequently occur at mid-elevation (our Hypothesis 4).

Not included in McCain's hypotheses is the notion that avian diversity is strongly associated with vegetation structure, and that this structure is associated with vegetation growth and air temperature. This structure or habitat complexity is then the causal ecological factor, the explanation being that complex habitats provide greater opportunities for resource subdivision (MacArthur *et al.*, 1966; Terborgh, 1977). This is related to the 'species energy hypothesis' in which it is proposed that the amount of available energy sets limits to the richness of the system and can be used to explain altitudinal and latitudinal reductions in species diversity (Currie *et al.*, 2004). It has been postulated that reduced moisture levels simplify vegetation structure (Tramer, 1974), but also that forest composition and canopy structure decrease with altitude (Terborgh, 1977). Along the elevational gradient of the Trans-Himalaya, there is a steady decline of woody species: between 3200 m a.s.l. and 3800 m a.s.l., Apple, Almond, Apricot, Poplar and Willow are planted in fields and along streams while uncultivated Sea Buckthorn *Hippophae rhamnoides* L. and *H. tibetana* Schltdl., wild Roses *Rosa ecae* Aitch., *R. webbiana* Wall. ex Royle and Tamarisk *Tamaricaria elegans* Royle also occur. Between 3800 m and 4000 m, Willow *Salix* spp., Tamarisk, Rose, Seabuckthorn and some Juniper *Juniperus indica* Bertol. still occur, but above 4000 m a.s.l., Poplar (with indigenous species *Populus ciliata* Wall. ex Royle and *P. pamirica* Komarov but exotic species occur too) and Willow disappear, to be replaced by the low Caragana shrub *Caragana versicolor* Benth., *C. brevifolia* Kom. Above 4400 m, Sea Buckthorn is no longer found, while above 4500 m, Roses and Tamarisk disappear from the landscape. Finally, above 4900 m Caragana disappears. The elevational gradient is thus also a strong vegetation structure gradient, with complex vegetation structure at the bottom and very simple vegetation at the top end. If vegetation structure is the defining ecological factor, then avian diversity should peak at low altitudes and decline steadily with increasing elevation (our Hypothesis 5). Plant species diversity also declines with altitude (cf. Tilman *et al.*, 1993), at least above 4000 m (Namgail *et al.*, 2012; cf. Vetaas & Grytnes, 2002), thus strengthening the case for declining resource availability for birds with increasing elevation. We know too little concerning the relationship between vegetation structure and altitude before the strong human influence of the present day to make any statement about the relation between avian diversity and vegetation structure in times when anthropogenic factors were less prevalent. For aquatic macroinvertebrate biomass (food for many waterbirds), we found significant decline with altitude but high biomass in saline lakes that occurred very high in the mountains, thus obfuscating a clear correlation with altitude (Prins *et al.*, Chapter 18). McCain's hypotheses, and ours in this chapter, deal primarily with 'terrestrial birds' (non-waterbirds), and we show that the difference in patterns of species richness along altitudinal gradient for birds that depend on wetlands and those that do not is substantial (Figure 20.5). We do not have data on terrestrial primary production, and so we have to rely on two proxy parameters, namely vegetation cover and water availability for plant growth.

Closely related to the notion of resource availability is that of resource predictability, and it has been postulated that animal diversity is well explained by resource predictability (see Lawton *et al.*, 1993). Yet we do not know enough about variability and predictability of resources in the Trans-Himalaya to test this. Indeed, resource predictability may well be related to temperature, or to vegetation structure, or to productivity. Resource predictability is also related to the (un)predictability of flash floods, since much bird habitat along streams can be washed away. These flash floods appear to be local events not related to monsoon conditions or western disturbances, but we do not know enough about their occurrence. We are thus not able to test for resource predictability.

The local distribution of species is not only determined by the abiotic environment (as in our Hypothesis 1), or by food resources (as in our Hypotheses 2, 4 and 5), but also by predation. Recently, much work has been done on the landscape of fear and its effects on bird behaviour and occurrence (Lank *et al.*, 2003; Dekker, 2009; Jonker *et al.*, 2010) (Ydenberg, Chapter 19). If top-down processes acting through predation are important in explaining avian diversity, then there is *no a priori* prediction concerning the relation between avian diversity and altitude, because predator pressure would not be expected to relate directly to altitude. This can, however, be investigated correlatively because a prediction would be that if the diversity of avian predators was high, then the diversity of the other bird species would be low (our Hypothesis 6).

Data sets: We conducted six expeditions to the areas roughly between Kargil and Manali, and between these two places and the border with China. The areas we surveyed included stretches along the Indus, Shayok, Zanskar and Nubra Rivers, the lakes Tsomoriri and Thatsangkasu and Tsokar. Data were collected during six expeditions to this region between 1998 and 2011. These expeditions were carried out during the summer (May–September) when all the migratory breeders were present in Ladakh. We walked along valley bottoms and high passes, recording birds observed on the way (daily absence and presence per elevation band of 100 m; see Prins & Van Wieren, 2004; Namgail *et al.*, 2012). Locations and altitudes were determined through a global positioning system (Garmin GPS 12) and an altimeter (Origo Accusense Multi-sensor). Observations were conducted with binoculars. For nomenclature of birds, we followed Grimmett *et al.* (1999). During some of the expeditions, we also estimated vegetation cover (in cover classes of 10%) in association with altitude and slope (reported in Namgail *et al.*, 2012). We walked at a constant speed (ca. 1 $km.h^{-1}$) during each of the expeditions to distribute sampling efforts amongst all altitudinal zones. On all the expeditions we used pack donkeys and horses to carry our gear.

We estimated that six expeditions would give us enough information to determine species diversity with some reliability. Following Harrison and Martinez (1995), we calculated the modified Shannon Index (H*) as

$$H^* = -\sum r_i \ln r_i$$

based on the view that r_i can be calculated as the number of records for species i divided by the total number of records for all species for all expeditions. We took the checklists resulting from each expedition as different checklists of the same area. Harrison and Martinez (*op. cit.*) found that five checklists gave 93% of the asymptotic value of the H*-index and 10 checklists gave 96% of that value. With six expeditions, we thus consider that we have measured approximately 95% of the expected diversity, and given the amount of strenuous labour required to obtain these data at altitudes between 3500 and 6000 m, we decided that this was sufficient.

McCain (2009) warned that adequate elevational analyses can only be conducted if at least 70% of the elevational gradient has been sampled. We only analyzed our data between 3200 m and 5700 m; altitudes between 5700 m and 6200 m were only visited rarely, and we did not collect data from below 3200 m because these areas do not occur within eastern Ladakh, so within the gradient, we intensely sampled more than 80%.

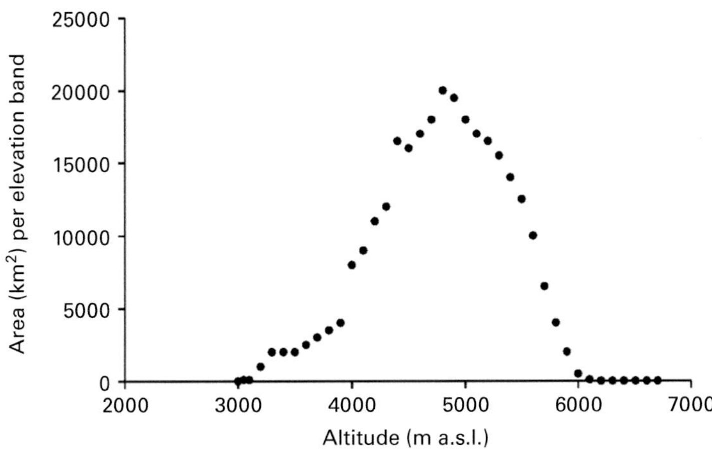

Figure 20.1. In eastern Ladakh, most land is located between 4900 m and 5000 m above sea level. If there is less land surface in a particular elevation band, then the terrain is steeper, and if there is more, then the terrain is flatter. This has repercussions for water infiltration and vegetation growth, and thus for distribution and density of biodiversity.

Because Tramer (1974) stated that 'since herons, ducks, shorebirds etc. are essentially aquatic organisms and as such are outside the scope of [his] paper' on patterns of species richness, we investigated patterns of species richness and diversity over an elevational gradient both with and without these 'aquatic organisms'.

On the basis of a Digital Elevation Model of eastern Ladakh (total area 28,040 km^2), we measured the total horizontally projected surface area per 100-m wide elevation band with a precision of a pixel size of one hectare. The lowest-altitude pixels were found at an altitude of 3000–3100 m a.s.l., and the highest in the band of 6600–6700 m. The surface area per elevation band shows a near-normal distribution with the largest area occurring at 4900–5000 m a.s.l. (19,850 km^2) and no areas occurring lower than 3000 m or higher than 6700 m (Figure 20.1).

The Avian Species Richness and Diversity Pattern in Ladakh

The relation between the cumulative number of species that we identified and the sequential number of expeditions forms a saturating curve (Figure 20.2). The regional species pool in a region of about 80,000 km^2 is estimated to be 294 species, since that is the number that has been recorded since the mid-nineteenth century (Pfister, 2004), or 308 species, according to Lepage (2015). Our sample area is much smaller, approximately 20,000 km^2, and on the basis of the curve in Figure 20.2, we estimate the local alpha diversity to be approximately 180 to 200 species instead of about 300.

The species diversity H* follows the same pattern as the species richness (Figures 20.3a and 20.3b): it is low at low elevations, then increases with altitude, showing two peaks,

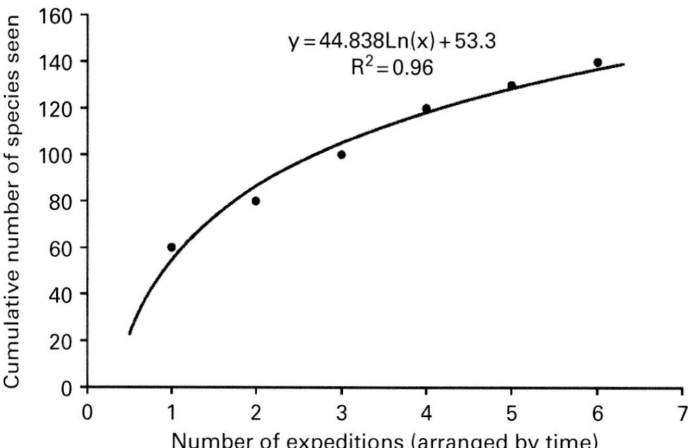

Figure 20.2. After six expeditions to East Ladakh (Indian Trans-Himalaya), approximately 50% of all species that have ever been recorded for East and West Ladakh since the mid-nineteenth century (cf. Pfister, 2004, who lists 276 species) were observed by us. On the basis of the curve, we would expect local alpha diversity to be approximately 180 species, while during the six expeditions 137 species were identified.

before decreasing precipitously above 5500 m a.s.l. Yet even at 6200 m two species were still observed, namely, Black Redstart *Phoenicurus ochruros* and Grey Wagtail *Motacilla cinerea*. The reason that the diversity pattern so closely follows the species richness pattern is because the number of bird observations closely reflected the number of species in a given 100 m-wide elevation band (number of observations/elevation band = 1.0046 $N^{1.1067}$ in which N is the number of species per elevation band; R^2 = 96%, n = 24). The relation between the species diversity H* and the number of species was very tight, and only when the number of species in an elevation band became very small (at altitudes greater than about 5500 m) did the diversity start dropping very fast (Figure 20.4). The species diversity pattern did not change if wetland birds (herons, ducks and shore-birds) were removed from the analysis (Figure 20.5). By and large, the avian diversity and species richness followed a bimodal (not unimodal) peak at mid-elevation.

The Robustness of the Data Set on Avian Species Richness and Diversity

Our results show that birds can occur at much higher altitudes than previously reported in Cramp (1988) and Cramp and Simmons (1985; cf. Pfister 2004), for example, Great Cormorant *Phalacrocorax carbo*, Garganey *Anas querquedula*, Mallard *Anas platyrhynchos*, Black-Winged Stilt *Himantopus himantopus* and Pied Avocet *Recurvirostra avosetta* (all between 4500 m and 4600 m a.s.l.). Voous (1960) even considered Garganey not to occur in mountains, and for Pochard *Aythya ferina*, he was of the opinion at the time that it exclusively occurred in lowlands. It remains unknown whether the local populations of these species, and

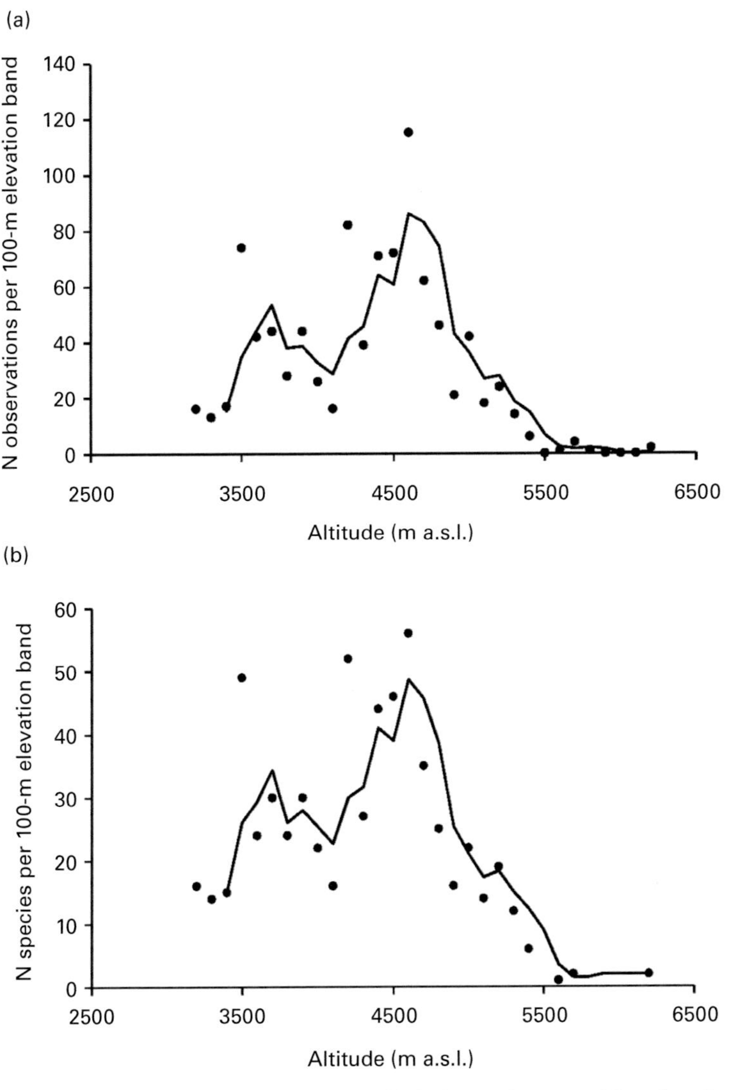

Figure 20.3. (a) Top panel: The number of bird observations per 100 m elevational band increases with altitude and peaks between 4500 m and 4800 m above sea level; the highest observations were at 6200 m. The peak around 3600 m reflects the effect of the mesic conditions along the Indus River. (b) Bottom panel: The number of species per 100 m elevational band. Lines are running means with n = 3.

especially those that raise young at this altitude of close to 5000 m a.s.l. namely, Great Crested Grebe *Podiceps cristatus*, Common Sandpiper *Tringa hypoleucos*, Common Redshank *Tringa totanus* and Ruddy Shelduck *Tadorna ferruginea*, but also Black Redstart, Saker Falcon *Falco cherrug*, Golden Eagle *Aquila chrysaetus* and several other species, have special physiological adaptations (cf. Hawkes *et al.*, Chapter 16). Since these bird species breed at extreme altitudes, we may safely assume that they are completely adapted to the very high elevational gradient that

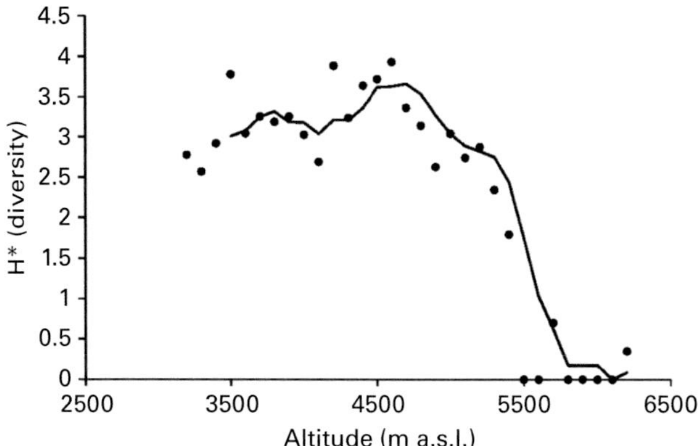

Figure 20.4. Species diversity (H*) peaks at about 4700 m above sea level, and then declines very fast at altitudes exceeding 5500 m. Line is running mean with n = 3.

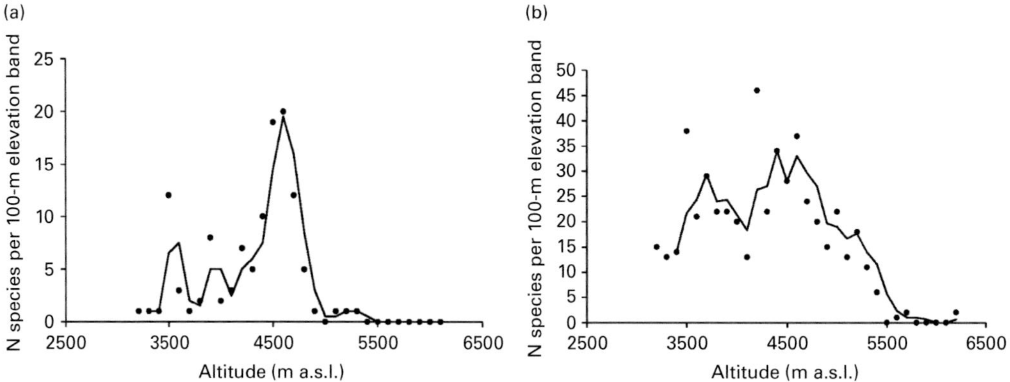

Figure 20.5. (a) Left panel: Wetland bird species richness (of ducks, geese, shorebirds, herons, coots, gulls etc.) peaks at 4600 m where the major wetlands of Tsomoriri and Tsokar can be found; the minor peaks at 3500 m and 3800 m are wetlands along the Indus River. (b) Right panel: If wetland birds are removed, then the pattern of species richness is basically unaffected as compared to Figure 20.3. Lines are running means with n = 3.

we studied. These individuals are not accidental stragglers, but animals that forage and reproduce here.

It is noteworthy that the relation between species diversity, as measured by the modified Shannon-index H* (Harrison & Martinez, 1995), is a direct continuation of the relation as determined by Harrison and Martinez (1995): their range of species numbers was between 30 and 300 per census block, while ours was between 2 and 55 per elevation band. This shows the utility of their approach for measuring species diversity, and preventing the effect of 'oversampling' and thus overestimating local species richness. Even though we are certain that after six expeditions we were not yet oversampling, it is indicative that the best-fitting curve for the relation between the number of expeditions and the number of species that we had observed (Figure 20.2) did show an ever-increasing number of species. Another problem may have been differential sampling efforts in the different elevation bands, the so-called sampling effect (McCain, 2009). Calculating H* on the basis of different censuses (different expeditions) takes this into account (Harrison & Martinez, 1995). For the region sampled, between the Greater Himalaya and Tibet, the patterns for species richness and species diversity are essentially the same, and even when we exclude wetland birds, which perhaps are differently dependent on the resource distribution and density compared to the terrestrial species, the pattern persists. The relation between elevation and bird diversity thus appears to have been robustly measured.

Hypotheses that Can Be Rejected on the Basis of the Patterns of Diversity

The patterns of both species diversity (H*) and species richness did not follow the pattern as predicted by Hypothesis 1 (a steady decrease of species richness because of decreasing temperatures with increasing elevation), and also not as predicted by Hypothesis 5 (a steady decrease of species richness because of decreasing vegetation structural complexity), and thus these two explanations can be rejected. The pattern also did not exactly conform to that expected if Hypothesis 2 were valid: then a unimodal peak was to be expected at around 4600–4800 m a.s.l. because vegetation grassland cover is highest at that elevation (Namgail *et al.*, 2012). The most important wetlands are also found at that elevation, but if wetland birds are removed from the data set, there are clearly two peaks: one at about 3700 m and one at about 4700 m a.s.l. Hypothesis 3 states that peak species richness is to be expected in elevation belts with the widest spatial extent: the so-called species-area effect. Our analysis of the area availability clearly did not support this hypothesis: the most spatially extensive altitude belt was found at a higher altitude than the peaks of species richness or species diversity, and also at low altitudes, far more species were found than would be the case if this prediction were true (compare Figure 20.3 with Figure 20.1). Area effects are highly variable for species groups in mammals, and the spatial extent of different altitude belts ('area size') in general seems to be a much better explanatory factor for terrestrial mammals than for bats, yet the maximum species diversity usually lies at a higher elevation than the midpoint of mountains (McCain, 2007). Area size in general was not the main driver for species diversity in most studies (McCain, 2007), although some bird studies found a strong effect in some mountain ranges (Rahbek, 1997; Kattan & Franco, 2004). In the Trans-Himalaya of Ladakh, there was thus no overriding effect of area on species diversity.

Our last hypothesis (6) stated that the species richness would be low where the number of predator species, or the number of predator observations, was highest. There is no evidence to support this hypothesis as there was a negative relationship between the species richness of predatory and non-predatory birds. In general, the diversity pattern of birds of prey (eagles, buzzards, falcons, Northern Goshawk, Eurasian Sparrowhawk, owls) closely followed the pattern of general avian diversity and species richness

Table 20.2. Birds of prey, including owls, and the maximum altitude at which they were observed.

Species	Maximum altitude (m a.s.l.)
***Owls*:**	
Little Owl *Athene noctua*	4400
Pallid Scops Owl *Otus brucei*	4500
Eurasian Eagle Owl *Bubo bubo*	4600
***Eagles, buzzards and falcons*:**	
Booted Eagle *Haeraaetus pennatus*	3500
Peregrine Falcon *Falco peregrinus*	3500
Northern Goshawk *Accipter gentilis*	3600
Laggar Falcon *Falco jugger*	3600
Eurasian Sparrowhawk *Accipter nisus*	3800
Eurasian Hobby *Falco subbuteo*	4200
Long-Legged Buzzard *Buteo rufinus*	4400
Common Buzzard *Buteo buteo*	4600
Saker Falcon *Falco cherrug*	4700
Common Kestrel *Falco tinnuculus*	5200
Golden Eagle *Aquila chrysaetos*	5400
***Vultures*:**	
Cinereous Vulture *Aegypius monachus*	4200
Egyptian Vulture *Neophron percnopterus*	4800
? Slender-Billed Vulture *Gyps tenuirostris*	5000
Himalayan Griffon *Gyps himalayensis*	5300
Lammergeyer *Gypaetus barbatus*	5300

(Figure 20.7a; Table 20.2), and the same held for birds that generally prey upon other birds or their nests (so birds of prey plus corvids, gulls and Egyptian Vulture) (Figure 20.7b). Yet proportionally, these birds that can become predatory become even more important with increasing elevation (Figure 20.8). However, no avian predators were observed above 5400 m a.s.l.; beyond that altitude, general species diversity (Figure 20.4) or species richness (Figure 20.3) also declined precipitously, and we assume that there was not enough food at these altitudes to sustain the predators. We conclude that avian species richness or diversity does not seem to be controlled by avian predators (George, 1987; see also Creswell, 2008; Ydenberg, Chapter 19).

A Causal Explanation of Avian Species Richness and Diversity

Terborgh (1977) noted that 'diversity is a complex community phenomenon not just a number which can be "explained" with one or another competing hypotheses.' We only partly agree: McCain's (2009) meta-analysis showed that competing hypotheses can be formulated to address the question of the relationship between bird diversity and landscape elevation. Our study of bird diversity in the dry, very high-altitude, cold system in the rain-

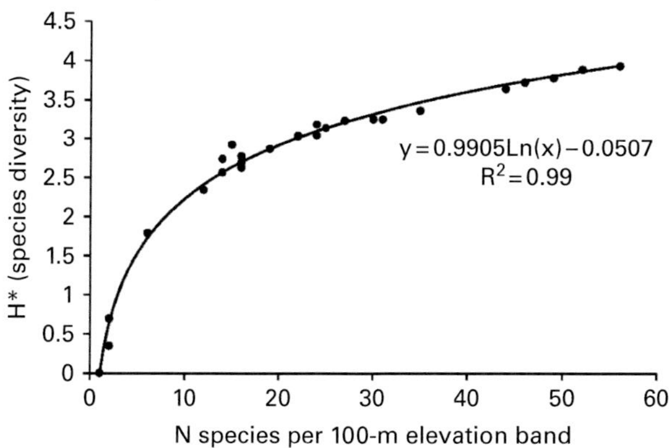

Figure 20.6. Species diversity as measured by a modified Shannon-index (H*; see text) is very well explained by species richness (i.e. the number of species) and slightly less so by the number of observations (H* = 0.8646 Ln(x) + 0.0395; R2 = 96%; n = 24) species diversity follows the same curve as described by Harrison and Martinez (1995).

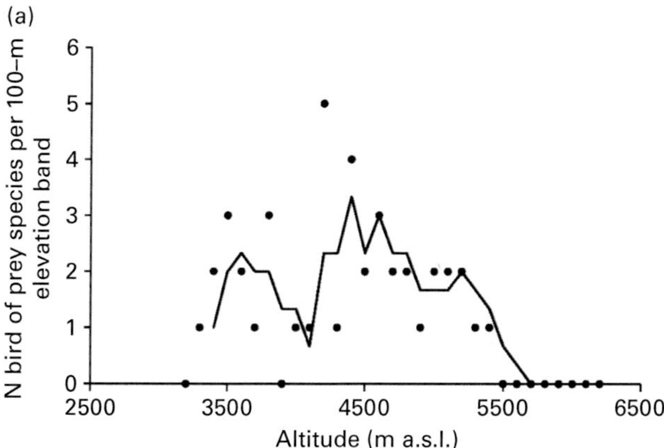

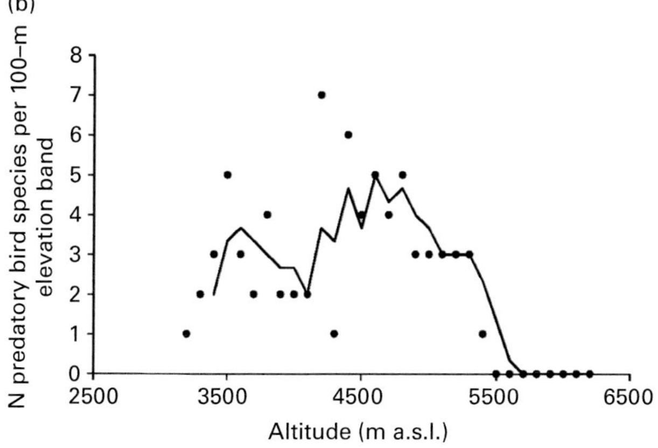

Figure 20.7. (a) Top panel. The number of birds of prey (falcons, eagles, goshawk, sparrowhawk and owls; excluding only vultures) in relation to altitude. These birds of prey show the same peaks at about 3600 m altitude and 4600 m as all bird species. (b) Bottom panel. The number of predatory bird species in the wide sense (so the same species as the 'proper' birds of prey but also including corvids, gulls and Egyptian Vulture) shows the same pattern.

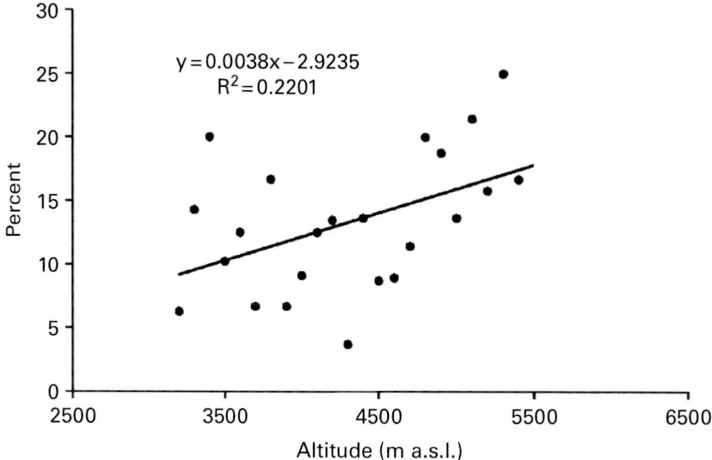

Figure 20.8. The proportion of birds that can be predatory on other birds or their nests increases with altitude. Yet above 5400 m above sea level, no such birds were seen, even though birds occur up to 6200 m. In this predator-free space, their numbers are, however, very low (cf. Figures 20.3a and 20.3b).

shadow of the Himalayas and towards the Tibetan Plateau shows that neither predation nor structural complexity are very important in explaining bird diversity at different altitudes. Our study sites are so high that Terborgh's (1977) anecdotal evidence supporting the notion that bird diversity is highest where productivity is highest, by noting that most farming takes place there, does not hold: farming is concentrated along rivers, particularly the Indus (3200–3700 m), and along streams originating from glaciers. Yet the highest farming of barley and green beans we encountered was at 4200 m altitude, while the avian diversity peaks at 4700 m. But the point is that the peak is bimodal and not unimodal, as expected. The bimodal species richness peak shows two things, namely, that the 'trough' around 4200 m altitude can be explained by the frequently occurring steep slopes at that altitude (see Hypothesis 4): steepness prevents water infiltration from rain- or meltwater, and on steep slopes there is no vegetation to produce seeds, berries, shoots or insects for birds to feed on (Namgail *et al.*, 2012). The general increase of bird diversity and richness from 3200 m altitude towards 4700 m can be explained by an ever-decreasing effect of the rain-shadow from the Himalayas, plus a general tendency for the terrain to become flatter with an increasing altitude towards 4900–5000 m (see Figure 20.1). Flatter areas provide more vegetation growth (Namgail *et al.*, 2012). We consider the decrease in diversity from 4700 m up to 6200 m to be an effect of an ever decreasing availability of free-flowing water to sustain plant growth due to lower and lower temperatures. Primary production ('productivity' *sensu* Terborgh, 1977) is thus the overriding causal explanatory factor predicting avian diversity, as was also deduced, but then rejected, by Kattan and Franco (2004) for the Colombian Andes. In their study of North American mammals and birds, Hawkins and Porter (2003) found that species diversity was very well explained by potential evapotranspiration (explaining 82% of the variation in case of birds), and this parameter is one of the best to explain primary productivity (Rosenzweig, 1968; Olff *et al.*, 2002).

Hawkins and Porter (*op. cit.*) prefer to link potential evapotranspiration to 'energy' as the causal factor, which may provide the explanation. We therefore suggest that primary production is the causal factor from an ecological point of view: it is the available resources that drive avian diversity.

Acknowledgements

The first expedition was only possible because of the men who took us over the Himalayan glaciers, namely, Namgyal, Lama and Sushil Dorjey. The second expedition would not have been possible without the support of Sonam Tashi, Samdup, Stanzin and Tashi Gyatso; we tremendously appreciated their loyalty when tracks and roads had been washed away. For the third expedition, we express thanks to Wangchuk Kalon, Sonam Rinchen and Lobzang Tsering. The fourth expedition was made possible through the assistance of Namgail Angchuk, Sonam Phunchok, Tashi Tobdan and Tsering Tenzin. We acknowledge Rinchen Namgial and his staff for organizing the fifth and sixth expeditions. We are appreciative to the Government of India for time and again restoring infrastructure in an area where the potential energy of rocks and water relentlessly translates into kinetic energy. We thank Herman van Oeveren for his literature searches.

References

Bishop, C.M., Spivey, R.J., Hawkes, L.A., *et al.* (2015). The roller coaster flight strategy of Bar-headed Geese conserves energy during Himalayan migrations. *Science*, **347**, 250–254.

Colwell, R.K. & Lees, D.C. (2000). The mid-domain effect: geometric constraints on the geography of species richness. *Trends in Ecology & Evolution*, **15**, 70–76.

Combreau, O., Launay, F., Al Bowardi, M. & Gubin, B. (1999). Outward migration of Houbara Bustards from two breeding areas in Kazakhstan. *The Condor*, **101**, 159–164.

Cramp, S. (1988). *Handbook of the Birds of Europe, the Middle East and North Africa: Ihe Birds of the Western Palearctic*. In S. Cramp, ed., Vol. IV *Flycatchers to Thrushes*. Oxford: Oxford University Press.

Cramp, S. & Simmons, K.E.L. (1985). *Handbook of the Birds of Europe, the Middle East and North Africa: The Birds of the Western Palearctic*. In S. Cramp & K.E.L. Simmons, eds., Vol. III *Waders to Gulls*. Oxford: Oxford University Press, pp. 1–913.

Cresswell, W. (2008). Non-lethal effects of predation in birds. *Ibis*, **150**, 3–17.

Currie, D.J., Mittelbach, G.G., Cornell, H.V., *et al.* (2004). A critical review of species-energy theory. *Ecology Letters*, **7**, 1121–1134.

Dekker, D. (2009). *Hunting Tactics of Peregrines and Other Falcons*. PhD thesis, Wageningen University.

Den Besten, J.W. (2004). Migration of Steppe Eagles Aquila nipalensis and other raptors along the Himalayas past Dharamsala, India, in autumn 2001 and spring 2002. *Forktail*, **20**, 9–13.

George, T.L. (1987). Greater land bird densities on island vs. mainland: relation to nest predation level. *Ecology*, **68**, 1393–1400.

Grimmett, R., Inskipp, C. & Inskipp, T. (1999). *Pocket Guide to the Birds of the Indian Subcontinent*. New Delhi: Oxford University Press.

Harrison, J.A. & Martinez, P. (1995). Measurement and mapping of avian diversity in southern Africa: implications for conservation planning. *Ibis*, **137**, 410–417.

Hawkins, B.A., Diniz-Filho, J.A.F., Jaramillo, C.A. & Soeller, S.A. (2007). Climate, niche conservatism, and the global bird diversity gradient. *American Naturalist*, **170**, S16–S27.

Hawkins, B.A. & Porter, E.E. (2003). Does herbivore diversity depend on plant diversity? The case of California butterflies. *American Naturalist*, **161**, 40–49.

Jonker, R.M., Eichhorn, G., Van Langevelde, F. & Bauer S. (2010). Predation danger can explain changes in timing of migration: the case of the Barnacle Goose. *PLoS ONE*, **5**(6), e1136g.

Kattan, G.H. & Franco, P. (2004). Bird diversity along elevational gradients in the Andes of Colombia: area and mass effects. *Global Ecology and Biogeography*, **13**, 451–458.

Lank, D.B., Butler, R.W., Ireland, J. & Ydenberg, R.C. (2003). Effects of predation danger on migration strategies of sandpipers. *Oikos*, **103**, 303–319.

Lawton, J.H., Lewinsohn, T.M. & Compton, S.G. (1993). Patterns of diversity for the insect herbivores on bracken. In Ricklefs, R. E. and Schluter, D. (eds.) *Species Diversity in Ecological Communities: Historic and Geographic Perspectives*. University of Chicago Press, Chicago, pp. 178–184.

Lepage, D. (2015). *Checklist of the birds of Ladakh. Avibase, the world bird data base*. Retrieved 9 April 2015, from http://avibase.bsc-eoc.org/checklist.jsp?lang=EN®ion=inwhjk01&list-howardmoore.

Lomolino, M.V. (2001). Elevation gradients of species density – historical and prospective views. *Global Ecology and Biogeography*, **10**, 3–13.

MacArthur, R., Recher, H. & Cody, M. (1966). On the relation between habitat selection and species diversity. *The American Naturalist*, **100**, 319–332.

MacGillivray, M. (1832). Remarks on the phenogamic vegetation of the river Dee, in Aberdeenshire. *Memoires of the Wernian Natural History Society*, **6**, 539–556.

Mani, M.S. (1978). *Ecology & Phytogeography of High-Altitude Plants of the Northwest Himalaya*. London: Chapman & Hall.

McCain, C.M. (2007). Area and mammalian elevational diversity. *Ecology*, **88**, 76–86.

McCain, C.M. (2009). Global analysis of bird elevational diversity. *Global Ecology and Biogeography*, **18**, 346–360.

McLure, H.E. (1974). *Migration and Survival of the Birds of Asia*. Bangkok.

Mishra, C. (2001). *High Altitude Survival: Conflicts between Pastoralism and Wildlife in the Trans-Himalaya*. PhD thesis, Wageningen University, Wageningen, The Netherlands.

Namgail, T., Rawat, G., Mishra, C., Van Wieren, S. & Prins, H.H.T. (2012). Biomass and diversity of dry alpine plant communities along altitudinal gradients in the Himalayas. *Journal of Plant Research*, **125**, 93–101.

Namgail, T. & Yom-Tov, Y. (2009). Elevational range and timing of breeding in the birds of Ladakh: the effects of body mass, status and diet. *Journal of Ornithology*, **150**, 505–510.

Olff, H., Ritchie, M.E. & Prins, H.H.T. (2002). Global environmental controls of diversity in large herbivores. *Nature*, **415**, 901–904.
Pfister, O. (2004). *Birds and Mammals of Ladakh*. New Delhi: Oxford University Press.
Prins, H.H.T. & Van Wieren, S.E. (2004). Number, population structure and habitat use of Bar-headed Geese Anser indicus in Ladakh (India) during the brood-rearing period. *Acta Zoologica Sinica*, **50**, 738–744.
Rahbek, C. (1997). The relationship among area, elevation, and regional species richness in neotropical birds. *The American Naturalist*, **149**, 875–902.
Rosenzweig, M.L. (1968). Net primary productivity of terrestrial communities: predictions from climatological data. *The American Naturalist*, **102**, 67–74.
Scott, G.R. & Milsom, W.K. (2007). Control of breathing and adaptation to high altitude in the Bar-headed Goose. *American Journal of Physiology Regulatory Integrative and Comparative Physiology*, **293**, R379–R391.
Scott, G.R., Schulte, P.M., Egginton, S., Scott, A.L.M., Richards, J.G. & Milsom, W.K. (2011). Molecular evolution of cytochrome c oxidase underlies high-altitude adaptation in the Bar-Headed Goose. *Molecular Biology and Evolution*, **28**, 351–363.
Takekawa, J.Y., Heath, S., David, C.D., *et al.* (2009). Geographic variation in Bar-Headed Geese Anser indicus: connectivity of wintering areas and breeding grounds across a broad front. *Wildfowl*, **59**, 100–123.
Terborgh, J. (1977). Bird species diversity on an Andean elevational gradient. *Ecology*, **58**, 1007–1019.
Tilman, D., Pacala, S.W., Ricklefs, R.T. & Schluter, D. (1993). The maintenance of species richness in plant communities. In R.E. Ricklefs & D. Schluter, eds., *Species Diversity in Ecological Communities*. Chicago: University of Chicago Press, pp. 13–25.
Tramer, E.J. (1974). On latitudinal gradients in avian diversity. *Condor*, **76**, 123–130.
Vetaas, O.R. & Grytnes, J.A. (2002). Distribution of vascular plant species richness and endemic richness along Himalayan elevation gradient in Nepal. *Global Ecology and Biogeography*, **11**, 291–301.
Voous, K.H. (1960). *Atlas of European Birds*. London: Nelson.
Williams C.T. & Delany, S.N. (1985). Migration through the north-west Himalaya – some results of the Southampton University Ladakh Expeditions. Part 1. *Oriental Bird Club Bulletin*, **2**, 10–14.
Williams C.T. & Delany, S.N. (1986). Migration through the north-west Himalaya – some results of the Southampton University Ladakh Expeditions. Part 2. *Oriental Bird Club Bulletin*, **3**, 11–16.

Part IV

People and Their Effects on the Himalayas

21 Evidence of Human Presence in the Himalayan Mountains: New Insights from Petroglyphs

Martin Vernier and Laurianne Bruneau

Ancient Peoples in the Himalayas

Birds have a subtle yet remarkable and ubiquitous presence in the Himalayan scenery. Just as high peaks, glaciers and barren valleys have never been a barrier to their migration, these geographical features never stopped people from moving across the Himalayas. The concept of mountains as an impassable obstacle is nothing more than imagery. Extensive archaeological surveys have been conducted in Ladakh and Himachal Pradesh (India), Mustang (Nepal) and western Tibet (Tibetan Autonomous Region, China) between 1995 and the present. These surveys have brought to light the antiquity of human occupation in the Himalayas, which, far from being an isolated and disconnected part of the settlement of secluded parts of Inner Asia, is, on the contrary, a crossroads.

Evidence of prehistoric human presence in the western Himalayas has been found in the form of lithic tools. For the protohistoric and early historic periods, the bulk of the remains consist of rock art. Although painted images, sometimes sheltered, were found, until now, petroglyphs along riverbanks and on plateaus are the salient feature of western Himalayan rock art. Tens of thousands of engraved images have been documented in Ladakh, Mustang, western Tibet and more recently in Spiti (Himachal Pradesh).

As with any archaeological material, when it comes to interpretation, rock art has its advantages as well as disadvantages. Apart from its sheer quantity, rock art is also characterized by its diversity. The subjects and themes represented are extremely varied, making the possibilities of analysis almost endless. Furthermore, the immobility of petroglyphs makes them the most reliable material for identifying and understanding the movement of ancient people. However, to date, no sound chronometric technique is available for dating petroglyphs. On the issue of chronology, experts depend on the combination of various parameters such as the degree of patina covering the images. Patina has been defined as 'visually obvious skin on rock surfaces which differs in colour or chemical composition from the unaltered rock and whose development is a function of time' (IFRAO, undated), Also important in chronology is knowledge of the techniques and tools used to produce the petroglyphs, as well as their thematic, stylistic and comparative analysis. Ideally, engraved images should be brought into perspective with representations found on other types of material (ceramics, textiles, metal objects, etc.)

through which cultural and chronological contexts can be asserted. However, archaeologists face another difficulty in the Himalayan region, that is, the lack of excavated materials from the area itself, and for this they have to look to sites further afield.

Within the Himalayas, rock art was first reported from Ladakh at the end of the nineteenth century, and this was when it was most thoroughly studied. In this chapter we, therefore, focuses on the petroglyphs of this western Himalayan region. More specifically we concentrate on the fauna of Ladakh through zoomorphic images engraved on rock surfaces. We provide further insights into the bird images, their spatial distribution and identification, and conclude with comments on their possible historical and cultural significance.

Animals in the Rock Art of Ladakh

Petroglyphs in Ladakh were first reported by Western explorers as early as the 1880s. In the first decade of the twentieth century, Moravian missionary and scholar A.H. Francke published several articles based on his discoveries along the Indus. Renowned geologists and Tibetologists H. De Terra and G. Tucci made further mention of petroglyphs in the 1930s. Ladakh attracted rock art researchers again only in the late 1980s and early 1990s from the region itself (T. Ldawa Tsangspa), India (R. Vohra, B.R. Mani) and abroad (P. Denwood, G. Orofino, H.-P. Francfort). During this period, specialized papers on Tibetan rock inscriptions and protohistoric zoomorphic images were published along with general presentations of rock art (for a detailed historiography of the rock art of Ladakh as well as all references, see Bruneau & Bellezza, 2013).

A systematic documentation of Ladakh's rock art was initiated in 1996 by M. Vernier in Central Ladakh and Zanskar. He recorded about 10,000 petroglyphs across the region (Vernier, 2007). Since 2006, we (the authors) have been working in collaboration and we have further investigated Lower and Upper Ladakh as well as the Nubra area. To date, we have identified 158 rock art sites (91 sites have been systematically documented and 67 surveyed only) totalling almost 20,000 petroglyphs. Most rock art sites are located on the banks of the Indus and its tributaries (Nubra, Shayok, Zanskar, Doda and Tsarab Rivers). The size of these sites varies from a single isolated rock to a cluster of up to about 1000 engraved boulders (for the location of rock art sites, their access to water and possible use as resting places, see Bruneau & Bellezza, 2013).

In total, we identified 74 motifs in the rock art of Ladakh, which were either figurative or nonfigurative. The content is mainly zoomorphic: out of 13,597 images systematically recorded 7270 (53%) were classified as animals. The various species depicted were mainly identified through their body shape, horns and tails. They were, in decreasing order of occurrence, Ibex, wild sheep, Yak, canids, indeterminate caprids, felids, equids, birds, deer, Markhor, Argali, camel and antelope. For about a quarter of zoomorphic images, the species was indeterminate.

The Ibex *Capra ibex sibirica*, usually recognizable by its long upward and backward-curving horns, accounted for almost half of the zoomorphic images

(47%). It is one of the most common wild ungulates found north of the Himalayan ranges (Himachal Pradesh, Kashmir) and in Central Asia (in the Pamirs, Tien Shan and Altai Ranges) (Schaller, 1998). In Ladakh, the Ibex currently extends east to the environs of Leh and north to the Nubra Valley (Namgail *et al.*, 2013). In engravings, it is represented all over the region and is found at 83 out of 91 sites. Although solitary images occur, we most commonly find several Ibexes engraved on the same boulder. Sometimes the intention of representing a herd, typically from 5 to 10 individuals, is unquestionable.

Apart from the Ibex, wild sheep were also commonly found in the rock art of Ladakh (8.4%). By this term we refer to images of Blue Sheep (Bharal, *Pseudois nayaur*) and Urial *Ovis vignei*: both species have massive, short horns sweeping up and out, but their representation in the petroglyphs is not distinctive enough to discern between them. Such images are found all over Ladakh. Nowadays, the Bharal occurs throughout the region, except the westernmost part, whereas the Urial has a linear distribution along the Indus and the Shayok river valleys (Namgail *et al.*, 2010). There are solitary carvings of wild sheep, but they most often appear in herds.

The Yak *Bos grunniens*, easily identifiable by long, curved horns, conspicuous hump and short tail terminating in a large bushy tuft, was third in abundance in the rock art of Ladakh (7%). Yaks are, more often than not, represented in isolation. When they are part of compositions, they occur along with other individuals of their species or with Ibexes. Representations are those of wild Yaks, now occurring only in the Chang Chenmo Valley of eastern Ladakh. We do not know of any irrefutable representation of domestic Yaks (loaded or on a lead for example).

Canines (accounting for 4.5% of zoomorphic images), recognizable by fairly short legs, short, upright, pointed ears and long tails, are difficult to identify with accuracy. Straight tails might be indicative of Red Foxes *Vulpes vulpes* or Grey Wolves *Canis lupus*, as both species inhabit Ladakh, whereas curved tails might point to dogs. Some compositions display packs of canines attacking herbivores.

Other carnivores, and in particular the Snow Leopard *Panthera uncia* (1.8%), easily identified by a long thin tail, curled at the tip, and a spotted body, were also depicted in the rock art of Ladakh. Such images were only documented in the Indus Valley, where the snow leopard is still found. These were most often represented in packs; only in two instances was the snow leopard shown attacking an Ibex or a deer.

The last group of animals of importance in the rock art of Ladakh belonged to the Equidae family (4.8% of zoomorphic images). Seventy-five per cent of these images included a rider, thus identifying the animals as horses. However, anatomical details are not sufficient to identify the type(s) of mounts. Unmounted animals with a long neck, long legs and tail may be either horses or the Tibetan Wild Ass '*Kiang' Equus kiang*. The Tibetan Wild Ass inhabits the entire Changthang (Tibetan Plateau) and was still common in eastern Ladakh at the beginning of the twentieth century (Schaller, 1998; Bhatnagar *et al.*, 2006). Images of the *kiang* are recognizable by their large head and robust body, but most of all by their short and bristly upright mane.

Other animals (birds, deer, Markhor *Capra falconeri*, Argali *Ovis ammon*, camel and antelope) accounted for about 1% (each) of zoomorphic engravings. Images of

Musk Deer *Moschus chrysogaster* and Tibetan Antelope *Pantholops hodgsonii* were of particular interest: both species currently inhabit areas of the Changthang, and their images suggest a wider distribution in the past, as do those of Markhor documented in the west of Ladakh. This wild goat is now found further west in the Gilgit-Baltistan province of Pakistan and in the Indian Himalayas.

Finally, of great significance are representations of double-humped Bactrian Camels *Camelus bactrianus*, sometimes mounted or on a lead. Most were documented in the Nubra Valley, where feral and domestic camels are still found, or in adjacent areas where they were used in the trans-Karakoram trade route until the middle of the twentieth century.

From this description of zoomorphic images, we note that the species depicted reflect a local fauna dominated by ibexes and wild sheep. These ungulates are still widespread in the region, whereas other species such as the wild Yak and antelope are now found only in eastern Ladakh. Their representation in rock art might reflect a narrowing of their habitat induced by a change of climate and by anthropogenic pressure. At the Neolithic site of Kiari (circa 900 BCE) in eastern Ladakh, bones of Himalayan Goral *Nemorhedus goral* were found, suggesting a modification in game and vegetation conditions, since this bovid lives in wooded environments (Ota, 1993). All the wild ungulates, except for camels, were represented in a cynegetic (i.e. hunting) context in the rock art of Ladakh. However, hunted animals accounted for a small percentage of zoomorphic representations. For example the Yak was being hunted in only a quarter of all images. In hunting scenes, the prey was often being chased or savaged by one or several dogs. Along with horses and camels, dogs were the only domestic species definitely represented in the rock art of Ladakh: all other species appear to be wild.

Because of the animals depicted, we suggest that the rock art of Ladakh was largely created by indigenous people familiar with the local fauna. Nevertheless, petroglyphs are not merely an inclusive natural history inventory: large animals such as the bear, and small mammals such as the marmot, hare and pika, which are common in nature, did not occur in rock art. The species represented in the rock art of Ladakh are influenced not only by the natural environment, but also by the cultural background of the engravers, the best example being that of the Ibex. The Ibex is still present in oral traditions and rituals of the Ladakhis as well as several peoples in the Pamirs, Hindu Kush, Iranian plateau and Caucasus, but is absent from other areas of Tibetan culture, thus underscoring the existence of ancient beliefs (Dollfus, 1988).

Bird Representations in the Rock Art of Ladakh

Among the zoomorphic petroglyphs of Ladakh, 130 images have been identified as representing birds. They were found at 22 rock art sites across Ladakh: all are indicated and labelled on the map (Figure 21.1).

Figure 21.1 shows that there were very few images in Zanskar (only six, and all located at one site: Zamthang, near Char village) or Nubra (13 images in total at four sites: Tsati, Murgi, Chomolung and Yulkam). A few bird images were located along the Indus (in central and lower Ladakh, from the site of Trishul in the east, to Tilichang in the

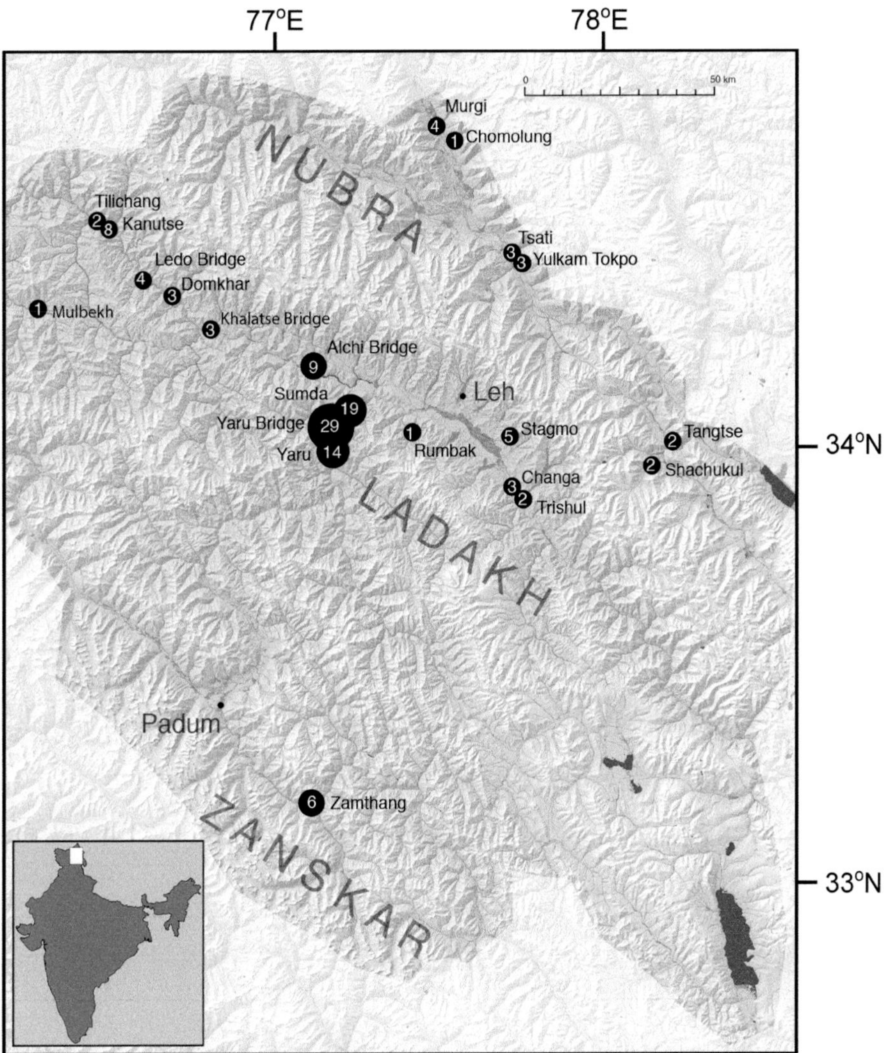

Figure 21.1. Location of rock art sites in Ladakh (India) with bird images and number of images at each site (Credit: M. Vernier / A. Pointet; reworked by H. van Oeveren). The white block in the top of the inset gives the location of the map.

west), but the main concentration of bird images was found in the Zanskar Gorge, on both banks of the river (Figure 21.2), about 25 km upstream of its confluence with the Indus. There, the three sites of Yaru, Yaru Bridge and Sumda accounted for half of all bird images recorded in the whole of Ladakh (62 images). It seems possible that this concentration could be partly explained by the Zanskar Gorge being an avian migration corridor. Except for a small number of images (one at Rumbak, two at Shachukul, one at Zamthang and one at Tangtse) displaying a white patina and a crude engraving technique, all bird images were characterized by a brown patina

Figure 21.2. View of the Zanskar Gorge upstream of the Zanskar River confluence with the Indus in Ladakh, India (Photo credit: M. Vernier). (A black-and-white version of this figure will appear in some formats. For the colour version, please refer to the plate section.)

ranging in colour from dark brown to medium brown, and they demonstrate a skilful engraving technique.

Birds in rock art were mainly represented in two different ways: flying (with one or both wings stretched) or standing on the ground (with wings folded back on their bodies) (Figure 21.3), respectively as seen from the front and side. Whilst more than 310 bird species have been reported from Ladakh (Pfister, 2004; Prins *et al.*, Chapter 20), the criteria of identification of the images are usually not very precise. Bird images tend to be schematic, and the presence of a beak only enables us to qualify the representation as that of a bird. In some cases the shape of the belly, wings and neck, the length and shape of the legs, as well as the presence of distinctive head or tail feathers, allows identification of the type of bird represented. As we will see, orders of birds can be identified through the images but not species (except in one case).

Many bird images are indeterminate: they show the bird as seen from the ground with outstretched and pointed wings depicted horizontally, downwards or upwards. The tail is generally quadrangular or V-shaped. A good example of this was documented at the site of Murgi (Figure 21.3: line A, number 7). A beak as seen in profile is sometimes, but not always, indicated. When it is depicted as hooked, we can identify the image as that of a raptor. In some instances, long tail feathers are clearly engraved, as on the images from Sumda (Figure 21.3: B16, C33–34). It is not possible to conclude whether buzzards,

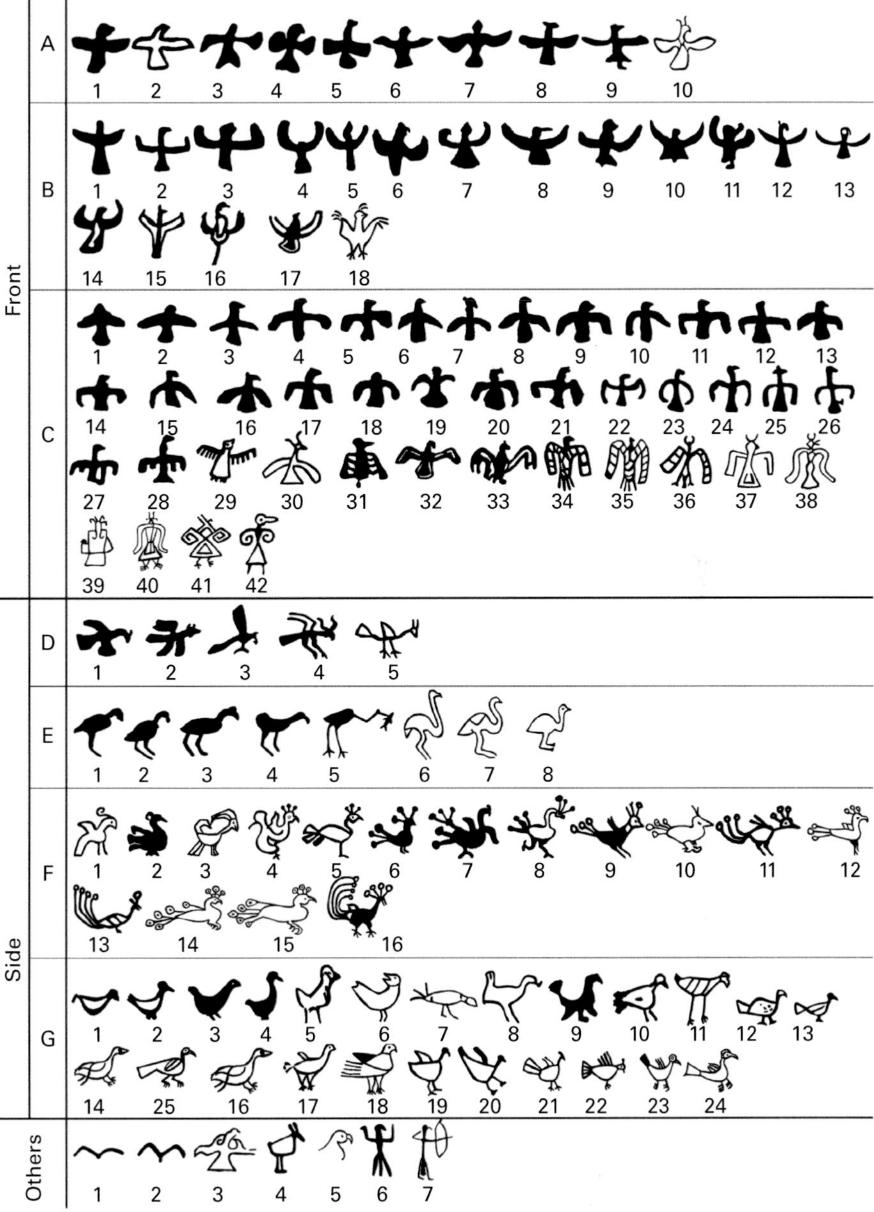

Figure 21.3. Engraved bird images from Ladakh, sorted by shapes and type. All images have been standardized (orientation and size) and redrawn according to photographic documentation for classification (Drawings: M. Vernier).

lammergeiers, kites, eagles or falcons, for example, are specifically represented. The bodies of raptors were usually fully engraved, whereas those of Galliformes or Anseriformes were, more often than not, outlined only (Figure 21.3: lines F and G). The rounded belly, rounded head and short legs enable us to hypothesize the representation of snowcocks, partridges, quails, geese or ducks, but without being able to be more specific. These birds were engraved as seen from the side and as if standing on the ground. Toes (usually three) were sometimes indicated and a wing was sometimes represented folded over their ovoid body shape or stretched above it (Figure 21.3: G14–23). The tail, pointing either downwards or upwards, was sometimes indicated by a triangle or signified by feathers. An eye, in shape of a circle or dot, was often engraved. The curved beak on some images may point to galliformes. Some engraved birds are distinguishable from these generic images by the depiction of a crest and long tail and may represent the Common Hoopoe *Upupa epops* that is widespread in Ladakh and is recognizable by its fanned erect crest (Pfister, 2004) (Figure 21.3: F4, F6, F8). However, some crested bird images display a long curved neck and long feathers, both from the crest and tail, ending in circles (Figure 21.3: F9–16). These recall the train and eyed feathers of peacocks. This species is indigenous to the Indian subcontinent and is found in moist and dry deciduous forests mainly below an altitude of 1800 m and in rare cases at about 2000 m. It is feasible that the climate of Ladakh was moister in the past (as suggested earlier by the bones of Himalayan goral excavated at the site of Kiari), but because of the high altitude it seems very unlikely that the peacock ever inhabited Ladakh.

Other images of relatively long-necked birds, but without crests, might refer to species found in the region: recognizable through their long legs, they are probably migratory birds of the Rallidae, Gruidae or Ardeidae (Figure 21.3: E5–8). This identification is proposed not only based on the way these birds were represented, but also on their context.

At the site of Yaru, four heron-like birds were represented in a group, arranged in a circle, a depiction of a water body, such as a pond or lake (Figure 21.4). This is confirmed by the image of a fish in its centre. This composition is surrounded by several other images of birds, represented as flying, as well as images of *khyungs*, the mythical horned eagle indigenous to the Tibetan cultural area, discussed separately later in this chapter. A unique petroglyph from Murgi (Nubra) shows a long-necked bird, most probably a crane, holding a fish in its beak. (Figure 21.3: E5). At Yulkam (Nubra), there are three heron-like birds associated with 22 other animal motifs, among which a lizard, a human hand, a mascoïd (a design representing a stylized human face or a mask), a bowman, Blue Sheep and Ibexes. *In situ* observation of the patina and engraving technique indicated that all images were contemporary with each other. The main image, by its relative large size (69 cm wide and 63 cm high) and central position, is that of an Ibex with bent legs, hindquarters and body ornamented by a volute and S motif, respectively. This type of zoomorphic representation, discussed at length in another paper, is typical of the 'animal style' or 'art of the steppes' (Bruneau & Vernier, 2010). Beyond its possible Iron Age (first millennium BCE) dating, what is of interest is the association of zoomorphic images in this specific style with birds (for a discussion of bird images in the 'animal style', see Kubarev & Zabelin, 2006). The same association (i.e. bird and Ibex) is found at the well-known and significant sites of Domkhar (lower Ladakh) and Zamthang (Zanskar) where birds display a fanned erect crest. The pair from the latter site was discussed at length by H.-P. Francfort, who proposed a dating, based on comparative analysis with petroglyphs from northern Pakistan and Chinese bronzes,

Figure 21.4. Rock engraved with four heron-like birds arranged around a circle, a possible pond or lake (left corner). This group is surrounded by several other images of flying birds and *khyungs*, Yaru rock art site (Ladakh) (Photo credit: L. Bruneau, Image enhanced using Adobe Photoshop©). (A black-and-white version of this figure will appear in some formats. For the colour version, please refer to the plate section.)

between the seventh and fourth centuries BCE (Francfort *et al.*, 1990). The association of bird and Ibex is not uncommon in the 'art of the steppes', as illustrated by a bronze piece from Kandia (Kohistan, northern Pakistan) representing a crouching Ibex with a bird's head attached to its horn (Jettmar, 1982). This piece was compared to similar bronzes from the Pamirs and dated to the fifth or third century BCE (Jettmar, 1982; Litvinsky, 1993; Parzinger, 2001). The Kandia plaque seems to have been locally produced because of the crested bird, most probably a Himalayan monal *Lophophorus impejanus*, a species not encountered in the Pamirs (Jettmar, 2002). We might also mention here a metal piece, acquired in Leh, in the shape of a bird of prey, comparable to pieces from the steppes, but unfortunately, its provenance being unknown, it does not provide much valuable information (Koenig, 1984). We should also mention here the images of four peacocks engraved at the site of Dachi (Dachi/Dartsigs, lower Ladakh), where stags, on the tips of their hooves and with their heads turned backwards, as well as felines with bodies ornamented with volutes and typical of the 'animal style' were found engraved nearby.

Other images of peacocks were found at the site of Yaru Bridge, but their cultural context is very different. One of these peacocks is standing with its beak

Figure 21.5. A bird holding a vase engraved next to a *stūpa* and an early Tibetan inscription, Alchi rock art site (Ladakh) (Credit: L. Bruneau & M. Vernier).

above a floral design. This association is found again at the same site with two birds (one peacock and one crested bird) holding the petals of a flower in their toes. We assume a Buddhist context for these bird images based on the fact that many of the petroglyphs at the site of Yaru Bridge are from the historic period and include *brāhmī* inscriptions (unpublished) as well as the historic composition of an elephant *Elephas maximus*, Yak and Blue Sheep (for more details about this composition, see Bruneau & Bellezza, 2013). This association (bird/flower) is also encountered at the rock art site at Alchi, where it is indubitably associated with Buddhist *stūpa* engravings. There, below the carving of a triple *stūpa* on a common platform, two birds (galliformes or anseriformes) are depicted standing and facing each other on the side of a lotus petal. The identification of the central motif as a lotus is made possible by the identical ornamentation of the *stūpa* platform. The identification of the lotus ornamentation in the carvings is confirmed by ruined *stūpas* all over Ladakh with similar decorations plastered on their bases (for a list of such *stūpas*, see Devers *et al.*, 2014). Below the *stūpa* and next to the birds is the representation of a vase, a well-known Buddhist motif.

Interestingly, on another boulder at Alchi, we encountered a bird holding a vase engraved next to a *stūpa* (Figure 21.5). This association of bird and *stūpa* is also found at the sites of Ledo Bridge (lower Ladakh) and Zamthang (lower Zanskar). It was therefore concluded that these bird images have a religious value. Indeed, the presence of birds in Buddhist iconography is attested in the earliest forms of Indian art (for example at Sanchi, Madhya Pradesh), and is well attested in the mural paintings of early Ladakhi temples, on the ceilings of the Alchi Sumtseg, for example. Many *stūpa* images from the rock art site of Alchi are accompanied by Tibetan inscriptions dated from the mid seventh to the early eleventh century CE, and the bird images may be attributed to the same period (Takeuchi, 2012).

Images of the *khyung* can also be assigned to the historic period (circa 100 BCE–1000 CE; for details of the historical division of Ladakh, see Bruneau & Bellezza, 2013). Notably, except for one representation at the site of Kanutse, the 10 images of the pre-Buddhist mythical horned eagle were all documented in the Zanskar Gorge, at the sites of Yaru and Yaru Bridge (Figure 21.3: A10; C30, C36–40; D1–2, D4–5). Apart from the presence of horns, the most distinctive feature of *khyung* images is the rendering of the tail in the shape of a triangle. Most *khyungs* in the rock art of Ladakh are represented as seen from the front with wings pointing downwards (Figure 21.3: C30, C36–40). There are Tibetan copper alloy bird talismans with downward-pointing wings, but these are not common (for such a Tibetan talisman, see Bruneau & Bellezza, 2013). In Tibet, the horned eagle in rock art appears to date as far back as the Iron Age, and the *khyung* remains an important religious and secular symbol down to the present day. According to the Bön religion, the *khyung* was the primary political emblem and genealogical symbol of Zhang Zhung, a prehistoric kingdom and culture based in Upper Tibet (for a preliminary survey of the horned eagle rock art of upper Tibet, see Bellezza, 2012; 2013; for some of the major cultural functions of the horned eagle, consult indexes in Nebesky-Wojkowitz, 1993; Bellezza, 1997 2005, 2008; and Norbu, 2013).

While the significance of the rock art *khyung* remains conjectural, its copper alloy counterpart clearly had a talismanic function (for images of *khyung* talismans, see Bellezza, 1998; Norbu, 2013). Whatever their significance, *khyung* rock images unquestionably have a symbolic value and underscore the cultural interconnections between Ladakh and western Tibet. This is reinforced by the association of the *khyung* images with *swastikas* (anti-clockwise) and the branched motif, engraved on the same boulders as the *khyung* images at Yaru and Yaru Bridge, which may refer to a Bön cultural background. About 100 swastikas were documented in the rock art of Ladakh, and motifs 10 ten branches (sometimes also referred to as 'fishbone' motifs). The branched motif appears to represent a tree and to have been imbued with symbolic or mythological meaning (for more details about the boulder from Yaru and the association of images, see Bruneau & Bellezza, 2013) (Figure 21.4). These motifs (*swastika*, branched motif and *khyung*) are only very exceptionally found in the rock art of Ladakh and noticeably not in association, except at the sites of Yaru.

Conclusion

The species engraved in the rock art of Ladakh are those of the local fauna, and this conveys information about the people who produced the images, that is, indigenous people, and environmental change, under the assumption that the petroglyphs represent a reasonably close relation between the motif and the local setting. It is clear that people did not make engravings of animals that they saw at faraway locations, nor were they influenced by exogenous imagery.

The images of mammals have a high preponderance of animals from a steppe environment (Yak, *Kiang*) intermixed with those from a montane environment (Ibex, Blue Sheep, Urial). Most representations of birds (cranes, galliform birds, Hoopoes, geese etc.) were those of species still occurring in the area or that might have occurred in the past. Representations of deer in rock art, but most of all, the excavated bones of a Himalayan Goral, may refer to more forested conditions and 'indicate that the area experienced better vegetation and game conditions than today for early humans to cope with the physiological and environmental constraints' in the first millennium BCE (Ganjoo & Ota, 2012). Since the images of peacocks are found in Buddhist and 'animal style' contexts, we cannot infer that Ladakh was more forested than it is now in the first millennium BCE or CE, but rather that this bird was introduced into Ladakhi imagery. Were these images a distant echo resulting from a fascination with neighbouring cultures, respectively Buddhist and steppic, or were they created by people originating from those cultures? In other words, were they directly or indirectly transmitted, and by what means: trade, military, religious or other links? We mentioned a small metal piece in the shape of a bird displaying steppic characteristics that was purchased in Leh but of which the provenance is unknown. This piece points to the existence of small and easily transportable artefacts that could have been seen in the region and served as models. Besides artistic copying from portable art, we should also mention here the possibility of copying pre-existing rock images. Thus, avian representations, and more generally zoomorphic images, may not only be naturalistic (i.e. result from the observation of nature), but also artistic, that is to say resulting from thematic and stylistic conventions. This may be especially true for bird images of line B (Figure 21.3) with their non-naturalistic wings pointing upwards.

Leaving the question of naturalistic versus artistic aside, quite uniquely, our iconographic work, which stands in a very different scientific tradition than ecology, and uses different sources of information, may provide sound evidence for animal behaviour which cannot otherwise be deduced from observations and which otherwise leaves few material traces. 'The avian fauna features of the past in any zone can be reconstructed through analysis of available paleontological data. . . . The main obstacle to collecting data on birds from the ancient past is the relatively poor preservation of feathers and bones. Determining species also presents considerable difficulty. . . . Bird images . . . serve as an important and in some cases even unique source of information on the bird taxa composition . . . in the remote past' (Kubarev & Zabelin, 2006).

As indicated by this chapter, rock art is crucial to the understanding of the ancient fauna of Ladakh and more generally its past. Yaru, Stakna and other nearby major rock art sites have already suffered severe damage in recent years, due to the construction of a road along the Zanskar River. We would therefore like to raise our reader's awareness of the need and urgency of preserving rock art in Ladakh and more generally in the Himalayas.

References

Bellezza, J.V. (1997). Notes on three series of unusual symbols discovered on the Byang Thang. *East and West*, **47**, 395–405.

Bellezza, J.V. (1998). Thogchags: talismans of Tibet. *Arts of Asia (The Asian Arts & Antiques Magazine)*, **28**, 44–64.

Bellezza, J.V. (2005). *Calling Down the Gods: Spirit-Mediums, Sacred Mountains, and Related Bon Textual Traditions in Upper Tibet*. Leiden: Brill (Brill's Tibetan Studies Library).

Bellezza, J.V. (2008). *Zhang Zhung: Foundations of Civilization in Tibet. A Historical and Ethnoarchaeological Study of the Monuments, Rock Art, Texts, and Oral Tradition of the Ancient Tibetan Upland*. Vienna: Verlag der österreichischen Akademie der Wissenschaften.

Bellezza, J.V. (2012). The horned eagle: Tibet's greatest ancestral and religious symbol across the ages. *Flight of the Khyung*, **January 2012**, www.tibetarchaeology.com/january-2012/.

Bellezza, J.V. (2013). High on the khyung. *Flight of the Khyung*, **January 2013**, www.tibetarchaeology.com/january-2013/.

Bhatnagar, Y.V., Wangchuk, R., Prins, H.H.T., Van Wieren, S.E. & Mishra, C. (2006). Perceived conflicts between pastoralism and conservation of the kiang Equus kiang in the Ladakh Trans-Himalaya, India, *Environmental Management*, **38**, 934–941.

Bruneau, L. & Bellezza, J.V. (2013). The rock art of upper Tibet and Ladakh: Inner Asian cultural adaptation, regional differentiation and the western Tibetan Plateau style, *Revue d'Études Tibétaines*, **28**, 5–161.

Bruneau, L. & Vernier, M. (2010). Animal style of the steppes in Ladakh: a presentation of newly discovered petroglyphs. In L.M. Olivieri, L. Bruneau & M. Ferrandi (Eds.) *Pictures in Transformation: Rock Art Researches between Central Asia and the subcontinent*. Oxford: Archeopress (BAR International Series 2167), 27–36.

Devers, Q., Bruneau, L. & Vernier, M. (2014). An archaeological account of ten early painted chortens in Ladakh. In E. Lo Bue & J. Bray (Eds.) *Art and Architecture in Ladakh, Cross-cultural Transmissions in the Himalayas and Karakoram*. Leiden: Brill (Brill's Tibetan Studies Library), pp. 100–140.

Dollfus, P. (1988). La représentation du bouquetin au Ladakh, région de culture tibétaine de l'Inde du Nord. In *Tibetan Studies, Proceedings of the 4th seminar of the International Association of Tibetan Studies-Münich 1985*, vol. **2**. Münich: Kommission für Zentralasiatische Studien, Bayerische Akademie der Wissenschaften, pp. 125–138.

Francfort, H.P. & Klodzinski, D., Mascle, G. (1990) Petroglyphes archaïques du Ladakh et du Zanskar. *Arts Asiatiques*, **45**, 5–27.

Ganjoo, R.K. & Ota, S.B. (2012). Mountain environment and early human adaptation in NW Himalaya, India: A case study of Siwalik Hill Range and Leh Valley, *Quaternary International*, **269**, 31–37.

IFRAO (International Federation of Rock Art Organisations), e-RockArtGlossary, www.ifrao.com/ifrao-glossary/

Jettmar, K. (1982). Petroglyphs and early history of the Upper Indus Valley: the 1981 expedition – a preliminary report. *Zentralasiatische Studien*, **16**, 293–308.

Jettmar, K. (2002). Rock art in northern Pakistan. Researches between 1979–1989. In E. Kattner (Ed.) *Beyond the Gorges of the Indus. Archaeology before Excavation*. Karachi: Oxford University Press, pp. 80–109.

Koenig, G. (1984). Skythen in Tibet? In C. Muller & W. Raunig (Eds.) *Der Weg zum Dach der Welt*. Innsbruck: Pinguin-Verlag, pp. 318–320.

Kubarev, V.D. & Zabelin, V.I. (2006). Avian fauna in Central Asian rock art: archaeological and ethnological evidence. *Archaeology, Ethnology & Anthropology of Eurasia*, **26**, 87–103.

Litvinsky, B.A. (1993). Pamir und Gilgit Kulturhistorische Verbindunge. In K. Jettmar (Ed.) *Antiquities of Northern Pakistan: Reports and Studies*, vol. **2**. Mainz: Philipp von Zabern, pp. 141–149.

Namgail, T., Van Wieren, S.E., Mishra, C. & Prins, H.H.T. (2010). Multi-spatial co-distribution of the endangered Ladakh urial and blue sheep in the arid Trans-Himalayan mountains. *Journal of Arid Environments*, **74**, 1162–1169.

Namgail, T., Van Wieren, S.E. & Prins, H.H.T. (2013). Distributional congruence of mammalian herbivores in the Trans-Himalayan mountains. *Current Zoology*, **59**, 116–124.

Nebesky-Wojkowitz, R. de (1993). *Oracles and Demons of Tibet. The Cult and Iconography of the Tibetan Protective Deities, first reprint*. Kathmandu: Tiwari's Pilgrims Book House.

Norbu, C.N. (2013). *A history of Zhang Zhung and Tibet, Volume One, The Early Period*. Berkeley, CA: North Atlantic Books.

Ota, S.B. (1993). Evidences of transhumance from Ladakh Himalayas, Jammu & Kashmir, India. In R.K. Ganjoo & S.B. Ota (Eds.) *Current Advances in Indian Archaeology*, vol. **1**. Nagpur: Dattsons, pp. 79–95.

Parzinger, H. (2001). Bemerkungen zu einigen Tierstillbronzen zwischen Karakorum und Pamir. In R.M. Boehmer & J. Maran (Eds.) *Lux orientis: Archäologie zwischen Asien und Europa, Festschrift für Harald Hauptmann zum 65. Geburtstag*. Rahden, Westf.: M. Leidorf, pp. 321–326.

Pfister, O. (2004). *Birds & Mammals of Ladakh*. New Delhi: Oxford University Press.

Schaller, G. (1998). *Wildlife of the Tibetan Steppe*. Chicago: University of Chicago Press.

Takeuchi, T. (2012). Old inscriptions near Alchi, in historical development of the Tibetan languages. *Journal of Research Institute*, **49**, 29–69.

Vernier, M. (2007). *Exploration et documentation des pétroglyphes du Ladakh. 1996–2006*. Como: Fondation Carlo Leone et Mariena Montandon

22 Pastoralism and Wetland Resources in Ladakh's Changthang Plateau

Sunetro Ghosal and Monisha Ahmed

Contested Ecologies of Changthang

The high-altitude plateau in eastern Ladakh, India, called Changthang (*byang-thang*: 'northern plains') is home to a rich assemblage of birds. The plateau is characterized by different habitats, including wetlands that serve as staging and breeding grounds for a diversity of migratory birds, and the wetland resources are shared with pastoral nomadic communities called the 'Changpas' (Northerners). The Changpas have traditionally relied on their herds of goat, sheep and yak for subsistence, by husbanding the pastures of the high-altitude steppe of Changthang and negotiating resource usage with the wild animals. However, changes over the past few decades have altered social relations within Changpa society, redistributed access to pastures, included new actors and introduced new economic pressures. In this chapter, we discuss the complexity of these changes and their impact on wetland resources in the region.

The Landscape of Changthang

Ladakhi Changthang is situated at an average height of 4500 m above mean sea level, which is around the upper limit for agriculture for this latitude. The short summer months (May/June to August) record maximum temperatures of 30°C and higher, while the temperature in the winter months (October to March/April) can drop to minus 45°C or lower.

Like the neighbouring regions of Tibet and the higher reaches of the Spiti Valley, high altitude and a short growing season limit cultivation on the alpine steppe in Changthang (Miller, 1999; Mishra *et al.*, 2003). However, high precipitation compared to other parts of Ladakh helps sustain a high-altitude steppe (Miller, 1999), on which a diversity of mammalian and avian fauna depend (see Pfister, 2004; Namgail *et al.*, 2007).

Changthang has historically been a complex natural, cultural and economic landscape. While some cultivation is practised at lower elevations of Changthang, nomadic pastoralism has historically been the main source of livelihood for local residents. Miller (1999) draws an 'ecological distinction' between the pastoral nomads of the Tibetan plateau in relation to those in other parts of the world. He argues that nomadic pastoral movements in Eurasia and Africa are driven by 'water or the lack of it', while the important factor on the Tibetan Plateau is altitude (Miller, 1999, p. 16) – which is an important determinant of vegetation growth across different seasons – in addition to availability of water. This is further supported by Namgail *et al.* (2012), who report a hump-shaped relationship between phytomass along altitudinal gradients in Changthang, which they link with precipitation and grazing pressure.

The landscape of Changthang is marked by high-altitude steppe, brackish lakes, wetlands and mountains, while the climate is characterized by extreme cold, strong winds, relative aridity and high radiation (Rizvi, 1996; Namgail *et al.*, 2007). The lakes, rivers and other wetlands serve as important breeding and feeding grounds for resident and migratory birds. In the past, Changpas collected salt and soda from lakes such as Tsokar for consumption and barter (Ahmed, 1997a). The lush meadows along the shores of the high-altitude lakes and the banks of the Indus and Hanle Rivers are also used as grazing pastures for domestic sheep, goats, yaks and horses.

Ladakh's Changthang hosts a rich assemblage of birds, including migratory breeders and passage migrants (Pfister, 2004). Many avian species use the wetlands to breed, while others use them as important staging sites during their migrations between South Asia and breeding grounds further north (Pfister, 2004). Tsomoriri and two wetlands nearby are the only known breeding sites in India for the Bar-headed Goose *Anser indicus* (Prins & Van Wieren, 2004). In 2002, Tsomoriri was declared a Ramsar site, a Wetland of International Importance (Ramsar Site No. 1213), while four other areas have been identified as potentially important sites for birds. The wetlands of Changthang are also used by Ruddy Shelduck *Tadorna ferruginea*, Great Crested Grebe *Podiceps cristatus*, Black-necked Crane *Grus nigricollis*, Common Redshank *Tringa totanus* and several other species of birds adapted to breed and survive in high-altitude wetlands (see Prins *et al.*, Chapter 26).

Pastoralism in Changthang

'Changpa' is a generic term used for resident pastoralists on the Changthang Plateau. However, they are not a single homogeneous community and are divided into groups, based on their place of origin, with each having its own social hierarchy and specified grazing areas. The estimated population of Changpas is around 9000 individuals (LAHDC-Leh, 2014).

The Changpas rear sheep, Changra goats (which yield *pashm* or *cashmere* wool) and yaks. A number of them also maintain horses and donkeys for transportation. According to official records, there were an estimated 168,000 goats and 46,000 sheep in Changthang in 2012 (LAHDC-Leh, 2014). While pastoralism is the main occupation for the Changpas, some do practise agriculture during the short summer, tending to some of the world's highest fields to cultivate barley and potatoes (Ahmed, 2009). In the context of the ecological dynamics and climatic variability in Changthang, the Changpas' main survival strategy revolves around tending to their herds and management of pastures. Milk is sourced from their livestock and is an important source of nutrition for the Changpas (Goodall, 2007). Meat is also an important part of the

Changpa diet, and sourced through periodic and often seasonal slaughter of their animals. The herds also supply wool, which is an important commodity that the Changpas use and sell. The wool (from sheep and yaks) and hair (from goats and yaks) are used to weave a range of fabrics (Ahmed, 2002). *Pashm*, the winter undercoat of Changra goats, forms the raw material to make *pashmina* cloths (Ahmed, 2004). This is generally traded and rarely used by the Changpas, who report that the fibre is difficult to weave as it breaks easily when stretched. Animal skins are used and also traded. Historically, these products were bartered for barley, tea and agricultural produce from other parts of Ladakh and neighbouring Himalayan regions.

Over several generations, Changpas have developed an intricate body of traditional ecological knowledge to manage resources and negotiate the unpredictability of climatic cycles. For instance, like other pastoral communities in the Trans-Himalayas, the Changpas practise overstocking as an insurance against the risks of unpredictable events such as extreme weather cycles that take a heavy toll on their herds (Mishra *et al.*, 2003). However, many changes have taken place in Changthang since 1947, when Ladakh became a part of independent India. One important change occurred with the Sino-Indian war of 1962. The immediate impact of this geopolitical event was the creation of a Line of Actual Control between India and China, which cut through the traditional migratory routes of several Changpa groups (Singh *et al.*, 2013). It also resulted in an increased presence of Indian security forces, influx of Tibetan pastoralists, introduction of formal government institutions and infrastructure development (Chaudhuri, 2000; Goodall, 2007). Since 1994, certain parts of Changthang have been opened for tourism – especially around the wetlands of Tsomoriri, Tso Kar and Pangong Tso – which was further expanded in May 2014 (Humbert-Droz, Chapter 23). Each of these changes has had a significant impact on the ecological processes in Changthang.

Pastoral Dynamics

'Today, the government brings rations to our doorstep. This has made life as a nomad much easier. In the past, we trekked deep into Tibet and to Lahul-Spiti to get these supplies," said a 78-year-old Changpa man near Korzok in 2005. The observation of the elderly Changpa highlights the impact of changes under way in Changthang. Unpredictability and variability are key features of Changthang's ecological systems, and the Changpas have developed relatively flexible strategies to adapt to this ecological variability. However, political and economic changes in the past few decades have had far-reaching impacts on Changthang's intricate socio-ecological systems. These changes include the loss of access to pastures and wetlands in western Tibet (Singh *et al.*, 2013; Angmo, 2014; Dolkar, 2014; Stobdan, 2015). In addition, the Changpas have stopped travelling to Himachal Pradesh and Zanskar, as they did in the past, and have lost access to the pastures they used along the way (Ahmed, 2004). The Changpas have responded to these losses by realigning their migratory routes. They have also reduced the number of yaks, which require large amounts of grass and are no longer essential for migrations with the development of roads.

After Tibet came under Chinese rule in 1959, hundreds of Tibetan pastoralists crossed into Ladakh along with their livestock (Ahmed, 2013). Some have settled in refugee camps closer to Leh, while others live in Changthang. It is estimated that a total of 20,000 Tibetans have settled in Ladakh, of which about 3500 have settled in Changthang (Ahmed, 2013). Initially, the Tibetan pastoralists did not follow a fixed migratory path. This changed in 1977, when the Central Tibetan Administration in Dharamsala (India), with help from the Government of India and the state governments of Jammu and Kashmir, organized them into groups with fixed migratory routes. In Changthang, they were settled in nine places: Nyoma, Kakjung, Koyul, Hanle, Sumdo, Samad, Kharnak, Chushul and Chumur, which changed the migration patterns and pasture use of Changpa groups whose traditional pastures were allocated to Tibetan pastoralists. The increased population of livestock and changes in resource management strategies had the direct impact of intensifying pressures on pastures and wetlands in Changthang.

The number of households in Changpa society has increased in the past few decades. This is largely driven by the availability of new economic opportunities, and by the shift from the tradition of polyandry, where one woman married two or more brothers, to monogamy, after polyandry was banned in 1941 (Goodall, 2007; Namgail, 2015). Each nuclear family tends to live in one large tent (*reb-chen*), with elderly parents residing in a smaller tent nearby (*reb-chung*). As children get married and start their own families, they prefer to set up their own tents in the vicinity of their parents' tents. This division of households was less common in the past, as making a new tent required large amounts of yak and goat hair. Now, new tents are made from thick, white canvas available in Leh. For instance, Namgail (2015) reports that the number of Changpa households in Sato-Kargyam has increased from 15 to 109 in the past two decades. However, reduced resource availability, shortage of manpower and new economic opportunities in urban areas have fuelled seasonal and permanent settlement of Changpas near Leh, the capital town of Ladakh, especially from Rupshu-Kharnak – an estimated 306 people or 25% of this community moved permanently to Leh between 1961 and 2001 (Goodall, 2004, 2007). At the same time, the Changpas in Changthang, especially the young people, have started exploring alternative livelihoods that are now accessible in different parts of Changthang. This includes jobs as drivers, tourist guides, porters, daily-wage labour, shopkeepers and government jobs (Namgail, 2015).

In 1983, the Changpas were integrated into the Public Distribution System by the Indian government. This had a significant impact on their diet and economic systems, which were earlier oriented towards subsistence and barter. In addition, Changpas source supplies from security agencies stationed in the region. In periods of distress – like the heavy snowfall in the winter of 2013 – the Ladakh Autonomous Hill Development Council, Leh, supplies animal feed sourced from outside the area. These changes have thus diversified the subsistence and barter economy of the Changpas and integrated them into a market economy with an increased dependence on government subsidies and requirement for money to purchase various supplies (Ghosal, 2005).

The New Changpa Economy and Wetland Resources

Despite these changes, nomadic pastoralism remains the cornerstone of Changpa society, though these practices have also changed significantly (Namgail, 2015; Namgail *et al.*, 2007). This is most evident in terms of herd composition, which now has more *pashm*-producing goats – the goats produce the fine quality of *pashm* only when they are exposed to the high altitude, cold and harsh winter conditions of Changthang (Rizvi, 1999; Ahmed, 2004).

The *pashm* combed from the goats is sold to traders from Leh, who in turn sell it to traders in Srinagar. A male goat yields around 300 g of *pashm*, though larger individuals may even yield 500 g, while a female goat yields around 200 g to 250 g (Ahmed, 2004; Namgail *et al.*, 2007; Namgail *et al.*, 2010). Trading practices have changed since the 1960s, when traders from Himachal Pradesh would collect most of the *pashm*, while the bulk of the *pashm* now goes to Kashmir. In addition, some Changpa have also turned middlemen and deal directly with traders from Srinagar.

Before 1960, most *pashm* reaching Kashmir originated in western Tibet with Ladakh serving as a conduit. This trade was controlled through tight regulations and treaties, with transgressors facing severe punishment, including death. The first agreement was the Treaty of Tingmosgang of 1684 that stipulated that Tibetan authorities would supply their entire wool and *pashm* to Ladakh, who would in turn supply it, along with their own produce, to Kashmir (Petech, 1977). This was reiterated in the 1842 Treaty of Leh. As a result of these treaties, *pashm* from western Tibet travelled through Ladakh to Kashmir. The amount of *pashm* produced in Ladakh's Changthang was meagre compared to that of western Tibet and as a result the Changpas got a much lower price for their *pashm* than their Tibetan counterparts (Ahmed, 2002). It was only after the closure of the trade routes to Yarkand and western Tibet that the weavers in Kashmir started depending on *pashm* from Changthang. The income earned from the sale of raw *pashm* ranged from Rs. 2 per kilo in the 1950s to Rs. 5000 per kilo in 2014 (Ahmed, 2004; Goodall, 2007).

Given the lack of economic research in Ladakh over the past few decades, it's difficult to accurately compare the value of a rupee across time. However, taking an annual average inflation rate of 7.8% in India, based on the Consumer Price Index between 1960 and 2015 (inflation.eu 2015), suggests that despite devaluation of the Indian rupee, the Changpas are earning significantly more cash in 2014 for *pashm* than they did in the 1960s.

The most evident impact of this integration of the Changpas to the market economy is on their herd composition. Earlier, sheep were favoured over goats, as they are better adapted to the cold, easier to herd and there was greater demand for their meat and wool. The Changpa herds used to have a ratio of three sheep for every goat in the early 1990s, which was equal by the end of the 1990s, and by 2009 it had changed to three goats for each sheep. The ecological impact of the change in herd composition remains unclear as there have been no long-term studies on this process in Changthang. Furthermore, the foraging habits of goats and sheep remain understudied. Some Changpas claim that goats damage rangelands in the long term, which is supported by biologists George Schaller and Martin Williams, who work in the Tibetan Changthang (pers. comm., 2009). The former argues that goats cause more damage as they are non-selective feeders and paw the ground to feed on roots and the base of grasses, which could be detrimental around wetlands and disturb breeding birds. However, he adds that the evidence is not conclusive and that climate change and overgrazing are important contributing factors to habitat degradation on the Tibetan Changthang. While the impact of goats in Changthang is unclear, a more important issue is that of overstocking, which has been a traditional practice to mitigate risks related to unpredictable climatic events (Fox *et al.*, 1994). As pasture management systems change, there has been a push from the Sheep Husbandry Department to increase herd sizes for higher *pashm* yield. Studies in Spiti suggest that overstocking, in the absence of social mechanisms to limit the impact on the pastures can lead to overgrazing and ecological degradation (Mishra *et al.*, 2001; Chandrasekhar *et al.*, 2007). Similar patterns are emerging in Changthang too, where a formerly negative attitude towards goats has been replaced by market-oriented pastoral practices, which has eroded earlier conservation practices and ecological knowledge. While there have been retaliatory killings of large carnivores in the past, these were done to control the population of predators and not to locally exterminate them, which seems to have been the objective of the anecdotal reports of wolf killings in 2013–2014. In 2014, a 75-year-old Changpa man from Hanle lamented these changes in Changpa relationships with wildlife:

> We no longer value wildlife such as *shangku* (wolves), *kiang* (Tibetan wild ass), *Cha thung thung* (Black-necked Crane) . . . [unless] they bring us monetary benefit . . . we have forgotten that without them, our goats and sheep will soon become wild.

This is further highlighted by the emerging conflict between wildlife and Changpas in Changthang (Bhatnagar *et al.*, 2006; Bhatnagar *et al.*, 2007). While the current data on the ecological impact of overstocking in Changthang are meagre, evidence from similar landscapes in Spiti does give cause for concern for wildlife conservation and the future of pastoralism in Changthang (Mishra *et al.*, 2001). The abundant livestock, apparently, deplete the wetland vegetation, which is an important food source for species such as Bar-headed Goose. Furthermore, the grazing livestock are known to disturb the foraging activities of breeding birds, and herding dogs are known to attack chicks and eggs of migratory breeders (Pfister, 2004). The inter-generational loss of ecological knowledge and erosion of institutional mechanisms of management are further accelerated by the current education system of centralized residential schools in Puga and Sato-Kargyam for Changpa children. As the children stay at these schools, they spend little time with their parents and grandparents to access their knowledge and experience, while learning skills required to practise pastoralism in Changthang (Namgail, 2015).

The Changpas' pasture management systems have historically included the wetlands of Changthang, which are crucial habitats for wildlife, especially migratory birds. The Changpas historically used the wetlands for grazing, as a source of fresh water and in some cases, to collect salt and soda. However, historically the Changpas also had access to several more wetlands, many of which are in western Tibet and no longer accessible to them. The wetlands are also sites for tourism, which in its current form has had a negative impact on the ecology (Humbert-Droz, Chapter 23). A large share of the revenue generated by tourism goes to Leh-based travel agents with the Changpas playing a peripheral role and deriving marginal benefits, despite the negative ecological impact of tourism (Ghosal, 2005; Geneletti & Dawa, 2009). Thus, in the context of tourism and market dynamics, Changthang remains a marginal economic landscape, with the Changpas being peripheral economic players. The current processes of change outlined earlier will probably result in ecological degradation and will marginalize the landscape and the Changpas further, and in turn undermine conservation of Changthang's rich ecosystem.

These changes are closely intertwined with each other, especially the socioeconomic and ecological processes, with wetlands serving as a barometer of their impacts. For instance, changes in social structure, migratory routes, marketization of pastoral production and herd composition have increased grazing pressure around wetlands and tested the socio-ecological resilience of resource management institutions. Namgail *et al.* (2012) and Raghuvanshi *et al.* (2016) also report structural changes in vegetation around Tangtse and Pangong Tso in the Durbuk block of Changthang with an observed increase of hardy plants such as creeping thistle *Cirsium arvense*, which are known to grow aggressively and outcompete others in terms of nutrients and moisture. Such changes are detrimental for pastoralism, as these plants are unpalatable, and also for wetland ecosystems, as they lead to reduced plant diversity. In addition, as alluded to earlier, the feral dogs around wetlands in Chushul and Korzok have been observed to cause nesting failures in breeding birds. While the Changpas do breed dogs to guard livestock, these feral dogs are primarily linked to the presence of security forces and waste generated by unregulated tourism.

At the same time, not all the change has been detrimental to wetlands. For example, in some places Changpas have asserted their community-based traditional rights over the land and its resources to check the ecological impact of unregulated tourism on pastures and wetlands. A good example is the Changpa community in the Tsokar Basin, who have reserved specific sites for tourist camps. Members of this community take turns to monitor the impact of these camp sites, collect money from travel agents to maintain the sites, impose fines for any violations and dispose of waste generated from the campsites to prevent it from entering the wetland ecosystem. This community has asserted its control over the wetlands and the surrounding pastures. In the past, this community has engaged in physical fights, and as recently as the 1980s taken recourse to the court of law to safeguard their control over the area and its resources (Ahmed, 1997b). Similarly, the Changpas in Tsokar and Korzok have established home-stays, with support from the Department of Wildlife Protection and NGOs such as the World Wildlife Fund, to redistribute the economic benefits of tourism and reduce the demand for tents for tourists.

References

Ahmed, M. (1997a). The salt trade: Rupshu's annual trek to Tsokar. In Van Beek, M., Bertelsen, B. & Pedersen, P. (eds.) *Ladakh: Culture, History and Development between Himalaya and Karakoram. Recent Research on Ladakh* **8**, pp. 32–48. Aarhus (Denmark): Aarhus University Press.

Ahmed, M. (1997b). *We Are Warp and Weft – Nomadic Pastoralism and the Tradition of Weaving in Rupshu (Eastern Ladakh)*. Oxford: Oxford University.

Ahmed, M. (2002). *Living Fabric: Weaving among the Nomads of Ladakh Himalaya*. Bangkok: Orchid Press.

Ahmed, M. (2004). The politics of pashmina: the Changpas of eastern Ladakh. *Nomadic Peoples*, **8**, 89–106.

Ahmed, M. (2009). Why are the Rupshupa leaving the Changthang? In Ahmed, M. & Bray, J. (eds.) *Recent Research on Ladakh 2009: Papers from the 12th Colloquium of the International Association for Ladakh Studies, Kargil*, International Association for Ladakh Studies, pp. 145–151.

Ahmed, M. (2013, January). Negotiating Place on Ladakh's Changthang – What Happened after 1959? The Himalayan Impasse, University of Bonn: Unpublished paper (presented at the conference).

Angmo, D.K. (2014). Making sense of the border incursions. *Stawa*, **1**, 8–9.

Bhatnagar, Y.V., Seth, C.M., Takpa, J., *et al.* (2007). A strategy for conservation of the Tibetan Gazelle Procapra picticaudata in Ladakh. *Conservation and Society*, **5**, 262–276.

Bhatnagar, Y.V., Wangchuk, R., Prins, H.H.T., Van Wieren, S.E. & Mishra, C. (2006). Perceived conflicts between pastoralism and conservation of the Kiang Equus kiang in the Ladakh Trans-Himalaya, India. *Environmental Conservation*, **38**, 934–941.

Chandrasekhar, K., Rao, K.S., Maikhuri, R.K. & Saxena, K.G. (2007). Ecological implications of traditional livestock husbandry and associated land use practices: a case study from the Trans-Himalaya, India. *Journal of Arid Environments*, **69**, 299–314.

Chaudhuri, A. (2000). Change in Changthang: to stay or to leave? *Economic and Political Weekly*, **35**, 52–58.

Dolkar, T. (2014). Unstable borders: India vs China? *Stawa*, **1**, 4–7.

inflation.eu. (2015). *Inflation India 1960:* www.inflation.eu/inflation-rates/india/historic-inflation/cpi-inflation-india-1960.aspx. (accessed: 6 Dec 2015).

Fox, J.L., Nurbu, C., Bhatt, S. & Chandola, A. (1994). Wildlife conservation and land-use changes in the Transhimalayan region of Ladakh, India. *Mountain Research and Development*, **14**, 39–60.

Geneletti, D. & Dawa, D. (2009). Environment impact assessment of mountain tourism in developing regions: a study in Ladakh, Indian Himalaya. *Environment Impact Assessment Review*, **29**, 229–242.

Ghosal, S. (2005). *The Nomad's Journey: From Barter to Cash*. London: University of London, School of Oriental and African Studies.

Goodall, S. (2004). Rural-to-urban migration and urbanization in Leh, Ladakh: a case study of three nomadic pastoral communities. *Mountain Research and Development*, **24**, 220–227.

Goodall, S. (2007). *From Plateau Pastures to Urban Fringe: Sedentarisation of Nomadic Pastoralists in Ladakh, North-West India*. Adelaide (Australia): University of Adelaide, Department of Geographical and Environment Studies.

LAHDC-Leh. (2014). *Statistical Handbook of Leh*. Leh-Ladakh: Ladakh Autonomous Hill Development Council, Leh.

Miller, D.J. (1999). Nomads of the Tibetan Plateau rangelands in western China part two: pastoral production practices. *Rangelands*, **21**, 16–19.

Mishra, C., Prins, H.H.T. & Van Wieren, S.E. (2001). Overstocking in the Trans-Himalayan rangelands of India. *Environmental Conservation*, **28**, 279–283.

Mishra, C., Prins, H.H.T. & Van Wieren, S.E. (2003). Diversity, risk mediation, and change in a Trans-Himalayan agropastoral system. *Human Ecology*, **31**, 595–609.

Namgail, S. (2015). Changes in the institution of family in the Changpa community. *Ladakh Studies*, **32**.

Namgail, T., Bhatnagar, Y.V., Mishra, C. & Bagchi, S. (2007). Pastoral nomads of the Indian Changthang: production system, land use and socioeconomic changes. *Human Ecology* **35**, 497–504.

Namgail, T., Rawat, G.S., Mishra, C., Van Wieren, S.E. & Prins, H.H.T. (2012). Biomass and diversity of dry alpine plant communities along altitudinal gradients in the Himalayas. *Journal of Plant Research*, **125**, 93–101.

Namgail, T., Van Wieren, S.E. & Prins, H.H.T. (2010). Pashmina production and socio-economic changes in the Indian Changthang: implications for natural resource management. *Natural Resources Forum*, **34**, 222–230.

Petech, L. (1977). *The Kingdom of Ladakh*. Rome: Istituto Italiano per il Medio ed Estremo Oriente.

Pfister, O. (2004). *Birds and Mammals of Ladakh*. New Delhi: Oxford University Press.

Prins, H.H.T. & Van Wieren, S.E. (2004). Number, population structure and habitat use of Bar-headed Geese Anser indicus in Ladakh (India) during the brood-rearing period. *Acta Zoologica Sinica*, **50**, 738–744.

Raghuvanshi, M.S., Mishra, A.K., Tewari, J.C., Landol, S., Stanzin, J. & Bhatt, R.K. (2016). The growing threat of creeping thistle in Changthang. *Stawa*, **3**, 11–12.

Rizvi, J. (1996). *Ladakh: Crossroads of High Asia*. New Delhi: Oxford University Press.

Rizvi, J. (1999). The trade in pashm and its impact on Ladakh's history. In Van Beek, M., Bertelsen, B. & Pedersen, P. (eds.) *Ladakh: Culture, History and Development between Himalaya and Karakoram. Recent Research on Ladakh* **8**, Aarhus: Aarhus University, pp. 317–338.

Singh, N.J., Bhatnagar, Y.V., Lecomte, N., Fox, J.L. & Yoccoz, N.G. (2013). No longer tracking greenery in high altitudes: pastoral practices of Rupshu nomads and their implications for biodiversity conservation. *Pastoralism: Research, Policy and Practice*, **3**, 1–15.

Stobdan, P. (2015). India and China: adversaries or partners? *Stawa*, **2**, 14–16.

23 Impacts of Tourism and Military Presence on Wetlands and Their Avifauna in the Himalayas

Blaise Humbert-Droz

The Himalayas form a unique and popular tourist destination. A growing number of tourists visits various Himalayan lakes for their outstanding scenery and to observe migratory waterbirds. The Himalayan range, which straddles the boundary of several countries, also witnesses a substantial army presence, and military activities around the lakes add to pressure on the wetland ecosystem. In view of its growing economic and strategic importance, the region is seeing the rapid development of new lines of communication and other infrastructure, which further impacts its wetlands.

Here I review these modern uses of Himalayan wetlands, with a focus on the lakes of the western Tibetan Plateau, which form part of the Trans-Himalayan region of Ladakh in northwest India. These wetlands, hosting a diverse avifauna characteristic of the Plateau – Black-necked Crane, Bar-headed Goose, more than 60 species of migratory waterbirds (Humbert-Droz, 2011; several chapters in the present book) – are seen as emblematic of tourism and other recent developments in the region.

Tsomoriri, Tsokar and Pangong Tso, the largest lakes of the region, are located close to the international border. Easily accessible from urban centres during the summer months, the lakes bear the brunt of modern development, a process I have documented from its inception in the mid-1990s through annual environmental and waterbird surveys.

Himalayan Tourism

Himalayan tourism is a comparatively recent industry. Nepal opened to tourists in 1949, Bhutan and most of Ladakh in 1974, the Tibet Autonomous Region in 1979. The Indian portion of the Tibetan Plateau in eastern Ladakh and the neighbouring Spiti Valley in the State of Himachal Pradesh were the last to open their doors starting in 1993, one year after the Mustang Valley, bordering Tibet in north-central Nepal (East *et al.*, 1998; Gurung & DeCoursey, 1999; Nepal, 2000).

Tourism development has been rapid throughout the region, though uneven due to recurring political and international tensions in several areas. Nepal, one of the most popular Himalayan destinations, which had a trickle of under 1000 arrivals per year until the early 1960s, has seen regular and sizable increase, barring years of peak

Maoist insurgency in the early 2000s. The country registered close to 800,000 arrivals per year in 2012–2014 (Nepal, 2000; Nepal Tourism Statistics, 2010–2014). Tibet has also witnessed a tremendous increase in tourist numbers – from barely 1000 in 1980 to 15 million in 2014 – but with significant drops in some years on account of political upheaval and subsequent restrictions on tourism (Lihua & Jingming, 2002; China Tibetology Research Centre, 2009; Xinhua, 2015). The Kashmir Valley had more than 700,000 visitors per year until tourism was hit by militancy in 1989. Numbers dropped to fewer than 70,000 per year in the early to mid-1990s, only to recover to pre-militancy levels 10 years later. From then onwards, with militancy at a generally lower level, tourism flow has increased steadily, reaching nearly 1.3 million in 2012 (East *et al.*, 1998; Islam, 2014; Malik & Bhatba, 2015).

Similarly, neighbouring Ladakh recorded more than 20,000 entries in 1988–1989, before these figures fell to a few thousand in the following years. It took 15 years for tourist numbers to return to their original level. From then on, numbers have increased at an exponential rate, reaching a (temporary?) plateau between 140,000 and 180,000 per year in 2011–2015 (Assistant Director of Tourism Leh, 2015). Although the number of foreign visitors rose during the period, the main driver for the increase has been the rapidly growing popularity of Ladakh with Indian tourists, their numbers multiplying six times between 2003 and 2011 (Figure 23.1). Thus, while in its first phase, tourism was dominated by overseas visitors, domestic tourism is now predominant, as is generally the case in the Himalayan region (Sharma, 1998; Lihua & Jingming, 2002; Xinhua, 2014; Malik & Bhatba, 2015).

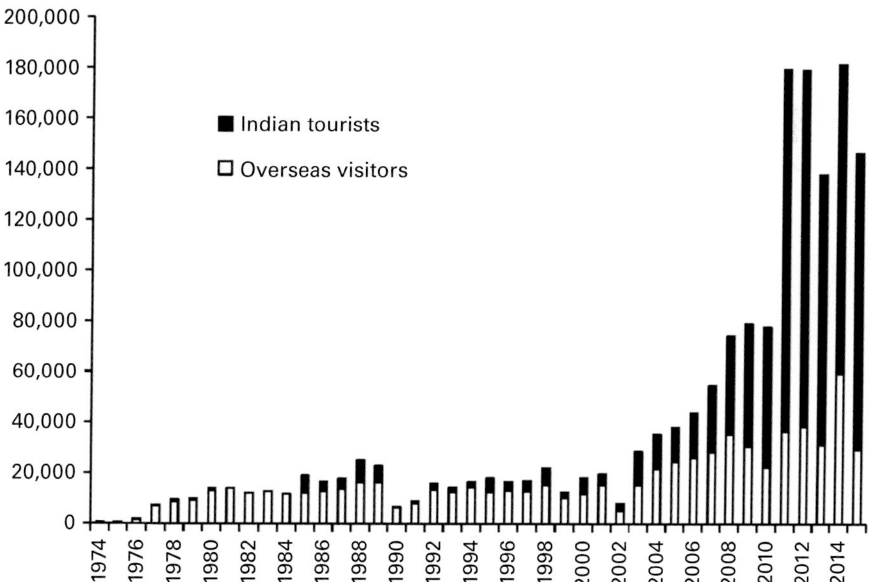

Figure 23.1. Tourist entries in Ladakh (based on data from Assistant Director Tourism, Leh, Ladakh). The 2015 figures do not include entries by road for October to December and by air for November–December.

Within Ladakh, the last area to open to visitors, the Indian portion of Pangong Lake has seen the fastest expansion. No visitors were allowed to stay at Pangong until 2001. Ten years later, there were an estimated 100,000 to 150,000 tourists spending anything from a few hours to a few days in the area. The main trigger for this sudden increase was a series of successful films shot around the lake, in particular the blockbuster *3 Idiots*. Within two years of its release in 2009, the number of domestic tourists had almost trebled.

In contrast with Ladakh's uncontrolled, low-cost tourism, Bhutan adopted from the start a policy of 'low volume–high yield' tourism, by imposing a substantial minimum daily tariff per visitor (€200–250 at present). This policy initially succeeded in controlling tourist numbers, but since visitors from the Indian subcontinent are excluded from the fee and have been growing the fastest in number, tourism development is currently comparable with other parts of the Himalayas: more than 130,000 people visited the country in 2014, half of them from the Indian subcontinent, essentially from India (more than 80%) and Bangladesh (more than 15%) (Dorji, 2001; Nyaupane & Timothy, 2010; Tourism Council of Bhutan, 2014, 2015).

Infrastructure Development

With its short development history, Himalayan tourism has seen an explosive, mostly uncontrolled and demand-led growth. Tourists have arrived in areas formerly visited by the occasional trader or pilgrim, and the regions have reacted to meet their needs (East *et al.*, 1998). In many areas, especially in remote locations, tourism has been a key factor of change, spurring infrastructure development from roads to housing to administrative and defence posts, which have, sometimes quite rapidly, transformed the landscape (Sharma, 1998, 2000; Cole & Sinclair, 2002). Popular wetlands of the Tibetan Plateau, the lakes of Ladakh in particular, are cases in point.

Within a few years of Lake Tsomoriri opening up to tourism, several camps and other constructions cropped up around Korzok, the only village of the area, spilling over into the wetlands (Mishra & Humbert-Droz, 1998). A first temporary camp for foreign devotees of a religious dignitary was set up in a lakeside meadow in 1996. A summer residence for the same dignitary was built in 1999, becoming the first permanent construction on the shore. The same year saw the establishment of a first seasonal 'drive-in' camp on the banks of one of the lake's main feeder streams. Next, in 2000, came the first hotel and a permanent camp of the border forces – also responsible for checking visitors' entry permits (Humbert-Droz & Dawa, 2004). In 2001, work started on a new link road to a main border settlement on the nomadic route, 70 km away. Once completed (2011), a defence post was erected, complete with check post and barracks on the road and a Tibetan-style villa, boathouses and slipways on the shore. A vehicle service area was developed above an adjoining bay. These became the first constructions around the lake away from the village. In 2014, dirt tracks were carved out of the hill slopes and lakesides to connect all major bays and the Norbu Sumdo bog to the new road.

Meanwhile, with visitor numbers rising rapidly, augmented by migrant workers and border forces, facilities catering to their needs started to bloom. By 2015, there were five seasonal camps, four campsites, four hotels and resorts, more than 20 home-stay guest-houses and two café-restaurants. The initial border force post had expanded and occupied, together with a mud brick extracting unit, a solar power plant, a health centre and other administrative buildings, what was formerly the northern meadow of the village.

Development around other wetlands has been similar or faster. At Pangong Lake, there was no infrastructure other than a small army camp next to the shore until 2001. Thirteen years later, we recorded, along the ca. 30 km coastline, 21 permanent camps, six hotel-resorts, 30 home-stay guesthouses and 17 café-restaurants. The defence outpost had expanded on the roadside with boathouses and slipways added on the shore. Rubber dinghies were frequently patrolling between the main tourist settlement and the riverine marshes at the northern end of the lake. The road had been enlarged as far as the popular film-shooting bay and the nearby hamlet of Spangmik, which had become the main tourist hub at Pangong Lake.

The example of the Ladakh lakes shows that tourism can act as a powerful trigger for rapid and generalized infrastructure development. However, the build-up of important infrastructure in the Himalayas well predates the advent of tourism, with geopolitics playing a crucial role. Border wars between Pakistan, India and China in the 1940s and 1960s sealed off major Trans-Himalayan trade routes and led to the construction of extensive civilian and military infrastructure, including roads and hydroelectric power projects, on both sides of the lines of control (Fox *et al.*, 1994; Humbert-Droz & Dawa, 2004; Rahman, 2014). In the case of Ladakh, Leh Airport was built in the wake of the first Kashmir war (1947). The Leh–Srinagar road opened in 1966, one year after the second Kashmir war, and the Leh–Manali road in 1976, five years after the Indo-Pakistan war of 1971 (Dame & Nüsser, 2008). In the aftermath of the Pakistan-India Kargil conflict of 1999, defence expenditure and military positions have increased exponentially in the border area (Aggarwal & Bhan, 2009). Significant development of road and other military infrastructure has also taken place along the Sino–Indian border, including in Ladakh and the eastern region (Rajagopalam & Prakash, 2013).

From the 1960s, a few traditional border trading posts reopened and work started on new strategic Trans-Himalayan roads linking the Indian subcontinent with Central Asia and China. The Kathmandu–Lhasa road opened in 1967, and the Karakorum Highway connecting Pakistan's Punjab with the Xinjiang Autonomous Region was completed in 1986 (Naim, 2010; Jha, 2013). The most recent road connection to be opened for trade (2006) and vehicle traffic (2014) is the Nathu La Pass linking Sikkim in eastern India to Lhasa (Jacob, 2011; India TV News Desk, 2014). Infrastructure development across the Himalayas is set to intensify as the new 'Silk Road Economic Belt' project championed by China's President Xi Jinping gets under way. Under the project, which aims at connecting China with Europe and South and West Asia, a vast network of roads, railways and pipelines is under construction or at an advanced stage of planning. On the northern Tibetan Plateau, work is under way to extend the rail and road links from Qinghai Province to Xinjiang Autonomous Region, and further west to Pakistan and Central Asia, while in the south-central region, the Tibet rail link is planned to be extended to Kathmandu and the Gangetic Plain in the southwest and Bhutan in the southeast (*Guardian Weekly*, 2014; Rahman, 2014; *Hindu*, 2015).

Impacts

Much of the infrastructure development in the Himalayas takes place along river valleys and at high-altitude lakes, the main points of passage for the military, tourists and trade. These activities are highly seasonal, due to inaccessibility of many areas in winter, which maximizes their impact during the sensitive bird migration and breeding periods. Wetland loss, pollution and threats to the avifauna are the most common negative consequences.

Wetland Loss

Landfill and construction of roads and commercial, residential and defence buildings are major causes of wetland loss (Mercer, 1990; Dugan, 2005). These developments are concentrated along transportation corridors and across high-altitude passes, also the main migration routes for birds and other wildlife. Typical examples are the Sutlej/Indus-Yarlung Tsangpo and Pangong-Wujang-Amdo east–west corridors in the southern portion of the Tibetan Plateau and the Yarlung Tsangpo/Lhasa-Qinghai Lake, south-north-east route. Wetlands of critical importance as bird breeding, staging and wintering grounds are located along these routes, including six Ramsar sites and several other Protected Areas (Lang *et al.*, 2007; Prosser *et al.*, 2011; Bishop *et al.*, 2012; Ramsar Information Service). The most accessible sites are the most impacted.

Within 20 years of its opening to modern development, Tsomoriri Lake, the first designated Ramsar site of the Tibetan Plateau (2002), has lost several hectares of grassland and wet meadows to private, tourism and defence constructions. These include more than four hectares of grassland and prime grazing habitat of Bar-headed Goose *Anser indicus* and Ruddy Shelduck *Tadorna ferruginea*. Most important losses are due to the setting up of semi-permanent camps and access routes in lakeside meadows. First established at Tso Moriri and Tsokar in the late 1990s, such camps have rapidly spread to other popular wetlands such as the Nubra Valley and Pangong Lake. With more than 20 camps and various other facilities, Pangong has lost several hectares of prime grassland and wetland in less than 10 years of intensive development. Filling for solid waste disposal further adds to these losses, as does the extraction of turf, sand and gravel as construction source materials. Infrastructure used by the border forces has a more severe impact as these facilities are permanent and used throughout the year. Slipways and boathouses built at Tsomoriri and Pangong represent the first loss of natural shoreline on Ladakh's lakes. Together with Hemis National Park in central Ladakh, the wetlands of the Changthang (Tsokar, Tsomoriri and Pangong) have been identified as two regions particularly affected by environmental degradation due to tourist activities (Geneletti & Dawa, 2009).

The less accessible wetlands of Tibet appear to have been less affected. With about 10,000 tourists per year, the Mapangyong Cuo lakes, a Ramsar site close to the Indo-Nepal-China border, 400 km southeast of Tsomoriri and 950 km west of Lhasa, has no permanent infrastructure other than access roads (Ramsar Information Service). However, with well over 10 million tourists now visiting Tibet every year, major wetlands with good road connections such as Yamdrok Co and Nam Co, as well as Ramsar sites of the Tibet-Qinghai Plateau, appear to have fared less well (Jianqiang, 2012; Yan *et al.*, 2013). Qinghai Lake, the largest saline wetland in China and the first Ramsar site on the Plateau (Niaodao), has seen a significant increase in human population, residential areas and reclamation of marshes since the 1970s, leading to net wetland loss and degradation (Duo *et al.*, 2014). During the same period, road construction, mining and shrub land slash and burn around Eling–Zahling lakes, key Ramsar sites for breeding and staging Bar-headed Geese and Black-necked Cranes, have led to water loss and soil erosion in upstream wetland habitats. Overall, the loss of Plateau wetlands due to human activities has been estimated at 186 km^2 per annum, that is, a total of some 3000 km^2 between 1990 and 2006 (Song *et al.*, 2014).

Pollution

Closely linked with infrastructure development, water pollution is a major cause of wetland degradation. Sewage discharge and garbage accumulation stem from inadequate sanitation and waste management in many tourist, civil and defence facilities (Nepal, 2000; Sharma, 2000; Malik & Bhatba, 2015). These facilities are mostly located close to wetlands and typically sited, in the semiarid Tibetan Plateau, on alluvial fans with perennial or seasonal sub-surface run off. These structures are generally equipped with water closets (WCs), which facilitate contamination of groundwater and water bodies. At Pangong Lake, for example, more than 200 WCs discharge their effluent into shoreline wetlands along the 30 km coastline opened to modern development. Trekking groups often dig their own temporary latrines also located close to water sources. Faecal pollution of water bodies, which are the main source of water for cooking and drinking, has become a common phenomenon around much-used campsites and other accommodation facilities, with potentially severe consequences for human health and for wildlife (Banskota & Sharma, 1998; Humbert-Droz & Dawa, 2004; Malik & Bhatba, 2015).

In popular wetlands located at lower elevations, such as Dal Lake next to Srinagar, capital of Kashmir, raw sewage discharge is leading to severe pollution, including by coliforms and streptococci, as well as eutrophication, with rapid expansion of macrophytes choking the lake's surface (Mercer, 1990). The picture is similar in other wetlands of the Jhelum floodplain in the Kashmir Valley, including its two Ramsar sites, Wular and Hokersar Lakes (Wetland international, 2007; Yousuf *et al.*, 2015; Ramsar Information Service). Oligotrophic lakes of the Tibetan Plateau, which are generally devoid of outside drainage, are also particularly at risk. Since faecal bacteria are known to persist in sediments at much higher concentration than in the

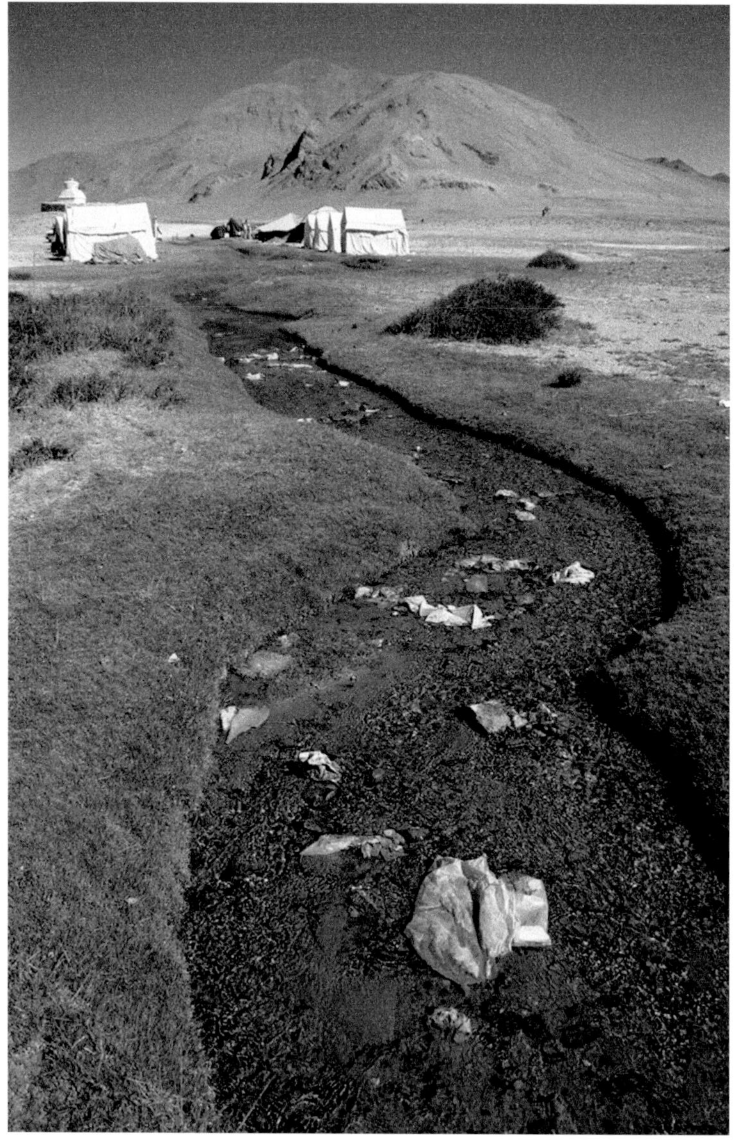

Figure 23.2. Point source pollution at Tsokar campsite.

water column, areas subject to intense use, for example, near tourist and military facilities, may present further health hazards when sediments are disturbed or dispersed (Ceballos-Lascuráin, 1996; Hammit *et al.*, 2015).

In the absence of solid waste management in many Himalayan destinations, litter accumulates in the environment, eventually contaminating water bodies. With an estimated daily generation of 150 to 300 grams of non-degradable garbage per person (Daga, Reliance plastic industry, pers. comm.), the problem has reached crisis proportions in different areas, including Everest National Park, the Annapurna region, Bhutan and several Himalayan lakes (Sharma, 1998, 2000; Dorji, 2001). While in most lakes of Ladakh such as Tsomoriri and Tsokar, pollution is still at a point-source stage (Figure 23.2), incipient generalized pollution can be observed at Pangong Lake, which is under more intense anthropogenic pressure. Pangong, located like most Plateau lakes in an enclosed basin, where 'what goes in, stays in', can be seen as an indicator of the likely evolution of many popular wetland destinations in the region.

Contamination by heavy metals, in particular cadmium and lead, a form of pollution that is often overlooked, has been reported at moderate to high levels along roads with heavy tourist traffic and in fish of popular lakes of the Tibet-Qinghai Plateau, including Nam Co, the largest salt lake of the plateau, Eling–Zhaling Lakes and Yamdrok Co near Lhasa (Yang *et al.*, 2007; Yan *et al.*, 2013). Combustion of fossil fuels is the main source of cadmium emission in the air, however, possible natural sources of cadmium, which may accumulate at high levels in sedimentary rocks and evaporites (ICdA, undated), are not factored into the Yan *et al.* study. In view of its high toxicity for the avifauna and human health, possible cadmium and other heavy metal contamination requires further monitoring, in particular around saline wetlands along the main transportation corridors.

Migratory Bird Disturbance and Population Decline

Wetland degradation, pollution and general disturbance by human activities can have severe impacts on the avifauna. Impacts in the Himalayan region are exacerbated by the fact that bird breeding and migration largely coincide with human activities, in place – along the main rivers and lakes – and time – the vegetation period during which the high passes are opened. In many wetlands, the birds' breeding, staging or wintering areas lie adjacent to major roads, trekking routes, campsites and other infrastructure (Humbert-Droz, 2011; Bishop *et al.*, 2012; Song *et al.*, 2014).

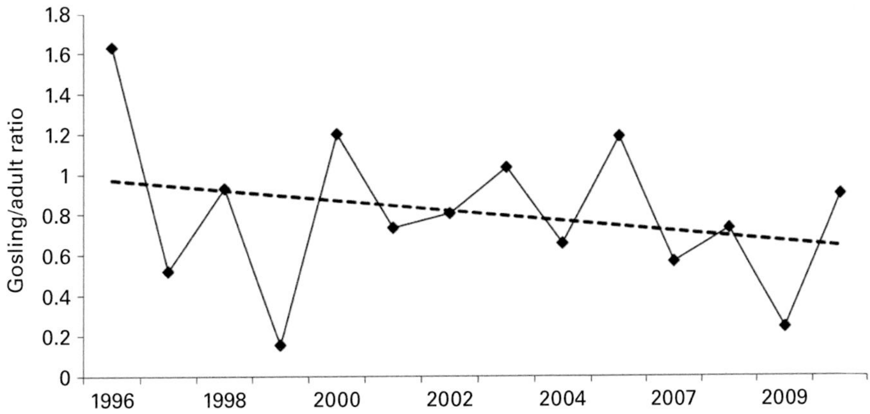

Figure 23.3. Bar-headed Goose young-to-adult ratios at Tsomoriri (redrawn from Humbert-Droz, 2011).

Disturbance at these sites, by vehicular traffic, movement of tourists or troops and construction activities, can lead to breeding failure and is particularly damaging to unfledged young, causing stress, increased visibility to predators and possible mortality (Ceballos-Lascuráin, 1996; Maitland & Morgan, 1997; Hammitt *et al.*, 2015). A case in point is the declining trend in young-to-adult ratio of Bar-headed Geese observed in Ladakh lakes – the sole breeding ground of this species in the Indian subcontinent (Prins & Van Wieren, 2004) – after the beginning of intensive tourism in the mid-1990s (Figure 23.3). Similar trends have been noted for the Great Crested Grebe *Podiceps cristatus* and Black-necked Crane *Grus nigricollis*, whose habitat, shared with the goose, has been increasingly disturbed by tourist and military activities. Most other waterbirds breeding in the area, such as the Goosander *Mergus merganser*, Lesser Sandplover *Charadrius mongolus*, Common Redshank *Tringa totanus* and Common Tern *Sterna hirundo*, also show population declines. These declines have been ascribed to growing anthropogenic pressures as well as an increased frequency of extreme weather events such as drought and floods over the period (Humbert-Droz, 2011). In China, breeding areas for both Black-necked Cranes and Bar-headed Geese appear to have suffered less damage from human activities, owing to their relative inaccessibility in the central and northern Tibetan Plateau (Song *et al.*, 2014). However, overwintering sites, which are almost all located in areas submitted to heavier anthropogenic pressure, including tourism, in the Lhasa River Basin, northern Yunnan, and Guizhou Provinces, as well as in Bhutan (Black-necked Crane), have been subjected to frequent disturbance, resulting in greater habitat damage (Bishop, 1996; Lang *et al.*, 2007; Fengshan *et al.*, 2011; Bishop *et al.*, 2012).

Destruction of the avifauna is in some cases directly associated with modern uses of wetlands. For example, intense poaching of geese, ducks and waterbird eggs is reported around Wular Lake, the main wetland of the Kashmir Valley, which has been subjected to rapid infrastructure development and urbanization (Wetlands International, 2007). In Ladakh, hunting by border forces appears to have declined in past decades due to effective government intervention (Mallon, 1991), but poaching by construction workers, who partly live from the resources of the land, is not uncommon (Fox *et al.*, 1994; Humbert-Droz & Dawa, 2004). Illegal hunting appears to be presently uncommon in Tibet, although incidents were reported in the past, while in neighbouring Arunachal Pradesh, it led to the extirpation of a small wintering population of Black-necked Crane (Bishop, 1996; Farrington & Xiulei, 2013). Poaching and egg collection are reported as prevalent in Qinghai, Xinjiang, Yunnan and most other provinces of China (Boer *et al.*, 2011; Zhang *et al.*, 2015). According to a recent incomplete estimate, and with the exception of Tibet, between 80,000 and 120,000 waterbirds belonging to ca. 40 species are illegally hunted every year in China, geese, ducks and swans accounting for the greatest numbers (Ma Ming *et al.*, 2012).

Certain seemingly innocuous activities like bathing and boating can have serious detrimental effects, especially on birds nesting on the shore (e.g. Sandplovers), on islands (e.g. Bar-headed Goose, Brown-headed Gull *Larus brunnicephalus*) and those that build floating nests (grebes, terns). Boating is especially disruptive for unfledged and moulting waterbirds (Ceballos-Lascuráin, 1996; Maitland & Morgan, 1997). Common in tourist destinations on the southern flank of the Himalaya like the Kashmir lakes, these activities were, until recently, totally absent from the Tibetan Plateau. Plateau lakes are generally left undisturbed by the local population, who holds them as sacred sites (Maharana *et al.*, 2000; Shen *et al.*, 2012; Ramsar Information Service, site 1213) and are advertised as such by the tourist industry (Kelly, 2015). With growing tourist and army presence, swimming and boating, though opposed by locals and religious authorities, are becoming increasingly common. Introduced in Ladakh lakes during scientific surveys in the 1990s and 2000s, boating activities have rapidly expanded with patrolling and building of landing facilities by the military from 2010 onwards. In Tibet, according to reports on Chinese websites, the tourism industry has sought, but apparently not been granted so far, the development of boating facilities in two large lakes of the Plateau (Jianqiang, 2012).

Pollution Impacts on Waterbirds

Waste accumulation near human settlements and its seepage into water bodies is a major anthropogenic impact that directly and indirectly affects wetland birds. Organic litter attracts scavengers like Brown-headed Gulls (Figure 23.4), but also, at least in Ladakh wetlands, Black-necked Cranes, which have been regularly observed near garbage dumps at popular campsites in recent years. Although I did not observe it directly, their feeding on garbage is strongly suspected and has been reported by visitors and fieldworkers (Heygers & Rigzin, WWF field officer, pers. comm.). The crane is one of the few species breeding in the area to have shown

Figure 23.4. Brown-headed Gull feeding among garbage at Tsomoriri (Photo credit: Blaise Humbert-Droz). (A black-and-white version of this figure will appear in some formats. For the colour version, please refer to the plate section.)

a steady increase in number – though not in breeding success – in the past 20 years (Humbert-Droz, 2011), a trend also reported from Tibet and neighbouring provinces of China (Farrington & Xiulei, 2013; Song *et al.*, 2014). Feeding at camp dumps, however, carries its own risks as organic matter is often mixed with glass, metal and plastic waste, which can be harmful to the birds when ingested along with food. A recent example of how significant the exploitation of refuse as a food source can be for large and long-lived migratory birds is provided by a population of White Storks *Ciconia ciconia*, which have become resident in Portugal over the past 30 years and now use their nests in the vicinity of landfill sites throughout the year (Gilbert *et al.*, 2016).

Rubbish also attracts stray dogs, which tend to congregate around military and tourist camps and may prey on birds and eggs (Humbert-Droz, 2011). In Ladakh, nearly 50% of Black-necked Crane chicks are estimated to be killed by feral dogs (Islam & Rahmani, 2008). Dog predation is also reported as a serious threat in Tibet and neighbouring provinces (Bishop, 1996; Farrington & Xiulei, 2013).

Increased water pollution levels are likely to have more serious indirect impacts as they affect the entire avifauna using the area. For example, nutrient enrichment in the littoral zone, through sewage and other organic discharges, is likely to induce changes in species composition and abundance of benthic flora and fauna as well as wetland aquatic insects (Nelson & Steinman, 2013). This in turn is bound to affect waterbird populations, for which they are a main food source.

Such habitat alteration added to other disturbances may lead birds to change their behaviour patterns and avoid affected areas. A graphic example of the combined effects on migratory birds of water pollution, infrastructure development and increased human disturbance is provided by the Indian portion of Pangong Lake. During two surveys conducted in the peak tourist season (August 2014 and 2015), we found all the portions of coastline lying directly adjacent to tourism structures and a popular shoreline track to be totally devoid of waterbird life except for scavengers (Brown-headed Gull and Arctic Jaeger *Stercorarius parasiticus*). This is in spite of the fact that most of these structures are established on lakeside meadows, which are, potentially, prime bird habitats. The loss of such key habitats assumes particular significance as the lake is one of the most important breeding grounds and migration corridors for Great Crested Grebes and Goosanders in Ladakh and India. Another example, from Tibet, is the displacement of small Black-necked Crane populations from wetlands along the Yarlung Tsangpo near Shigatse after construction of an expressway (Bishop *et al.*, 2012). Injury and death of cranes and Bar-headed Geese following collision with newly built power lines have also been reported in the area and in Qinghai Province (Fengshan *et al.*, 2011; Farrington & Xiulei, 2013). The region is the main wintering ground for the crane and the goose in Tibet.

Ultimately, altered behaviour patterns and shifts in distribution in response to severe disturbance may lead to greater population fluctuations and decline (Song *et al.*, 2013). Based on earlier work on recreation impacts (Ceballos-Lascuráin, 1996; Hammit *et al.*, 2015), the ecological impacts of modern development on the region's avifauna could be summarized as follows: intrusion into bird habitat by development and recreational activities leads to disturbance, alteration of habitat and increased mortality (of young in particular, e.g. Bar-headed Goose). These in turn lead to adaptation (e.g. increased scavenging by Brown-headed Gulls, cranes feeding on crop residues and also possibly on garbage), displacement from severely disturbed habitats and reduced reproduction levels (e.g. lower recruitment in geese and cranes). Displacement and lower reproduction levels induce population changes and decline, which may ultimately result in altered species composition in bird communities.

Perspectives on Protection

Although many key bird habitats are afforded a degree of protection by their inclusion in conservation areas, they also face the direct onslaught of modern development, since the wetlands in which they are located are often prime tourism, defence and road construction sites. The severity of the threat that this entails for the survival and maintenance of migratory bird populations calls for focussed and robust conservation action. Protective measures are specifically required to curb infrastructure development and prevent human disturbance and pollution in key habitats along the Trans-Himalayan migratory routes.

Several such measures were identified in a resolution of the Goose Specialist Group and IUCN Species Survival Commission in Leh in 2008. They include: the diversion of vehicular traffic away from bird breeding grounds, the restriction of trekking to designated paths, a ban on off-road driving and permanent camps and infrastructure in wetlands, the promotion of tourist accommodation in local homes, a ban on sewage discharge into water bodies and the promotion of local-technology dry toilets, the setting-up of adequate garbage management including removal of waste and recycling and the control of feral dogs (IUCN, 2008).

Such protection work faces immense challenges. The combined pressure on wetlands of the development of tourism, civilian and military infrastructure is enormous. There are now well over 20 million tourists visiting the Himalayas and Tibet every year, and although political and border tensions have periodically reduced tourist flows, numbers have rebounded every time and are rising. These tensions have also given a boost to military and civilian construction along and across Himalayan borders.

To succeed, the conservation work on hand needs to carry the active support of all those concerned: the local people, the administration, the tourism industry and the military. There are several ways in which the main stakeholders can directly contribute to wetland protection. Here are a few examples.

Policies aimed at regulating tourist flows such as the trekking permit system in force in Nepal and Sikkim, in particular the 'low volume–high yield' approach adopted in Bhutan and Nepal's Mustang Valley, should be replicated elsewhere, especially in eco-sensitive and newly opened areas. These systems cannot on their own limit visitor numbers as the comparatively high daily fees (~€70 in Mustang, €200–250 in Bhutan) paid by international visitors are not levied on domestic or regional tourists – which comprise the fastest-growing tourist segment (Gurung & DeCoursey, 1999; Nepal,

2000; Dorji, 2001). However, they do exert a degree of control, limit impact through the provision of trained guides and provide additional resources that can be invested in conservation and programmes benefitting local communities.

With their regular presence in many remote and wetland areas, the defence forces can directly help control poaching as well as stray dogs, often loosely attached to military camps. Such an active anti-poaching role, currently played by the military in Nepal (Herbert Prins, pers. comm.) and also, in the past, in Ladakh, could be replicated on a wider scale. Defence forces also need to ensure that motor-boating only takes place in border lakes when necessary and as far as possible from waterbird habitats.

Local administration across the Himalayan region needs to enlist the support of the local population and tourism industry to strengthen conservation efforts and ensure, as they have apparently managed to do so far in Tibet, that no recreational swimming and boating – a curse for breeding waterbirds – takes place in Plateau lakes. A fundamental task, implemented with little success as yet, is to site camps, infrastructure and sanitation facilities away from wetlands, as is the traditional practice of local people.

Appropriate disposal of rubbish, through composting, removal from wetlands and recycling, is another key task requiring the cooperation of all stakeholders. Positive examples exist such as the 'if you bring it in, bring it out' garbage bag deposit system in force in Everest National Park. Such approaches, which directly involve tourist company staff and the tourists themselves in the clean-up effort, must be adopted on a much larger scale.

Partnerships should be the order of the day, as no one organization can address the conservation challenge on its own, nor can conservation succeed without local community participation. The pitfall of too lengthy and elaborate planning processes should be avoided, as the example in Ladakh of more than 10 years of participative planning and very little implementation emphatically shows (Humbert-Droz & Dawa, 2004; IUCN, 2008). Several other examples exist, however, of successful conservation partnerships *in action* that could be emulated in Ladakh and elsewhere.

At Caohai, on the Yunnan-Guizou Plateau, southeast of Tibet, a key Black-necked Crane wintering site has been regenerated over the past 20 years through a partnership between the local administration, international NGOs and local communities. Local people are provided with grants to develop businesses non-detrimental to the wetland and play a central role in restoration. The project has now become a model for nature reserve management in China (Bishop, 1996; Wagner 2012; International Crane Foundation, undated). Another innovative partnership tested successfully in Bhutan involves the country's main conservation society, a local community development organisation and the International Crane Foundation (ICF), in which the foundation organises eco-tours, whose participants pay an extra fee, which is invested in local area development and maintenance of the reserve (Dorji, 2001; ICF, undated). Other successful conservation partnerships include those formed between local communities and support organizations like the Annapurna Conservation Area Project and the Everest Pollution Control Committee (Sharma, 2011). Ultimately, partnerships such as these, which benefit both local communities and habitat restoration efforts, offer the best prospect for long-term conservation of critical bird habitats in Himalayan wetlands.

References

Assistant Director Tourism (2015). Year-wise tourist arrival by air/road to Ladakh. Unpubl. report. Office of the Assistant Director Tourism, Leh.

Aggarwal R. & Bhan, M. (2009). 'Disarming violence': development, democracy, and security on the borders of India. *The Journal of Asian Studies*, **68**, 519–542.

Banskota, K. & Sharma, B. (1998). Understanding sustainability in mountain tourism: case study of Nepal. In East, P., Luger, K., Inmann, K., eds., *Sustainability in Mountain Tourism: Perspectives for the Himalayan Countries*. Delhi: Book Faith India and Insbruck-Vienna: Studienverlag, pp. 111–126.

Bishop, M.A. (1996). Back-necked Crane. In *The Cranes: – Status Survey and Conservation Action Plan*, Gland, Switzerland, and Cambridge, UK: IUCN, pp. 184–194.

Bishop, M.A., Tsamchu, D. & Li, F.X. (2012). Number and distribution of Black-necked Cranes wintering in Shigatse prefecture, Tibet. *Chinese Birds*, **3**, 191–198.

Boer, W.F. de, Cao, L., Barter, M., *et al.* (2011). Comparing the community composition of European and Eastern Chinese waterbirds and the influence of human factors on the China waterbird community. *AMBIO*, **40**, 68–77.

Ceballos-Lascuráin, H. (1996). *Tourism, Ecotourism, and Protected Areas: The State of Nature-Based Tourism around the World and Guidelines for Its Development*. Gland, Switzerland and Cambridge, UK: IUCN.

China Tibetology Research Center (2009). Report on the Economic and Social Development of Tibet, Zheng Limin Ed. www.cctv.com/english/20090330/106989.shtml.

Cole, V. & Sinclair, A.J. (2002). Measuring the ecological footprint of a Himalayan tourist center. *Mountain Research and Development*, **22**, 132–141.

Dame, J. & Nüsser, M. (2008). Development perspectives in Ladakh, India. *Geographische Rundschau International Edition Review*, **4**, 20–27.

Dorji, T. (2001). Sustainability of tourism in Bhutan. *Journal of Bhutan Studies*, **3**, 84–104.

Dugan, P. (ed.) (2005). *Guide to Wetlands*. London: Philip's Octopus Publishing.

Duo, H., Shi, L. & Lei, G. (2014). Remote sensing applications to monitoring wetland dynamics: a case study on Qinghai lake Ramsar site, China. In Wang, G. & Weng Q., eds., *Remote Sensing of Natural Resources*, Boca Raton, FL, CRC Press, pp. 479–493.

East, P., Luger, K. & Inmann, K. (1998). Himalayan tourism on the sustainable trail? In East, P., Luger, K., Inmann, K., eds., *Sustainability in Mountain Tourism: Perspectives for the Himalayan Countries*. Delhi: Book Faith India and Insbruck-Vienna: Studienverlag, pp. 1–14.

Farrington, J.D. & Xiulei, Z. (2013). The Black-necked Cranes of the Longbao National Nature Reserve, Qinghai, China: current status and conservation issues. *Mountain Research and Development*, **33**, 305–313.

Fengshan L.I., Bishop, M.A. & Drolma, T. (2011). Power line strikes by Black-necked Cranes and Bar-headed Geese in Tibet Autonomous Region. *Chinese Birds*, **2**, 167–173.

Fox, J.L., Nurbu, C., Bhatt, S., Chandola, A. (1994). Wildlife conservation and land-use changes in the Trans-Himalayan region of Ladakh, India. *Mountain Research and Development*, **14**, 39–60.

Geneletti, D. & Dawa, D. (2009). Environmental impact assessment of mountain tourism in developing regions: a study in Ladakh, Indian Himalaya. *Environmental Impact Assessment Review*, **29**, 229–242.

Gilbert, N.I., Correia, R.A., Silva, J.P., *et al.* (2016). Are white storks addicted to junk food? Impacts of landfill use on the movement and behaviour of resident white storks (*Ciconia ciconia*) from a partially migratory population. *Movement Ecology*, **4**(7), 1–13.

Gurung. P. & DeCoursey, M.A. (1999). Too much too fast: lessons from Nepal's lost kingdom of Mustang. In Godde, P.M., Price, M.F., Zimmermann, F.M., eds., *Tourism and Development in Mountain Regions*. Wallingford (UK) & New York: CAB International, pp. 239–254.

Hammitt, W.E., Cole, D.N. & Monz, C.A. (2015). *Wildland Recreation: Ecology and Management*, 3rd edition. Hoboken, NJ: John Wiley & Sons.

Humbert-Droz, B. (2011). Trends in waterfowl populations and development of tourism in high-altitude wetlands of Ladakh, north-western India. *Casarca*, **14**, 184–203.

Humbert-Droz, B. & Dawa S. (Eds.) (2004). *Biodiversity of Ladakh: Strategy and Action Plan*. New Delhi: Sampark.

India TV News Desk (2014). China announces opening of new route to Kailash Manasarovar. www.indiatvnews.com/news/india/chinese-prez-may-announce-opening-new-route-to-kailash-manasarov-42109.html.

International Cadmium Association (ICdA) (undated). *Environment: Cadmium emissions*. www.cadmium.org/environment/cadmium-emissions.

International Crane Foundation, (undated). Cao Hai Project. www.savingcranes.org/cao-hai-project/.

Islam, A. (2014). Impact of armed conflict on economy and tourism: a study of state of Jammu and Kashmir. *IOSR Journal of Economics and Finance*, **4**, 55–60.

Islam, M.Z. & Rahmani, A.R. (2008). *Potential and Existing Ramsar Sites in India*. New Delhi: Oxford University Press.

IUCN, Wetlands International (2008). Resolution 1 of the 11th Meeting of the Goose Specialist Group of Wetlands International & the Species Survival Commission of IUCN concerning the protection of wetlands, grasslands and their water bird populations in Ladakh. Unpublished resolution, dated 29 May 2008, submitted to the Department of Wildlife Protection, Jammu and Kashmir State.

Jacob, J.T. (2011). The Sino-Indian boundary dispute: sub-national units as ice-breakers. *Eurasia Border Review*, **2**, 35–45.

Jha, H.B. (2013) Nepal's border relations with India and China. *Eurasia Border Review*, **4**, 63–75.

Jianqiang, L. (2012). Locals fight tourism on Tibet's holy lakes. *The Third Pole*. www.thethirdpole.net/locals-fight-tourism-on-tibets-holy-lakes.

Kelly (2015). Nine sacred lakes in Tibet. www.chinahighlights.com/tibet/sacred-lake.htm.

Lang, A., Bishop, M.A. & Le Sueur, A. (2007). An annotated list of birds wintering in the Lhasa river watershed and YamzhoYumco, Tibet Autonomous Region, China. *Forktail*, **23**, 1–11.

Lihua, L. & Jingming, H. (2002). Sustainable rural tourism and its implications for poverty alleviation in Tibet Autonomous Region, P.R. China. In Jodha *et al.*, eds., *Poverty Alleviation in Mountain Areas of China*. Feldafin, Germany: InWent, pp. 209–220.

Maharana, I., Rai, S.C. & Sharma, E. (2000). Valuing ecotourism in a sacred lake of the Sikkim Himalaya, India. *Environmental Conservation*, **27**, 269–277.

Maitland, P.S. & Morgan, N.C. (1997). *Conservation Management of Freshwater Habitats*. London: Chapman & Hall.

Malik, M.I. & Bhatba, M.S. (2015). Sustainability of tourism development in Kashmir – Is paradise lost? *Tourism Management Perspectives*, **16**, 11–21.

Mallon, D.P. (1991). Status and conservation of large mammals in Ladakh. *Biological Conservation*, **56**, 101–119.

Ma Ming, R., Zhang, T., Blank, D., Ding, P. & Zhao, X. (2012). Geese and ducks killed by poison and analysis of poaching cases in China. *Goose Bulletin*, **15**, 2–11.

Mercer, D.C. (1990). Recreation and wetlands: impacts, conflict and policy issues. In Williams, M., ed., *Wetlands: A Threatened Landscape*. Oxford, UK & Cambridge, MA: Blackwell, pp. 267–295.

Mishra, C. & Humbert-Droz, B. (1998). Avifaunal survey of Tso Moriri Lake and adjoining Nuro Sumdo wetland in Ladakh, Indian Trans-Himalaya. *Forktail*, **14**, 65–67.

Naim, T. (2010). South Asia. In *UNESCO Science Report 2010*. UNESCO Publishing, pp. 322–347.

Nelson, W.A. & Steinman, A.D. (2013). Changes in the benthic communities of Muskegon Lake, a Great Lakes Area of Concern. *Journal of Great Lakes Research*, **39**, 7–18.

Nepal, S.K. (2000). Tourism in protected areas: the Nepalese Himalaya. *Annals of Tourism Research*, **27**, 661–681.

Nyaupane, G.P., &Timothy, D.J. (2010). Power, regionalism and tourism policy in Bhutan. *Annals of Tourism Research*, **37**, 969–988.

Prins, H.H.T. & Van Wieren, S.E. (2004). Number, population structure and habitat use of Bar-headed Goose, Anser indicus in Ladakh (India) during the brood-rearing period. *Acta Zoologica Sinica*, **50**, 738–744.

Prosser, D.J., Cui, P., Takekawa J.Y., *et al.* (2011). Wild bird migration across the Qingzang-Tibetan Plateau: a transmission route for highly athogenic H5N1. *PLoS one*, **6**(3), e17622.

Rahman M.Z. (2014). Territory roads and trans-boundary rivers: an analysis of Indian infrastructure building along the Sino-Indian border in Arunachal Pradesh. *Eurasia Border Review*, **5**, 59–75.

Rajagopalan, R.P. & Prakash, R. (2013). Sino-Indian border infrastructure: an update. *ORF Occasional Paper* #**42**. New Delhi: Observer Research Foundation.

Ramsar Sites Information Service. Tso Moriri Ramsar Site Number 1213, Wular 461, Hokera 1570, Mapangyong Cuo 1439, Maidika 1438, Eling 1436, Zhaling 1442, Niaodao 552. https://rsis.ramsar.org/ris/ <Ramsar Site Number>.

Sharma, P. (1998). Sustainable tourism in the Hindu Kush-Himalayas: issues and approaches. In East, P., Luger, K., Inmann, K., eds., *Sustainability in Mountain Tourism*. Delhi: Book Faith India and Insbruck-Vienna: Studienverlag, pp. 47–69.

Sharma, P. (2000). Tourism and livelihood in the mountains regional overview and the experience of Nepal. In Banskota *et al.*, eds., *Growth, Poverty Alleviation and Sustainable Resource Management in the Mountain Areas of South Asia*. Proceedings of the International Conference 31.1. – 4.2. 2000, Kathmandu. Feldafing (Germany): InWEnt, pp. 349–376.

Sharma, P. (2011). Sustainable mountain tourism development in Nepal: an historical-perspective. In Kruk *et al.*, eds., *Integrated Tourism Concepts to Contribute to Sustainable Mountain Development in Nepal*. Feldafing (Germany): GIZ, pp. 40–47.

Shen, X., Lu, Z., Li, S., & Chen, N. (2012). Tibetan sacred sites: understanding the traditional management system and its role in modern conservation. *Ecology and Society*, **17**, 1–13.

Song, H., Zhang, Y., Gao, H., Guo, Y. & Li, S.(2014). Plateau wetlands, an indispensable habitat for the Black-necked Crane (Grus nigricollis). *Wetlands*, **34**, 629–639.

The Guardian Weekly (2014). China hopes to revive the Silk Road with bullet trains to Xinjiang. *The Guardian Weekly*. www.theguardian.com/world/2014/sep/30/china-bullet-high-speed-train.

The Hindu (2015). Nepal to join Silk Road Economic Belt through Tibet. *The Hindu*. www.thehindu.com/news/international/south-asia/nepal-to-join-silk-road-economic-belt-through-tibet/article6749342.ece.

Tourism Council of Bhutan (2014, 2015). *Bhutan Tourism Monitor Annual Reports 2013&2015*. http://tcb.cms.ebizity.net/attachments/tcb_060514_bhutan-tourism-monitor-2013http://tcb.img.ebizity.bt/attachments/tcb_081415_btm-2015–booklet-(web).

Wagner, E. (2012). Flight plan. *Earth Island Journal Summer*, **27**. www.earthisland.org/journal/index.php/eij/article/green_dragon/.

Wetlands International – South Asia (2007). *Comprehensive management action plan for Wular Lake, Kashmir*. Final report. http://sites.wetlands.org/reports/ris/2IN003_mgtplan.pdf.

Xinhua (2014). Tourists to Tibet surge in 2013. http://news.xinhuanet.com/english/china/2014-02/02/c_133089541.htm.

Xinhua (2015).Tibet sees record high tourist arrivals in 2014. http://news.xinhuanet.com/english/china/2015-01/11/c_133911210.htm.

Yan, X., Zhang, F., Gao, D., *et al.* (2013). Accumulations of heavy metals in roadside soils close to Zhaling, Eling and Nam Co Lakes in the Tibetan Plateau. *International Journal of Environmental Research and Public Health*, **10**, 2384–2400.

Yang, R., Yao, T., Xu, B., Jiang, G. & Xin, X. (2007). Accumulation features of organochlorine pesticides and heavy metals in fish from high mountain lakes and Lhasa River in the Tibetan Plateau. *Environment International*, **33**, 151–156.

Yousuf, T., Yousuf, A.R. & Mushtaq, Ba. (2015). Comparative account on physico-chemical parameters of two wetlands of Kashmir Valley. *International Journal of Recent Scientific Research*, **6**, 2876–2882.

Zhang, Y., Jia, Q., Prins, H.H.T., Cao, L. & Boer, W.F. de (2015). Effect of conservation efforts and ecological variables on waterbird population sizes in wetlands of theYangtzeRiver. *Scientific Reports*, **5**, 17136.

24 Birds in Relation to Farming and Livestock Grazing in the Indian Trans-Himalayas

T.R. Shankar Raman, Kulbhushansingh R. Suryawanshi and Charudutt Mishra

Birds and Man in the Trans-Himalayas

The Trans-Himalayan region comprises a vast rangeland system in the rain-shadow of the Himalayan Mountains, extending north onto the Tibetan Plateau and its marginal mountains (Mishra *et al.*, 2010). The region harbours a diversity of wildlife, including a rich assemblage of mountain ungulate species, carnivores such as the Snow Leopard *Panthera uncia* and the Grey Wolf *Canis lupus* and a diversity of alpine and migratory birds (Pfister, 2004). This region is also home to several high-altitude fresh- and salt-water wetland systems, which are breeding grounds for wetland avifauna (Mishra & Humbert-Droz, 1998; Namgail *et al.*, 2009). The Trans-Himalayan mountains have a long history of human presence (Vernier & Bruneau, Chapter 21). Seasonal human presence on the Tibetan Plateau, for example, is recorded as early as 30,000 years ago, and more permanent pastoral habitation around 8200 years ago (Brantingham *et al.*, 2007). Extensive nomadic pastoralism, and, to some extent, agro-pastoralism, have doubtlessly influenced these systems for several millennia.

The relationship between the ecology of birds and human land use in the Trans-Himalayan region presumably strengthened with the advent of settled agriculture (cf. Ghosal & Ahmed, Chapter 22). Many bird species can consume large quantities of invertebrate prey, acting as potential pest control agents (Kirk *et al.*, 1996; Johnson *et al.*, 2010). Other species consume grain and fruit, potentially damaging agriculture and farm production (de Mey *et al.*, 2012; Canavelli *et al.*, 2013). There are also environmental and cultural connections between birds and humans. Birds are useful as indicators of environmental change and ecosystem health, including pollution (Gregory & Van Strien, 2010; Smits & Fernie, 2013). Species such as the Black-necked Crane *Grus nigricollis* have been a part of the traditional Tibetan Buddhist culture and folklore that dominates the Trans-Himalayan region (Shen *et al.*, 2012; C. Mishra, pers. obs.).

Bird habitats such as grasslands and wetlands have long been important for the grazing of livestock and for agriculture, respectively. Traditional methods of management of these habitats may have incidental or intentional benefits for bird conservation (Harris, 1994). Archaeological evidence has shown that the Amdo Tibetan people used land clearing, grazing and fire to maintain meadows which are today important for conservation (Urgenson *et al.*, 2014). Many other Tibetan societies are known to be tolerant towards wildlife and to have actively encouraged wildlife conservation in their rangelands and wetlands (Shen & Tan, 2012). Traditional Buddhist values of no hunting or disturbance of wildlife have assisted in the conservation of birds such as pheasants and cranes (Woodhouse, 2012). In some parts of the Tibet-Qinghai Plateau, pastoralists have rejected state-sponsored mass poisoning programmes for pikas (Ochotonidae or 'whistling hares') and rodents, and have instead created perches for raptors for the control of small herbivore populations (C. Mishra, pers. obs.). Songs have been composed about species such as the Black-necked Crane. The relationship between birds and people in the Trans-Himalayan region is extensive and culturally significant.

Conservation Management Efforts in the Trans-Himalayas

International and national efforts such as the Ramsar Convention on Wetlands, ratified by 168 countries, have played a role in the protection and conservation of wetland bird habitats in the region. The Ramsar Convention provides a framework for the conservation and wise use of wetlands and their resources as a contribution towards achieving sustainable development throughout the world. There are several important Ramsar Wetlands such as Mapangyong Cuo in the Tibetan Autonomous Region, China, Niaodao (Bird Island) wetland and Zhaling Lake in Qinghai Province China, and the Tsomoriri wetland in Ladakh, India (Mishra & Humbert-Droz, 1998; Prins *et al.,* Chapter 26), although their actual protection status varies. National-level protected area status has also helped secure important bird habitats across the Trans-Himalayan region. Large, state-controlled protected areas such as the Changthang Wildlife Sanctuary, Ladakh, India, extend protection to important bird habitats, but again, the actual on-the-ground status is debatable because most nature reserves in the Tibetan region are poorly managed due to insufficient staff and low financial support (Shen & Tan, 2012). Species-focussed initiatives such as the Black-necked Crane Nature Reserve in China have intended to extend protection to all the breeding sites of this vulnerable species. However, much is yet to be achieved with regard to actual, effective protection of wetlands (Zou & Liu, 2005; Zhang, 2016).

Human Sedentarization and Birds: Bottom-Up and Top-Down Effects

While some practices of the local communities and efforts of the state may have benefitted bird conservation in the region, other state policies and patterns of local land use, while deemed important for economic development, may have been counter-productive for bird conservation. One of the most significant recent changes in the Trans-Himalayan region is the ongoing transition from nomadic pastoralism into agro-pastoralism (Namgail *et al.*, 2007). This has been especially vigorously facilitated since the end of the past century. Sedentarization of nomadic livestock-rearing peoples, which is still continuing, has been accompanied by intensification of grazing pressure in many regions, leading to degradation of many rangelands (Mishra *et al.*, 2001; Namgail *et al.*, 2007; Berger *et al.*, 2013). Privatization and fencing of pastures has led to further fragmentation and degradation (Ning & Richard, 1999; Bauer, 2005). Promotion of agriculture in dry steppe areas of the Trans-Himalayas is accompanied by the need to draw water, especially from snow melt and natural springs, away from the pastures to the fields, reducing pasture productivity (Mishra, 2001). The government's active assistance in building channels to harvest water for agriculture and also as a means to provide employment to rural people in the Indian Trans-Himalayas has increased the diversion of water for agriculture (Mishra, 2001). Similarly, there are inputs in the form of fertilizers and pesticides. For example, the Indian state of Himachal Pradesh, where our long-term monitoring of birds has been conducted, provides a 100% subsidy on the transport of fertilizers anywhere in the state, in addition to directly subsidizing the cost of certain fertilizers (Himachal Pradesh State Government Economics and Statistics Department, 2013).

The impacts of sedentarization of pastoral peoples and intensification of agriculture on bird communities of the Trans-Himalayan region are poorly understood. The issue is topical in the Trans-Himalayas, but also has wider relevance to bird conservation and ecology in larger parts of the world, as the conversion of pastoralism to agriculture is a global issue (Fratkin, 1997). To understand the impacts of agricultural intensification on bird communities of the Trans-Himalayan mountains, we examined patterns in bird species richness and abundance over a nine-year period (2002 to 2010) in a landscape mosaic which comprised agricultural fields, steppes and meadows intensively grazed by livestock, and steppes from which livestock grazing had been curtailed since 1998. Our study was conducted in the Trans-Himalayan Spiti Valley in the state of Himachal Pradesh.

Bird Surveys and Methods

Bird surveys were carried out across the four main habitats: agricultural fields, grazed meadows, grazed steppe and protected ungrazed steppe (Table 24.1). For fields and meadows, we established a single site each in the vicinity of Kibber Village, Spiti Valley, Himachal Pradesh, India. For grazed steppe, we selected three sites on different aspects: Bandifarah (west facing), Chugecha (north facing) and Bandang (northwest facing). In ungrazed steppe, we had two sites: Rongolong (north facing) and Tsankar (west facing). In each of these four sites, we established two replicate line transects for bird surveys. The pair of transects within each site was less than 1 km apart. In total there were 14 transects, in seven sites, across four habitats.

We used variable-width line transects to survey birds (Buckland *et al.*, 1993; Buckland *et al.*, 2005). Each transect was marked from a randomly selected starting point within each site as a 500 m long straight line using a compass. We attempted to uniformly sample all sites between the months of May and September in each year. For

Table 24.1. Bird density in the four main habitat strata estimated using distance sampling methods with data pooled across years.

Stratum	Agricultural fields	Grazed meadow	Grazed steppe	Ungrazed steppe
Period	2002–10	2002–10	2004–10	2002–10
Transect surveys	59	55	135	126
Total line length (km)	29.5	27.5	67.5	63
Number of observations	551	310	886	947
Model selected	Half-normal	Half-normal	Hazard-rate	Half-normal
Density (individuals.ha^{-1})	8.8	4.25	4.78	7.17
Standard error	0.75	0.52	0.28	0.42
% coefficient of variation	8.5	12.3	6.0	5.9
95% confidence interval	7.43–10.43	3.33–5.43	4.24–5.38	6.38–8.06

Table 24.2. Density of birds (number.ha^{-1}) grouped by their feeding habits or guilds across the four habitat strata. CI stands for 95% confidence interval.

Diet (number of species)	Agriculture field	CI	Grazed steppe	CI	Grazed meadow	CI	Ungrazed steppe	CI
Granivores (15)	6.2	5.7–6.7	3.4	3.0–3.8	2.9	1.9–3.9	3.9	3.3–4.5
Insectivore (14)	1.5	1.0–2.0	1.9	1.6–2.2	1.1	0.9–1.2	2.6	2.2–3.0
Omnivore (2)	1.8	1.2–2.4	1.1	0.8–1.4	0.9	0.6–1.2	1.0	0.6–1.4
Grazers (1)	0.2	0.1–0.3	0.1	0.0–0.3	0.3	0.0–0.7	0.8	0.6–1.0

logistical reasons, it was not possible to ensure the same distribution of sampling effort across months in different years; most of the sampling was restricted to the period June to August (Table 24.2). This period spanned the major breeding season of resident birds and summer visitors (Ali & Ripley, 1983; pers. obs.).

Birds were surveyed along transects in the morning hours between 06.30 h and 11.00 h on clear, relatively sunny days. The agricultural fields, meadows and steppe vegetation allowed for excellent visibility, and virtually all birds detected were seen (including birds first detected by call and subsequently sighted). We noted the species and number of individuals of all birds detected on the ground or perched on vegetation while walking at a slow, uniform pace to complete each transect survey in around 45–50 minutes. Birds seen flying low over the ground (< 5 m from the surface) were also noted as flying birds; birds flying at greater heights were ignored as they could not be unambiguously assigned as belonging to the general area of that transect. The perpendicular distance to each detection was measured (by pacing) or visually estimated and recorded in the following distance classes: 0–5 m, 5–10 m, 10–20 m, 20–30 m, 30–50 m and 50–100 m. The distance intervals were wider further away from the centre line to minimise errors in distance estimation. Detections that were greater than 100 m away were discarded as outliers in all analyses.

Birds were identified using standard references and field guides (Ali & Ripley, 1983; Grimmett *et al.*, 2011). Bird taxonomy (including common and scientific names) in this chapter follows the 2014 eBird/Clements checklist (www.birds.cornell.edu/clement schecklist/) as implemented in eBird (http://ebird.org; Sullivan *et al.*, 2014). Transects were surveyed by trained observers well versed in identifying the local avifauna by sight and call.

Bird species richness was a fundamental assemblage variable of interest. We used rarefaction analysis to estimate individual-based species richness for a standardized sample of individuals (Gotelli & Colwell, 2001) using the computer program EcoSim (Gotelli & Entsminger, 2001). The number of species for each sample of individuals was obtained through simulations involving the random sampling of individuals from the data set of individuals detected in each site (pooled across both transects and across years in each site).

Patterns of change in bird assemblage composition were analyzed by constructing matrices of pair-wise dissimilarities between samples using the Bray-Curtis dissimilarity index to analyze bird assemblage compositional variation across habitats and years. Input data were strip transect bird abundance (estimated as described earlier) from data gathered between June and August. The dissimilarity matrices were subjected to non-metric multidimensional scaling (NMDS) ordination. NMDS was used to provide a convenient two-dimensional representation of similarities among sites in bird assemblage composition. Data were pooled across years (within transects) to assess spatial variation and pooled across transects (within habitats and years) to assess inter-annual variation and temporal trajectory of change.

To obtain measures of bird density (individual birds per hectare) we used distance sampling density estimation of variable-width line transect data using the package Distance (Miller, 2015) in statistical software R (2.3.4, R Core Team, 2016). Distance sampling explicitly models variation in detectability as a function of distance from the transect line, and provides accurate and robust estimates of density and its variance. Detection was modelled using the uniform, half-normal and hazard-rate models with cosine adjustment terms, and half-normal with simple polynomial adjustment terms (Buckland *et al.*, 2005). The shape of the detection function (or how detection probability varies as a function of distance from the transect line) is likely to vary by habitat type and species. Larger, more vocal and visible birds may have higher detectability which is better represented by models such as uniform and hazard rate, while half-normal models may better represent smaller and more cryptic species whose detectability falls more sharply with distance from the line (Thomas *et al.*, 2010). To estimate density at the level of individual species, feeding guilds and entire community, we compared uniform, half-normal and hazard rate models without additional parameters to determine the model that best fitted the data for each density estimate. Model selection, assessment of fit and inference followed standard guidelines such as using the model with the lowest Akaike Information Criterion (AIC, Buckland *et al.*, 1993; Thomas *et al.*, 2010). Bird densities (all species) were first estimated using data pooled across all years within each stratum, then with data pooled across transects in each year for each stratum, and finally at the level of individual transect samples. In the latter case, due to lower detection, it was not possible to estimate the detection function sample-wise; therefore, detection functions estimated stratum-wise were used to estimate sample-wise densities. Individual bird species densities were also similarly estimated using the package Distance in R, with habitat-specific detection functions and appropriate pooling of data to assess variations in density across the four main habitats.

To examine patterns at the level of dietary guilds, we estimated the density of bird groups classified based on their primary diet type as granivore, insectivore, omnivore, grazers, carnivores or scavengers. We did not record any bird species with nectar as the primary diet. We could not estimate densities for carnivores and scavengers due to insufficient observations. Although there was only one species in our data set that is primarily a grazer (Himalayan Snowcock *Tetraogallus himalayensis*), we chose to analyze it as we expected a significant response from this species to protection (ungrazed).

Bird Assemblage Variation across Habitats

During nine years of sampling along line transects, we recorded around 7162 individual birds of 40 species. Of these, five species were recorded as single detections overall: White-winged (Guldenstadt's) Redstart *Phoenicurus erythrogastrus* (single bird in agricultural fields), European Goldfinch *Carduelis carduelis* (three birds in agricultural fields), Tibetan Partridge *Perdix hodgsoniae* (one pair in meadows), Tibetan Sandgrouse *Syrrhaptes tibetanus* and Golden Eagle *Aquila chrysaetos* (single bird in ungrazed steppe). The number of individuals seen of the remaining 35 species ranged from four individuals (Eurasian Kestrel *Falco tinnunculus* and Common House Martin *Delichon urbicum*) to more than 850 in the case of the two most abundant species, Black Redstart *Phoenicurus ochruros* (1024 individuals) and Horned Lark *Eremophila alpestris* (897 individuals).

Rarefaction analysis indicated that bird species richness in different habitats in comparable samples of 700 individuals pooled across nine years of study ranged narrowly between 22 and 27 species (Figure 24.1). The meadows had significantly higher bird species richness than all other habitats (mean = 26.8, 95% CI = 25.7 – 27.9). Agricultural fields had the next highest estimate of species richness at a sample size of 700, but with increasing samples, the estimate for agricultural fields flattened out while that for ungrazed areas continued to increase. At sample 700, livestock grazed sites had the lowest estimate of bird species richness 21.7 (95% CI = 18.9 – 24.5).

Bird assemblage composition showed a clear pattern of variation across habitats as visualized in the ordination plot (Figure 24.2a). Agricultural fields and meadow transects appear separate from the ungrazed steppe, while grazed sites are intermediate in composition. Across years, differences are still noticeable among habitats with some exceptions: (i) grazed meadow bird assemblage composition in 2006 was substantially altered due to disturbance associated with construction of a new village road in the transect area, (ii) grazed and ungrazed steppes did not show substantial variation, although a diverging trajectory is noticeable between 2005 and 2008 (Figure 24.2b).

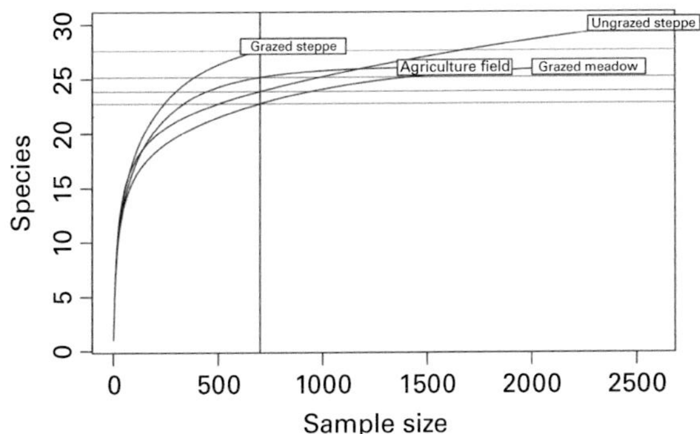

Figure 24.1. Rarefaction bird species richness across the habitats. Vertical line indicates the minimum number of individuals used to estimate species richness in all the four habitats. The horizontal lines indicate the species richness at the comparable sample size of 700 individuals.

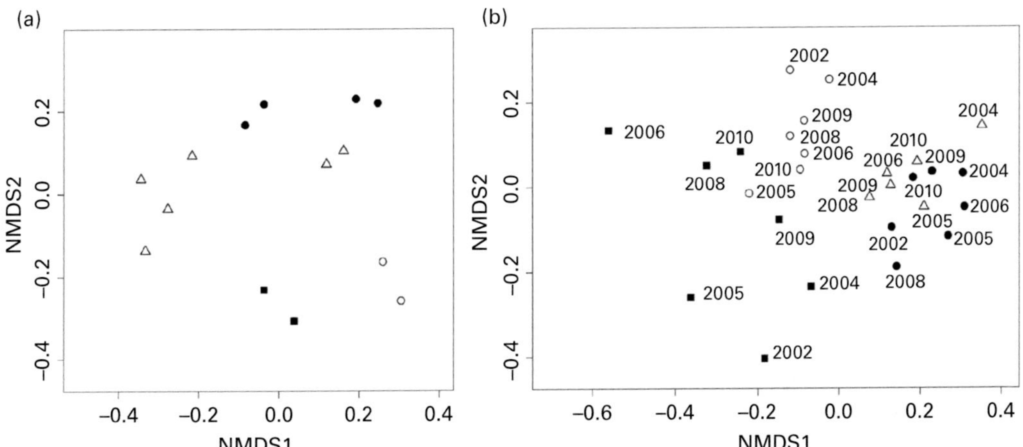

Figure 24.2. Variation in bird assemblage composition across sites: results of non-metric multidimensional scaling (NMDS) using the Bray-Curtis dissimilarity index. The NMDS plot represents the bird assemblage dissimilarity among sites in reduced (two) dimensions for illustrative purposes, such that points that are further apart are also more dissimilar in the bird assemblage composition as measured by the Bray-Curtis dissimilarity index. (a) Ordination of 14 study transects using data pooled across years. (b) Bird assemblage change in the four habitats across years using data pooled across all transects. The open triangles indicate grazed areas, closed circles indicate ungrazed, open circles indicate agriculture fields and the closed squares indicate meadows. Numbers on the graph indicate year of sampling.

The average density across years ranged between 5 and 10 birds per hectare in the four habitat types (Table 24.1). Comparison of 95% confidence intervals indicated that bird density was significantly higher at 8.8 birds.ha^{-1} in the agricultural fields as compared to meadows (4.3 birds.ha^{-1}) and grazed steppes (4.8 birds.ha^{-1}), but was not significantly higher than in ungrazed steppe (7.2 birds.ha^{-1}).

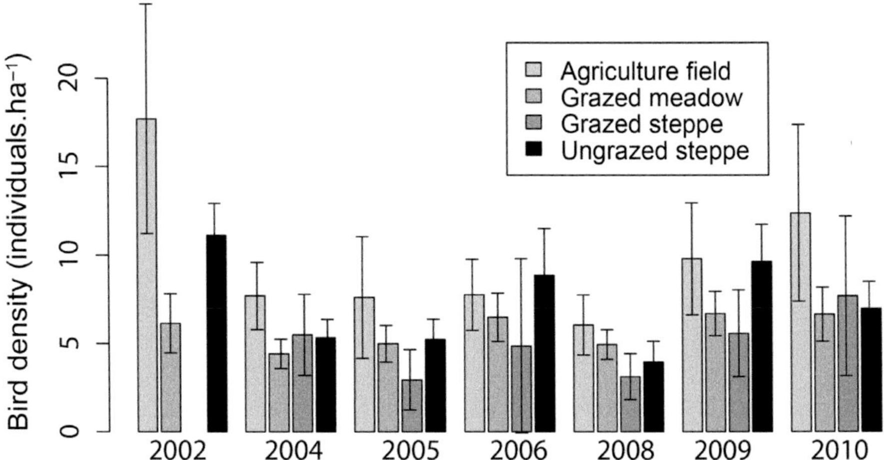

Figure 24.3. Inter-annual variation in bird density across the study habitats. Bird densities were estimated using distance sampling methods; vertical bars represent 95% confidence intervals.

Inter-annual variation in bird densities was substantial. Bird densities were highest in agricultural fields in 2002, but declined by more than 50% in succeeding years (Figure 24.3). Bird densities in agricultural fields between 2004 and 2008 did not appear to be significantly different from the other habitats. The mean density across all years in ungrazed steppes was higher than in grazed steppes, however, overlap of confidence intervals indicated that the difference was not statistically significant.

Granivores were the most abundant birds in all the habitats. Granivore density was highest in the agricultural fields (mean = 6.2 birds.ha^{-1}; 95% CI = 5.7–6.7 birds.ha^{-1}). However, about one-third of these were contributed by house sparrows (*Passer domesticus*) (2.0; CI 1.5–2.5 birds.ha^{-1}). The ungrazed steppe habitat had the next highest density of granivores (3.9; CI 3.3–4.5 birds.ha^{-1}) followed by grazed steppe (3.4; CI 3.0–3.8 birds.ha^{-1}) and grazed meadow (2.9; CI 1.9–3.9 birds.ha^{-1}) habitats.

Insectivore density appeared highest in the ungrazed steppe habitat (2.6; CI 2.2–3.0 birds.ha^{-1}), followed by grazed steppe (1.9; CI 1.6–2.2 birds.ha^{-1}), agricultural fields (1.5; CI 1.0–2.0 birds.ha^{-1}) and meadows (1.1; CI 0.9–1.3 birds.ha^{-1}). The density of the only grazing bird species, Himalayan Snowcock, was highest in the ungrazed steppe habitats (0.8; CI 0.6–1.0 birds.ha^{-1}), and much lower in grazed meadows (0.3; CI 0.0–0.7 birds. ha^{-1}), agricultural fields (0.2; CI 0.1–0.3 birds.ha^{-1}) and grazed steppe (0.1; CI 0.0–0.3 birds.ha^{-1}) habitats. Omnivore density was highest in the agricultural fields (1.8; CI 1.2–2.4 birds.ha^{-1}) and comparable in grazed meadows (0.9; CI 0.6–1.2 birds.ha^{-1}), grazed (1.1; CI 0.8–1.4 birds.ha^{-1}) and ungrazed steppe (1.0; CI 0.6–1.4 birds.ha^{-1}) habitats.

Densities of the 17 most common species estimated in each of the four main habitats along the gradient of land-use intensification are shown in Table 24.3. As indicated by the variation in mean density, 8 of the 17 species appeared to be negatively associated with intensification and sedentarization: Himalayan Snowcock, Black-headed Mountain Finch *Leucosticte brandti* and Plain Mountain Finch *Leucosticte nemoricola*, Black Redstart, Desert Wheatear *Oenanthe deserti*, Robin Accentor *Prunella rubeculoides*,

Table 24.3. Density of individual bird species per hectare in the four main habitat strata. Detection function was estimated by aggregating at the level of strata. Half-normal detection function for all except grazed strata for which hazard rate was used. CI stands for 95% confidence interval.

Species	Scientific names	Agriculture field	CI	Grazed steppe	CI	Grazed meadow	CI	Ungrazed steppe	CI
Black Redstart	*Phoenicurus ochruros*	0.9	0.8–1.1	1.1	0.9–1.2	0.8	0.6–0.9	1.2	1.1–1.4
Blue Rockthrush	*Monticola solitarius*		0–0		0–0	0.6	0–1.3	0.4	0.0–1.6
Black-headed Mountain Finch	*Leucosticte brandti*	1	0.5–1.4	1.7	1.3–2.1	1.3	0.1–2.4	2.5	1.8–3.1
Brown Accentor	*Prunella fulvescens*		0–0	0.8	0.0–2.5		0–0		0–0
Citrine Wagtail	*Motacilla citreola*	0.4	0.4–0.4	0.2	0.0–0.6		0–0	0.4	0.4–0.4
Eurasian Hoopoe	*Upupa epops*	0.7	0.3–1.1	0.4	0.2–0.6	0	0–0	0.4	0.4–0.4
Eurasian Kestrel	*Falco tinnunculus*		0–0	0	0–0		0–0	0	0–0
Common Raven	*Corvus corax*		0–0		0–0	0.3	0.3–0.4		0–0
Common Rosefinch	*Carpodacus erythrinus*		0–0	0.8	0.7–0.8		0–0	0.8	0.7–0.9
Desert Wheatear	*Oenanthe deserti*	0.6	0.4–0.7	0.8	0.7–0.8	0.8	0.5–1	0.9	0.8–1.1
Eurasian Crag-Martin	*Ptyonoprogne rupestris*	0.3	0.0–0.7	0	0–0	0.3	0.0–1.4	0	0–0
European Goldfinch	*Carduelis carduelis*	1.1	0.0–3.2		0–0		0–0		0–0
Fire-Fronted Serin	*Serinus pusillus*	0.6	0.5–0.8	2.3	0.8–3.7	0.6	0.3–0.9	0.6	0.2–1
Golden Eagle	*Aquila chrysaetos*		0–0		0–0		0–0	0	0–0
Grandala	*Grandala coelicolor*		0–0	0.4	0.0–1.6		0–0	0.4	0.0–1.6
Grey Wagtail	*Motacilla cinerea*	1.1	0.4–1.8	0.8	0.5–1	0.4	0.2–0.6	1	0.8–1.3
Great Rosefinch	*Carpodacus rubicilla*	0.8	0.6–1	0.8	0.6–0.9	0.6	0.3–0.9	1.1	0.9–1.2
Hill Pigeon	*Columba rupestris*	2.1	1.3–2.8	0.7	0.5–0.9	0.7	0.4–1	0.8	0.7–1
Himalayan Griffon	*Gyps himalayenis*	0	0–0		0–0	0.3	0.0–0.9		0–0
Himalayan Snowcock	*Tetraogallus himalayensis*	0.2	0.1–0.3	0.1	0–0.2	0.3	0.0–0.7	0.8	0.6–1
Horned Lark	*Eremophila alpestris*	1.4	1.1–1.6	1.3	1.2–1.5	0.8	0.5–1	1.2	0.9–1.4
House Sparrow	*Passer domesticus*	2	1.5–2.5		0–0	2.2	0.7–3.7		0–0
Hume's Lark	*Calandrella acutirostris*	1	0.7–1.2		0–0	0.6	0.1–1.2	0.8	0.0–2.6
Lammergeyer	*Gypaetus barbatus*		0–0	0.8	0.0–2.5		0–0	0	0–0
Common House-Martin	*Delichon urbicum*	0.6	0.2–0.9		0–0	0.3	0.0–1.4		0–0
Plain Mountain-Finch	*Leucosticte nemoricola*	1.3	0.7–1.8	1.2	0.9–1.5	0.9	0.2–1.6	1.3	0.9–1.6

Table 24.3. (cont.)

Species	Scientific names	Agriculture field	CI	Grazed steppe	CI	Grazed meadow	CI	Ungrazed steppe	CI
Red-billed chough	*Pyrrhocorax pyrrhocorax*	1.4	0.1–2.6	1.3	0.5–2.1	0.6	0.0–2.2	0.5	0.3–0.8
Red-Fronted Rosefinch	*Carpodacus puniceus*		0–0		0–0	1	0.0–3	0.9	0.7–1
Robin Accentor	*Prunella rubeculoides*	0.7	0.7–0.8	0.6	0.5–0.7	0.4	0.3–0.6	0.6	0.5–0.7
Rock Bunting	*Emberiza cia*	0.4	0.1–0.6	0.5	0.2–0.8	0.6	0.4–0.7	0.6	0.3–0.8
Rock Pigeon	*Columba livia*	1.3	0.3–2.3	0.6	0.4–0.9	1.3	0.0–3.1	0.3	0–0.7
Snow Pigeon	*Columba leuconota*		0–0	0.4	0.2–0.7	0.3	0.0–1.4	0.4	0.3–0.6
Sulphur-Bellied Warbler	*Phylloscopus griseolus*	0.5	0.4–0.7	0.9	0.7–1	0.6	0–1.3	1.2	1.1–1.4
Black-winged Snowfinch	*Montifringilla adamsi*	1.2	0.9–1.5	1.1	0.8–1.4	1.8	0.9–2.7	1.1	0.8–1.4
Tickell's Leaf-Warbler	*Phylloscopus affinis*	0.6	0.4–0.9		0–0	1	0.0–2.2	1	0.6–1.4
Yellow-billed Chough	*Pyrrhocorax graculus*	1.8	1.2–2.3	1.1	0.8–1.4	1	0.6–1.3	1	0.7–1.4

Snow Pigeon *Columba leuconota* and Sulphur-bellied Warbler *Phylloscopus griseolus*. Species that were more abundant in intensively used areas included Hill and Rock Pigeons *Columba rupestris, C. livia*, House Sparrow, Hume's Lark *Calandrella acutirostris*, Black-winged Snowfinch *Montifringilla adamsi* and Yellow-billed Chough *Pyrrhocorax graculus*. Variations in abundance of Great Rosefinch *Carpodacus rubicilla*, Horned Lark and Red-billed Chough *Pyrrhocorax pyrrhocorax* appeared idiosyncratic and did not show any obvious pattern in relation to land use classes that we discerned (Table 24.3).

Bird Responses to Agricultural Intensification

Our long-term data set from the Spiti Valley provides pointers towards the possible future of bird communities in the Trans-Himalayas in the face of sedentarization and agricultural intensification. While removal of livestock in grazing habitats can lead to reductions in bird species richness and abundance (Garcia *et al.*, 2008), in our case, both species richness and abundance were higher in the livestock-free habitats compared to grazed ones. This is presumably a reflection of the high intensity of livestock grazing in our study site (Mishra *et al.*, 2001). A further increase in livestock grazing intensity in future is likely to cause more declines, while a reduction in people's reliance on livestock and a shift to agriculture could increase bird species richness and abundance in the rangelands.

In other parts of the world, agricultural intensification, accompanied by modifications such as irrigation and reduction in landscape heterogeneity, has caused declines in bird diversity and abundance of ground-nesting, grassland and Mediterranean shrub-steppe birds (Brotons *et al.*, 2004; Guerrero *et al.*, 2012; da Silva *et al.*, 2015). In contrast, in our study area, while there were compositional differences, agricultural fields supported a relatively high abundance of birds and bird species richness, comparable to those of ungrazed steppes. High inputs of water and manure into agricultural fields in the Trans-Himalayas lead to greater primary productivity compared to the surrounding rangelands (Mishra, 2001). Because the natural vegetation of the region was already an open habitat (Mishra, 2001), some conversion of rangelands into agricultural land seems not to have affected the bird assemblage drastically, allowing for persistence, especially of granivores and omnivores. Together with minimal use of pesticides and near absence of hunting or killing (at least in our study site), these factors presumably help in maintaining high bird abundance.

At the same time, agricultural fields showed the most different bird species composition compared to all the other habitats, and especially compared to ungrazed steppes. The species composition appeared to diverge in relation to the level of land use intensification, with ungrazed steppes at one extreme, agricultural fields at the other and the grazed areas lying in between. Granivores, while generally being the most abundant guild, also attained the highest densities in agricultural fields, while insectivores occurred at lower abundance. The only grazing species recorded was significantly less abundant in agricultural fields.

In the current situation, a matrix of agricultural fields, rangelands and ungrazed steppe can allow continued persistence of birds in the Trans-Himalayan landscape. However, the future trajectory for bird conservation in such multiple-use landscapes remains uncertain. The use of pesticides in our study site was minimal, despite the fact that a government scheme is available offering farmers the possibility of acquiring pesticides and chemicals for 50% of the market cost (Himachal Pradesh State Government Economics and Statistics Department, 2013). There is a strong likelihood of increased use of chemicals in future, which would be detrimental for bird conservation, unless greater incentives can be provided to farmers for continued organic cultivation. Similarly, greater recognition, appreciation and reinforcement of peoples' willingness to coexist with wildlife would secure a future for birds in the face of rapid lifestyle changes in the Trans-Himalayas.

Acknowledgements

We are thankful to the Whitley Fund for Nature and Fondation Segré for their continued support to our field research and conservation programmes.

References

Ali, S. & Ripley, S.D. (1983). *Handbook of the Birds of India and Pakistan (compact edition)*. Oxford University Press and BNHS, Mumbai.

Ali, S. & Ripley, S.D. (1995). *The Pictorial Guide to the Birds of Indian Sub-continent*. Oxford University Press and BNHS, Mumbai.

Baskaran, S.T. (1992). Sighting of Dusky Horned Owl. *Newsletter for Birdwatchers*, **32**, 10.

Bauer, K. (2005). Development and the enclosure movement in pastoral Tibet since the 1980s. *Nomadic Peoples*, **9**, 53–81.

Berger, J., Bayarbaatar, B. & Mishra, C. (2013). Globalization of the cashmere market and the decline of large mammals in Central Asia. *Conservation Biology*, **27**, 679–689.

Brantingham, P.J., Gao, X., Olsen, J.W., *et al.* (2007). A short chronology for the peopling of the Tibetan Plateau. *Developments in Quaternary Sciences*, **9**, 129–150.

Brotons, L., MaÑosa, S. & Estrada, J. (2004). Modelling the effects of irrigation schemes on the distribution of steppe birds in Mediterranean farmland. *Biodiversity & Conservation*, **13**, 1039–1105.

Buckland, S.T., Anderson, D.R., Burnham, K.P. & Laake, J.L. (1993). Assumptions and modelling philosophy. In *Distance Sampling*. Springer Netherlands, pp. 29–51.

Buckland, S.T., Anderson, D.R., Burnham, K.P. & Laake, J.L. (2005). *Distance Sampling*. John Wiley & Sons, Ltd.

Canavelli, S.B., Swisher, M.E. & Branch, L.C. (2013). Factors related to farmers' preferences to decrease monk parakeet damage to crops. *Human Dimensions of Wildlife*, **18**, 124–137.

da Silva, T.W., Dotta, G. & Fontana, C.S. (2015). Structure of avian assemblages in grasslands associated with cattle ranching and soybean agriculture in the Uruguayan savanna ecoregion of Brazil and Uruguay. *The Condor: Ornithological Applications*, **117**, 53–63.

De Mey, Y., Demont, M. & Diagne, M. (2012). Estimating bird damage to rice in Africa: evidence from the Senegal River Valley. *Journal of Agricultural Economics*, **63**, 175–200.

Fratkin, E. (1997). Pastoralism: governance and development issues. *Annual Review of Anthropology*, **26**, 235–261.

García, C., Renison, D., Cingolani, A.M. & Fernández-Juricic, E. (2008). Avifaunal changes as a consequence of large-scale livestock exclusion in the mountains of Central Argentina. *Journal of Applied Ecology*, **45**, 351–360.

Gotelli, N.J. & Colwell, R.K. (2001). Quantifying biodiversity: procedures and pitfalls in the measurement and comparison of species richness. *Ecology Letters*, **4**, 379–391.

Gotelli, N.J. & Entsminger, G.L. (2001). EcoSim: Null models software for ecology. www.uvm.edu/~ngotelli/EcoSim/EcoSim.html.

Gregory, R.D. & Van Strien, A. (2010). Wild bird indicators: using composite population trends of birds as measures of environmental health. *Ornithological Science*, **9**, 3–22.

Grimmett, R., Inskipp, C., Inskipp, T. & Allen, R. (2011). *Birds of the Indian Subcontinent*. Second edition. Oxford University Press, London.

Guerrero, I., Morales, M.B. & Oñate, J.J. (2012). Response of ground-nesting farmland birds to agricultural intensification across Europe: landscape and field level management factors. *Biological Conservation*, **152**, 74–80.

Harris, J. (1994). Cranes, people and nature: preserving the balance. *The future of cranes and wetlands*. Proceedings of the International Symposium, Wild Bird Society of Japan, Tokyo, pp, 1–15.

Himachal Pradesh State Government Economics and Statistics Department (2013). Economic Survey of Himachal Pradesh. Public report.

Johnson, M.D., Kellermann, J.L. & Stercho, A.M. (2010). Pest reduction services by birds in shade and sun coffee in Jamaica. *Animal Conservation*, **13**, 140–147.

Kirk, D.A., Evenden, M.D. & Mineau, P. (1996). Past and current attempts to evaluate the role of birds as predators of insect pests in temperate agriculture. *Current Ornithology*, **13**, 175–269.

Miller, D.L. (2015). Distance: Distance Sampling Detection Function and Abundance Estimation. R package version 0.9.4. https://CRAN.R-project.org/package=Distance.

Mishra, C. (2001). High altitude survival: conflicts between pastoralism and wildlife in the Trans-Himalaya. PhD thesis, Wageningen University, The Netherlands.

Mishra, C., Bagchi, S., Namgail, T. & Bhatnagar, Y.V. (2010). Multiple Use of Trans-Himalayan Rangelands: Reconciling Human Livelihoods with Wildlife Conservation. In J.T. du Toit, R. Kock and J.C. Deutsch, eds., *Wild Rangelands: Conserving Wildlife while Maintaining Livestock in Semi-arid Ecosystems*. Chichester: Wiley-Blackwell, pp. 291–311.

Mishra, C. & Humbert-Droz, B. (1998). Avifaunal survey of Tsomoriri Lake and adjoining Nuro Sumdo Wetland in Ladakh, Indian Trans-Himalaya. *Forktail*, **14**, 65–68.

Mishra, C., Prins, H. H. T. & Van Wieren, S. E. (2001). Overstocking in the Trans-Himalayan rangelands of India. *Environmental Conservation*, **28**, 279–283.

Namgail, T., Bhatnagar, Y.V., Mishra, C. & Bagchi, S. (2007). Pastoral nomads of the Indian Changthang: production system, landuse, and socio-economic changes. *Human Ecology*, **35**, 497–504.

Namgail, T., Mudappa, D. & Raman, T.R.S. 2009. Waterbird numbers at high altitude lakes in eastern Ladakh, India. *Wildfowl*, **59**, 137–144.

Ning, W. & Richard, C.E. (1999, July). The privatisation process of rangeland and its impacts on the pastoral dynamics in the Hindu Kush Himalaya: the case of Western Sichuan, China. In *People and Rangelands*. Proceedings of VI International Rangelands Congress, Townsville, Australia, pp. 14–21.

Pfister, O. (2004). *Birds and Mammals of Ladakh*. New Delhi: Oxford University Press.

R Core Team (2016). *R: A language and environment for statistical computing*. R Foundation for Statistical Computing, Vienna, Austria. www.R-project.org/.

Shen, X., Li, S., Chen, N., Li, S., McShea, W.J. & Lu, Z. (2012). Does science replace traditions? Correlates between traditional Tibetan culture and local bird diversity in Southwest China. *Biological Conservation*, **145**, 160–170.

Shen, X. & Tan, J. (2012). Ecological conservation, cultural preservation, and a bridge between: the journey of Shanshui Conservation Center in the Sanjiangyuan region, Qinghai-Tibetan Plateau, China. *Ecology and Society*, **17**, 38.

Smits, J.E.G. & Fernie, K.J. (2013). Avian wildlife as sentinels of ecosystem health. *Comparative Immunology, Microbiology and Infectious Diseases*, **36**, 333–342.

Sullivan, B.L., Aycrigg, J.L., Barry, J.H., *et al.* (2014). The eBird enterprise: an integrated approach to development and application of citizen science. *Biological Conservation*, **169**, 31–40.

Thomas, L., Buckland, S.T., Rexstad, E.A., *et al.* (2010). Distance software: design and analysis of distance sampling surveys for estimating population size. *Journal of Applied Ecology*, **47**, 5–14.

Urgenson, L., Schmidt, A.H., Combs, J., *et al.* (2014). Traditional livelihoods, conservation and meadow ecology in Jiuzhaigou National Park, Sichuan, China. *Human Ecology*, **42**, 481–491.

Woodhouse, E. (2012). The role of Tibetan Buddhism in environmental conservation under changing socio-economic conditions in China (Doctoral dissertation, Imperial College London).

Zhang, Y. (2016). Wild geese of the Yangtze River: their ecology and conservation. PhD thesis submitted to Wagningen University. The Netherlands.

Zhou, Z.Q., & Liu, T. (2005). The current status, threats and protection way of Sanjiang Plain wetland, Northeast China. *Journal of Forestry Research*, **16**, 148–152.

25 Migratory Ducks and Protected Wetlands in India

Tsewang Namgail, John Y. Takekawa, Sivananinthaperumal Balachandran, Taej Mundkur, Ponnusamy Sathiyaselvam, Diann J. Prosser, Tracy McCracken and Scott H. Newman

Wetlands and Migratory Ducks in India

India is the most important country for wintering migratory ducks in the Central Asian Flyway (Rahmani & Islam, 2008). Because of its latitudinal and climatic extent, the country provides a diversity of wetland habitats for migratory ducks (Ali & Ripley, 1978; Kumar *et al.*, 2005). India is the seventh largest country in the world with an area of about 3.3 million km^2 or 2.4% of the world's land area. Mainland India stretches nearly 3200 km from north to south (6° to 36° N), and 3000 km from west to east (68° to 98° E). Given this huge geographical extent and the migration strategies of ducks, migratory ducks wintering in the southern part of the country need to refuel at several wetlands before they cross the Himalayas on their way to breeding sites in Central Asia and Siberia. There is, however, no proper regulatory framework for conservation of wetlands in India (Bassi *et al.*, 2014).

India has a variety of wetlands, ranging from high-altitude lakes in the Himalayas to mangroves in West Bengal and coastal plains in the southeastern part of the country. These wetlands encompass an estimated area of about 153,000 km^2, constituting 4.63% of the geographical area of the country (Panigrahy *et al.*, 2012). Rivers and streams form the most extensive wetlands, however, encompassing an area of about 53,000 km^2. More than 60 wetlands, including lakes, rivers and reservoirs in the country, are protected (Gopal & Sah, 1995). Twenty-six of these wetlands with a total area of about 7,000 km^2 are protected as Wetlands of International Importance under the Ramsar Convention on Wetlands (Islam & Rahmani, 2008). A further 130 areas have been identified as potential Ramsar sites (Islam & Rahmani, 2008).

The utilization pattern of these protected areas by migratory ducks is, however, not completely understood. Even if the ducks use these protected wetlands, it is not known whether the wetlands adequately fulfil their habitat requirements. This information is needed for the proper management of migratory ducks and is also important for developing conservation strategies for these ducks, as the populations of several duck species are declining in India (Wetlands International, 2016).

Recognizing this, we assessed the extent of overlap between Ramsar sites and migratory routes of Northern Pintail *Anas acuta* and Garganey *A. querquedula*. Subsequently, we checked for spatial correspondence between protected wetlands and high utilization areas of these two species in India. We also examined the habitat preferences of seven species of ducks: Garganey *Anas querquedula*, Gadwall *Anas strepera*, Northern Pintail *Anas acuta*, Common Teal *Anas crecca*, Eurasian Wigeon *Anas penelope*, Northern Shoveler *Anas clypeata* and Ruddy Shelduck *Tadorna ferruginea*, to see if the current network of protected areas provides suitable habitat for these migratory ducks in India.

Correspondence between Protected Wetlands and Migratory Routes of Ducks

To evaluate spatial correspondence between the migratory paths of ducks and Ramsar sites, we assigned 50 km-wide buffers along the migratory paths of Garganey and Northern Pintail, obtained during the present study, and overlaid these on a layer of Ramsar sites using ArcGIS version 9.3 (Environmental Systems Research Institute, Inc., Redlands, CA, USA). We then calculated the percentages of protected wetlands that intersected the migratory paths of these species. The results showed that 16 Ramsar sites (61%, n = 26) intersected the 50-km buffer along the migratory routes of Garganey and Northern Pintail (Figure 25.1). This does not mean that the ducks necessarily stop at these protected wetlands, and further analysis showed that this is indeed not the case. For example, the two species marked in Tamil Nadu stopped at several wetlands on the Deccan Plateau, but there are no Ramsar sites in this region (Figure 25.1). The Srisailam Backwater near Kurnool and Moosi Reservoir in Andhra Pradesh were used rather extensively by these species as staging sites, but currently they do not have any protection status (Islam & Rahmani, 2008).

Similarly, the migratory ducks used several unprotected wetlands on the Indo-Gangetic Plain, just south of the Himalayas, as staging sites during their spring migration. The network of wetlands on this plain is very important for migratory ducks, because the birds have to cross the formidable Himalayas after feeding in these wetlands (Namgail *et al.*, Chapter 2). Their migratory success over this mountain range is therefore determined to a large extent by the quality of habitat around these wetlands. Given the rapid economic development in the country, it is crucial to ensure protection of these wetlands before they are modified or degraded beyond recovery. Building of dams to meet the energy needs of the country is further impacting the wetland habitats of migratory birds in the Himalayas (Grumbine & Pandit, 2013)

Do Migratory Ducks Use Protected Wetlands?

To examine whether migratory ducks use the Ramsar sites, we assessed the spatial correspondence between these protected wetlands and the ranges of distribution (utilisation distribution) of Garganey and Northern Pintail. For this, we overlaid a map of Ramsar sites on the distribution ranges of the ducks that we studied. We digitized Ramsar sites in India, as listed on the Ramsar Convention Website (http://ramsar.wetlands.org). Utilization distributions were determined by the Brownian bridge movement model (BBMM) in R (Horne *et al.*, 2007; R Development Core Team, 2008). Brownian bridge is

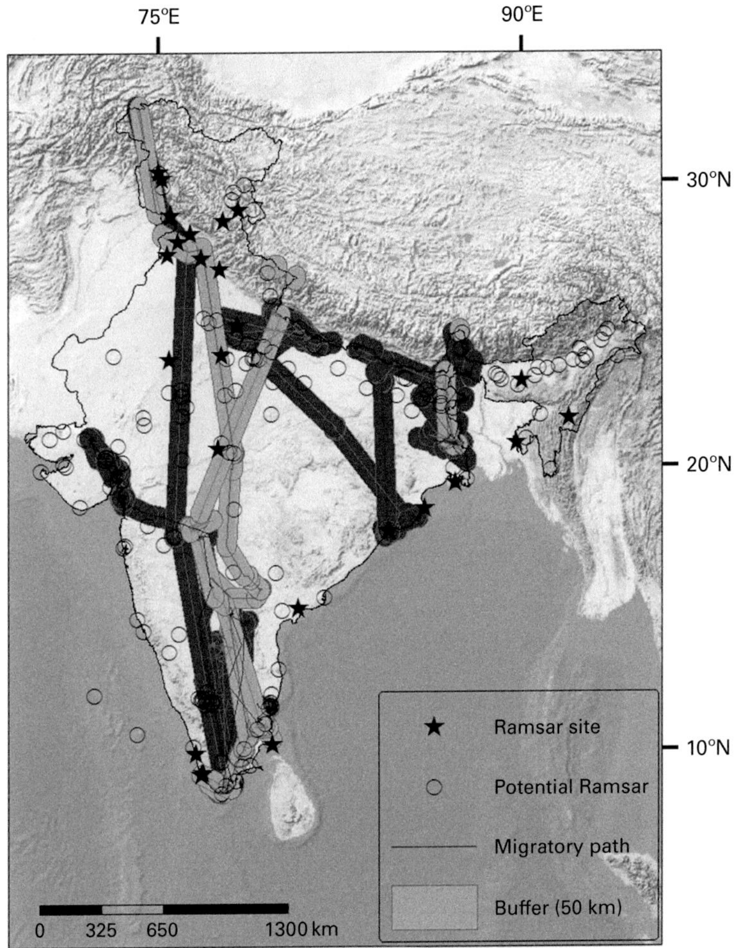

Figure 25.1. Northward migration routes (lines and buffers) of Northern Pintail (light grey) and Garganey (dark grey) within India, plotted in relation to existing Ramsar sites (stars) and potential Ramsar sites (circles). (A black-and-white version of this figure will appear in some formats. For the colour version, please refer to the plate section.)

a continuous-time stochastic model of movement, in which the probability of an individual duck being in an area is conditioned on starting and ending locations (Horne *et al.*, 2007).

For estimating utilization distribution, we used location coordinates from 13 individuals of Garganey and nine individuals of Northern Pintail. The utilization distribution of the two species highlighted the use of wetlands on the Deccan Plateau and the Indo-Gangetic Plain as the main wintering areas (Figure 25.2), but there was little correspondence between the Ramsar sites and the utilization distributions of these ducks. In the case of Garganey, only 8% of the Ramsar sites overlapped with the species' high-use areas (Figure 25.2), while in the case of the Northern Pintail, 36% of the Ramsar sites overlapped with its high-use areas. The utilization distributions of the two species in this study corresponded well with their distribution maps provided by the Asian Waterbird

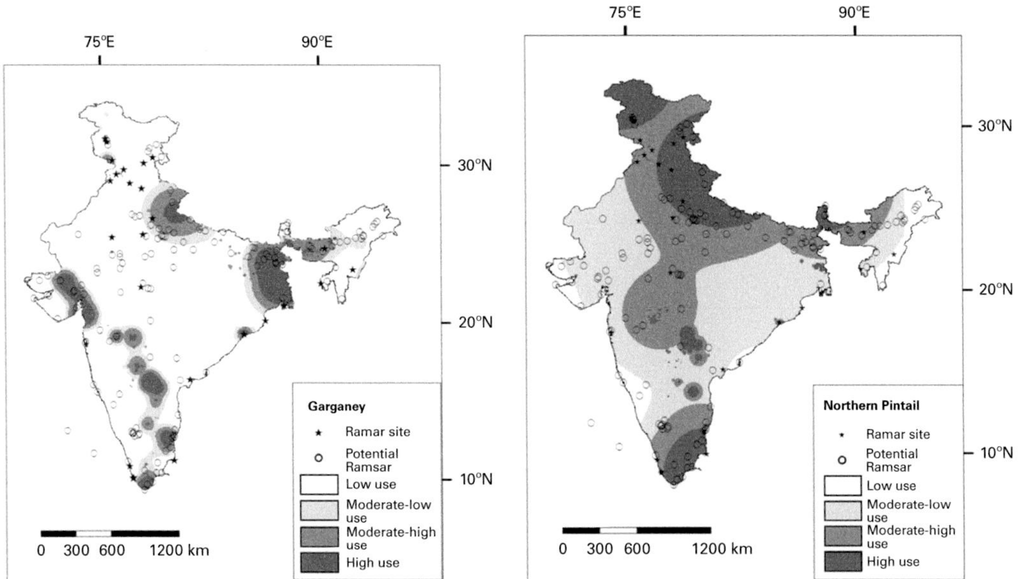

Figure 25.2. Brownian Bridge Utilization Distributions of Garganey and Northern Pintail in relation to Ramsar and potential Ramsar sites (wetlands that fulfil the Ramsar criteria but not yet declared as a Ramsar site).

Census (AWC, Wetlands International, 2002), indicating that our samples were representative of the populations of these species in India. Thus, the migratory dabbling ducks apparently use the Ramsar sites rather sparingly, which raises the question as to whether migratory dabbling ducks receive adequate protection in India, their most important wintering ground in the Central Asian Flyway. There was, however, a greater degree of overlap between the potential Ramsar sites and the species' utilization distributions. For example, 36% of the potential Ramsar sites overlapped with the high-use areas of Garganey, while 87% of the potential Ramsar sites overlapped with the high use areas of Northern Pintail (Figure 25.2). The declaration of these potential sites as Ramsar sites might therefore be crucial to the conservation of these migratory ducks in India.

Do Protected Wetlands Provide Adequate Resources for Migratory Ducks?

The existing Ramsar sites also provided inadequate habitats for migratory ducks in India. This was indicated by a habitat analysis of seven species of ducks. We analyzed the habitat use of these species by comparing their habitat use with the habitat available at the Ramsar sites. For estimating habitat utilization, we calculated the third-order selection ratios (i.e. use and available habitats were measured for each bird; Thomas & Taylor, 1990). We used χ^2 tests to statistically test significant departures of use from the available habitat. A habitat was used selectively, if the confidence limit for that habitat excluded 1. If selected, a habitat was preferred if the interval was > 1, and avoided if it was < 1 (Manly *et al.*, 1993).

Most of the ducks: Northern Pintail (χ^2 = 90.82, $p < 0.001$), Northern Shoveler (χ^2 = 59.54, $p < 0.001$), Common Teal (χ^2 = 77.03, $p < 0.001$) and Ruddy Shelduck (χ^2 = 118.92, $p < 0.001$), used lakes less than expected from their availability, but this wetland class was the predominant habitat type in the protected wetlands, constituting more than 70% of the total area of the Ramsar site network. Dabbling ducks used coastal habitats more extensively. For example, Northern Shoveler (χ^2 = 59.54, $p < 0.001$), Eurasian Wigeon (χ^2 = 59.54, $p < 0.01$) and Common Teal (χ^2 = 77.03, $p < 0.001$) used this habitat type more than would be expected from its availability. Yet the Ramsar sites encompassed very little of this habitat type.

Northern Pintail also preferred intermittent or seasonal wetlands (χ^2 = 90.82, $p < 0.001$), but the network of existing Ramsar sites included only a small proportion of this habitat type. Similarly, Ruddy Shelduck preferred flood plains (χ^2 = 118.92, $p < 0.001$) as well as rivers, but both these habitat types were less represented in the existing network of Ramsar sites in India (flood plain < 10% and river < 1%). Even if the migratory ducks stop at the existing Ramsar sites, these protected wetlands do not provide enough suitable habitats for migratory dabbling ducks. Thus, it is important to confirm the adequacy of the other wetlands identified as potential Ramsar sites, and if they are found suitable for migratory ducks, then upgrade their protection status as a high priority.

Recommendations for the Conservation of Ducks in India

Although 16 of the Ramsar sites coincided with the migratory routes of the ducks, only small percentages of the areas used by these ducks overlapped with the Ramsar sites. This indicates that although the migratory ducks fly over some of the Ramsar sites, they do not make much use of these wetlands. Furthermore, most of the protected wetlands do not include habitats such as floodplains, rivers and coastal wetlands that are preferred by the migratory ducks. Most of the ducks used lakes less than expected from their availability, but this wetland type is the predominant habitat in the network of protected wetlands. The reason for under-utilization of lakes by these ducks is not known. Perhaps these dabbling ducks cannot find much food in the deep water of lakes. Less than 5% of India's wetlands are protected, and the overrepresentation of lakes among these areas apparently reduces the benefits to the conservation of dabbling ducks. Halting and reversing the decline of wildfowl populations in India requires the designation and management of additional protected areas, including the potential Ramsar sites containing habitats preferred by the ducks.

Most of the species studied were found to prefer coastal habitats. Thus, these habitat types need to be conserved to benefit the long-term survival of migratory ducks in India. Nearly 250 million people live within 50 km of the Indian coast. Therefore, the pressure on coastal areas is increasing as more people move closer to the coast due to climate change–related droughts in the interior parts of India. Currently, the coastal wetlands face many threats, including those from shrimp aquaculture (Nagarajan & Thiyagesan,

2006). Coastal development initiatives are further putting coastal wetlands at great risk (Venkataraman, 2008).

Wetlands in the Himalayas are rapidly becoming degraded, and in some areas as many as 30% of lakes and marshes have disappeared because of over-exploitation and climate change (Kanwal *et al.*, 2013). The migratory ducks in this study used many small, seasonal wetlands on the Indo-Gangetic Plain before crossing the Himalayas. Although crucial for bird migration across the Himalayas, these wetlands are individually too small to be considered in the network of protected areas. But since they are important staging sites for birds taking the arduous journey across the Himalayas, it may be beneficial to collectively consider these as Ramsar sites (Prins *et al.*, Chapter 26), so that the great migration system across the world's highest mountain range can be conserved now and for posterity.

Acknowledgements

We thank the UN Food and Agriculture Organization (FAO) and US Geological Survey (USGS) for funding this programme. Thanks are due to partners and government officials in India (Yogendrapal Singh, R. Sundararaju, P. Gangaimaran, Dr Anmol Kumar, Dr Lal Krishna, Devojit Das, A.K. Pattnaik, Dr S. Nagarajan, Dr G. Chetri, D.M. Singh, Mr Mukherji and V. Thirunavukarasu). We thank William Perry, Ben Gustafson, David Douglas, Shane Heath and Bridget Collins of USGS for their help and support. We thank M.V. Subba Rao (FAO) and Sourabh Sen (US State Department, US Embassy in Kolkata). We also thank Asad Rahmani at the Bombay Natural History Society. Capture, handling and marking procedures were approved by the US Geological Survey Patuxent Wildlife Research Center Animal Care and Use Committee (2007–01). The use of trade, product or firm names in this publication is for descriptive purposes only and does not imply endorsement by the US government.

References

Ali, S. & Ripley, D.S. (1978). *Handbook of the Birds of India and Pakistan Together with those of Bangladesh, Nepal, Bhutan and Sri Lanka*. New Delhi, India: Oxford University Press.

Bassi, N., Kumar, M.D., Sharma, A. & Pardha-Saradhia, P. (2014). Status of wetlands in India: a review of extent, ecosystem benefits, threats and management strategies. *Journal of Hydrology: Regional Studies*, **2**, 1–19.

Gopal, B. & Sah, M. (1995). Inventory and classification of wetlands in India. *Vegetatio*, **118**, 39–48.

Grumbine, R.E. & Pandit, M.K. (2013). Threats from India's Himalaya dams. *Science*, **339**, 36–37.

Horne, J.S., Garton, E.O., Krone, S.M. & Lewis, J.S. (2007). Analyzing animal movements using Brownian bridges. *Ecology*, **88**, 2354–2363.

Islam, M.Z. & Rahmani, A.R. (2008). *Existing and Potential Ramsar Sites of India*. Mumbai, India: Oxford University Press.

Kanwal, K.S., Samal, P.K., Lodhi, M.S. & Kuniyal, J. (2013). Climate change and high-altitude wetlands of Arunachal Pradesh. *Current Science*, **105**, 1037–1038.

Kumar, A., Sati, J.P., Tak, P.C. & Alfred, J.R.B. (2005). *Handbook on Indian Wetland Birds and Conservation*. Dehradun, India: Shiva Printing Press.

Li, D.Z.W., Bloem, A., Delany, S., Martakis, G. & Quintero, J.O. (2009). *Status of Waterbirds in Asia. Results of the Asian Waterbird Census: 1987–2007*. Wetlands International, Wageningen, the Netherlands.

Manly, B.F.J., McDonald, L.L. & Thomas, D.L. (1993). *Resource Selection by Animals*. London, UK: Chapman & Hall.

Nagarajan, R. & Thiyagesan, K. (2006). The effects of coastal shrimp farming on birds in Indian mangrove forests and tidal flat. *Acta Zoologica Sinica*, **52**, 541–554.

Panigrahy, S., Murthy, T.V.R., Patel, J.G. & Singh, T.S. (2012) Wetlands of India: inventory and assessment at 1:50,000 scale using geospatial techniques. *Current Science*, **102**, 852–856.

R Development Core Team (2008). *R: A Language and Environment for Statistical Computing*. Vienna, Austria: R Foundation for Statistical Computing.

Rahmani, A.R. & Islam, M.Z. (2008). *Ducks, Geese and Swans of India: Their Status and Distribution*. Mumbai, India: Oxford University Press.

Ramsar Convention Secretariat (2010). Wise use of wetlands: concepts and approaches for the wise use of wetlands. *Ramsar Handbooks for the Wise Use of Wetlands*, 4th edition, vol. **1**. Ramsar Convention Secretariat, Gland, Switzerland.

Singh, M.P. (2009). Rice Productivity in India under Variable Climates. Proceedings of the MARCO Symposium 2009 'Next Challenges of Agro-Environmental Research in Monsoon Asia'. www.niaes.affrc.go.jp/marco/marco2009/english/program/W2-02_Singh_P.pdf

Thomas, D. and Taylor, E. (1990). Study designs and tests for comparing resource use and availability. *Journal of Wildlife Management*, **54**, 322–330.

Venkataraman, K. (2008). Coastal and marine wetlands in India. Proceedings of the Taal 2007: the 12th World Lake Conference, pp. 392–400.

Wetlands International (2002). Waterbird population estimates. Wetlands International, Wageningen, The Netherlands.

Wetlands International (2016). 'Waterbird Population Estimates'. Retrieved from wpe.wetlands.org on 1 February 2016.

26 A Network of Small, Dispersed Himalayan Wetlands Suitable for Designation under the Ramsar Convention

Herbert H.T. Prins, Sipke E. van Wieren and Tsewang Namgail

The Need for Fuelling Stations for Migratory Birds

The Himalayas form a massive and globally unique barrier to avian migration. We use the term 'Himalayas' to refer collectively to the Pamirs, Hindu Kush, Karakoram, Zanskar, Great Himalaya and Kailas Ranges, and the Trans-Himalaya ranges behind them. Migrants in the New World do not face east–west mountain ranges of nearly the same extent or altitude, and in Europe, the Alps and Pyrenees can be circumvented relatively easily because of their much smaller spatial extent. Migration in other areas such as the Middle East, the Far East and Africa does not necessitate high-altitude crossings. However, millions of birds migrate from Tibet, Mongolia and Siberia to the Indian subcontinent in the autumn, and return in spring. Already, decades ago, ring recovery data from Bharatpur in northern India (McClure, 1974), for example, unambiguously showed that many bird species cross these ranges to breed in Siberia, roughly in the triangle from Taymyr in the north, to Lake Baikal in the southeast and Lake Balkash in the southwest, with many Ruffs *Philomachus pugnax* migrating to east Siberia (see also Rogacheva, 1992). The wetlands of Pakistan, India and Bangladesh (which must have been of enormous extent before humans transformed them) lie closer to these vast breeding areas than to the east Atlantic wetlands or to those along the west Pacific Flyway. The price of this reduced distance, however, is a crossing of very high mountains exceeding 8000 m above sea level (a.s.l.) in altitude, in which even many of the mountain passes exceed 6000 m (Groen *et al.*, Chapter 17). The mean recorded altitude of migrants globally is in the order of 1300 m, with fewer than 10% flying higher than 3000 m, and a typical reported maximum of only 4800 m (Alerstam & Gudmundsson, 1999) to 5400 m over the Alps (Bruderer, 1999). However, Liechti and Schaller (1999) in a radar study over Israel, found that nocturnal migrants on spring migration flew at altitudes between 5000 m and nearly 9000 m a.s.l. to take advantage of optimal wind conditions, indicating that the ecological challenges of a barrier as high as the Himalayas, while formidable, are within the range of possibility for many species.

In other areas, migrants are known to skirt rather than cross mountains (Bruderer & Jenni, 1990; Bruderer, 1999). The Himalayas and associated ranges, however, stretch over about 35° of longitude (~4000 km), which would necessitate very long detours. The crossing is also very wide. The mountain ranges extend over approximately 10° of latitude (1100 km), and are bordered to the north by the extensive arid lands of the Taklamakan and Gobi Deserts. The Trans-Himalayan ranges themselves are as dry as these deserts. For example, the annual precipitation in Leh, the capital of Ladakh (3620 m a.s.l. in the State of Jammu Kashmir, India), is about 80 mm. The recorded precipitation in the village of Kibber (4200 m a.s.l. in Lahaul-Spiti, part of the State of Himachal Pradesh, India), just to the south of the Greater Himalaya range, is about 50 mm (Charudutt Mishra, pers. comm.) (see Bookhagen, Chapter 11 for further details). Summer temperatures may reach 35°C while snowfall in winter is low (in the order of 50 cm, which is equivalent to 50 mm of rain). At higher altitudes, temperatures and hence evaporation are much lower, while both summer rainfall and winter snowfall are higher. The net result is that small, dispersed wetlands have developed at altitudes between 4500 m and 5200 m a.s.l. At even higher altitudes, the summer season is so short that even though water may be available, there is no appreciable plant cover, and wetlands in the normal sense of the word do not exist (see Mani, 1978; Mishra, 2001; Rawat, Chapter 12).

In this high, arid and barren landscape, these wetlands are the only places where wetland birds can rest and forage, and sometimes even breed and reproduce. Very little has been published about these very high-altitude wetlands of the northwestern Himalayas and associated ranges, and this chapter offers insights into the importance of the different types of wetland. We specifically aim at understanding which types of wetland are important for resident and migrating birds because it is well known that stop-over sites for migrating birds are essential for the success of their migrations (Chapters 1–8, 16–20, 25).

We describe our results based on our expeditions between 1998 and 2011 to Ladakh (India). We estimated that six of these expeditions would give us enough information to determine species diversity with some reliability. For details, see Prins *et al.*, Chapter 20. For nomenclature of birds, we follow Grimmett *et al.* (1999) and Pfister (2004).

Wetland Types

We distinguished seven different types of wetland (Table 26.1). In all cases, their existence is due entirely to snowmelt from glaciers on the surrounding high mountains, rising up to more than 7500 m. Bird densities are low (Table 26.2), and the number of species is modest. This appears to be caused by limited food resources. Generally, densities of invertebrates are very low (see Prins *et al.*, Chapter 18); the hostile environment during the winter, with temperatures as low as −50° C and very little snow cover, restricts the distribution of invertebrate life. Streams and rivulets above 5000 m a.s.l. did not support any observable fish or invertebrates, and their banks were devoid of vegetation. The only birds observed along these extreme high-altitude streams were Grey Wagtails *Motacilla cinerea*, which we even observed at 6200 m. Away from the streams, we saw a number of other species (Table 26.3), none of

Table 26.1. Seven different types of wetland were distinguished in the Himalayas of northwestern India.

- *high-altitude stream*: streams above 5200 m a.s.l. but lower than 6000 m; the area is too high for plant growth, fish or crustaceans and these rivers are thus apparently devoid of life;
- *high-altitude marsh*: a marshy area with much cryoturbation and impeded drainage due to permafrost, between 4800 and 5200 m;
- *rivulets through grassland*: two types of rivulets could be discerned, namely, fast-running (faster than 1 m), steep rivulets and much slower-flowing rivulets (slower than 0.2 m); both types have grassy banks (between 2 m and 20 m wide) with shortly cropped sedge vegetation and some grasses;
- *large rivers*: the Indus, with its steep banks, is approximately 50 m wide and very deep; it generally does not have any vegetation growing along its banks in the range between 4400 m above sea level and 3800 m. In lower-lying areas, it has agriculture on its banks, or willow plantations and thickets of Sea Buckthorn *Hippopahe rhamnoides*;
- *freshwater lake*: Lake Tsomoriri, with a total bank length of some 75 km, is filled with good, potable water. Fish occur, and the lake is some 40 m deep, but it has no outflow. It lies about 4560 m above sea level. The banks are rocky (approximately 40 km), covered in shingle (some 25 km) or have grass and sedges growing (some 10 km);
- *brackish lake*: Lake Thatsangkaru, with a total bank length of some 15 km, has slightly brackish water. The lake has no outflow, and is situated at 4720 m above sea level. Its banks are grass and sedge covered over a distance of some 8 km and the remainder is formed by a shingle beach;
- *saline lake*: Lake Tsokar consists of two lakes which are both saline. The two lakes together have a bank length of some 35 km, and they are surrounded by raised beaches situated some 80 m above the present-day lake level. Their borders consist of saltmarsh (some 8 km), freshwater marsh (some 7 km), sand and gravel (some 10 km) and salty clay deposits (10 km). These lakes lie at 4580 m above sea level, and are without outflow.

which is normally dependent on water or wetlands except Brown-headed Gull *Larus brunnicephalus*, which we encountered flying over a mountain pass towards the wetland of Tsokar; Bar-headed Geese *Anser indicus* while we crossed a mountain pass at 4900 m, which may have been flying at an altitude of 5100 m. In other words, above 5000 m altitude, wetland-associated birds do not occur (Figure 26.1) except as flying migrants or stragglers.

Large Rivers

The banks of the Shayok, Zanskar and Nubra Rivers, and of the River Indus too, are devoid of vegetation for long stretches. In the area between Panamik, Diskit, Khalsar and Tsati (each about 40 km apart), at the confluence of the Shayok and Nubra Rivers at 3300 m a.s.l., we did not observe any waders, terns, gulls or ducks. Apart from some barley fields, the riverbanks are desert-like and sand dunes occur in the rather wide plain between the river and the high and very steep mountains. Along the River Indus, from Phey (3600 m a.s.l., about 25 km downstream of Leh) to Mahe (close to the border with China and some 100 km upstream of Leh) at about 4400 m a.s.l., there are some grassy banks and thickets of willow (*Salix* spp.) and seabuckthorn *Hippophae rhamnoides* L. and *H. tibetana* Schltdl. The riverbanks are generally steep, and the shale and granite

Table 26.2. Wetland birds species recorded in the different wetland types distinguished in Table 26.1. Only the densities of typical wetland birds were recorded, although absence and presence of other birds, such as raptors and songbirds were recorded too. Densities are expressed in totals per 100 km shoreline. For streams and rivers we calculated these per 100 km of river length (shore line was not double counted). July surveys were in 2000 and 2006, August in 2002 and 2006, and September in 1998.

Species	High-altitude streams 5200–6000 m		High-altitude marsh 4800–5200 m		Rivulets with grassy borders 4500–5200 m		Large rivers (Indus, Shayok, Zanskar) 3800–4400 m			Freshwater lake 4500 m				Brackish-water lake 4700 m	Saline lake 4600 m
Month of survey	Aug.	Sep.	Aug.	Sep.	July	Aug.	July	Aug.	Sep	July	Aug	Sep	July	July	Aug.
Great Crested Grebe	0	0	0	0	0	0	0	0	0	29	70	48	0	0	0
Great Cormorant	0	0	0	0	0	0	0	0.5	4	0	2	0	0	0	0
Bar-headed Goose	0	0	75	190	0	0	0	0	0	619	1172	391	0	0	0
Ruddy Shelduck	0	0	15	30	6	0	0	0	0	109	82	65	325	3440	300
Mallard	0	0	0	0	0	0	0	0	0	9	0	0	0	0	0
Northern Shoveler	0	0	0	0	0	0	0	0	0	0	2	0	0	0	0
Garganey	0	0	20	43	0	0	0	0	0	0	8	75	0	0	0
Gadwall	0	0	0	0	0	0	0	0	0	0	18	0	0	0	0
Common Pochard	0	0	0	0	0	0	0	0	0	0	6	0	0	0	0
Red-crested Pochard	0	0	0	0	0	0	0	0	0	4	0	0	0	0	0
Goosander	0	0	0	0	0	0	0	0	0	9	42	12	0	0	0
Brown-headed Gull	0	0	0	0	3	0	0	1	6	1	4	5	0	510	1360
Whiskered Tern	0	0	0	0	1	0	0.5	1.5	4	17	0	0	0	0	0
Common Tern	0	0	0	0	0	0	0.5	0	2	3	0	0	0	0	0
Black-necked Crane	0	0	0	7	0	0	0	0	0	3	0	0	0	0	0
Black-winged Stilt	0	0	0	3	0	0	0	0	0	0	22	16	0	60	0
Pied Avocet	0	0	0	0	0	0	0	0	0	0	2	3	0	20	0
Kentish Plover	0	0	0	0	0	0	0	0	0	1	0	0	50	550	0
Lesser Sandplover	0	0	5	0	1*	0	0	0	0	127	6	26	25	60	0
Temminck's Stint	0	0	0	0	0	2	0	0	0	0	20	0	0	0	0
Common Redshank	0	0	35	10	3	0	0	0	0	189	38	37	13	740	10
Common Greenshank	0	0	0	0	0	0	0	0	0	0	22	0	0	0	0
Green Sandpiper	0	0	20	0	3	0	0	0	0	21	0	13	0	20	0

Table 26.2. (cont.)

Species	High-altitude streams 5200–6000 m		High-altitude marsh 4800–5200 m		Rivulets with grassy borders 4500–5200 m		Large rivers (Indus, Shayok, Zanskar) 3800–4400 m			Freshwater lake 4500 m				Brackish-water lake 4700 m	Saline lake 4600 m
Month of survey	Aug.	Sep.	Aug.	Sep.	July	Aug.	July	Aug.	Sep	July	Aug	Sep	July	July	Aug.
Common Sandpiper	0	0	5	7	0	2	1	0	0	63	30	51	150	0	0
Ruff	0	0	0	0	0	0	0	0	0	0	0	0	350	0	0
White-throated Dipper	0	0	0	0	5	2	0	0	0	0	0	0	0	0	0
Total number of birds per 100 km shore	0	0	175	2900	22	6	2	3	16	1204	1546	742	913	5400	1670
Length of survey	40 km	35 km	20 km	30 km	78 km	44 km	250 km	190 km	50 km	75 km	50 km	75 km	4 km	10 km	10 km

* The Lesser Sandplover also occurred in low densities in arid grasslands quite far removed from any wetland.

Table 26.3. The maximum altitude (≥ 5000 m above sea level) at which bird species were observed in Ladakh during the expeditions. None of these species (except the Brown-headed Gull and Green Sandpiper; see text) can be considered birds that are dependent on wetlands.

5000 m	Rock Pigeon
5100 m	Tibetan Partridge Brown-headed Gull Green Sandpiper Desert Wheatear Pied Wagtail Great Rosefinch
5200 m	Tibetan Snowcock Common Kestrel Lesser Sandplover Tibetan Sandgrouse Common Hoopoe Twite
5300 m	Lammergeier Himalayan Griffon Red-billed Chough Common Raven Tibetan Snowfinch (= Black-winged Snowfinch) Plain Mountain Finch
5400 m	Golden Eagle Horned Lark White-winged Redstart (= Güldenstadt's Redstart) Brown Accentor Plain-backed Snowfinch (Blanford's Snowfinch) Brandt's Mountain Finch
5600 m	Hill Pigeon
5700 m	Yellow-billed Chough Robin Accentor
5800 m	Rock Bunting
6200 m	Black Redstart Grey Wagtail

is bare; perhaps this bareness is a natural phenomenon, although severe overgrazing by sheep and goats is noteworthy. In many places, the river is deeply incised into ancient, arid, level riverbeds or into the rock. The same holds for the Zanskar, the Tsarab and the Kargyak Chu Rivers between Lamayuru and Padum (a distance of about 200 km), and the Sangpo and Jankar Rivers on the 100 km stretch between Padum and Darcha (on the Manali–Leh road). This explains why the banks of this river are of low importance as wetland habitat for birds (Table 26.2); the density of aquatic macroinvertebrates is low

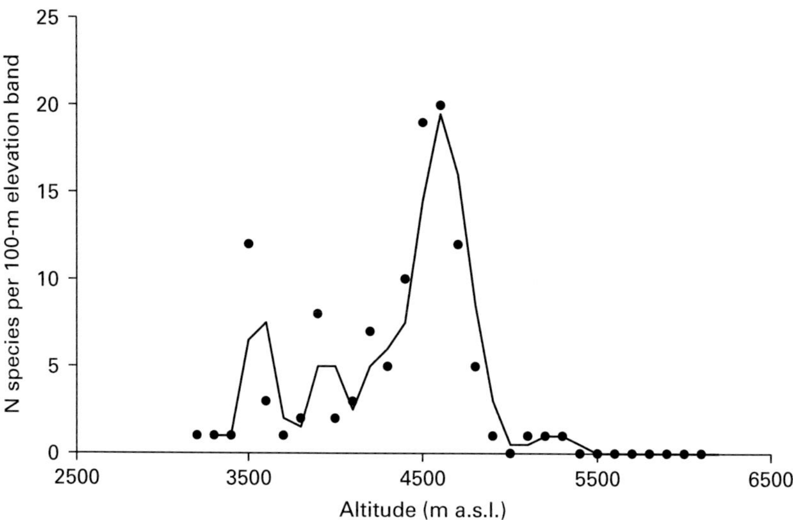

Figure 26.1. Wetland-associated birds (see Table 26.4 for species) are mainly associated with high-altitude lakes in the arid Trans-Himalaya of east Ladakh. Lower species diversity peaks are associated with the wetlands along the Indus and Zanskar Rivers. The lines are running means with n = 3.

(Prins *et al.*, Chapter 17). In September 1998 and August 2002, the Indus was fast flowing in its usual channel, but in July 2000 and July 2006, the monsoon had been severe and the Indus and other rivers were in full spate and very turbid. Common Terns *Sterna hirundo* and Whiskered Terns *Chlidonias hybridus* could be observed only where clear spring water flowed into the river, creating small patches of crystal-clear water in an otherwise muddy river. In August 2002, we observed Great Cormorants *Phalocrocorax carbo* at 3600 m a.s.l. near Phey. In July 1998, we had seen them at 4200 m a.s.l. near Mahe, which is higher than previously recorded (Ali & Ripley, 1995). In August 2002, we even observed this species on Lake Tsomoriri at 4560 m a.s.l.. Other birds on the banks of the Indus and Zanskar included Chukar *Alectoris chukar* and Citrine Wagtail *Motacilla citreola*.

Along some stretches, the Indus (at 3300 m a.s.l.) and Zanskar Rivers (at 3650 m) form braided channels (multi-stream networks), with bordering vegetation dominated by seabuckthorn or willow interspersed with meadows. The water usually remains turbid in summer because these rivers are so silt-laden. Rarely, backwaters have developed where the water is stagnant and clearer. In these areas, we observed Grey Heron *Ardea cinerea* and Indian Pond Heron *Ardeola grayii* in August 2006 and again in May 2011; Pfister (2004) also reported Black-crowned Night Heron *Nycticorax nycticorax*, Greylag Goose *Anser anser*, Ruddy Shelduck *Tadorna ferruginea* and Garganey *Anas querquedula* in these areas. We did not include our observations of herons in the calculations of Table 26.2 because of the limited extent of these multi-stream areas. In May 2008, we observed a pair of the Central Asian endemic Ibisbill *Ibidorhyncha struthersii* near their nest. On the basis of our records, it appears as if in eastern Ladakh, these areas are less important than the high-altitude lakes and marshes (Figure 26.1).

High-Altitude Marsh

Intense cryoturbation, coupled with drainage impeded by permafrost and water accumulation because of snowmelt, has led to suitable conditions for wetland development at altitudes between 4800 m and 5200 m. These high-altitude marshy areas show closed vegetation cover dominated by sedges, and are interspersed with numerous small waterholes. This habitat is suitable for a small number of waders (Table 26.2), namely Redshank *Tringa totanus*, Green Sandpiper *T. ochropus* and Common Sandpiper *Acitis hypoleucos*. Redshanks were observed with young in July 1998 and in August 2002. In May 2011, they arrived at the breeding grounds in the last days of May. We noticed some insect life here, and also saw small crustaceans, but did not see fish or molluscs in these high-altitude marshy areas, which may explain the absence of diving ducks, Goosanders *Mergus merganser* or grebes. Ducks and geese were typically herbivorous, namely Bar-headed Goose, Ruddy Shelduck and Garganey only (Table 26.2). We also sighted the very rare Black-necked Crane *Grus nigricollis*, which breeds in the area. In 2011, we observed them feeding on spilt barley, hours after nomads struck their camp. Bar-headed Geese and Ruddy Shelducks arrived at the breeding grounds in the last 10 days of May 2011, and started laying eggs within days of arrival; they nested on small islands (for further details, see Prins & Van Wieren, 2004), but fed in adjacent marshes. Pfister (2004, p. 54) also mentions Wood Sandpiper *Tringa glareola* and Temminck's Stint *Calidris temminckii* for this area; ring recoveries for Wood Sandpiper show that they originate from central Siberia (Rogacheva, 1992, p. 243; and see Delany *et al.*, Chapter 5).

Rivulets through Grassland

Between 4500–5200 m a.s.l., there are quite numerous small streams and rivulets, sometimes flowing very fast with water speeds above 1 $m.s^{-1}$ but often flowing slower than 0.2 $m.s^{-1}$. They invariably have grassy borders, sometimes only a few metres wide but often much wider. Sedges and herbs frequently dominate the vegetation, but grasses also occur. The vegetation is always intensely cropped by horses, donkeys and smaller stock, because nomads and trekkers alike camp along these streams with their domestic herbivores. This may explain why bird life is scanty (Table 26.2) (cf. Humbert-Droz, Chapter 23). However, aquatic macroinvertebrate densities are low here at this altitude: the explanation appears to be that there is not enough oxygen at these high altitudes to sustain much aquatic life (Prins *et al.*, Chapter 18).

Different Types of Lakes

The most important wetland habitats are the freshwater Lake Tsomoriri, the saline Lake Tsokar and the brackish Lake Thatsangkaru (see also Pfister, 2004). Bird density was high along these lake shores, and many bird species were seen with young (Bar-headed Goose, Ruddy Shelduck, Redshank, Whiskered Tern, Great Crested Grebe *Podiceps cristatus*) or were showing behaviour as if they were forming breeding colonies (Brown-headed Gull *Larus brunnicephalus*). Lake Tsomoriri, situated at 4560 m a.s.l., is a freshwater lake, some 30 km long and 6 km wide. Even though it has no connection with other river systems, because it lies in a natural depression, this holy lake has good potable water and contains freshwater fish. This explains why Great Crested Grebes and Goosanders live on this lake. The percentage of juvenile Great Crested Grebes observed in 1998 was 50% (total n = 46), in 2000 this percentage was 41% (total n = 22), and in 2002 it was 30% (total n = 27). Humbert-Droz (Chapter 23) suggests that this apparent decrease in breeding success has been caused by disturbance related to increased tourism. Among pairs observed with young, in 2000, the average number of young per pair was 2.4 (n pairs = 5) and in 2002 it was 2.4 again (n pairs = 8). In 2011, we arrived before the breeding season. The first grebes arrived during the night of 24 or 25 May and started courting behaviour on 25 May. The lake was nearly ice-free by that date, although there were still ice banks along the shore. The Great Crested Grebe is considered a lowland species in the western Palearctic, although it has been observed to breed well above 4000 m in 'exceptional cases' (Cramp & Simmons, 1977, p. 79), as confirmed by our observation on Lake Tsomoriri at about 4600 m a.s.l. The saline Lake Tsokar and the brackish Lake Thatsangkaru do not support fish, and piscivorous ducks and grebes were not seen. These lakes were still frozen over at the very end of May in 2011. Lake Tsomoriri is apparently not even too high for Great Cormorants. The absence of herons on all these lakes is noteworthy (*pace* Pfister, 2004). Goosander occurs, according to Cramp and Simmons (1977, p. 681), 'to at least at 4000 m', but we saw it even higher at 4600 m (Table 26.2) and flying through a pass towards Lake Tsomoriri at 4900 m in May 2011.

The density of duck species on these lakes is quite high (Table 26.2) and on a par with those reported from Scotland for Red-breasted Merganser *Mergus serrator* (Marquiss & Duncan, 1993) or for a number of duck species in Finland (Nummi & Poysa, 1995). Lake Tsomoriri holds a notable proportion of the world population of Bar-headed Geese and on this basis meets the criteria for designation as a Wetland of International Importance under the Ramsar Convention (Prins & van Wieren, 2004), to which India is a Contracting Party. Breeding success of Bar-headed Goose was high: at the beginning of September 1998 and at the end of August 2002, the juvenile percentage was close to 60%, and at the end of July 2000 it was about 55% (Prins & Van Wieren, 2004). For the Ruddy Shelduck, the number of observed juveniles was low in 1998 (total n = 57, 13% juveniles) and also in 2000 (total n = 44, 9% juveniles). Again, in 2002 very few juveniles were seen (total n = 71, 10% juveniles). Breeding of this species was only recorded near the saline lakes.

Other species of duck did not show any evidence of breeding, but their absolute numbers were very low and juveniles may have been overlooked (Common Pochard *Aythya farina* in 2002, n = 3; Gadwall *Anas strepera* in 2002, n = 9; Mallard *Anas platyrhynchos* in 2000, n = 7; Red-crested Pochard *Netta rufina* in 2000, n = 3; Northern Shoveler *Anas clypeata* in 2002, n = 1; Goosander in 1998, n = 9; in 2000 n = 7; in 2002 n = 6). In 2011, there were no ducks (yet) in May. Red-crested Pochard breeds in the steppes of southern Siberia and winters in India (McClure, 1974, p. 120; Cramp & Simmons, 1977, p. 554; pers. obs. in Bardiya NP in the Terai of Nepal in March 2015, together with Northern Pintail *Anas acuta*, Garganey and Goosander), and we did not expect breeding in Ladakh. Of Mallard, Cramp and Simmons (1977, p. 508) wrote that this is mainly a lowland species, but it 'will extend to 2000 m or higher on occasion'. We encountered the species at much higher altitudes (Table 26.2). Except for Mallard, ring recovery data of ringed birds from Bharatpur in northern India show all these species to migrate between Siberia and the Ganges plains (Common Pochard: McClure, 1974, p. 116; Gadwall: *ibid.*, p. 115; Northern Shoveler: *ibid.*, p. 106; Northern Pintail: *ibid.*, p. 101; Common Teal (*A. crecca*): *ibid.*, p. 107; Eurasian Wigeon *A. penelope*: *ibid.*, p. 111, and Tufted Duck *Aythya fuligula*: *ibid.*, p. 118) (see Namgail *et al.*, Chapter 2). The Common Teal is the most common duck species in central Siberia (Rogacheva, 1992, p. 132), together with Eurasian Wigeon (*ibid.*, p 137) and Northern Pintail (*ibid.*, p. 137, with some 920,000 birds, *ibid.*, p. 139). Mallard appear to be declining in central Siberia, for which causes are unknown (*ibid.*, p. 130). The Gadwall is decreasing rapidly in the same region (*ibid.*, p. 135), as is the Northern Shoveler (*ibid.*, p. 141), but Garganey may be increasing (*ibid.*, p. 140). Tufted Duck, with a population of some 440,000 in Central Siberia has also been declining (*ibid.*, p. 143). The Scaup *Aythya marila* population, which does not migrate over the Himalayas, stands at approximately 530,000 and appears to be stable (*ibid.*, p. 144). The same applies to Goldeneye *Bucephala clangula*, with a population of close to 700,000 (*ibid.*, p. 147), and Long-tailed Duck *Clangula hyemalis*, with a population of 1.6 million (*ibid.*, p. 145), so the decline of a number of duck species may find its cause on their migration towards India and Pakistan.

The numbers of Garganey in September 1998 were reasonably high (n = 82), but we did not see any in May 2011 or July 2000, and in August 2002 we only saw four. Garganeys breed in the middle Siberian steppe and winter in the Gangetic plains and lowland India, and further away in Southeast Asia (Cramp & Simmons, 1977, p. 531). Cramp and Simmons (1977, p. 531) state that 'it tends to avoid high upland or mountain regions,' but this clearly does not include the migration routes. Ring recovery data show that this species uses the Ganges plain in winter and the Siberian wetlands in summer (McClure, 1974, pp. 113, 114). We saw Garganey at altitudes between 4500 m and 4700 m a.s.l. (Table 26.2). Local people reported that Northern Shoveler and Northern Pintail also use the high-altitude lakes of Tsokar and Tsomoriri in quite large numbers, and we saw a single Northern Shoveler in 2002. Both these species also have to cross the high mountains of the Himalayas.

Likewise, Ruffs were observed in one group only (July 2000, n = 14), indicative of migration. Ruff breed on the Siberian tundra and winter inland in Pakistan, India and Bangladesh (Cramp & Simmons, 1983, p. 388; see also McClure, 1974, p. 155; Rogacheva, 1992, p. 269) and thus also have to cross the high mountains. They also migrate via India to eastern and southern Africa (Rogacheva, 1992, p. 269; Delany *et al.*, 2009, p. 411). Kentish Plover *Charadrius alexandrines* was encountered singly (n = 3) and in one flock (n = 52) in July 2000. Cramp and Simmons (1983, p. 153) state that Kentish Plover occurs 'generally in lowland, but occasionally ascends river valleys or settles on plateau areas up to 2000 m or more'. Ascending, for example, the side valleys of the Indus Valley would lead to altitudes of more than 4500 m, such as the sites where we observed Kentish Plovers. Temminck's Stint *Calidris temminckii* were observed in two small flocks of four and six individuals, respectively, along Lake Tsomoriri in August 2002, and a single individual on the grassy fringe of a small backwater near the village of Pistu along the Zanskar River at 3500 m in August 2006. Single individuals of Common Sandpiper were encountered at many localities in July 2000 and 2006, August 2002 and 2006 and September 1998, but in July 2000 we also saw a flock of 22 at the southwestern end of Lake Tsomoriri. The individuals in this flock were feeding and flying together, which suggested that they were migrating through. Cramp and Simmons (1983, pp. 595, 596) recorded the species as breeding in Ladakh, at altitudes exceeding 4000 m, but we did not observe nests or young.

Lesser Sandplover *Charadrius mongolus* showed nest-defence behaviour on the arid grasslands close to Lake Tsokar, and we suspect it was breeding in the area. Redshank were also observed with young. Of Redshank it is reported that it 'dislikes frost and arid conditions', even though it ranges up to 4000 m in the Pamirs on the border between Afghanistan and Kazakhstan (Cramp & Simmons, 1983, p. 526). Redshank ringed in the Altai and near Novosibirsk have been recovered from India (Rogacheva, 1992, p. 247). The area where we saw Redshank breeding in Ladakh is even as high as 5100 m (Table 26.2): frost may occur in the middle of July and air humidity may be very low indeed. Redshanks only arrived at the breeding grounds on 26 May in 2011.

Pied Avocet *Recurvirostra avosetta* and Black-winged Stilt *Himantopus himantopus* occurred in pairs in 1998, 2000, 2002 and 2011, but young were not observed, even though Pied Avocet made calls very typically given when engaged in nest defence along the North Sea coast (pers. obs.). Cramp and Simmons (1983) does not give Ladakh as part of the Black-winged Stilt's range (*ibid.*, p. 38) and states that the species avoids mountains in the former Soviet Union, although it has been reported from mountain lakes at about 2000 m a.s.l. in Turkey (*ibid.*, p. 37, and see Delany *et al.*, Chapter 5). Likewise, for Pied Avocet, Cramp and Simmons (1983, p. 48) state that the species is absent from mountains during the breeding season, even though it may be observed at altitudes exceeding 3000 m. Both species were observed by us at altitudes exceeding 4500 m, but neither the avocet nor the stilt is reported to breed in Ladakh (Pfister, 2004). The density of aquatic macroinvertebrates is high in the saline lakes, but not in the freshwater lakes at these extreme altitudes (Prins *et al.*, Chapter 18), possibly explaining why these birds can occur here in sizable numbers.

Most Wetlands Are not Individually Important but Are Collectively Crucial

Natural selection must have had an enormous pressure for high-altitude adaptations when mountain ranges such as the Pamir, Karakoram, Hindu Kush, Himalayas and Kailash Ranges all formed one gigantic barrier between the northern breeding grounds and the wintering areas on the Indian subcontinent. Ecologists have given some attention to the physiological requirements of very high-altitude flight (e.g. Hawkes *et al.*, Chapter 16), but ecological requirements are less well studied. For wetland-associated birds, the zone above about 5000 m is basically a desert (Figure 26.1) in which there are

Table 26.4. Wetland-associated birds in Ladakh that typically depend on animal food; only Bar-headed Geese are totally herbivorous, but juveniles also eat insects, like other geese.

Food resource	Insects	Invertebrates	Crustaceans	Fish	Seeds	Shoots	Other
Species							
White-throated Dipper	*						
Brown Dipper	*						
Yellow Wagtail	*						
Citrine Wagtail	*						
Grey Wagtail	*						
White Wagtail	*						
Ruddy Shelduck	*		*			*	
Mallard	*		*			*	
Great Crested Grebe	*		*	*			
Common Tern	*		*	*			
Indian Pond Heron	*			*			
Whiskered Tern	*			*			
Brown-headed Gull	*			*		*	
Common Black-headed Gull	*			*		*	
Gadwall	*					*	
Red-crested Pochard	*					*	
Northern Shoveler	*						**plankton**
Common Greenshank	*	*		*			
Garganey	*	*				*	
Ibisbill	*	*					
Pied Avocet	*	*					
Common Redshank	*	*					
Green Sandpiper	*	*					
Common Sandpiper	*	*					
Temminck's Stint	*	*					
Ruff	*	*			*		
Black-winged Stilt	*	*			*		
Black-necked Crane	*	*			*		
Lesser Sandplover		*	*				
Common Pochard		*				*	
Common Moorhen		*				*	
Common Coot		*				*	
Goosander				*			
Great Cormorant				*			
Grey Heron				*			**rodents**
Bar-headed Goose						*	

no oases where animal food on which they all depend (Table 26.4) can be found. In this zone, only flying migrants or stragglers can be encountered, and wetland species in this zone are fully dependent on their body reserves. We think it is no coincidence that we observed no raptors above 5400 m (Prins *et al.*, Chapter 20): prey is basically absent there.

Yet many of our observations show that birds can occur and some can breed at much higher altitudes than previously reported by Cramp and Simmons in the western Palearctic (1977, 1983), for example, Great Cormorant, Kentish Plover, Garganey, Mallard, Black-Winged Stilt and Pied Avocet as confirmed by Pfister (2004). These species, and especially those that raise young at this altitude of close to 5000 m a.s.l., probably possess special physiological adaptations, as has been found for other species and at other high-altitude locations (Hiebl & Braunitzer, 1988; Fedde *et al.*, 1989; Velarde *et al.*, 1991; Velarde, 1993; Weber *et al.*, 1993; Velarde *et al.*, 1997a; Dragon *et al.*, 1999). One such adaptation is to lay larger eggs (Liu, 2000) with increased eggshell permeability (Velarde, 1997b). Other adaptations to extremely high altitude among thrushes and finches are lower wing-loading factors, longer and more pointed wings and stronger hind limbs with longer claws (Landmann & Winding, 1993, 1995a, 1995b). Natural selection must have favoured these adaptations, since without being able to make use of the very high-altitude wetlands (up to 4800 m), the barrier of the combination of the Himalayas in the broad sense and the deserts to the north of it (Gobi, Taklamakan, Betpak-Dala, Peski Muyunkum) would be as formidable as the Sahara.

The second conclusion from our expeditions is that bird life is scarce even at elevations below 5000 m altitude. That is not only because the wetlands are few, and widely dispersed, but also because they show very low productivity (see Prins *et al.*, Chapter 18 for aquatic macroinvertebrates) because either they are frozen for so many months of the year, or because their waters are low in available oxygen (Dillon *et al.*, 2006; Jacobsen, 2008; Verberk *et al.*, 2011; cf. Figure 0.3).

This poses a particular problem for the protection of these wetlands, especially the small ones: individually they may appear insignificant, but collectively they are essential because, although few and far between, these wetlands must have great significance for the success of migrations through the highest mountains on Earth. Network analysis shows how the disappearance of one node in a network can unravel the whole of it, making migration next to impossible (Shimazaki *et al.*, 2004). Even though these small wetlands are remote from villages and fixed human settlement, they are suffering increasing human use. The upward shift of permanent agriculture towards the highest sources of meltwater is incessant, and barley cultivation is subsidized by government in extremely high and remote places (see also Mishra, 2001). Drainage of marshes for irrigated agriculture of barley and peas takes place, while other areas are being planted with willow shrubs. Remaining wetlands are intensely utilized by domestic stock, and very few areas are effectively protected against use and over-utilization (Veer *et al.*, 2006; Namgail *et al.*, 2010); yet there is indication that the pressure may now be declining: (Singh *et al.*, 2015). All these activities threaten these high-altitude wetlands and thus the migration of central Asian and Siberian birds. None of these wetlands will individually meet the Ramsar Convention criteria (except Lake Tsomoriri; Prins & Van Wieren, 2004; Humbert-Droz, Chapter 23) simply because they are too small and too unproductive, and harbour too few birds. Yet, taken together they are all indispensable, and losing one more site is further weakening the whole of the migratory network across the highest mountain range on Earth.

The Ramsar Convention on Wetlands includes clusters of small wetlands in arid areas in its *Strategic Framework and guidelines for the future development of the List of Wetlands of International Importance of the Convention on Wetlands* (Ramsar, Iran, 1971). The fourth edition of this document, approved in 2012 in Resolution XI.8 and included in Annex 2 of that Resolution, includes the following as point 91:

> '**Site Clusters**. Clusters of small sites, or individual small "satellite" sites associated with larger areas, should be considered for listing where these are: . . .
>
> v) found in arid or semi-arid zones, where complexes of dispersed wetlands (sometimes of a non-permanent nature) can both individually and collectively be of very great importance for both biological diversity and human populations (e.g. essential links in incompletely known chains).'

The small, high-altitude lakes in eastern Ladakh appear to meet these guidelines very closely, and it would be highly appropriate for the Indian and Chinese Governments to propose them for listing as wetlands of international importance under the Ramsar Convention.

Terrestrial bird migration can nearly always be thought of as following a 'braided network' in which staging grounds form the essential nodes (Si *et al.*, 2015). Shimazaki *et al.* (2004) drew attention to the inadequacy of protection measures when their network analysis identified crucial staging grounds for Oriental White Storks *Ciconia boyciana* without a protection status in eastern China. Similarly, we discovered that the highest concentration of food on offer (in the form of macro-invertebrates) is not in the freshwater Lake Tsomoriri, which falls under the Ramsar Convention, but in the adjacent much smaller saline lakes such as Tsokar, which are not protected under the Convention. Ydenberg *et al.* (2004) drew attention to the fact that simple censuses do not adequately measure the biological importance of staging grounds for migrating birds; the Ramsar Convention criteria, however, still mainly reflect the importance of census figures and not of migration ecology. Strongly improved protection of the small dispersed wetlands in the Himalayas, and beyond, is therefore an urgent necessity. Designation of the small, dispersed, high-altitude wetlands in eastern Ladakh under the Ramsar Convention, and the implementation of the Convention at these sites, could play an important part in this protection. The weakness of the present definition of site clusters ('Clusters of small sites, or individual small "satellite" sites associated with larger areas') is the emphasis on the association with larger areas (that meet the Ramsar Criteria). Indeed, the small dispersed wetlands in the Himalayas, Tibetan Plateau or Taklamakan Desert form a network of which the integrity for migration depends on that collective of small wetlands and not necessarily on their association with larger areas.

Acknowledgements

For acknowledgements, see Prins, Van Wieren and Namgail, Chapter 20.

References

Alerstam, T. & Gudmundsson, G.A. (1999). Migration patterns of tundra birds: tracking radar observations along the Northeast Passage. *Arctic*, **52**, 346–371.

Ali, S. & Ripley, S.D. (1995). *A Pictorial Guide to the Birds of the Indian Subcontinent* (2nd ed.). Oxford: Oxford University Press.

Bhatnagar, Y.V., Wangchuk, R., Prins, H.H.T., Van Wieren, S.E. & Mishra, C. (2006). Perceived conflicts between pastoralism and conservation of the kiang *Equus kiang* in the Ladakh Trans-Himalaya. *Environmental Management*, **38**, 934–941.

Bruderer, B. (1999). Three decades of tracking radar studies on bird migration in Europe and the Middle East. In Y. Leshem, Y. Mandalik, J. Shamoun-Barangs (Eds.) *Migrating Birds Know No Boundaries: Proceedings International Seminar on Birds and Flight Safety in the Middle East, Israel*. Tel Aviv: Tel Aviv University, pp. 107–141.

Bruderer, B. & Jenni, L. (1990). Migration across the Alps. In E. Gwinner (ed.). *Bird Migration*. Berlin Heidelberg: Springer, pp. 60–77.

Cramp, S. & Simmons, K.E.L. (eds.) (1977). *Handbook of the Birds of Europe, the Middle East and North Africa. The Birds of the Western Palaearctic, Vol. I. Ostrich to Ducks*. Oxford: Oxford University Press.

Cramp, S. & Simmons, K.E.L. (eds.) (1983). *Handbook of the Birds of Europe, the Middle East and North Africa. The Birds of the Western Palaearctic, Vol. III. Waders to Gulls*. Oxford: Oxford University Press.

Delany, S., Scott, D., Dodman, T. & Stroud, D. (Eds.) (2009). *An Atlas of Wader Populations in Africa and Western Eurasia*. Wageningen: Wetlands International.

Dillon, M.E., Frazier, M.R. & Dudley, R. (2006). Into thin air: physiology and evolution of alpine insects. *Integrative and Comparative Biology*, **46**, 49–61.

Dragon S., Carey, C., Martin, K. & Baumann, R. (1999). Effects of high altitude and in vivo adenosine/beta-adrenergic receptor blockade on ATP and 2,3BPG concentrations in red blood cells of avian embryos. *Journal of Experimental Biology*, **202**, 2787–2795.

Fedde, M.R., Orr, J.A., Shams, H. & Scheid, P. (1989). Cardiopulmonary function in exercising Bar-headed Geese during normoxia and hypoxia. *Respiration Physiology*, **77**, 239–252.

Grimmett R., Inskipp, C. & Inskipp, T. (1999). *Pocket Guide to the Birds of the Indian Subcontinent*. New Delhi: Oxford University Press.

Hiebl, I. & Braunitzer, G. (1988). Anpassungen der Haemmoglobine von Streifengans (Anser indicus), Andengans (Chloephaga melanoptera) und Sperbergeier (Gyps rueppellii) an hypoxische Bedingungen. *Journal fuer Ornithologie*, **129**, 217–226.

Jacobsen, D. (2008). Low oxygen pressure as a driving factor for the altitudinal decline in taxon richness of stream macro invertebrates. *Oecologia*, **154**, 795–807.

Landmann, A. & Winding, N. (1993). Niche segregation in high-altitude Himalaya chats (Aves, Turdidae): does morphology match ecology? *Oecologia*, **95**, 506–519.

Landmann, A. & Winding, N. (1995a). Adaptive radiation and resource partitioning in Himalayan high-altitude finches. *Zoology (Jena)*, **99**, 8–20.

Landmann, A. & Winding, N. (1995b). Guild organisation and morphology of high-altitude granivorous and insectivorous birds: convergent evolution in an extreme environment. *Oikos*, **73**, 237–250.

Liechti, F. & Schaller, E. (1999). The use of low-level jets by migrating birds. *Naturwissenschaffen*, **86**, 549–551.

Liu, N. (2000). Breeding of Beicki's blood pheasant (Ithaginis cruentus beicki) in northwestern Gansu, China. *Game and Wildlife Science*, **17**, 17–27.

Mani, M.S. (1978). *Ecology & Phytogeography of High-Altitude Plants of the Northwest Himalaya*. London: Chapman & Hall.

Marquiss, M. & Duncan, K. (1993). Variation in the abundance of Red-breasted Mergansers Mergus serrator on a Scottish river in relation to season, year, river hydrography, salmon density and spring culling. *Ibis*, **135**, 33–41.

McLure, H.E. (1974). *Migration and Survival of the Birds of Asia*. United States Army Medical Component, South East Asia Treaty Organisation (SEATO) Medical Project, Bangkok.

Mishra, C. (2001). *High Altitude Survival: Conflicts between Pastoralism and Wildlife in the Trans-Himalaya*. PhD thesis, Wageningen University

Namgail, T., van Wieren, S.E. & Prins, H.H.T. (2010). Pashmina production and socio-economic changes in the Indian Changthang: implications for natural resource management. *Natural Resources Forum*, **34**, 222–230.

Nummi, P. & Poysa, H. (1995). Breeding success of ducks in relation to different habitat factors. *Ibis*, **137**, 145–150.

Pfister O. (2004). *Birds & Mammals of Ladakh*. New Delhi: Oxford University Press.

Prins H.H.T. & van Wieren, S.E. (2004). Number, population structure and habitat use of Bar-headed Geese Anser indicus in Ladakh (India) during the brood-rearing period. *Acta Zoologica Sinica*, **50**, 738–744.

Rogacheva, H. (1992). *The Birds of Central Siberia*. Husum: Husum Druck- und Verlaggesellschaft.

Shimazaki, H., Tamura, M., Darman, Y., *et al.* (2004). Network analysis of potential migration routes for Oriental White Storks (Ciconia boyciana). *Ecological Research*, **19**, 683–689.

Si, Y., Xin, Q., de Boer, W.F., Gong, P., Ydenberg, R.C. & Prins, H.H.T. (2015). Do Arctic breeding geese track or overtake a green wave during spring migration? *Scientific Reports*, **5**, 8749.

Singh, R., Sharma, R.K. & Babu, S. (2015). Pastoralism in transitions: livestock abundance and composition in Spiti, Trans-Himalayan Region. *Human Ecology*, 43: 799–810.

Velarde F.L., Espinoza, D., Monge, C.C. & de Muizon, C. (1991). A genetic response to high altitude hypoxia: high hemoglobin-oxygen affinity in chickens (Gallus gallus) from the Peruvian Andes. *Comptes Rendus de l'Academie des Sciences, serie III Sciences de la Vie*, **313**, 401–406.

Velarde F.L., Mejia, O., Palacios, J.A. & Monge, C.C. (1997a). Changes in whole blood oxygen affinity and eggshell permeability in high altitude chickens translocated to sea level. *Comparative Biochemistry and Physiology B*, **118**, 53–57.

Velarde F.L., Monge, C.C. & Carey, C. (1997b). Physiological strategies of oxygen transport in high altitude bird embryos. *Comparative Biochemistry and Physiology A*, **118**, 31–37.

Velarde F.L., Sanchez, J., Bigard, A.X., Brunet, A., Lesley, C. & Monge, C.C. (1993). High altitude tissue adaptation in Andean coots: capillary, fibre area, fibre type and enzymatic activities of skeletal muscles. *Journal of Comparative Physiology B Biochemical Systemic and Environmental Physiology*, **163**, 52–58.

Verberk, W.C.E.P., Bilton, D.T., Calosi, P. & Spicer, J.I. (2011). Oxygen supply in aquatic ectotherms: partial pressure and solubility together explain biodiversity and size patterns. *Ecology*, **92**, 1565–1572.

Weber R.E., Jessen, T.H., Malte, H. & Tame, J. (1993). Mutant hemoglobins (alpha-119-Ala and beta-55-Ser): functions related to high altitude respiration in geese. *Journal of Applied Physiology*, **75**, 2646–2655.

Ydenberg, R.C., Butler, R.W., Lank, D.B., Smith, B.D. & Ireland, J. (2004). Western sandpipers have altered migration tactics as peregrine falcon populations have recovered. *Proceedings of the Royal Society of London, Series B: Biological Sciences*, **271**, 1263–1269.

Part V

Conclusions

27 Bird Migration across the Himalayas and Beyond: The Need for Better Conservation and Management of a Natural Wonder

Herbert H.T. Prins and Tsewang Namgail

> 'Somehow the word flyway seems somewhat magical; conjuring up the idea of invisible flightlines for birds to follow on their seasonal movements back and forth between their wintering and breeding grounds. . . . Flyways were defined as rather broad arterial boulevards, to which migration routes of individual birds act as tributaries.'
>
> (Johnsgard, 2012, p. 15)

Bird Migration across the Himalayas

The bird life of the Himalayas is spectacular. Altogether, 783 species (80%) of the birds found on the Indian subcontinent occur in the Himalayas (Price *et al.*, 2003). Of these, at least 35 species are endemic to the Himalayas (Thakur & Negi, 2015). The question of whether migratory birds cross over or migrate around the Himalayas is a long-standing one. All the chapters in Part I indicate that most of the migratory birds, ranging from passerines to cranes, do cross the Himalayas. Some individuals (but not all) of species such as Peregrine Falcon *Falco peregrinus* and Demoiselle Crane *Anthropoides virgo* circumvent the Himalayas, however (Higuchi & Minton, Chapter 3; Dixon *et al.*, Chapter 8). Birds adopt different strategies to cross this mountainous barrier. For example, raptors with high wing loadings soar with thermals and updraughts (Juhant & Bildstein, Chapter 6). Raptor migration across the Himalayas has also been reported by several observers, counting birds from raptor watch-sites across the Himalayas (e.g. Den Besten, 2004). The arid slopes of the northern flank of the Himalayas create numerous thermals that are readily used by birds during the autumn migration (Ohlmann, Chapter 14; Heise, Chapter 15). Ducks cross relatively low passes to get through the Himalayas (Namgail *et al.*, Chapter 2), but these passes, at 5000 m above sea level (a.s.l.), are still high by global standards. Passerines such as the Hume's Lesser Whitethroat *Sylvia althaea* probably pass directly over the Zanskar and Great Himalaya ranges to winter in lowland India (Delany *et al.*, Chapter 4).

Chapters in Part II highlight the physiography of the Himalayan barrier. The Himalayas form the longest mountain range in the world that stands at right angles to the routes of migratory birds in their flyway. There are, however, rivers that cut through the mountain range: the Indus and the Sutlej in the west and the Brahmaputra in the east. The source of the Sutlej, however, is Lake Rakshastal, still standing at a respectable altitude of 4580 m and surrounded by mountains. The source of the Indus is in Gêgyai District in Tibet with an elevation of about 5500 m. The Brahmaputra originates at the Angsi glacier in Tibet at 5200 m a.s.l.

These rivers existed before the rise of the Himalayas (Searle, Chapter 9). The rise of the Himalayas had an enormous influence on the climate of Asia. In fact, the Asian monsoon may have originated between 25 million and 22 million years ago as result of the rise of the Himalayas (Searle, Chapter 9). In due course, glaciers formed on the high mountains, the extent of which fluctuated widely throughout the Quaternary ice age. This in turn had profound influence on the region's hydrology (Owen, Chapter 10). The Himalayas form a climatologically and hydrologically diverse system reflecting the scale and diversity of their geology and topography (Searle, Chapter 9; Bookhagen, Chapter 11). This diverse climatic and hydrological system has led to the growth of diverse vegetation ranging from tropical forests to alpine meadows (Rawat, Chapter 12), supporting a rich assemblage of migratory birds, including altitudinal migrants.

Now the question is: why do some birds fly over and not around the Himalayas? Chapters in Part III address this important question. Groen and Prins (Chapter 17) suggest that this is simply a trade-off between distance and altitude. The main constraint for birds is the cost of ascending rather than the cost of sustaining flight once at high altitude. Under such a scenario, raptors that benefit from thermals have an edge over other groups of birds. Forty-five raptor species are known to cross the Himalayas (Juhant & Bildstein, Chapter 6; see also Batbayar & Lee, Chapter 7), of which eagles, vultures and buzzards typically use soaring flight. In fact, the Himalayan region hosts one of the greatest assemblages of migratory raptors on any flyway, perhaps because it forms the junction of four zoogeographical realms, of which the two eastern ones are the most raptor-diverse realms in the world (Ferguson-Lees & Christie, 2001; cf. for owls, Romulo, 2012). Yet two other issues crop up, namely food supply and wind conditions. In the autumn, the food supply is reasonably good, even at high altitudes (see Introduction; Prins *et al.*, Chapter 18). Yet also in autumn, the subtropical jet stream in the Himalaya region starts developing again. The exact timing of that development depends on the cooling of the Indian subcontinent. When the jet stream is present in full force again, strong winds develop and the air shows much turbulence, which makes crossing dangerous (Ohlmann, Chapter 14; Heise, Chapter 15). The spring migration season occurs before the maximum development of food supply for migrants, especially at higher altitudes (see Introduction; Prins *et al.*, Chapter 18) and the possibility of safely crossing the Himalayas after the influence of the subtropical jet stream has died off now depends on the heating up of the Indian subcontinent (Ohlmann, Chapter 14; Heise, Chapter 15). In other words, the right conditions for safe crossing and the availability of food resources for refuelling migrants may be more constrained than we thought. Warming up of the climate over the Indian subcontinent may thus perhaps be favourable for migration because of the potential for reducing the influence of the subtropical jet stream and a longer season higher up in the mountains.

Two chapters (Takekawa *et al.*, Chapter 1; Hawkes *et al.*, Chapter 16) focus on the migration routes and the strategies used by the Bar-headed Goose *Anser indicus*, which has often been touted as the highest-flying (migrant) bird. This goose is a relatively large bird, and it crosses the Himalayas without the help of thermals and updraughts (Takekawa *et al.*, Chapter 1). Instead it has evolved special physiological systems to cross high mountains (Hawkes *et al.*, Chapter 16). Others assert, however, that these adaptations are common to all birds (Ruben *et al.*, 1997; see also Cieri & Farmer, 2016). Perhaps other birds such as ducks that cross the Himalayas regularly (Namgail *et al.*, Chapter 2) or small passerines that were observed at great altitudes (Prins *et al.*, Chapter 20) have similar physiological adaptations. This needs to be investigated. Indeed the flight paths of Bar-headed Geese and Mallards show that the energetics of the migrations of both species are basically identical (Groen & Prins, Chapter 17). Yet the Bar-headed Geese are better able to cope with hypoxia than another goose species (the Barnacle Goose *Branta leucopsis*; Hawkes *et al.*, 2014), and also better than domesticated Pekin Ducks (Faraci *et al.*, 1984).

In any case, several chapters in this book show that birds crossing the Himalayas fly very high. An expedition to Mount Everest found skeletons of Northern Pintail *Anas acuta* and Black-tailed Godwit *Limosa limosa* at 5000 m a.s.l. on the Khumbu Glacier of Mount Everest (Geroudet, 1954); we ourselves found carcasses of Golden Oriole *Oriolus oriolus* and Common Kestrel *Falco tinnunculus* in areas higher than 4500 m a.s.l. and found many species to live and sometimes even to breed at these enormous elevations (Prins *et al.*, Chapter 18; Prins *et al.*, Chapter 20).This book shows that the adaptations of the Bar-headed Goose may not be unique and that some 100-odd species are known to migrate over the extremely high Himalayas (134 long-distance migrant species for Ladakh only: Delany *et al.*, Chapter 4). This includes Mallard (Groen & Prins, Chapter 17), which has been observed to fly at some 6400 m a.s.l. in Nevada in the United States, contrary to expectations suggested by results reported by Black and Tenney (1980). In other words, even though the Himalayas almost straddle the Central Asian Flyway, and even though these mountains and the associated Tibetan Plateau are nowhere lower than 4500 to 5000 m a.s.l. (Figure 0.1 in the Introduction), and even though these vast stretches of land are generally arid, cold and often devoid of vegetation, many individual birds of many species are able to cope with this apparent barrier, and have done so for millions of years. This biannual migration is thus one of the wonders of the natural world.

The birds crossing the Himalayas between their breeding and non-breeding areas face a plethora of threats ranging from raptors (Ydenberg, Chapter 19) to intensive modern pastoralism (Ghosal & Ahmed, Chapter 22). Pastoralism in several parts of the Himalayas is intensifying in response to the market demands for *Pashmina* or cashmere wool. The very high grazing pressures in turn imperil the high-altitude lakes, the refuelling stations for migratory birds (Prins *et al.*, Chapter 18). Intensification of agriculture with increased use of subsidized artificial fertilizers and pesticides in the Himalayan region is also negatively affecting bird populations in agricultural areas

(Raman *et al.*, Chapter 24). With globalization, many nomadic herders look down at nomadism as a primitive way of life (Namgail *et al.*, 2010). Many of them therefore increasingly settle near wetlands and build permanent houses, which put additional pressures on the wetland resources (Namgail *et al.*, 2007). Furthermore, the rapidly growing tourism industry is also threatening the wetland ecosystems all across the Himalayan and the Tibetan Plateau region (Humbert-Droz, Chapter 23). The other important threats are damming of rivers for hydroelectricity (Pandit & Grumbine, 2012) and irrigation. Both China and India have enormous demand for energy to fuel their economic growth. Damming of rivers for irrigation is of major concern in these countries, and also in Pakistan and Nepal (see Tockner *et al.*, 2016).

The number of migratory birds along certain routes such as the Kali Gandaki in the Himalayas has already decreased in recent years (Higuchi & Minton, Chapter 3). Many of these birds depend on wetlands in the otherwise arid and barren landscape. Therefore, it is imperative to protect the wetlands along the routes of migratory birds in the Himalayan region, including small, dispersed wetlands (Prins *et al.*, Chapter 26). Although there is a network of wetlands in India protected under the Ramsar Convention, Namgail *et al.* (Chapter 25) pointed out that even these wetlands often do not support habitats preferred by ducks. Therefore, it is crucial to identify wetlands that are suitable for particular groups of birds, and manage and protect them for that particular group. For example, shallow wetlands with extensive mudflats should be protected and managed for waders.

History of a Barrier and Implications for Understanding Birds

Most of the present-day bird species that migrate over the Himalayas and the Tibetan Plateau date back about 2.5 million years to the beginning of the Pleistocene (e.g. Greylag Goose *Anser anser*, Gadwall *A. strepera*, Eurasian Wigeon *Anas penelope*, Northern Pintail, Goosander *Mergus merganser*, Red-breasted Merganser *M. serrator*, Red-crested Pochard *Netta rufina*, Common Pochard *Aythya farina*, Black-tailed Godwit, Spotted Redshank *Tringa erythropus* or Green Sandpiper *T. ochropus*. Some species that migrate across the Himalayas are older, for example, Mallard *A. platyrhynchos* (Late Pliocene or early Pleistocene), Common Redshank *Tringa tetanus* (Late Pliocene), Northern Shoveler *A. clypeata* and Garganey *Anas querquedula* (Middle Miocene) or very old, such as Common Teal *A. crecca* (Late Miocene, 5.3 million years ago) (Mlíkovský, 2002).

Of course, from an evolutionary perspective, species lineages are much older than this approximate 2 to 5 million years (cf. White, 2014), but the point is that present-day birds that brave the deserts and high mountains of the Central Asian Flyway have already been exposed to a very high and extensive barrier for an enormous period of time. Indeed, Searle (Chapter 9) points out that this barrier has existed for some 120 million years along the western side of what are now the Himalayas. Further east, the uplift started some 65 million years ago with peak elevation some 30 million years ago. Hence, the ancestors of many species that now migrate to the Indian subcontinent were exposed to natural selection to cope with the massive barrier for tens of millions of years. One of the messages of the present book is that well over 130 species have been recorded crossing the Tibetan Plateau and the Himalayas. In other words, it is not the *rara avis* that crosses the high mountains, but it appears that many bird species are able to cope with elevations close to 6000 m a.s.l. This in itself is not strange because it is likely that the Himalayan ecosystem has not changed dramatically in the past hundreds (Vernier & Bruneau, Chapter 21) or thousands of years (Owen, Chapter 10). Apart from local deforestation and overgrazing, the vegetation is still very much the same as it used to be (Rawat, Chapter 12), and there is still little evidence for the impacts of climate change on the phenology of the vegetation (Bagchi *et al.*, Chapter 13).

The Central Asian Flyway and Its Estimated Number of Birds

Worldwide, eight major migratory flyways for birds have been recognized (Boere & Stroud, 2006). A *flyway* is defined as a geographical region within which one or more migratory species, or a distinct population of a given migratory species, complete all components of their annual cycle (breeding, moulting, migrating, staging, nonbreeding etc.). Flyways are often distinct 'pathways' linking a network of key sites, but for some species/groups, flyways are more dispersed (Galbraith *et al.*, 2014). There are many limitations to the flyway concept (Boere & Stroud 2006), but especially in North America, it has high utility for the concerted management actions by states in the United States and provinces in Canada for sustainable hunting, especially of waterfowl (Johnsgard, 2012, p. 18). The eight global flyways are the East Atlantic, the Mediterranean-Black Sea, the East Asia–East Africa, the Central Asia, the East Asia–Australasia, and three flyways in the Americas and the Neotropics. The total number of birds migrating in these flyways is estimated to be billions (BirdLife International n.d., b). This book focusses on the Central Asian Flyway (Figure 0.2 in the Introduction).

Globally, about 2000 species of birds make regular seasonal movements between their breeding and non-breeding areas (Birdlife, n.d., a). Kirby *et al.* (2008) believe that out of 9856 extant species of birds there are 1855 migrants, but Fuller (2016) contends there are only about 1200 migratory species of birds in the world. A migratory species is defined as a species of which a substantial proportion of the global or a regional population makes regular cyclical movements beyond the breeding range, with predictable timing and destinations (Kirby *et al.*, 2008). Most of these migrations entail journeys of thousands of kilometres twice a year between their breeding and wintering areas. More than 450 of these migratory species are known to be declining, and nearly 200 are now classified as Globally Threatened (BirdLife International, n.d., a). Partial migrants are notably less likely to be declining than are full migrants (Gillroy *et al.*, 2016; Government of Canada, n.d.), suggesting a clear disadvantage to this strategy of full migration at present.

The causes of the present negative selection against migration are all man-induced and range from agricultural practices to climate change. Naturally, migration must have been beneficial in the past so that it could evolve, and the causal explanations centre on climate, food, predation and breeding as factors favouring the evolution of migration, but even disease control has been shown to be a possible factor of importance (Johns & Shaw, 2016; see also Rotics *et al.*, 2016). Contrary to what one may have expected, it has recently been shown in at least one group of birds (the American sparrows) that long-distance migration evolved primarily through an evolutionary shift in geographic range to the south from the northern temperate zone in winter, and not the other way around (Winger *et al.*, 2014).

If flyways were used at similar levels, then a null hypothesis would be that the Central Asian Flyway would be important for several hundreds of millions of or even a billion birds. Perhaps that has been the case in the past, but currently there is no evidence that such an enormous number of birds uses this flyway. On the contrary, Delany *et al.* (Chapter 5) make a rough estimate of the total number of waders migrating across the Himalayas of about 1.4 million per season, based on passage rates 'guesstimated' from a limited number of observations. Since waders also migrate over the Pamirs and Hindu Kush (see for instance the many ring recoveries between southwest Siberia and the Indian subcontinent presented in Veen *et al.*, 2005), the total figure for waders may be in the order of 2 to 3 million in the Central Asian Flyway. This tallies well with results from hundreds of censuses collated and analyzed by Li *et al.* (2009). For example, 100 sites in India recorded more than 20,000 waterbirds, and 13 extremely important sites recorded 100,000 waterbirds. Although many of the birds at these wetlands are not migratory and are properly thus not part of the Central Asian Flyway (cf. Table 27.1). In Pakistan, 56 sites each supported counts of more than 20,000 waterbirds at some point in the year.

Table 27.1. The most numerous waterbird species for the major South Asian countries in the past 25 years. The most numerous species (> 100,000 individuals or > 50,000 individuals, depending on country) recorded are given in order of decreasing abundance. Nepal had one wetland with > 15,000 waterbirds in the early 1990s, but since that time numbers have declined seriously to a few thousand (from Li *et al.*, 2009). Species that do not cross the Himalayas or the Tibetan Plateau appear in italics. Common Coots are mostly resident.

Bangladesh	India	Pakistan	Sri Lanka
> 50,000	> 100,000	> 50,000	> 100,000
Northern Pintail	Northern Pintail	Common Coot	Northern Pintail
Ferruginous Duck	Common Coot	Common Teal	*Lesser Whistling Duck*
Gadwall	Northern Shoveler	Common Pochard	Garganey
Fulvous Whistling Duck	Gadwall	Northern Pintail	
Eurasian Wigeon	Tufted Duck	Northern Shoveler	
Lesser Whistling Duck	*Lesser Whistling Duck*	Eurasian Wigeon	
	Eurasian Wigeon	Gadwall	
	Garganey	*Greater Flamingo*	
	Common Pochard	Black-Tailed Godwit	
	Demoiselle Crane		
	Black-tailed Godwit		
	Ruff		

However, only eight sites were reported to hold more than 20,000 waterbirds after the year 2000. In Nepal, only one site yielded a count of more than 20,000 waterbirds during 1992–1994, but recent counts were much lower (Li *et al.*, 2009).

The most numerous waterbird species that cross the Himalayas are, in decreasing order, Northern Pintail, Common Teal, Northern Shoveler, Common Pochard, Garganey, Gadwall, Tufted Duck *Aythya fuligula*, Eurasian Wigeon, Black-tailed Godwit, Demoiselle Crane and Ruff *Philomachus pugnax*. Common Coot *Fulica atra* is one of the most numerous birds recorded by counts in South Asia in winter, but the proportion which are resident and the number of additional birds which cross the Himalayas to winter in the subcontinent remain unknown (Table 27.1). The combined number of all these species may be in the order of 1 million birds (Li *et al.*, 2009). It is the order of magnitude that is of importance here.

A total estimate for waterbirds in the Central Asian Flyway of 2 to 3 million is, we think, defensible. This contrasts strongly with North America, where wildlife biologists study waterfowl populations intensively to set hunting seasons and limits. They have a good idea of how many waterfowl head south each autumn, namely about 100 million. Out of these, about 40 million return each spring; hunters kill about 20 million and about 40 million fall victim to predation, accidents, environmental factors and disease (Shackelford *et al.*, 2005). Recent counts for North America stand at about 50 million ducks (breeding population) and 16 million geese and swans (mid-winter count) (US Fish & Wildlife Service, 2015), thus indicative for the same size of waterfowl population. The total number of waterfowl in Europe and Southwest Asia is substantially lower than in North America and is in the order of 26 million only (namely, 11 million ducks and coots, 3 million geese and swans, 5 million waders, 2.5 million gulls and 4.5 million grebes, flamingos, cranes etc.; Gilissen *et al.*, 2002).

Each autumn, the skies above Canada and the northern United States become engulfed with birds. Three to five billion birds emerging from the boreal forest after each summer breeding season fly towards their wintering grounds. For landbirds alone, the top five destinations are (estimated number of wintering birds): the United States (1150×10^6), Mexico (680×10^6), Brazil (200×10^6), Colombia (110×10^6) and Venezuela (60×10^6) (Boreal Songbird Initiative, n.d.). The estimate for the number of passerines and near-passerines that yearly migrate from Europe to sub-Sahara Africa is in the same order of magnitude as in North America, namely, about 2.1 billion (ranging between 1.52 and 2.91×10^9) (Hahn *et al.*, 2009) and a similar number must come from west Siberia.

In North America, there is thus roughly a ratio of 40:1 for the number of boreal birds vs. the number of waterfowl. In Europe and Southwest Asia, this is roughly 80:1. If a ratio of 60:1 (the average of the two estimates) holds true for the central Siberian taiga, then one could roughly expect that the total number of birds in the Central Asian Flyway would be some 100 million individuals. Delany *et al.* (Chapter 4), however, were not able to make an estimate of the number of passerines or other bird species that make use of the Central Asian Flyway. We, at this stage of research in Eurasia and South Asia, would only dare to present a ballpark figure of 100 million migratory birds, but perhaps it is as low as 20 million.

Many birds may circumvent the high mountains but still migrate between central Siberia and the Indian subcontinent, as indicated by morphological differences (Irwin & Irwin, 2005), but neither satellite tracking nor ring recoveries give much evidence for this (Dixon *et al.*, Chapter 8). Since an equitable allotment of the billions of migratory birds that make use of the eight global flyways would lead to an estimate of about 1 billion birds using the Central Asian Flyway (the null hypothesis referred to earlier), it is reasonable to conclude that this flyway is 'under-used' (because we guestimate it is at least one order of magnitude lower). This may be explained in different ways, namely, by natural causes such as (a) the existence of the enormous barrier of the Himalayas and Tibetan Plateau, the Taklamakan Desert and several other mountain ranges, (b) the relatively small size of the Indian subcontinent as compared to South America and especially Africa and (c) a relative lack of upwelling cold seawater near the subcontinental coasts, resulting in less food for shorebirds as compared to that on shores of other continents (Butler *et al.*, 2001). Or perhaps even high depredation by predators is a factor. The Central Asian Flyway includes the Palearctic, Sino-Japanese, Saharo-Arabian and Oriental terrestrial zoogeographical realms (Holt *et al.*, 2013), which are the most raptor-diverse realms in the world (Ferguson-Lees & Christie, 2001). Perhaps, however, the relatively small number of birds in the Central Asian Flyway is also explained by recent onslaughts by humans on their habitat and food supplies (see later in this chapter). India and China are the most populous countries in the world (1.2 billion and 1.4 billion, respectively) with high average human population densities (365 and 144 people.km^{-2}, respectively), but the area north of the Himalayas is generally not densely populated (both Siberia and the Tibet Autonomous Region have a human density of only 3 people.km^{-2}). Thus, if people are a major cause of decline in the Central Asian Flyway, we most likely have to take a hard look at the Indian subcontinent.

How Is the Central Asian Flyway Faring?

When we started working on birds in the Himalayas in the 1990s, we had the feeling that we were entering a fairy-tale land. Birds were everywhere, in valleys and on mountain passes. Since that time, we have been travelling from the southern cape of India, through the Indian subcontinent, the Terai of Nepal, the Himalayas, along the Pamirs, through the Taklamakan Desert, the Tien Shan, the Gobi Desert and the Altai Mountains. We have been travelling the great Yenisei River from close to its source all the way to the Arctic Ocean, and likewise the Ob River. These long journeys ended last year, and we finished with the startling insight that birds are becoming scarcer in the Central Asian Flyway. Gone are the circling vultures and eagles. The paddy fields and wheat fields of India are nearly devoid of songbirds. The oases in the Taklamakan and Gobi Deserts are shockingly empty of birds. This is partly because this flyway does not appear to hold many birds for natural reasons: the landmass of the Indian subcontinent (Bangladesh, Bhutan, India, Nepal, Pakistan, Sri Lanka and the Maldives) is only 4.5 x 10^6 km^2, while that of South America is 17.8 x 10^6 km^2, and that of Africa without the Sahara even

20.8×10^6 km^2. So merely based on the size of the available habitat during the winter season, one would expect the Central Asian Flyway to contain fewer birds. And it should not be forgotten that along the coasts of the Indian subcontinent there appear to be less favourable conditions due to less upwelling of cold seawater than elsewhere in the tropics (Butler *et al.*, 2001).

The Central Asian Flyway encompasses some of the harshest tracts of land, with extreme deserts (Taklamakan, Gobi, Badain Jaran), barren plateaus (Tibetan Plateau, Mongolian Plateau) and the highest mountains (Himalayas, Tian Shan, Kunlun Shan, Karakoram, Hindu Kush, Pamirs) on Earth. But the truly worrying fact is that the number of birds is declining rapidly: there are far fewer birds now than 30 years ago (see later in this chapter). Critically, we need to identify which migratory species are declining in which regions and the principal reasons for their declines. This is particularly important in South Asia (including the Indian subcontinent), where information is generally poor or non-existent (Kirby *et al.*, 2008). The regular wetland counts coordinated by Wetlands International give some good insights, but are becoming increasingly difficult to sustain.

Coastal birds are becoming scarce at many key sites in India, which may be caused by habitat loss, harmful industrial developments, unregulated agricultural practices, human disturbance or climatic and environmental changes (Balachandran, 2012). Some of these declines are frightening. For example, at Point Calimere, situated on a low promontory on the Coromandel Coast of India, 1 million migratory waders and ducks found in the 1980s declined to fewer than 150,000 in the late 1990s. However, in recent years the total number of waterbirds at this site increased to 350,000, but that may be due to the shifting of populations from neighbouring coastal wetlands, where coastal bird populations declined correspondingly (Balachandran, 2012; see Table 27.2). Indeed, wader and duck population declines (from 300,000 to 150,000) have been reported at the Pulicat wetlands on the eastern coast of India, and at Chilika Lake, the world's second largest brackish-water lagoon (from 10 million shorebirds, flamingos and ducks to 4.5 million) (Balachandran, 2012).

The causes of these appalling declines are not known, but can be surmised. Often the root causes are not mentioned in the Indian subcontinent, and much hot air has been vented on climate change as a cause of the population declines. A case in point is

Table 27.2a. Counts of the most common waterbird species at Point Calimere (S. India): the decline of the number of individuals has been rapid (based on Balachandran, 2012).

Species	1980s	1990s	2000–2008
Curlew	> 150,000	> 80,000	< 25,000
Little Stint	> 200,000	> 100,000	< 30,000
Lesser Sandplover	> 100,000	> 75,000	< 40,000
Ruff	> 100,000	> 30,000	< 10,000
Black-tailed Godwit	> 50,000	> 40,000	> 15,000
Black-winged Stilt	> 15,000	> 3000	> 1000
Pied Avocet	> 7000	> 500	< 100

Table 27.2b. Bird counts in the Gulf of Mannar some 300 km south of Point Calimere (see Table 1a; on basis of Balachandran, 2012).

Species	1985–1988	1993–2001	2005–2007
Curlew Sandpiper	> 10,000	> 8000	< 1000
Little Stint	> 8000	> 3000	< 1000
Lesser Sandplover	> 13,000	> 8000	< 2500
Crab Plover	900	150	< 20
Eurasian Curlew	450	450	55
Grey Plover	970	970	100
Great Knot	350	140	< 200
Red Knot	300	90	< 50
Common Greenshank	250	180	< 50

Pulicat wetland: 'Integrated studies of the lagoon environment may reveal vital scientific information on its environmental status in changing climatic scenarios. This will help the authorities to identify small changes in environmental quality so that appropriate actions can be taken for remedial work before the lake degrades' (Kannan, 2014). Yet the same source states that about 50,000 plus families of fishermen depend on this lake for livelihood! Also Namgail *et al.* (Chapter 25) point out that Ramsar sites in India are either selected without habitat assessment or are not effectively protected against overexploitation by people. On India's west coast, along the Arabian Sea, there are fewer wetlands of importance. The Kadalundi-Vallikkunnu Community Reserve, the only community reserve in Kerala, is one of the most important wintering grounds and stopover sites for migratory waterbirds on the west coast of India. Serious disturbances have been noticed here, including sand extraction and dumping of organic and other solid waste: disposal of large quantities of solid waste especially of plastics, clothes, carrier bags, bottles, food wrappers, domestic waste, poultry farm waste and slaughterhouse waste takes place in this reserve (Aarif & Prasadan, 2015). This and similar observations have been made by Humbert-Droz in the Himalayas, where wetlands and even Ramsar sites are becoming overwhelmed by tourism and all its associated pollution and habitat destruction (Humbert-Droz, Chapter 23). Wetlands International has made an extensive inventory of the threats to the South Asian wetlands and their birds (Table 27.3): the list of threats to the remaining birds in the Central Asian Flyway is long and the leadership shown to date in solving these issues by the authorities over territories within this flyway appears minimal.

Plausible Impact of Modern Agriculture on Bird Numbers

Pesticide use in agriculture in South Asia in the 1990s was still low as compared to that of the middle-income countries: average pesticide use for India, Nepal, Pakistan and Sri Lanka was, respectively, 0.3 < 0.1, 0.9 and 0.8 $kg.ha^{-1}$. This was in the same order of

Table 27.3. Major threads to wetlands and their birds, as reported in order of importance for different countries of South Asia (from Li *et al.*, 2009). There is major convergence between the threats as assessed in the different countries.

Bangladesh	Agricultural development, wetland reclamation, clearance of vegetation, waterbird hunting and poaching.
India	Agricultural development, excessive cattle grazing and fishing, illegal hunting of waterbirds.
Nepal	Habitat loss and degradation, reduced food availability for many wetland-dependent birds due to over-fishing, water pollution, river poisoning and dynamite blasting to obtain fish, increased use of pesticides – particularly on the paddy crop, hunting and associated human disturbance, the spread of invasive alien plant species in wetland areas.
Pakistan	Failure of rainfall, drainage, reclamation, water diversion for irrigation and other uses, pollution, hunting, trapping, poaching, overgrowth of vegetation, eutrophication, over-fishing and tourism/recreation.
Sri Lanka	Fishing, agricultural development, pollution from agricultural and industrial sources, infilling and other anthropogenic transformation, changes of water level; hunting (despite a total ban), raiding of heronries.

magnitude as the use intensity in the United States of America (1.7 $kg.ha^{-1}$) and lower than that in the Netherlands (6.5 $kg.ha^{-1}$) or that of the world record user the Bahamas (63 $kg.ha^{-1}$) (Schreinemachers & Tipraqsa, 2012). However, pesticide use in South Asia is on the rise, while in the rich countries it is declining (Schreinemachers & Tipraqsa, 2012). Today, the biggest producer and consumer of pesticides is China (Zhang *et al.*, 2011), and India is the second largest producer of pesticides in Asia, after China. It ranks already as the fourth largest pesticide-producing nation in the world (Yadav *et al.*, 2015). Generally, water bodies of croplands are frequently polluted in China (and, we suspect, by analogy on the Indian subcontinent too), where the pesticide concentration of water bodies can attain the magnitude of dozens of milligrams per litre. In China, the levels of water pesticide pollution can be ranked as: cropland water > field ditch water > runoff > pond water > groundwater > river water > deep groundwater > seawater. Residues of DDT, dieldrin, endrin and so forth were detected in most of the water bodies measured in China (Zhang *et al.*, 2011). Data on recent insecticide use in the Indian subcontinent are not widely available but are reported to be high; Bangladesh and Pakistan use even more per hectare than India (Liu *et al.*, 2015). The Indian subcontinent has experienced heavy use of organochlorine pesticides, such as DDT, while India continues the use of DDT to combat vector-borne diseases, which are prevalent in the monsoon season (Wang *et al.*, 2010). All along the border of south Tibet, traces of pesticide are measured in the monsoon air, which is indicative of heavy use in the Indian subcontinent (Wang *et al.*, 2010). It was even concluded that India is currently one of the major contributors of the global persistent organic pesticide (POP) distribution; every environmental component in India is contaminated with POPs, and residue levels of pesticides are high in air, water and soil despite low consumption (Yadav *et al.*, 2015). This is a problem in Pakistan too (Mehmood *et al.*, 2015).

Of course, it is understandable that farmers use pesticides if they can afford it. Farmers benefit by an estimated €3 to €5 in crops for every €1 that they invest in pesticides (Raven *et al.*, 2008), but several studies argue that pesticide use produces overall low economic returns if social and ecosystem health impacts are accounted for (Atreya *et al.*, 2011). Indeed, every year, pesticide use causes 3 million human poisonings, 220,000 deaths and about 750,000 chronic illnesses worldwide. The majority of these are reported in developing countries (Atreya *et al.*, 2011). The widely used pesticides and their persistent residues are also likely to be taking a heavy toll on migratory birds in the Central Asian Flyway: bird numbers are rapidly declining, not only on the Indian subcontinent, but also in central Siberia (Rogacheva, 1992). Of course, birds still occur in agricultural areas in India (Laxmi Narayana *et al.*, 2015), but a very insightful paper by Amjath-Babu and Kaechele (2015) shows how many parts of India show a rapid transition to types and intensities of agriculture that have hardly supported biodiversity in recent times. Outside the protected areas, a 'silent spring' (Carson, 1962) may change into a 'silent monsoon'. The study of Amjath-Babu and Kaechele did not take into account the northern states close to or in the Himalayas. But Ghosal and Ahmed (Chapter 22) and Raman *et al.* (Chapter 24) also show how the pastoral economies in the high mountains are changing, and how bird species are negatively affected.

Plausible Impact of Deforestation on Bird Populations

In lowland India, 25% of the land surface is covered with forest, but it is declining; analysis of MODIS satellite data shows that between 2001 and 2009, the forest cover declined all across India by a total of 28,000 km^2 (so forest disappears at a rate of about 0.5 per cent.y^{-1}). Similarly, the areas under woodlands and barren lands have also decreased by 32,000 km^2 each (Mythili & Goedecke, 2015). At the current level of deforestation, by 2100 only about 10% of the land area of the Indian Himalayas will still be covered by dense forest (Pandit *et al.*, 2007). Pakistan currently has 2.5% of its land area under forest cover (20,000 km^2), with a 2.1% annual rate of deforestation (420 $km^2.y^{-1}$). In fact, Pakistan has the highest deforestation rate in Asia. According to the Pakistani Federal Bureau of Statistics, there was a 3% decline in forest area (600 km^2) between 2000 and 2005. The plausible causes are urbanization, overgrazing, farming methods, climate change and the use of wood as a major source of energy in rural areas (Ahmed *et al.*, 2015). The data showing that forests constitute about 46,000 km^2 of Pakistan's total land area and that in total, 610 km^2 of forest region have been transformed to non-forest use in the country ever since independence (in 1947) (Khalid *et al.*, 2015) is thus erroneous, and may reflect official land status only. Sri Lanka lost 4500 km^2 of forest since 1990 (from 2.9 x 10^4 km^2 in 1990) and Bangladesh lost 1500 km^2 over the same period (from 2 x 10^4 km^2) (Kim *et al.*, 2015). Nepal lost 25% of its forest cover, or around 12,000 km^2, between 1990 and 2010 (FAO, 2010). Thus, in total, the Indian subcontinent must have lost some 60,000 to 70,000 km^2 of forest since the 1990s. Note that deforestation in the humid tropics is apparently still accelerating (Kim *et al.*, 2015). These enormous losses of forest must be affecting the number of wintering birds, especially passerines, in the Central Asian Flyway (cf. Pandit *et al.*, 2007).

Land degradation is another major issue (Mythili & Goedecke, 2015). Remotely sensed data revealed that about 16% of the Indian territory, that is, about 470,000 km^2, showed declining normalized difference vegetation index (NDVI) trends between 1982 and 2006 (Le *et al.*, 2014). Fertilizer subsidy turns out to be a major determinant of land degradation. However, due to political pressure in response to lobbying by farmers' groups, the government has not been able to cut subsidies on fertilizer in a desirable manner. According to calculations, a 1% reduction in subsidy is likely to reduce land degradation by nearly 3%; also, the use of pesticides increases the occurrence of soil erosion (Mythili & Goedecke, 2015). Also in the Himalayas, the Government of India is subsidizing local pastoralists for various reasons (Ghosal & Ahmed, Chapter 22) and is engaged in road building and supporting local tourism (Humbert-Droz, Chapter 23). This has negative effects on natural habitats, including bird life (Raman *et al.*, Chapter 24).

What Conservation Measures must Be Taken to Maintain the Bird Migration across the Himalayas?

To save the natural splendour of the biannual bird migrations across the Himalayas for future generations, the range states of the Central Asian Flyway (Afghanistan, Bangladesh, Bhutan, China, India, Kazakhstan, Kyrgyzstan, Mongolia, Myanmar, Nepal, Pakistan, Russian Federation, Sri Lanka, Tajikistan and Turkmenistan) need to take a number of measures. These measures can be taken as sovereign states and do not need further international treaties or conventions. Indeed, the Ramsar Convention on Wetlands and the Convention on the Conservation of Migratory Species ('Bonn Convention') already exist for this purpose. The problem with these conventions is that they do not have 'teeth' (Wilcove, 2008). Measures therefore, have to be taken and upheld by national governments.

The Central Asian Flyway Action Plan for the Conservation of Migratory Waterbirds and Their Habitats is a fine piece of text which was provisionally endorsed by the range states falling within that flyway in New Delhi in 2005 (CMS, 2005). The website of the Convention on Migratory Species states that 'The Action Plan . . . was finally adopted in January 2008 after incorporation of further technical comments received following the Meeting.' An updated version has not been posted until now (May 2016: www.cms.int/en/legalinstrument/central-asian-flyway), so one must assume that the New Delhi 2005 version is the official version. If the range states were really implementing what they had agreed (see Appendix 1), to conserve and protect bird species, wetlands and grasslands, then one would have hope for the birds of the Central Asian Flyway. There are excellent provisions and promises for the protection of wetlands, the water flowing into these wetlands and the control of pollution and pesticides. Of course, this Action Plan does not cover non-wetlands, and as indicated earlier, there are major problems for other (migratory) birds, especially when it comes to the agricultural areas and forests, plus rangelands of the Indian subcontinent, the oases in the Taklamakan and Gobi Deserts and the grasslands of Tibet and Mongolia. Creating 'paper parks' is not difficult, but effectively managing and protecting them is. Indeed, a recent analysis of protected areas in China shows that waterbirds are declining in all protected areas, but more so in reserves with a provincial status than in reserves that are protected at national level (Zhang *et al.*, 2015). Wetlands in India also continue to disappear (Foote *et al.*, 1996). And it is the bigger, more charismatic species that are lost before the small ones, possibly due to illegal hunting (de Boer *et al.*, 2011).

Ineffective protection plagues many of the reserves in the Central Asian Flyway, thus not only in China (Wandesforde-Smith *et al.*, 2014; but the rate of deforestation seems to have been effectively lowered to about 1% per year: Ren *et al.*, 2015), but also in Pakistan (Khan, 2003), Bangladesh (Chowdhury *et al.*, 2014), Nepal and Bhutan (e.g. Oli *et al.*, 2013; Uddin *et al.*, 2015), India (Kannan *et al.*, 2013) and South Asia in general (Clark *et al.*, 2013). Conservation and protection also often fail, or have difficulties in delivering positive output for biodiversity in Siberia (e.g. Fiorino & Ostergren, 2012; Santangeli *et al.*, 2013; Anthony & Shestackova, 2015), Tajikistan (e.g. Michel *et al.*, 2015), Kazakhstan (Kamp *et al.*, 2016) and Mongolia (e.g. Bedunah & Schmidt, 2004). Indeed, Rogacheva (1992) already drew attention to the observation that many bird species in central Siberia were declining as long ago as the 1980s. A recent tragic case in point is the Yellow-breasted Bunting. *Emberiza aureola*. This species is currently showing a population collapse in Siberia, perhaps due to uncontrolled catching in China (Kamp *et al.*, 2015): its populations consisted of hundreds of millions of birds during the 1980s. Already in 2004, the IUCN conservation status of this species became 'less favourable', and in 2008 it was classified as 'Vulnerable' and in 2014 as 'Endangered'. Now the species is showing a very strong decline in the foothills of the Himalayas during winter (Kamp *et al.*, 2015). Over-exploitation and consumption in China are considered the most important drivers of population decline of this bunting species, but 53% of all endangered bird species are threatened by harvesting for food (Yiming & Wilcove, 2005; cf. Liang *et al.*, 2013). Similarly, it is probable that other bunting species are negatively affected by trapping in China.

Many threats to the birds in the Central Asian Flyway are so widespread, persistent and perhaps hopeless that one may wonder whether they can be tackled at all (cf. Appendix). It is very clear from this review that one can only succeed in halting the precipitous decline of the bird numbers in this flyway if governments start acting at local scales, tackling local issues such as over-hunting and pollution. One of the tools that governments have at their disposal is the selection of sites to be designated under the Ramsar Convention on wetlands. However, after that is done, it is of paramount importance to base such decisions on the right premises and the correct assessment of the effect on bird populations (see Namgail *et al.*, Chapter 25). We also recommend that wetlands should be conserved even if they are individually small in extent. Such wetlands may form a punctuated networks of sites that collectively need conservation; the text of the Ramsar Convention should be slightly modified so that it can serve governments better to protect such networks of small, dispersed wetlands on the migration routes high up in the Himalayas (Prins *et al.*, Chapter 26).

The very existence of a number of bird species in the Himalayas and beyond is now threatened, but the natural splendour is still there. Wetlands are still functioning between the glaciers and the deserts as natural oases for the birds to use on migration. We sincerely hope that this will continue for endless times to come. But people with the power to change the current state of affairs under which that splendid biodiversity is neglected must act now.

References

Aarif, K.M. & Prasadan, P.K. (2015). Migrant shorebirds in the southwest coast of India. *Biosystematica*, **8**, 51–58.

Ahmed, K., Shahbaz, M., Qasim, A. & Long, W. (2015). The linkages between deforestation, energy and growth for environmental degradation in Pakistan. *Ecological Indicators*, **49**, 95–103.

Amjath-Babu, T.S. & Kaechele, H. (2015). Agricultural system transitions in selected Indian states: what do the related indicators say about the underlying biodiversity changes and economic trade-offs? *Ecological Indicators*, **57**, 171–181.

Anthony, B.P. & Shestackova, E. (2015). Do global indicators of protected area management effectiveness make sense? A case study from Siberia. *Environmental Management*, **56**, 176–192.

Atreya, K., Sitaula, B.K., Johnsen, F.H., & Bajracharya, R.M. (2011). Continuing issues in the limitations of pesticide use in developing countries. *Journal of Agricultural and Environmental Ethics*, **24**, 49–62.

Balachandran, S. (2012). Avian diversity in coastal wetlands of India and their conservation needs. *Proceedings of International Day for Biological Diversity May*, 22.

Bedunah, D.J. & Schmidt, S.M. (2004). Pastoralism and protected area management in Mongolia's Gobi Gurvansaikhan National Park. *Development and Change*, **35**, 167–191.

Birdlife (n.d., a). www.birdlife.org/worldwide/programmes/migratory-birds.

Birdlife (n.d., b). www.birdlife.org/datazone/userfiles/file/sowb/flyways/2_Central_Americas_Factsheet.pdf.

Black, C.P., & Tenney, S.M. (1980). Oxygen transport during progressive hypoxia in high-altitude and sea-level waterfowl. *Respiration Physiology*, **39**, 217–239.

De Boer, W.F., Cao, L., Barter, M., Wang, X., Sun, M., Van Oeveren, H., de Leeuw, J., Barzen, J. & Prins, H.H.T (2011). Comparing the community composition of European and Eastern Chinese waterbirds and the influence of human factors on the China waterbird community. *Ambio*, **40**, 68–77.

Boere, G.C. & Stroud, D.A. (2006). *The Flyway Concept: What It Is and what It Isn't. Waterbirds around the World*. Eds. G.C. Boere, C.A. Galbraith & D.A. Stroud. The Stationery Office, Edinburgh, UK. pp. 40–47.

Boreal Songbird Initiative, n.d. www.borealbirds.org/boreal-bird-migrations?gclid=CIm8yNHFw8wCFUgq0wodel4CvA.

Butler, R.W., Davidson, N.C. & Morrison, R.G. (2001). Global-scale shorebird distribution in relation to productivity of near-shore ocean waters. *Waterbirds*, **24**, 224–232.

Carson, R. (1962). *Silent Spring*. Houghton Mifflin, New York.

Chowdhury, M.S.H., Nazia, N., Izumiyama, S., Muhammed, N. & Koike, M. (2014). Patterns and extent of threats to the protected areas of Bangladesh: the need for a relook at conservation strategies. *Parks*, **20**, 91–104.

Cieri, R.L. & Farmer, C.G. (2016). Unidirectional pulmonary airflow in vertebrates: a review of structure, function, and evolution. *Journal of Comparative Physiology B*, **186**, 541.

Clark, N.E., Boakes, E.H., McGowan, P.J., Mace, G.M. & Fuller, R.A. (2013). Protected areas in South Asia have not prevented habitat loss: a study using historical models of land-use change. *PloS one*, **8**(5), e65298.

CMS (2005). Central Asian Flyway Action Plan for the Conservation of Migratory Waterbirds and their Habitats, as finalised by Range States of the Central Asian Flyway at their second meeting in New Delhi, 10–12 June 2005 CMS/CAF/Report, Annex 4: www.cms.int/en/legalinstrument/central-asian-flyway.

Den Besten, J.W. (2004). Migration of Steppe Eagles Aquila nipalensis and other raptors along the Himalayas past Dharamsala, India, in autumn 2001 and spring 2002. *Forktail*, **20**, 9–13.

Faraci, F.M., Kilgore, D.L. & Fedde, M.R. (1984). Oxygen delivery to the heart and brain during hypoxia: Pekin duck versus Bar-headed Goose. *American Journal of Physiology-Regulatory Integrative and Comparative Physiology* **247**, R69–R75.

FAO (2010). The global forest resources assessment 2010. Food and Agriculture Organisation of the United Nations, Rome, Italy.

Ferguson-Lees, J. & Christie, D. (2001). *Raptors of the World*, Houghton Mifflin Company, New York.

Fiorino, T. & Ostergren, D. (2012). Institutional instability and the challenges of protected area management in Russia. *Society & Natural Resources*, **25**, 191–202.

Foote, A.L., Pandey, S. & Krogman, N.T. (1996). Processes of wetland loss in India. *Environmental Conservation*, **23**, 45–54.

Fuller, R.A. (2016). Animal migration: dispersion explains declines. *Nature*, **531**, 451–452.

Galbraith, C.A., Jones, T., Kirby, J. & Mundkur, T. (2014). *A Review of Migratory Bird Flyways and Priorities for Management. UNEP/CMS Secretariat*, Bonn, Germany. CMS Technical Series No. 27.

Geroudet, P. (1954). Des oiseaux migrateurs trouvés sur la glacier de Khumbu dans l'Himalaya. *Nos Oiseaux*, **22**, 254.

Gilroy, J.J., Gill, J.A., Butchart, S.H., Jones, V.R. & Franco, A. (2016). Migratory diversity predicts population declines in birds. *Ecology Letters*, **19**, 308–317.

Government of Canada (n.d.). Trends in Canada's Migratory Bird Populations. www.ec.gc.ca/indicateurs-indicators/default.asp?lang=en&n=F2BDAA4C-1.

Hahn, S., Bauer, S., Liechti, F. (2009). The natural link between Europe and Africa – 2.1 billion birds on migration. *Oikos* **118**, 624–626.

Hawkes, L.A., Butler, P.J., Frappell, P.B., Meir, J.U., Milsom, W.K., Scott, G.R., & Bishop, C.M. (2014). Maximum running speed of captive Bar-headed Geese is unaffected by severe hypoxia. *PloS one*, **9**(4), e94015.

Holt, B.G., Lessard, J.P., Borregaard, M.K., Fritz, S.A., Araújo, M.B., Dimitrov, D. & Nogués-Bravo, D. (2013). An update of Wallace's zoogeographic regions of the world. *Science*, **339**, 74–78.

Irwin, D.E. & Irwin, J.H. (2005). Siberian migratory divides: the role of seasonal migration in speciation. In R. Greenberg and P.P. Marra, eds., *Birds of Two Worlds*. Baltimore: John Hopkins University Press, pp. 27–40.

Fiorino, T., & Ostergren, D. (2012). Institutional instability and the challenges of protected area management in Russia. *Society & Natural Resources*, **25**, 191–202.

Gilissen, N., Haanstra, L., Delany, S., Boere, G. & Hagemeijer, W. (2002). Numbers and distribution of wintering waterbirds in the western palearctic and southwest Asia in 1997, 1998 and 1999 – results from the International Waterbird Census. Wetlands International Global Series No. 11. Wageningen, the Netherlands.

Johns, S., & Shaw, A.K. (2016). Theoretical insight into three disease-related benefits of migration. *Population Ecology*, **58**, 213–221.

Johnsgard, P.A. (2012). *Wings over the Great Plains: Bird Migrations in the Central Flyway*. Zea E-Books. University of Nebraska–Lincoln.

Kamp, J., Koshkin, M.A., Bragina, T.M., *et al.* (2016). Persistent and novel threats to the biodiversity of Kazakhstan's steppes and semi-deserts. *Biodiversity and Conservation*, 25: 2521.

Kamp, J., Oppel, S., Ananin, A.A., *et al.* (2015). Global population collapse in a super-abundant migratory bird and illegal trapping in China. *Conservation Biology*, **29**, 1684–1694.

Kannan, R., Shackleton, C.M. & Shaanker, R.U. (2013). Playing with the forest: invasive alien plants, policy and protected areas in India. *Current Science*, **104**, 1159–1165.

Kannan, V. (2014). Lakes are natural treasures – a look at Pulicat Lake. *Journal of Science, Technology and Management*, **7**: 109–116.

Khalid, N., Ahmad, S.S., Erum, S., Butt, A. (2015). Monitoring forest cover change of Margalla Hills over a period of two decades (1992–2011): a spatiotemporal perspective. *Journal of Ecosystem and Ecography* **6**, 174.

Khan, Z.I. (2003). Protected areas in Pakistan: management and issues. *Journal of the National Science Foundation of Sri Lanka*, **31**: 239–248.

Kim, D.H., Sexton, J.O. & Townshend, J.R. (2015). Accelerated deforestation in the humid tropics from the 1990s to the 2000s. *Geophysical Research Letters*, **42**, 3495–3501.

Kirby, J.S., Stattersfield, A.J., Butchart, S.H., *et al.* (2008). Key conservation issues for migratory land- and waterbird species on the world's major flyways. *Bird Conservation International*, **18(S1)**, S49–S73.

Laxmi Narayana, B., Vasudeva Rao, V. & Pandiyan, J. (2015). Avifaunal diversity in different croplands of Nalgonda District, Telangana, southern India. *International Journal of Current Research*, **7**, 17677–17682.

Le Q.B., Nkonya, E. & Mirzabaev, A. (2014). *Biomass Productivity-Based Mapping of Global Land Degradation Hotspots*, ZEF Discussion Paper on Development Policy No. 193, Centre for Development Research, University of Bonn.

Li, Z.W.D., Bloem, A., Delany S., Martakis G. & Quintero J.O. 2009. *Status of Waterbirds in Asia – Results of the Asian Waterbird Census: 1987–2007*. Wetlands International, Kuala Lumpur, Malaysia.

Liang, W., Cai, Y. & Yang, C.C. (2013). Extreme levels of hunting of birds in a remote village of Hainan Island, China. *Bird Conservation International*, **23**, 45–52.

Liu, Y., Pan, X. & Li, J. (2015). A 1961–2010 record of fertilizer use, pesticide application and cereal yields: a review. *Agronomy for Sustainable Development*, **35**, 83–93.

Mehmood, A., Mahmood, A., Eqani, S.A.M.A.S., *et al.* (2015). A review on emerging persistent organic pollutants: current scenario in Pakistan. *Human and Ecological Risk Assessment: An International Journal* **23**, 1–13.

Michel, S., Michel, T.R., Saidov, A., *et al.* (2015). Population status of Heptner's markhor Capra falconeri heptneri in Tajikistan: challenges for conservation. *Oryx*, **49**, 506–513.

Mlíkovský, J. (2002). *Cenozoic Birds of the World. Part 1: Europe*. Prague: Ninox Press.

Mythili, G. & Goedecke, J. (2015). Economics of land degradation in India. In A. Mirzabaev & J. von Braun, eds., *Economics of Land Degradation and Improvement – A Global Assessment for Sustainable Development*. Heidelberg, Springer, pp. 431–469.

Namgail, T., Bhatnagar, Y.V., Mishra, C. & Bagchi, S. (2007). Pastoral nomads of the Indian Changthang: production system, landuse, and socio-economic changes. *Human Ecology* **35**, 497–504.

Namgail, T., Van Wicren, S.E. & Prins, H.H.T. (2010). Pashmina production and socio-economic changes in the Indian Changthang: implications for natural resource management. *Natural Resources Forum*, **34**, 222–230.

Oli, K.P., Chaudhary, S. & Sharma, U.R. (2013). Are governance and management effective within protected areas of the Kanchenjunga landscape (Bhutan, India and Nepal)? *Parks*, **19**, 1–12.

Pandit, M.K. & Grumbine, R.E. (2012). Potential effects of ongoing and proposed hydropower development on terrestrial biological diversity in the Indian Himalaya. *Conservation Biology*, **2**, 1061–1071.

Pandit, M., Sodhi, N.S., Koh, L.P., Bhaskar, A. & Brook, B.W. (2007). Unreported yet massive deforestation driving loss of endemic biodiversity in Indian Himalaya. *Biodiversity and Conservation*, **16**,153–163.

Price, T., Zee, J., Jamdar, K. & Jamdar, N. (2003). Bird species diversity along the Himalaya: a comparison of Himachal Pradesh with Kashmir. *Journal of the Bombay Natural History Society*, **100**, 394–410.

Raven, P.H., Berg, L.R. & Hassenzahl, D.M. (2008). *Environment*. New Jersey, USA: John Wiley & Sons Inc.

Ren, G., Young, S.S., Wang, L., *et al.* (2015). Effectiveness of China's National Forest Protection Program and nature reserves. *Conservation Biology*, **29**, 1368–1377.

Rogacheva, H. (1992). *The Birds of Central Siberia*. Husum Druck- und Verlaggesellschaft, Husum.

Romulo, C.L. (2012). Geodatabase of global owl species and owl biodiversity analysis. Master of Natural Resources Capstone Paper. Virginia Polytechnic Institute and State University, Falls Church, Virginia.

Rotics, S., Kaatz, M., Resheff, Y.S., *et al.* (2016). The challenges of the first migration: movement and behavior of juvenile versus adult white storks with insights regarding juvenile mortality. *Journal of Animal Ecology*, **85**, 938–947.

Ruben, J.A., Jones, T.D., Geist, N.R. & Hillenius, W.J. (1997). Lung structure and ventilation in theropod dinosaurs and early birds. *Science*, **278**, 1267–1270.

Santangeli, A., Wistbacka, R., Hanski, I.K. & Laaksonen, T. (2013). Ineffective enforced legislation for nature conservation: a case study with Siberian flying squirrel and forestry in a boreal landscape. *Biological Conservation*, **157**, 237–244.

Sarkar, S.U. (1986). Population dynamics of raptors in the Sundarban forests of Bangladesh. *Birds of Prey Bulletin*, **3**, 157–162.

Schreinemachers, P. & Tipraqsa, P. (2012). Agricultural pesticides and land use intensification in high, middle and low income countries. *Food Policy*, **37**, 616–626.

Shackelford, C.E., Rozenburg, E.R., Hunter, W.C. & Lockwood M.W. (2005). *Migration and the Migratory Birds of Texas: Who They Are and Where They Are Going*. Houston, Texas Parks and Wildlife.

Thakur, M.L. & Negi, V. (2015). Status and phylogenetic analyses of endemic birds of the Himalayan region. *Pakistan Journal of Zoology*, **47**, 417–426.

Tockner, K., Bernhardt, E.S., Koska, A. & Zarfl, C. (2016). A global view on future major water engineering projects. In R.F. Hüttl, O. Bens, C. Bismuth & S. Hoechstetter, eds., *Society-Water-Technology: A Critical Appraisal of Major Water Engineering Projects*. Heidelberg: Springer International Publishing, pp. 47–64.

Uddin, K., Chaudhary, S., Chettri, N. *et al.* (2015). The changing land cover and fragmenting forest on the Roof of the World: a case study in Nepal's Kailash sacred landscape. *Landscape and Urban Planning*, **141**, 1–10.

US Fish and Wildlife Service (2015). Waterfowl Population Status, 2015. Division of Migratory Bird Management, U.S. Fish and Wildlife Service, Laurel, MD. pp. 1–75.

Veen, J., Yurlov, A.K., Delany, S.N., Mihantiev, A.I., Selivanova, M.A. & Boere, G.C. (2005). *An Atlas of Movements of Southwest Siberian Waterbirds*. Wageningen, The Netherlands: Wetlands International.

Wandesforde-Smith, G., Denninger Snyder, K. & Hart, L.A. (2014). Biodiversity conservation and protected areas in China: science, law, and the obdurate party-state. *Journal of International Wildlife Law & Policy*, **17**, 85–101.

Wang, X.P., Gong, P., Yao, T.D. & Jones, K.C. (2010). Passive air sampling of organochlorine pesticides, polychlorinated biphenyls, and polybrominated diphenyl ethers across the Tibetan Plateau. *Environmental Science & Technology*, **44**, 2988–2993.

White, T.D. (2014). Delimitating species in paleoanthropology. *Evolutionary Anthropology: Issues, News, and Reviews*, **23**, 30–32.

Wilcove, D.S. (2008) Animal migration: an endangered phenomenon? *Issues in Science and Technology* **24**, 71–78.

Winger, B.M., Barker, F.K. & Ree, R.H. (2014). Temperate origins of long-distance seasonal migration in New World songbirds. *Proceedings of the National Academy of Sciences*, **111**, 12115–12120.

Yadav, I.C., Devi, N.L., Syed, J.H., *et al.* (2015). Current status of persistent organic pesticides residues in air, water, and soil, and their possible effect on neighboring countries: a comprehensive review of India. *Science of The Total Environment*, **511**, 123–137.

Yiming L. & Wilcove, D.S. (2005). Threats to vertebrate species in China and the United States. *BioScience*, **55**, 147–153.

Zhang, W., Jiang, F. & Ou, J. (2011). Global pesticide consumption and pollution: with China as a focus. *Proceedings of the International Academy of Ecology and Environmental Sciences*, **1**, 125–144.

Zhang, Y., Jia, Q., Prins, H.H.T., Cao, L. & de Boer, W.F. (2015). Effect of conservation efforts and ecological variables on waterbird population sizes in wetlands of the Yangtze River. *Scientific Reports*, **5**:17136.

Appendix: Selected Articles of the 'Central Asian Flyway Action Plan'

Some selected articles of the 'Central Asian Flyway Action Plan for the Conservation of Migratory Waterbirds and their Habitats', as finalized by Range States of the Central Asian Flyway at their second meeting in New Delhi, 10–12 June 2005, CMS/CAF/Report, Annex 4: www.cms.int/en/legalinstrument/central-asian-flyway. The text was accepted in January 2008.

In the Action Plan, there are a number of proposed legal measures focusing on prohibition to take or trade birds or eggs. But the excerpted articles focus on habitat protection. The Table 2 referred to is a table listing all waterbird species and their vulnerability status in the Central Asian Flyway.

3.2.1 Range States shall endeavour to take decisions and implement measures to ensure:
   (a) adequate and timely supply of water required to maintain natural functions of wetlands and other important habitats known to be of importance for migratory waterbirds (especially in arid areas);
   (b) maintain and sustainably manage wetlands and other habitats important to migratory waterbirds (e.g. steppe grasslands); . . .

3.2.2 Range States shall endeavour to avoid degradation and loss of habitats that support populations listed in Table 2 through the introduction of appropriate regulations or standards and control measures.

3.2.3 Range States shall provide official support to designate, conserve and manage all important breeding, moulting, staging and non-breeding (wintering) sites for populations listed in Table 2, by establishing national networks of all important sites under appropriate national and international conservation categories (e.g. nature reserves, national parks, wildlife reserves, sanctuaries, non-hunting areas, Ramsar Sites, World Heritage Sites, Important Bird Areas and special conservation areas). These sites should be considered for inclusion within the Central Asian Flyway Site Network outlined in section 3.3.

3.2.4 Range States shall endeavour to give special protection to existing designated sites, including Ramsar Sites and World Heritage Sites, which meet internationally accepted criteria of international importance for populations listed in Table 2.

3.3.1 Range States shall actively support the establishment of the Central Asian Flyway Site Network, as a mechanism for linking national networks of waterbird sites of international importance across the flyway. Based on the principle of establishing an ecological network of internationally important sites through promotion of conservation and sustainable management of wetlands and other habitats. . . . This site network will extend to and integrate with the East Asian-Australasian Site Networks established under the Asia-Pacific Migratory Waterbird Conservation Strategy and the proposed West/Central Asian Site Network for the Siberian Crane (and other waterbirds).

3.4.1 Range States shall endeavour to rehabilitate or restore, where feasible and appropriate, sites which were previously important for the populations listed in Table 2.

3.5.1 Range States shall cooperate to determine and monitor the impacts of climate change on migratory waterbirds and their habitats and where appropriate respond to the threats.

4.1.6 Range States shall develop and implement necessary measures to eliminate, or reduce, as far as possible, illegal taking, poaching, and unsustainable hunting.

4.3.1 Range States shall, as far as possible, undertake an Environmental Impact Assessment, adopting international best practice methodologies, of human activities that could have impact on migratory waterbird populations and their habitats.

4.4.4 Range States shall ensure that adequate statutory pollution controls are in place, including those relating to the use of agricultural chemicals, pest control procedures, oil spills and the disposal of waste-water, which are in accordance with international norms, for the purpose of minimizing their adverse impacts on the populations listed in Table 2.

Gazetteer

Gazetteer of geographic localities (places, rivers, mountain ranges, etc.) mentioned in *Bird Migrations across the Himalayas*. Orthography follows The Times Comprehensive Atlas of the World, 11th edition. If we could not find a name there, we chose to follow Wikipedia. If we could not find it there, we chose to follow names as found on the internet through the Google search engines (last access 3rd January 2017). Tibetan names with their English spelling for localities in Tibet were found by using different search engines and through personal correspondence. Names of localities or their spelling do not reflect a political or legal point of view. Most research for this Gazetteer was carried out by Mr. Herman van Oeveren.

	Coordinates	
	North	East
Abakan Mountains	53°40'N	89°16'E
Ahiron	24°31'N	88°02'E
Ahmedabad	23°02'N	72°37'E
Ajmer	26°16'N	74°25'E
Akhnoor	32°54'N	74°44'E
Alake, Lake	35°34'N	97°06'E
Alchi	34°14'N	77°09'E
Allahabad	25°27'N	81°51'E
Almaty	43°14'N	76°53'E
Altai Mountains	49°N	89°E
Altun Mountains (= Altyn Tsan)	38°36'N	89°00'E
Amdo	33°N	90°E
Andhra Pradesh	14°48'N	79°18'E
Angsi Glacier	30°23'N	82°00'E
Annapurna	28°35'N	83°49'E
Argog, Lake	30°58'N	82°15'E
Arun	27°26'N	87°07'E
Arunachal Pradesh	28°N	93°E
Assam	26°08'N	91°46'E
Badain Jaran Desert	40°04'N	102°12'E
Bagmati River	27°26'N	85°21'E
Baikal, Lake	53°30'N	108°00'E

	Coordinates	
	North	East
Baikal Mountains	54°0’N	108°00’E
Balkhash, Lake	21°17’N	72°45’E
Balochistan	27°25’N	64°30’E
Baltistan	35°18’N	75°37’E
Baltoro Glacier	35°44’N	76°22’E
Bangong Tso (= Pangong Lake = Spang gong mtsho)	33°44’N	78°40’E
Bardiya National Park	28°23’N	81°30’E
Barkol, Lake	43°36’N	93°01’E
Baspa River	31°25’N	78°16’E
Bayan Har Mountains (= Bayen-káras = Bayan Har Shan)	34°N	98°E
Baytag Bogd Uul	45°12’N	91°38’E
Beas Reservoir	31°59’N	76°02’E
Beas River	31°49’N	76°45’E
Betpak-Dala	46°N	70°E
Bhagalpur	25°15’N	87°00’E
Bhagirathi (=Mount Bhagirathi)	30°16’N	78°29’E
Bharatpur	27°41’N	84°26’E
Bhasawar	27°02’N	76°49’E
Bihar	25°22’N	85°07’E
Brahmaputra	47°54’N	95°23’E
Broad Peak (was known as K3)	35°48’N	76°34’E
Buyant River	47°50’N	91°36’E
Calimere, Point	10°17’N	79°51’E
Caohai, Lake	26°51’N	104°14’E
Char	33°14’N	77°09’E
Chainjoin, Lake	35°54’N	87°00’E
Ch’a-k’o-ch’a-mu, Lake	33°08’N	85°44’E
Changthang (= Jangthang)	30°00’N	90°00’E
Chatyr, Lake	40°37’N	75°18’E
Chenab River	33°06’N	74°48’E
Chilika, Lake	19°38’N	85°11’E
Chitwan National Park	27°30’N	84°20’E
Choglamsar	34°06’N	77°35’E
Chohal Dam	31°36’N	75°57’E
Chomo Yummo	28°02’N	88°32’E
Chomolung (= Mount Everest = Jomo Langma)	27°59’N	86°55’E
Chomolung (in Ladakh)	34°23’N	76°22’E
Chorabari Glacier	30°45’N	79°03’E
Chorten Nyima Ri	27°57’N	88°10’E
Chuadanga	23°38’N	88°51’E
Chuma’er River	35°00’N	93°38’E
Chumba Yumco	28°12’N	89°37’E
Chumur	32°40’N	78°36’E
Chushul	33°33’N	78°43’E
Chusur	29°21’N	90°44’E
Co Ngoin (= Ngoin Lake = Ngoin Cho)	31°28’N	91°30’E

	Coordinates	
	North	East
Como Chamling	28°24'N	88°12'E
Communism (= Kommunizma) Peak (= Ismoil Somoni Peak)	38°55'N	72°01'E
Concordia	35°44'N	76°31'E
confluence of rivers Murti, Sipsu and Jaldhaka	26°42'N	88°52'E
Corbett National Park	29°32'N	78°56'E
Coromandel Coast	14°00'N	80°10'E
Cuolongka, Lake	32°22'N	92°27'E
Dachi	34°37'N	76°22'E
Dachigam National Park	34°08'N	75°02'E
Dadu Co	28°23'N	89°06'E
Dagze Co	31°54'N	87°29'E
Dahinsara	22°57'N	70°37'E
Dal Lake	34°08'N	74°52'E
Dalhousie	32°31'N	75°58'E
Damxung	30°26'N	90°59'E
Dangla Mountains (= Tanggula Shan)	34°13'N	92°26'E
Dangquezangbu River	30°04'N	83°22'E
Darcha	32°40'N	77°12'E
Darkhad Valley	51°10'N	99°30'E
Dartsigs	34°37'N	76°22'E
Dawei River mouth	13°43'N	98°13'E
Debiganj	26°07'N	88°45'E
Deccan Plateau	17°N	77°E
Deepor Beel Wildlife Sanctuary	26°07'N	91°39'E
Degh River	32°26'N	75°11'E
Dharamsala	32°13'N	76°19'E
Dhasan River	25°45'N	79°24'E
Dhauladhar Range	32°17'N	76°23'E
Dhaulagiri	28°41'N	83°29'E
Dibru-Saikhowa Wildlife Sanctuary	27°40'N	95°23'E
Diskit	34°33'N	77°32'E
Dochen, Lake	28°08'N	89°21'E
Doda River (= Stod River)	33°31'N	76°47'E
Dolpa Airfield	28°59'N	82°50'E
Doman River	34°00'N	80°15'E
Domkhar	19°10'N	73°03'E
Donggeicuona, Lake	35°17'N	98°31'E
Dou, Lake (= Dug Tso)	35°18'N	99°10'E
Dreku, Lake	28°41'N	91°40'E
Dudh Kosi (= Kosi Barrage)	27°35'N	86°40'E
Dug Tso (= Dou Lake)	35°18'N	99°10'E
Dung Co	31°44'N	91°10'E
Dzarangpu (= Rongbuk = Rongpu)	26°06'N	86°51'E
Dzungaria	45°00'N	85°00'E
East Tingri (= East Dingri)	28°34'N	86°37'E
Ejin	41°44'N	100°19'E

	Coordinates	
	North	East
Elephant Lake (= Kupup Lake = Tsongmo = Changu Lake)	27°20'N	88°51'E
Eling-Zahling Lakes	35°54'N	97°42'E
Etawah	26°46'N	79°01'E
Everest, Mount (= Jomo Langma = Chomolung)	27°59'N	86°55'E
Farakka Barrage	24°48'N	87°56'E
Gajoldoba (=Gajoldoba Reservoir)	26°46'N	88°30'E
Galalacuo, Lake	34°29'N	97°43'E
Gandaki River (see also Kali Gandaki Gorge)	25°39'N	85°11'E
Ganga River (= Ganges)	30°01'N	78°14'E
Gangdese Range (Gangdese batholit)	30°N	91°E
Ganges	22°05'N	90°50'E
Gangs Rin-po-che (Mount Kailash = Kangrinboqe)	31°04'N	81°18'E
Gar Tsangpo River (tributary of Indus in Tibet)	32°21'N	79°50'E
Garhwal	30°22'N	78°28'E
Gayaik	33°33'N	78°08'E
Gegeen, Lake	46°42'N	96°46'E
Gêgyai District	31°19'N	81°49'E
Gemang Co	31°34'N	87°03'E
Gilgit	35°55'N	74°17'E
Gilgit-Baltistan	35°35'N	74°37'E
Gnyan-chen-thang-lha (= Nyaingentanglha Shan)	30°30'N	94°30'E
Gobi Desert	42°35'N	103°25'E
Gomang Co	31°12'N	89°11'E
Great Himalaya Range	28°N	84°E
Guangxi	23°36'N	108°18'E
Guizhou Province	26°50'N	106°50'E
Gujarat	23°13'N	72°41'E
Güncang	29°48'N	94°10'E
Gurla Mandhata (= Memo'nani = Naymona'nyi)	30°26'N	81°17'E
Guwahati	26°11'N	91°44'E
Gyala Peri	29°49'N	94°58'E
Gyangtse	28°57'N	89°38'E
Gyayik	33°33'N	78°09'E
Gydan Peninsula	70°N	79°E
Gyirong (= Jilong = Kyirong)	28°51'N	85°18'E
Hanko-la	36°27'N	80°40'E
Hanle	32°47'N	79°00'E
Har Us, Lake	48°00'N	92°10'E
Haramosh	35°50'N	74°53'E
Hardoi	27°25'N	80°07'E
Haryana	30°44'N	76°47'E
Hemis National Park	33°59'N	77°26'E
Hengduan Shan	27°30'N	99°00'E
Himachal Pradesh	31°06'N	77°10'E
Hindu Kush	35°N	71°E
Hokersar, Lake	34°06'N	74°42'E

	Coordinates	
	North	East
Huangheyuan	36°40'N	101°16'E
Indus	23°59'N	67°25'E
Irrawaddy	25°42'N	97°30'E
Ismoil Somoni Peak (= Kommunizma/Communism Peak)	38°55'N	72°01'E
Jaldhaka River	27°01'N	88°53'E
Jammu	32°43'N	74°52'E
Jammu and Kashmir	32°43'N	74°52'E
Janakpur	26°43'N	85°55'E
Jangthang (= Changthang)	30°00'N	90°00'E
Jankar River	32°41'N	77°11'E
Jeu-ch'ing Hsiu-pu Tso	31°05'N	83°03'E
Jharkand	23°21'N	85°19'E
Jhatingri	31°56'N	76°53'E
Jhelum River	31°12'N	72°08'E
Jilong (= Kyirong = Gyirong)	28°51'N	85°18'E
Jodhpur	26°16'N	73°01'E
Jodia (=Jodiya)	22°41'N	70°18'E
Jomo Langma (= Chomolung = Mount Everest)	27°59'N	86°55'E
Jomsom	28°47'N	83°44'E
K2 (= Mount Chhogori)	35°52'N	76°30'E
Kabul	34°32'N	69°10'E
Kadalundi-Vallikkunnu Community Reserve	11°07'N	74°50'E
Kailash (= Mount Kailas = Kangrinboqe = Gangs Rin-po-che)	31°04'N	81°18'E
Kakjung	34°35'N	77°37'E
Kali Gandaki Gorge (= see Gandaki River =Andha Galchi)	28°42'N	83°38'E
Kandia	35°27'N	73°11'E
Kang Yatse	33°44'N	77°33'E
Kangchenjunga (= Mount Kanchenjunga)	27°42'N	88°08'E
Kangra	32°06'N	76°16'E
Kanutse	34°34'N	76°32'E
Kaphrauri	29°10'N	78°53'E
Karakoram	36°N	76°E
Karashura	39°20'N	71°47'E
Kara-tagh Pass	35°43'N	78°19'E
Karatoya River	24°13'N	89°36'E
Kargil	34°31'N	76°13'E
Kargyak Chu River	33°03 N	77°13'E
Karnataka	12°58'N	77°30'E
Kashmir	34°30'N	76°00'E
Kasuganji	27°59'N	78°38'E
Kathmandu	27°42'N	85°20'E
Kedarnath	30°43'N	79°04'E
Keoladeo (National Park)	27°10'N	77°31'E
Kerala	08°30'N	77°00'E

	Coordinates	
	North	East
Keriya River	38°36'N	82°05'E
Keshar	33°26'N	78°12'E
Khalsar	34°29'N	77°42'E
Khangai Mountains	47°30'N	100°00'E
Khare	28°20'N	83°40'E
Kharmang	34°56'N	76°13'E
Kharnak	33°15 N	77°42'E
Khart	30°22'N	69°59'E
Khatanga	26°06 N	83°57'E
Khatanga River	73°11'N	106°12'E
Khentii Mountains	48°47'N	109°10'E
Khentii Province	47°19'N	110°39'E
Khovd	48°00'N	91°38'E
Khulna	22°49'N	89°33'E
Khumbu	27°49'N	86°43'E
Khurkh River	48°21'N	110°25'E
Kiagar Tso (= Thatsangkuru Tso)	33°06'N	78°18'E
Kiari	33°28'N	78°07'E
Kibber	32°19'N	78°00'E
Kohistan	35°15'N	73°30'E
Kokkaymar	43°21'N	74°05'E
Koko Nor (= Tso Ngonpo = Qinghai Lake)	37°00'N	100°08'E
Kolhuwa	27°34'N	84°03'E
Kommunizma / Communism Peak (= Ismoil Somoni Peak)	38°55'N	72°01'E
Koonthankulum	08°34'N	77°45'E
Kopa	43°30'N	75°46'E
Korzok	32°58'N	78°15'E
Kosi Barrage	27°35'N	86°40'E
Koyul	32°53'N	79°14'E
Krasnoyarsk Krai	59°53'N	91°40'E
Kuhai, Lake	35°18'N	99°10'E
Kulong Chhu Valley	27°34'N	91°34'E
Kumaon	29°36'N	79°42'E
Kunlun Mountains (= Kunlun Shan)	36°N	84°E
Kupup, Lake (=Tsongmo = Elephant Lake = Changu Lake)	27°20'N	88°51'E
Kurnool	15°49'N	78°03'E
Kurram River	33°49'N	69°58'E
Kyagar Tso	33°06'N	78°18'E
Kyrgyzstan	41°N	75°E
Kyirong (= Gyirong = Jilong)	28°51'N	85°18'E
Ladakh	34°10'N	77°34'E
Ladakh Range	34°10'N	77°34'E
Lahul	32°30'N	77°50'E
Lahul-Spiti	32°30'N	77°50'E
Lamayuru	34°10'N	76°49'E

	Coordinates	
	North	East
Langtang Lirung (Mount Langtang Lirung)	28°15'N	85°30'E
Laptev Sea	76°16'N	125°38'E
Lawala Pass	27°28'N	90°10'E
Ledo Bridge	34°43'N	76°19'E
Leh	34°09'N	77°36'E
Lena Delta	72°52'N	123°21'E
Lhasa	29°39'N	91°07'E
Lhasa River	29°39'N	91°07'E
Lilung [on Tsangpo River in SW Tibet; q.v.]		
Loma	33°10'N	78°49'E
Lop Nur	40°10'N	90°35'E
Lukla	27°41'N	86°43'E
Machu (= Amo Chhuu = Torsa River)	26°49'N	89°20'E
Magadan	59°34'N	150°48'E
Maharashtra	18°57'N	72°49'E
Mahe	33°16'N	78°30'E
Mahe-Puga	33°14'N	78°22'E
Makalu (= Mount Makaru)	27°53'N	87°05'E
Manali	32°16'N	77°10'E
Manaslu (= Mount Kutang)	28°32'N	84°33'E
Mandi	30°50'N	80°40'E
Mannar, Gulf of	09°08'N	79°28'E
Mansarovar, Lake (= Mapham Yutso)	30°39'N	81°27'E
Mapangyong Cuo	30°41'N	81°23'E
Mapham Yutso (= Mansarovar Lake)	30°39'N	81°27'E
Mathabhanga River	23°24'N	88°43'E
McLeod Ganj	32°25'N	76°32'E
Mekong	11°00'N	105°11'E
Memo'nani (= Naymona'nyi = Gurla Mandhata)	30°26'N	81°17'E
Meru (= Meru Peak)	30°52'N	79°01'E
Mindam, Lake	33°12'N	81°21'E
Miram Shah	33°00'N	70°04'E
Mohanpur	28°16'N	80°15'E
Mokalsar	25°37'N	72°30'E
Moosi Reservoir	17°15'N	79°30'E
Mugqu Co	31°02'N	89°01'E
Murgi	34°45'N	77°31'E
Murti River	26°50'N	88°50'E
Mustang	28°47'N	83°44'E
Muztagh Pass	35°52'N	76°31'E
Muzzafarpur	26°00'N	85°27'E
Nagqu	31°29'N	92°03'E
Nainital	29°21'N	79°28'E
Nam Co	30°43'N	90°33'E
Namarodi River	30°29'N	81°25'E
Namchak Barwa (= Namche)	29°37'N	95°03'E
Namche (Namche Barwa)	29°37'N	95°03'E

	Coordinates	
	North	East
Nanga Parbat	35°14'N	74°35'E
Narayani River	27°35'N	84°10'E
Nathu Pass (= Nathoe La)	27°22'N	88°48'E
Nau Co	32°50'N	82°10'E
Naymona'nyi (= Gurla Mandhata = Memo'nani)	30°26'N	81°17'E
New Delhi	28°36'N	77°12'E
Ngangla Rinco, Lake	31°34'N	83°06'E
Ngoin, Lake (= Ngoin Cho = Co Ngoin)	31°28'N	91°30'E
Niaodao	36°50'N	100°10'E
Nilgiri, Mount	28°39'N	83°43'E
Nimaling	33°48'N	77°32'E
Niti Pass	30°58'N	79°52'E
Norbu Sumdo	32°29'N	78°19'E
Novaya Zemlya	59°19'N	142°48'E
Novosibirsk	55°03'N	82°57'E
Nubra River	34°36'N	77°42'E
Nuptse (= Nubtse)	27°57'N	86°53'E
Nuruchen	33°14'N	77°59'E
Nyaingentanglha Shan (= Gnyan-chen-thang-lha)	30°30'N	94°30'E
Nyang Chu River	29°47'N	93°50'E
Nyenchen Thanglha (= Nyaingentanglha Shan)	30°30'N	94°30'E
Nyoma	33°12'N	78°39'E
Ob	66°32'N	71°23'E
Odisha (formerly Orissa)	20°09'N	85°30'E
Oksu River	37°56'N	71°34'E
Olenyok River	72°55'N	120°14'E
Oma, Lake	27°45'N	91°38'E
Oma River	33°45'N	76°42'E
Omnogovi	43°00'N	104°15'E
Ondorkhaan City	47°19'N	110°40'E
Ordos Desert	39°35'N	109°43'E
Padum	33°27'N	76°52'E
Pamirs	38°30'N	73°30'E
Panamik	34°46'N	78°00'E
Pangong Tso (= Bangong Co = Spang gong mtsho)	33°44'N	78°40'E
Paro River	27°30'N	89°20'E
Periche (= Pheriche)	27°54'N	86°52'E
Peski Muyunkum	45°48'N	79°07'E
Phewa, Lake	28°12'N	83°56'E
Phey	34°08'N	77°28'E
Phobjika Valley	27°30'N	90°10'E
Pir Panjal (Range)	33°53'N	74°29'E
Pistu	34°08'N	77°15'E
Pokhara	28°15'N	83°58'E
Pokhara Airfield	28°12'N	83°59'E
Pong Dam	31°58'N	75°56'E
Popigai River	72°53'N	106°35'E

	Coordinates	
	North	East
Puga	33°14'N	78°19'E
Pulicat, Lake	13°30'N	80°15'E
Pung Co	31°29'N	90°57'E
Punjab	31°N	74°E
Purbasthali	23°27'N	88°21'E
Puthalam	08°06'N	77°28'E
Pyandsh River	37°35'N	71°30'E
Qiangzuo Co	28°18'N	88°06'E
Qilian Mountains (= Qilian Shan = Tsilian Mts.)	38°09'N	99°46'E
Qinghai, Lake (= Tso Ngonpo = Koko Nor)	37°00'N	100°08'E
Qinghai (= Tibetan) Plateau	33°N	88°E
Qinghai Province (= Tsongon Shingchen)	35°N	96°E
Quo Yangqu	32°08'N	88°17'E
Rabang	33°02'N	80°32'E
Raipur	21°15'N	81°37'E
Rajaji National Park	30°03'N	78°10'E
Rajasthan	26°34'N	73°50'E
Rajshahi	24°29'N	88°40'E
Rakshastal, Lake	30°39'N	81°15'E
Rangdum	34°03'N	76°21'E
Rann of Kutch	24°05'N	70°38'E
Rann of Kutch Lake	24°01'N	70°07'E
Rapti River	26°17'N	83°40'E
Rateu	26°13'N	71°42'E
Ravi River	30°35'N	71°49'E
Ringco Gongma	30°57'N	89°39'E
Ringco Yokma	30°55'N	89°50'E
Riyul Tso	32°55'N	78°36'E
Rongbuk (= Rongpu = Dzarangpu)	26°06'N	86°51'E
Rumbak	34°03'N	77°25'E
Rupsa River	22°45'N	89°33'E
Rupshu	33°05'N	78°11'E
Rutog	33°23'N	79°43'E
Sagsay River	48°55'N	89°38'E
Sakti	34°00'N	77°49'E
Salang Pass	35°18'N	69°02'E
Salween River	17°26'N	97°45'E
Samad	33°07'N	78°15'E
Sanchi	23°28'N	77°44'E
Sangpo River (=Tsangpo River)	29°09'N	93°58'E
Sangpo River (in Ladakh, India)	33°26'N	76°53'E
Sanjiangyuan Nature Reserve	34°00'N	96°12'E
Sayan Mountains (or Range)	53°15'N	94°58'E
Selincuo Reserve	31°31'N	89°09'E
Shachukul	33°59'N	78°07'E
Shanxi	37°42'N	112°24'E
Shayok (=Shyok)	34°12'N	78°10'E

	Coordinates	
	North	East
Shey	34°04'N	77°38'E
Shijianlame Co	29°46'N	92°23'E
Shigatse (= Shikatse)	29°15'N	88°53'E
Shikatse Air Base (= Xigazê Air Base)	29°20'N	89°18'E
Shillong Plateau	25°34'N	91°52'E
Shisha Pangma (= Mount Gosainthān = Shishapangma)	28°21'N	85°46'E
Shivling (= Shivling Peak)	30°52'N	79°03'E
Shyok (=Shayok)	34°12'N	78°10'E
Sichuan	30°08'N	102°56'E
Sikandra Rao	27°41'N	78°22'E
Sikkim	27°19'N	88°37'E
Siling Co	31°47'N	88°58'E
Simbalbara National Park (= Simbalwara N.P.)	30°25'N	77°29'E
Simi-la, Lake	28°50'N	89°52'E
Sipsu River	27°01'N	88°53'E
Siwalik Hills	27°46'N	82°24'E
Son, Lake	41°50'N	75°10'E
South Siberian Mountains (collective for Baikal, E and W Sayan, Yablonoi Mts: q.v.)		
Spangmik	33°54'N	78°27'E
Spang gong mtsho (= Pangong Tso = Bangong Co)	33°44'N	78°40'E
Spiti	32°17'N	78°00'E
Spontang	33°29'N	76°50'E
Srinagar	34°05'N	74°48'E
Srisailam Backwater	16°05'N	78°53'E
Stakna	33°52'N	77°48'E
Startsapuk Tso	33°15'N	78°02'E
Stod River (= Doda River)	33°31'N	76°47'E
Sulaiman Mountains	30°30'N	70°10'E
Sumda (=Sumda Cho)	34°08'N	77°09'E
Sumdo	29°17'N	100°04'E
Sundarbans	21°56'N	89°10'E
Suru River	34°34'N	76°06'E
Sutlej River and Gorge	29°21'N	71°01'E
Taklamakan Desert	38°54'N	82°12'E
Tamil Nadu	13°05'N	80°16'E
Tamuyen [on Tsangpo River in SW Tibet; q.v.]		
Tanggula Shan (= Dangla Mountains)	34°13'N	92°26'E
Tangtse	34°01'N	78°10'E
Tanintharyi	12°05'N	99°01'E
Tarim Basin	37°30'N	82°15'E
Taro Tso	31°08'N	84°07'E
Taymyr (= Taimyr Peninsula)	74°N	98°E
Teesta River	25°30'N	89°39'E
Tel River	19°54'N	82°46'E
Tengiz – Korgalzhyn Lake System	50°26'N	68°54'E
Terai Region	26°27'N	87°16'E

	Coordinates	
	North	East
Terhiyn Tsagaan Nuur	48°08'N	99°36'E
Thar Desert	39°N	83°E
Tharad	24°23'N	71°37'E
Thatsangkaru Tso (= Kiagar Tso)	33°07'N	78°16'E
Thikse	34°02'N	77°40'E
Thimphu	27°28'N	89°38'E
Thula Kharka (= Thoolokharka)	28°18'N	83°49'E
Tibet and Tibetan Plateau	32°N	88°E
Tien Shan	42°N	80°E
Tilaiya Dam Reservoir	24°19'N	85°30'E
Tilichang	34°34'N	76°32'E
Tomsk	56°30'N	84°58'E
Torsa National Reserve	27°21'N	89°03'E
Torsa River (= Torsha River = Machu = Amo Chhuu)	26°49'N	89°20'E
Trishul	30°18'N	79°46'E
Tsagaan, Lake (=Tsagaan Nuur)	49°58'N	105°21'E
Tsangpo River (= Zangbo River)	29°11'N	93°58'E
Ts'ao, Lake	41°31'N	84°36'E
Tsarab River	32°41'N	77°53'E
Tsati	34°28'N	77°43'E
Tsilian Mountains (= Qilian Tsan = Qilian Mts.)	38°09'N	99°46'E
Tsokar	33°19'N	77°59'E
Tsomoriri	32°54'N	78°18'E
Tsongmo (= Elephant Lake = Kupup Lake = Changu Lake)	27°20'N	88°51'E
Tso Ngonpo (= Qinghai Lake)	37°00'N	100°08'E
Tsongon Shingchen (= Qinghai Province)	35°N	96°E
Tungabhadra River	15°14'N	76°20'E
Turpan Basin (= Turpan Depression)	43°02'N	89°30'E
Tuul River	47°33'N	105°56'E
Upper Indus River	33°11'N	78°38'E
Upper Indus Valley	35°24'N	74°20'E
Upper Kali Gandaki	28°42'N	83°38'E
Ural Mountains	60°00'N	60°00'E
Urru, Lake (= Urru Cuo = Ur ru mtsho)	31°44'N	87°59'E
Uttar Pradesh	26°N	80°E
Uttarakhand	30°19'N	78°03'E
Vakhsh river	38°22'N	69°21'E
Waxxari	38°41'N	87°22'E
West Bengal	22°34'N	88°22'E
Wujang	33°37'N	79°55'E
Wular, Lake	34°20'N	74°36'E
Xigazê Air Base (= Shikatse Air Base)	29°20'N	89°18'E
Xinjiang	41°N	85°E
Yablonoi Mountains	52°01'N	113°40'E
Yakatograk	38°29'N	86°43'E
Yakutia	66°24'N	129°10'E

	Coordinates	
	North	East
Yamdrok, Lake (= Yamdrok Yutso = Yar-'brog g.yu-mtsho)	28°58'N	90°43'E
Yamuna River	25°25'N	81°53'E
Yar-'brog g.yu-mtsho (= Yamdrok Lake = Yamzhog Yumco)	28°58'N	90°43'E
Yarkand	38°25'N	77°14'E
Yarlung Tsangpo River (= Tsangpo River)	29°09'N	93°58'E
Yaru (and Yaru Bridge)	34°02'N	77°11'E
Yellow River	37°46'N	119°15'E
Yenisei	71°50'N	82°40'E
Yeru River	28°22'N	87°49'E
Yeyik	36°45'N	83°11'E
Yoqag, Lake	30°28'N	88°37'E
Yonghong, Lake	35°16'N	89°57'E
Yulkam	34°41'N	77°35'E
Yumthang River	27°49'N	88°42'E
Yungui	27°N	105°E
Yunnan	25°03'N	101°52'E
Yuru Bridge	34°16'N	76°47'E
Zamthang (in Ladakh, not China)	33°14'N	77°09'E
Zangbo River (= Tsangpo River)	30°00'N	90°26'E
Zanskar (= Zangskar)	33°29'N	76°50'E
Zanskar Range	33°29'N	76°50'E
Zanskar River	33°36'N	76°59'E
Zemithang River	27°42'N	91°43'E
Zemu River	27°46'N	88°28'E
Zhaling, Lake	34°55'N	97°31'E
Zhamu Co	33°00'N	93°25'E
Zhamucuo Wetland	33°03'N	94°04'E
Zhari Namco, Lake	30°55'N	85°34'E

Index

Printed in the United States
By Bookmasters